AAPG Treatise of Petroleum Geology

The American Association of Petroleum Geologists
gratefully acknowledges and appreciates the leadership and support
of the AAPG Foundation in the development of the
Treatise of Petroleum Geology

STRUCTURAL TRAPS VII

COMPILED BY
EDWARD A. BEAUMONT
AND
NORMAN H. FOSTER

TREATISE OF PETROLEUM GEOLOGY
ATLAS OF OIL AND GAS FIELDS

PUBLISHED BY
THE AMERICAN ASSOCIATION OF PETROLEUM GEOLOGISTS
TULSA, OKLAHOMA 74101, U.S.A.

Copyright © 1992
The American Association of Petroleum Geologists
All Rights Reserved

ISBN: 0-89181-589-9
ISSN: 1043-6103

Available from:
The AAPG Bookstore
P.O. Box 979
Tulsa, OK 74101-0979

Phone: (918) 584-2555
Telex: 49-9432
FAX: (918) 584-0469

Association Editor: Susan Longacre
Science Director: Gary D. Howell
Publications Manager: Cathleen P. Williams
Special Projects Editor: Anne H. Thomas
Science Staff: William G. Brownfield
Project Production: Custom Editorial Productions, Inc.

AAPG grants permission for a single photocopy of an item from this publication for personal use. Authorization for additional copies of items from this publication for personal or internal use is granted by AAPG provided that the base fee of $3.00 per copy is paid directly to the Copyright Clearance Center, 27 Congress Street, Salem, Massachusetts 01970. Fees are subject to change. Any form of electronic or digital scanning or other digital transformation of portions of this publication into computer-readable and/or transmittable form for personal or corporate use requires special permission from, and is subject to fee charges by, the AAPG.

TABLE OF CONTENTS

1. **Tip Top Field.** J. D. Edman and L. Cook 1
2. **Rangely Field.** T. A. Hefner and K. T. Barrow 29
3. **Matzen Field.** N. Kreutzer 57
4. **Cortemaggiore Field.** M. Pieri 99
5. **Malossa Field.** L. Mattavelli and V. Margarucci 119
6. **Penal/Barrackpore Field.** B. L. Dyer and P. Cosgrove 139
7. **Laojunmiao Field.** Zhai Guangming and Song Jianguo 159
8. **Raman, Bati-Raman, and Garzan Fields.** R. Salem, H. Ozbahceci, A. Ungor, and M. Isbilir 173
9. **Messoyakh Field.** J. Krason and P. D. Finley 197
10. **South Belridge Field.** D. D. Miller and J. G. McPherson 221
11. **Portachuelo Field.** H. Hay-Roe and P. M. Miller 245
12. **Raguba Field.** P. Brennan 267
13. **Tintaburra Field.** C. B. Newton 291
14. **La Cira-Infantas Field.** P. A. Dickey 323

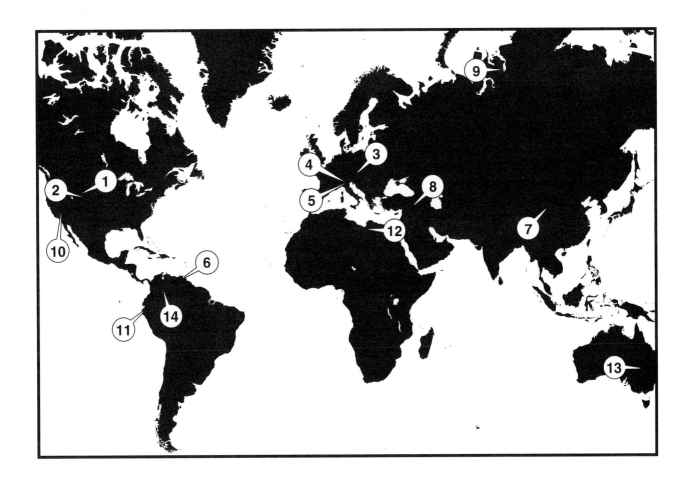

TREATISE OF PETROLEUM GEOLOGY
ADVISORY BOARD

Ward O. Abbott
Abdulaziz A. Al-Laboun
John J. Amoruso
George G. Anderman
John D. Armstrong
George B. Asquith
Donald W. Axford
Colin Barker
Ian R. Baron
Ted L. Bear
Edward A. Beaumont
Robert R. Berg
Steve J. Blanke
Richard R. Bloomer
John F. Bookout, Jr.
Louis C. Bortz
Donald R. Boyd
Robert L. Brenner
William D. Brumbaugh
Raymond Buchanan
Daniel A. Busch
Nick Cameron
David G. Campbell
J. Ben Carsey*
Duncan M. Chisholm
H. Victor Church
Don Clutterbuck
J. Glenn Cole
J. Frank Conrad
Robert J. Cordell
Robert D. Cowdery
Marshall C. Crouch, III
William H. Curry, III
Doris M. Curtis*
Graham R. Curtis
Clint A. Darnall*
Patrick Daugherty
Edward K. David
Herbert G. Davis
Gerard J. Demaison
Parke A. Dickey
Fred A. Dix, Jr.
Charles F. Dodge
Edward D. Dolly
James C. Donaldson
Ben Donegan
Robert H. Dott*
John H. Doveton
Marlan W. Downey
Bernard M. Durand
Richard Ebens
Joel S. Empie

Charles T. Feazel
William L. Fisher
Norman H. Foster
James F. Friberg
Richard D. Fritz
Lawrence W. Funkhouser
William E. Galloway
Donald L. Gautier
Lee C. Gerhard
James A. Gibbs
Melvin O. Glerup
Arthur R. Green
Donald A. Greenwood
R. R. Robbie Gries
Richard W. Griffin
Robert D. Gunn
Alfredo Eduardo Guzman
Merrill W. Haas
Cecil V. Hagen
J. Bill Hailey
Michel T. Halbouty
Bernold M. Hanson
Tod P. Harding
Donald G. Harris
Paul M. Harris
Frank W. Harrison, Jr.
Dan J. Hartmann
John D. Haun
Hollis D. Hedberg*
James A. Helwig
Thomas B. Henderson, Jr.
Neville M. Henry
Francis E. Heritier
Paul Hess
Lee Higgins
Mason L. Hill*
David K. Hobday
David S. Holland
Myron K. Horn
Gary D. Howell
Michael E. Hriskevich
Joseph P. D. Hull, Jr.
H. Herbert Hunt
Norman J. Hyne
J. J. C. Ingels
Russell W. Jackson
Michael S. Johnson
David H. Johnston
Bradley B. Jones
Peter G. Kahn
John E. Kilkenny
H. Douglas Klemme

Allan J. Koch
Raden P. Koesoemadinate
Hans H. Krause
Naresh Kumar
Susan M. Landon
Kenneth L. Larner
Rolf Magne Larsen
Roberto A. Leigh
Jay Leonard
Raymond C. Leonard
Howard H. Lester
Christopher J. Lewis
Donald W. Lewis
James O. Lewis, Jr.
Detlev Leythaeuser
Robert G. Lindblom
Roy O. Lindseth
John P. Lockridge
Anthony J. Lomando
John M. Long
Susan A. Longacre
James D. Lowell
Peter T. Lucas
Andrew S. Mackenzie
Jack P. Martin
Michael E. Mathy
Vincent Matthews, III
Paul R. May
James A. McCaleb*
Dean A. McGee*
Philip J. McKenna
Jere W. McKenny
Richard M. Meek
Robert E. Megill
Robert W. Meier
Fred F. Meissner
Robert K. Merrill
David L. Mikesh
Marcus Milling
George Mirkin
Michael D. Mitchell
Richard J. Moiola
Francisco Moreno
D. Keith Murray
Grover E. Murray
Norman S. Neidell
Ronald A. Nelson
Charles R. Noll
Clifton J. Nolte
David W. Organ
John C. Osmond
Philip Oxley

Susan E. Palmer
Arthur J. Pansze
John M. Parker
Stephen J. Patmore
Dallas L. Peck
William H. Pelton
Alain Perrodon
James A. Peterson
R. Michael Peterson
Edward B. Picou, Jr.
Max Grow Pitcher
David E. Powley
William F. Precht
A. Pulunggono
Bailey Rascoe, Jr.
R. Randy Ray
Dudley D. Rice
Edward P. Riker
Edward C. Roy, Jr.
Eric A. Rudd
Floyd F. Sabins, Jr.
Nahum Schneidermann
Peter A. Scholle
George L. Scott, Jr.
Robert T. Sellars, Jr.
Faroog A. Sharief
John W. Shelton
Phillip W. Shoemaker

Synthia E. Smith
Robert M. Sneider
Frank P. Sonnenberg
Stephen A. Sonnenberg
William E. Speer
Ernest J. Spradlin
Bill St. John
Philip H. Stark
Richard Steinmetz
Per R. Stokke
Denise M. Stone
Donald S. Stone
Douglas K. Strickland
James V. Taranik
Harry Ter Best, Jr.
Bruce K. Thatcher, Jr.
M. Ray Thomasson
Jack C. Threet
Bernard Tissot
Don F. Tobin
Don G. Tobin
Donald F. Todd
Harrison L. Townes
M. O. Turner
Peter R. Vail
B. van Hoorn
Arthur M. Van Tyne
Kent Lee Van Zant

Ian R. Vann
Harry K. Veal*
Steven L. Veal
Richard R. Vincelette
Fred J. Wagner, Jr.
William A. Walker, Jr.
Carol A. Walsh
Anthony Walton
Douglas W. Waples
Harry W. Wassall, III
W. Lynn Watney
N. L. Watts
Koenradd J. Weber
Robert J. Weimer
Dietrich H. Welte
Alun H. Whittaker
James E. Wilson, Jr.
Thomas Wilson
John R. Wingert
Martha O. Withjack
P. W. J. Wood
Homer O. Woodbury
Walter W. Wornardt
Marcelo R. Yrigoyen
Mehmet A. Yukler
Zhai Guangming
Robert Zinke

* Deceased

American Association of Petroleum Geologists Foundation
Treatise of Petroleum Geology Fund*

Major Corporate Contributors
($25,000 or more)

Amoco Production Company
BP Exploration Company Limited
Chevron Corporation
Exxon Company, U.S.A.
Mobil Oil Corporation
Oryx Energy Company
Pennzoil Exploration and Production Company
Shell Oil Company
Texaco Foundation
Union Pacific Foundation
Unocal Corporation

Other Corporate Contributors
($5,000 to $25,000)

ARCO Oil & Gas Company
Ashland Oil, Inc.
Cabot Oil & Gas Corporation
Canadian Hunter Exploration Ltd.
Conoco Inc.
Marathon Oil Company
The McGee Foundation, Inc.
Phillips Petroleum Company
Transco Energy Company
Union Texas Petroleum Corporation

Major Individual Contributors
($1,000 or more)

John J. Amoruso
Thornton E. Anderson
C. Hayden Atchison
Richard A. Baile
Richard R. Bloomer
A. S. Bonner, Jr.
David G. Campbell
Herbert G. Davis
George A. Donnelly, Jr.
Paul H. Dudley, Jr.
Lewis G. Fearing
Lawrence W. Funkhouser
James A. Gibbs
George R. Gibson
William E. Gipson
Mrs. Vito A. (Mary Jane) Gotautas
Robert D. Gunn
Merrill W. Haas
Cecil V. Hagen
Frank W. Harrison
William A. Heck
Roy M. Huffington
J. R. Jackson, Jr.
Harrison C. Jamison
Thomas N. Jordan, Jr.
Hugh M. Looney
Jack P. Martin
John W. Mason
George B. McBride
Dean A. McGee
John R. McMillan
Lee Wayne Moore
Grover E. Murray
Rudolf B. Siegert
Robert M. Sneider
Estate of Mrs. John (Elizabeth) Teagle
Jack C. Threet
Charles Weiner
Harry Westmoreland
James E. Wilson, Jr.
P. W. J. Wood

The Foundation also gratefully acknowledges the many who have supported this endeavor with additional contributions.

*Based on contributions received as of September 30, 1992.

PREFACE

The Atlas of Oil and Gas Fields and the Treatise of Petroleum Geology

The *Treatise of Petroleum Geology* was conceived during a discussion held at the annual AAPG meeting in 1984 in San Antonio, Texas. This discussion led to the conviction that AAPG should publish a state-of-the-art textbook in petroleum geology, aimed not at the student, but at the practicing petroleum geologist. The textbook gradually evolved into a series of three different publications: the Reprint Series, the Atlas of Oil and Gas Fields, and the Handbook of Petroleum Geology. Collectively these publications are known as the *Treatise of Petroleum Geology*, AAPG's Diamond Jubilee project commemorating the Association's 75th anniversary in 1991.

With input from the Advisory Board of the Treatise of Petroleum Geology, we designed this set of publications to represent, to the degree possible, the cutting edge in petroleum exploration knowledge and application: the Reprint Series to provide useful and important published literature; the Atlas to comprise a collection of detailed field studies that illustrate the many ways oil and gas are trapped and to serve as a guide to the petroleum geology of basins where these fields are found; and the Handbook as a professional explorationist's guide to the latest knowledge in the various areas of petroleum geology and related disciplines.

The Treatise Atlas is part of AAPG's long tradition of publishing field studies. Notable AAPG field study compilations include *Structure of Typical American Fields*, published in 1929 and edited by Sidney Powers; and Memoir 30, *Giant Fields of 1968-1978*, published in 1981 and edited by Michel T. Halbouty. The Treatise Atlas continues that tradition but introduces a format designed for easier access to data.

Hundreds of geologists participated in this first compilation of the Atlas. Authors are from all parts of the industry and numerous countries. We gratefully acknowledge the generous contribution of their knowledge, resources, and time.

Purpose of the Atlas

The purpose of the Atlas is twofold: (1) to help exploration and development geologists become more efficient by increasing their awareness of the ways oil and gas are trapped, and (2) to serve as a reference for both the petroleum geology of the fields described and the basins in which they occur.

Imagination is the primary tool of the explorationist. Wallace E. Pratt once said that the unfound field must first be sought in the mind. In part, what is imagined is based on what is remembered; memory is the direct link to what is created in the mind. To create ideas that lead to the discovery of new fields, the mind of the geologist builds from its knowledge of petroleum geology. To that end, the Atlas of field studies will be a primary source for locating much of the information necessary for creating prospects and will provide a connection to the phenomenon of oil and gas traps.

Next to the firsthand experience of having prospects tested with the drill bit, studying the many facets and concepts of developed fields is perhaps the best way for the geologist to develop the ability to create plays and prospects. Also, familiarity with the many ways oil and gas are trapped allows the geologist to see through the noise inherent to exploration data and to close gaps in that data.

Format of the Atlas

To facilitate data access, all field studies in the Atlas follow the same format. Once users become familiar with this format, they will know where to look for the information they seek. Different fields from different parts of the world can be easily compared and contrasted.

The following is a generalized format outline for field studies in the Atlas:
Location
History
   Pre-Discovery
   Discovery
   Post-Discovery
Discovery Method
Structure
   Tectonic History
   Regional Structure
   Local Structure
Stratigraphy
Trap
   General Description
   Reservoir(s)
   Source(s)
Exploration Concepts

Criteria for Inclusion of a Field

Fields described in the Atlas are selected using two main criteria: (1) trap type, and (2) geographic distribution. Our ultimate goal for the Atlas is to include a field study from each major petroleum-producing province and to include an example of each known trap type. Size or economic importance are not, of themselves, criteria. Many fields that are not giants are included because they are geologically unique, because they are significant examples of geological investigation and original thinking, or because they are historically important, having led to the discovery of many other fields.

Grouping of Fields into Separate Volumes

We considered several ways to group fields in these volumes. We chose trap type because the purpose of the Atlas is to make exploration geologists more effective oil and gas trap finders, regardless of where they search for traps.

Grouping oil and gas field studies into separate volumes by trap type is a difficult exercise. We decided to group the fields into volumes by designating them as structural or stratigraphic traps. Most traps are a combination of both structure and stratigraphy. Some traps are obviously more a consequence of one than the other, but many are not. The continuum that exists between purely stratigraphic and purely structural traps is what makes grouping difficult. A further complication is that many fields contain more than one trap type.

Papers selected for *Structural Traps VII*

This volume in the Atlas of Oil and Gas Fields series contains studies of fields with traps that are mainly structural in nature. Like many structural traps, there is a strong component of stratigraphic control in most of the traps of these fields.

Distribution of the fields is worldwide. Eleven are outside the United States, and three are in the United States.

All of the fields described in this volume have anticlinal traps. The first seven fields have compressional anticlines caused by thrust faulting. Tip Top, Rangely, Matzen, Cortemaggiore, Penal/Barrackpore, and Laojunmaio have traps in thrust plates detached from basement. Malossa's trap is in a basement-involved thrust plate.

The traps of Raman/Bati-Raman/Garzan fields are in compressional anticlines broken up by high-angle reverse faults. The anticlinal trap of Messoyakh produces gas from hydrates and from that gas sealed beneath shales and sandstones plugged by gas hydrates. South Belridge field of California has a trap that is mainly anticlinal in nature with porosity pinch-outs, unconformity truncation, fractured chert, and tar seal.

The oil of Portachuelo field was emplaced before numerous low-angle faults broke up the anticlinal trap and left fluid contacts perched at different levels in the broken segment. Tintaburra field has an anticlinal trap with multiple pays. Raguba and La Cira/Infantas are giant fields in faulted anticlines.

Another aspect described in each field is the history of the field's exploration and development. What seems obvious today usually was not obvious when these fields were discovered. Many times what was expected was not what was found. Geologists discovered these fields by creating concepts based on information sometimes limited by the technology available at the time, and at other times limited by economics and perceptions.

Drilling and discovery show how closely concept matches reality. Knowing the history of discovery may help explorationists realize that problems, seemingly insoluble at one time, were eventually solved. It is also instructive to learn of the sequence of thinking that solved these problems, as well as the role of serendipity in discovery. If luck has a role in successful exploration, we still come away with the sense that creativity, logic, and knowledge guide the explorationist to those circumstances where the serendipity may occur.

Careful study of these fields will enhance the propect generator's knowledge toward future prospecting.

Edward A. Beaumont
Norman H. Foster, Editors

Tip Top Field—U.S.A.
Wyoming-Utah-Idaho Overthrust Belt, Wyoming

JANELL D. EDMAN
Marathon Oil Company
Littleton, Colorado

LANCE COOK
Union Pacific Resources
Fort Worth, Texas

FIELD CLASSIFICATION

BASIN: Utah-Wyoming-Idaho Overthrust Belt
BASIN TYPE: Thrust Belt
RESERVOIR ROCK TYPE: Sandstone and Subarkose/Quartzarenite
RESERVOIR ENVIRONMENT OF DEPOSITION: Marine Bars (and Possible Deltaic) and Eolian
RESERVOIR AGE: Cretaceous and Triassic-Jurassic
PETROLEUM TYPE: Oil, Condensate, Gas
TRAP TYPE: Anticlinal and Stratigraphic/Diagenetic
TRAP DESCRIPTION: Multiple-pay stratigraphic/diagenetic trap; anticline created by forward splay of thrust fault and antithetic backthrust

LOCATION

Tip Top field is located in the western United States on the easternmost edge of the Wyoming-Utah-Idaho Overthrust belt, adjacent to the Green River basin to the east (Figure 1). Specifically, this field, which has an estimated ultimate recovery of 584 BCFG and 6.1 MMBO and natural gas liquids, is located in T28-29N, R113-114W, Sublette County, Wyoming (Figure 2). The reserves volumes cited here are for the Triassic-Jurassic Nugget and Cretaceous reservoirs only (Figures 3-6). Tip Top is unitized from 1500 ft (458 m) above the Cretaceous Frontier on down, thus excluding the shallower production (Figures 2-5), and Mobil has never produced gas out of the deep Paleozoic reservoirs because of problems with associated CO_2 and H_2S. Although Tip Top itself is not a giant field, it is within the giant La Barge platform field complex (Munsart, personal communication, 1988).

Tip Top field is significant from an exploration perspective for two reasons. First, it is one of the few fields located in a frontal thrust system, in this case the Darby (Figure 1), of the Wyoming-Utah-Idaho Overthrust belt. In contrast, the majority of fields in this region are in older thrust systems to the west toward the internal side of the thrust belt. More important, Tip Top and two fields to the south (Hogsback and Dry Piney) are among the only thrust belt fields that produce Phosphoria-sourced liquids.

In the remainder of the region, particularly to the west where the original sediment pile was thicker, the Phosphoria was probably already overmature prior to trap formation (Warner, 1982; Valenti, 1987); consequently, no Phosphoria-sourced liquids are preserved. The presence of Phosphoria-sourced liquids in the younger Darby system thus suggests a distinctive play within a region where the majority of production is from traps associated with older thrusts charged by Cretaceous-sourced hydrocarbons.

EXPLORATION HISTORY AND DISCOVERY OF HYDROCARBONS

Pre-Discovery

Exploration activity began much earlier in the Tip Top/La Barge area than in most of the thrust belt. Activity may have been initiated due to Schultz's 1907 discovery of oil seeps east of La Barge (Hogsback) ridge in T27N, R113W (Hodgden and McDonald, 1977). Schultz went on to map the Tip Top anticline, which he called the Dry Piney anticline (shown on the Intrasearch map in Figure 7), in T28N, R113W and projected this structure through Tertiary strata

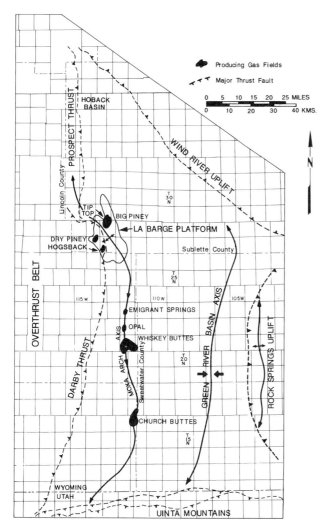

Figure 1. Green River basin index and tectonic map illustrating the location of Tip Top field in the transition zone between the Green River basin and the Wyoming-Utah-Idaho Overthrust belt of the western United States.

in T26N and 27N, R113W, encompassing the seeps. His 1914 map (Schultz, 1914) showing the superposition of the Tip Top anticline with surface seeps was probably a major factor in General Petroleum Corporation's decision to explore in the area during the 1940s and 1950s.

The next major publication documenting early activity is a March 1921 geological report by Charles Lackey that describes drilling on the Dry Piney anticline that led to the discovery of oil in upper Hilliard sands. The Texas Production Company well in Lot 7, Sec. 3, T26N, R113W, which was the La Barge field discovery well (Hodgden and McDonald, 1977), was also drilled in the early 1920s. This well was completed for 10 BOPD from an Almy (Paleocene) sand at 568 ft (173 m). By the end of 1924, there were nine producing wells in the La Barge field, but Texas Company remained the only major oil company interested in the area until the late 1940s. One exception was a deep test drilled in 1943 by Wasatch Producing Company in Sec. 8, T27N, R113W.

Discovery

In the late 1940s, General Petroleum Corporation became interested in the area, based in part on the surface anticline (Figure 7) initially mapped by Schultz in 1914 coupled with the presence of nearby production and plane table mapping undertaken by company geologists. The operating assumption was that the surface anticline out in front of the Darby thrust must also extend under the thrust. General Petroleum formed the Tip Top Unit (Figure 2) after conducting a seismic program over the area. The first unit test was the General Petroleum 81-7 Government drilled in NE NE NE Sec. 7, T28N, R113W. It was abandoned in the Triassic-Jurassic Nugget Sandstone at a total depth of 10,680 ft (3257 m) in December 1948.

A second well, in SW SW NW Sec. 12, T28N, R114W (Figure 2), encountered oil in the Triassic-Jurassic Nugget Sandstone. This well was completed in May 1951, flowing 266 BOPD from an open hole between 9678 and 9807 ft (2952-2991 m). The discovery did not create a great deal of excitement at the time, however, because of the relatively thin oil column, high drilling costs, complex geology, and production rates that were not particularly impressive (Hodgden and McDonald, 1977). Figure 8 is a photograph taken by Jack Rathbone showing the location of two early wells drilled at Tip Top, and Figure 9 is an early structural cross section drawn by General Petroleum geologists.

The recognized discovery well at Tip Top field (Webel, 1979) was not completed until three months later in September 1951 and was actually a recompletion of a Texaco well that found gas in the Frontier in 1939. Texaco had plugged the well because there was no market for the gas, but General Petroleum re-entered the well in 1951 and completed it as a Tip Top Unit well from the Frontier. This well, the 38-1G, in SE SW Sec. 1, T28N, R114W (Figure 2), had an initial potential of 1.8 MMCFGD on a ½-in. choke from 120 shots in the interval 5343-5373 ft (1630-1639 m). The discovery of the Tip Top field therefore occurred 24 years before the 1975 Pineview discovery that began the largest and most recent wave of exploration activity in the Wyoming-Utah-Idaho Overthrust belt.

Post-Discovery

Arthur B. Belfer, founder of Belco Petroleum (now Enron Oil and Gas Company), began developing the Tertiary gas in the Big Piney-La Barge area in 1952. By 1954, when he committed the gas to Pacific Northwest Pipeline, Belfer had 33 shut-in gas wells (Hodgden and McDonald, 1977). Development of the Frontier gas by both Belco and Mobil (Mobil had purchased General Petroleum prior to the discovery

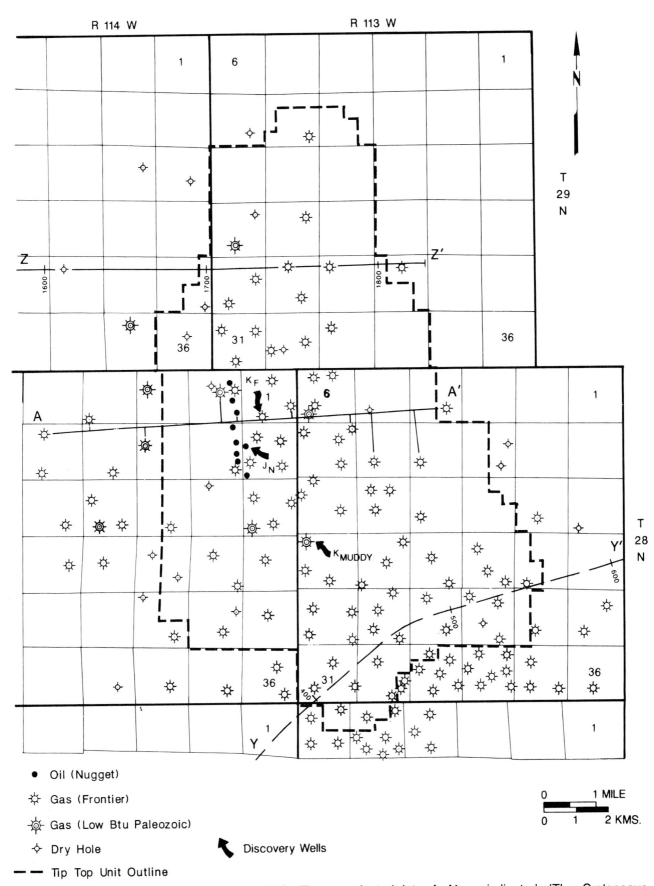

Figure 2. Detailed map of the Tip Top Unit. The locations of seismic lines Y-Y' and Z-Z' are shown as well as that of cross section A-A'. Well control is projected into A-A' as indicated. (The Cretaceous Muddy Formation in the discovery well indicated is the equivalent of the Bear River Formation. See Figure 3.)

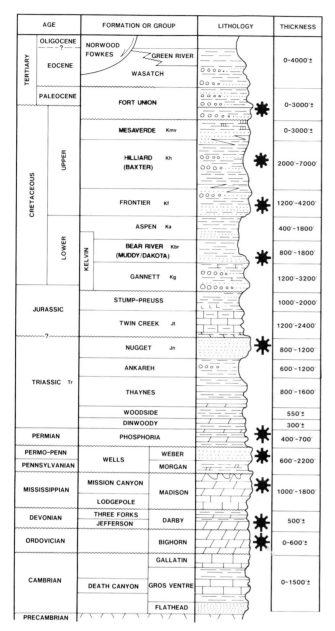

Figure 3. Stratigraphic column for Tip Top area. (Modified from Lamerson, RMAG, 1982.) Symbol indicates producing zones.

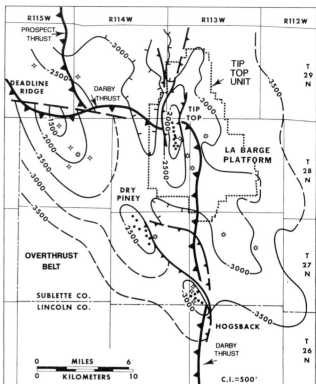

Figure 4. Regional Triassic–Jurassic Nugget structure. Where overlain by the Wyoming-Utah-Idaho Overthrust belt, autochthonous structure is shown. (Modified from Webel, 1977.) Contour interval, 500 ft (153 m).

of Tip Top but did not drop the General Petroleum name until later) soon followed, and by 1956 when Pacific Northwest completed their pipeline, a number of other operators were also involved in the play. The Big Piney–La Barge producing complex has been a major factor in Wyoming gas production since that time.

In 1962, the Mobil 22-19G well, NW NW Sec. 19, T28N, R113W, in the Tip Top field became the first well to penetrate lower Paleozoic reservoirs in the basin block of the western Green River basin (Marzolf, 1965). The high volume, low BTU gas initially detected in the Paleozoics by this well, however, would not be marketed for almost another 25 years (Munsart, personal communication, 1988), and Exxon rather than Mobil would tap this resource. Exxon began evaluating a large lease block with Paleozoic potential in the northwestern portion of the La Barge platform during the 1970s, but the first sales of Paleozoic gas did not occur until September 1986 (Hunter and Bryan, 1987). The delay was due primarily to environmental, government, and technical concerns.

Mobil has never produced deep gas out of the Paleozoic reservoirs because of problems related to removal of the associated CO_2 and H_2S, and Nugget condensate production was shut in in 1986. Thus, current production is limited to gas from the Cretaceous Frontier and Muddy formations. The 147 production wells in the combined Tip Top/Hogsback Unit are generally spaced 160 to 320 ac (65 to 130 ha) apart, and when a market exists, gas from these wells is transported by Northwest Pipeline to Opal, Wyoming, where it is processed and distributed to various markets.

DISCOVERY METHODS AND CONCEPTS

A surface anticline associated with oil seeps was described by Schultz in 1914. This surface feature,

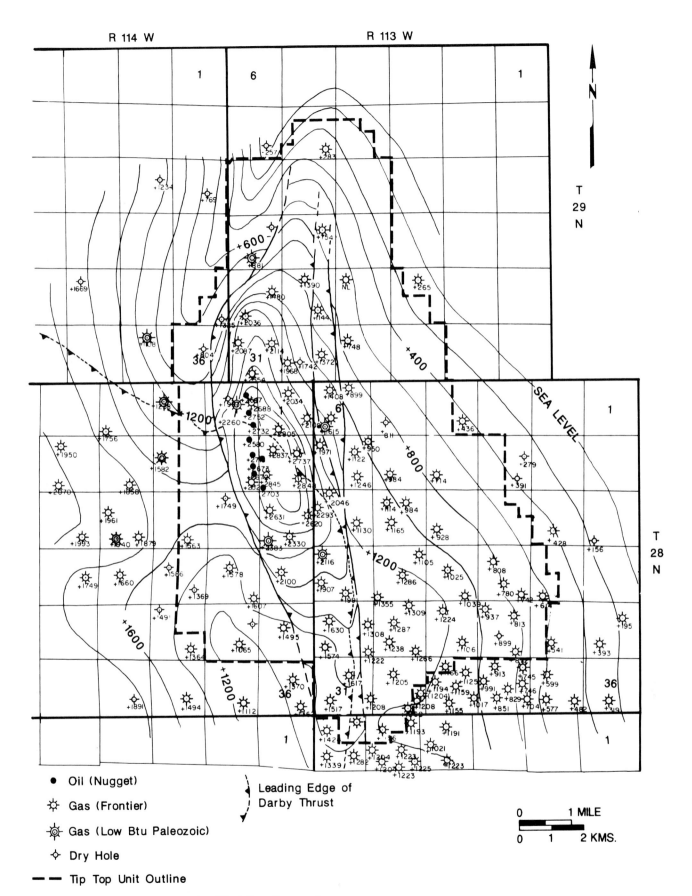

Figure 5. Structure map on the First Frontier marker. The surface trace of the Darby thrust fault cuts across the Tip Top Unit outline. Contour interval, 200 ft (61 m).

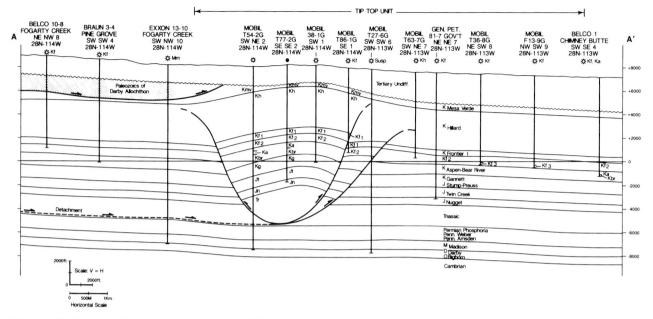

Figure 6. Structural cross section A-A'. Well control is projected. See Figure 2 for location and Figure 3 for formation abbreviations.

the Tip Top anticline, is seen in Figure 7 (labeled Dry Pine anticline). It is, however, fortuitous that the surface anticline at Tip Top coincides with the deeper Nugget structure. Webel (1977) shows the surface Tip Top anticline was created by shallow thrust faults rooted in the Hilliard shales. These shallow faults are not related to the deeper faults that created the Nugget trap. The shallow thrusts are clearly seen on seismic line Z-Z', as discussed later in the paper (under *Local Structure*).

Modern extension of the Tip Top play would involve searching for low magnitude structures created by forward splays of major thrust faults along strike to the north and south. Similar décollement-related structures could reasonably be anticipated to exist along trend. Because such low relief structures are not always expected to have a surface expression, seismic data would be essential. Owing to the limited area of closure associated with these structures, close seismic line spacing would be necessary to define a drillsite. In addition, any acquired seismic data would need to be accurately migrated and integrated with surface and subsurface geology. Migration is particularly important because the seismic expression of Tip Top is obscured by diffractions originating from faults bounding the structure.

The geologic concept of low magnitude structures in front of major thrusts created by forward splays or blind thrusts, such as the Tip Top Nugget structure shown in Figure 6, has long been recognized and actively used in overthrust belt exploration. Dry Piney and Hogsback fields in Wyoming are discoveries that were based on this concept (Webel, 1977). Johnson (1984) shows that Knowlton field in the Montana Disturbed belt is another structural analog. Canadian explorationists have used this concept successfully in the discovery of several fields, including Stolberg, Jumping Pound, Pincher Creek, and Sarcee in Alberta (Bally et al., 1966). The Canadian structures, while similar in structural style, are much larger in relief than the U.S. fields and were more easily identified.

STRUCTURE

Tectonic History

Western Wyoming was largely quiescent during the Paleozoic and was the site of shelf carbonate deposition (Peterson, 1977). Although there are numerous unconformities in the Paleozoic section, only the Ordovician-Devonian unconformity represents a significant hiatus (Figure 3). This event results in the omission of the Silurian section, which is commonly missing through much of the Rocky Mountain region. The cyclical shelf carbonate environment was interrupted more frequently in the early Mesozoic (Triassic-Early Jurassic) by periods of sabkha and eolian deposition (Armstrong and Oriel, 1965).

By Late Jurassic, compression resulting from subduction at the western craton margin became expressed as thin-skinned thrusting of the Sevier orogeny. The older cyclical shelf carbonate environment was disrupted by two coincident events, and the depositional pattern shifted to a dominantly clastic system. As the region moved north of 30°N latitude by continental drift (Peterson, 1977), the rate of carbonate deposition was significantly reduced. At about the same time, initial movement on the Paris-Willard fault system (Armstrong and Oriel, 1965) elevated the clastic source areas of the Ephraim conglomerate of the lower Gannett Group (Figures 3, 10). As the structurally uplifted highlands impinged further upon the craton, new thrusts propagated in an easterly direction (Oriel and Armstrong, 1966), with each successive thrust sourcing a new pulse of clastic sedimentation (Figure 10). Generally, as Sevier thin-skinned thrusting diminished at the close of the Mesozoic, Laramide (basement-involved) deformation began. Overlap of the Sevier and Laramide events has been recognized by Chapin and Cather (1981) and Jordan (1981) among others. The Moxa arch (Figure 1) is an early Laramide structure (Gries, 1983) that experienced major movement at the end of the Cretaceous as demonstrated by a regional unconformity at the Cretaceous-Paleocene boundary. After the structural climax of the overlapping Sevier and Laramide orogenies, western Wyoming became relatively quiet, although a period of normal faulting followed cessation of compression (Royse et al., 1975).

Regional Structure

Tip Top field sits in a transition zone between the Sevier orogenic belt, marked by eastward vergence of thin-skinned décollement-style thrusting, and the Rocky Mountain foreland, characterized by basement-involved reverse faulting and uplift (Brown, 1988). Figure 10 shows that the Sevier highlands were active in the Late Jurassic or earliest Cretaceous, based on the relationship of the Paris-Willard thrust and the Ephraim conglomerate of the Gannett Group (Royse et al., 1975). Movement on Sevier thrust faults continued through the Paleocene (Dorr et al., 1977). The Laramide orogeny overlaps and then postdates the Sevier event. Gries (1983) shows the time span of Laramide events ranging from Late Cretaceous to middle Eocene.

In the Tip Top area, two main overthrust faults, the Darby and the Prospect, are involved with the development of the Nugget structure (Figures 1 and 4). The Moxa arch (Figure 1) is a Laramide uplift that also plays a role in the evolution of the Tip Top Nugget structure. Kraig et al. (1987) have examined the interaction of the Moxa arch and the Overthrust belt.

Sevier structures are involved in the development of the productive Frontier sands in the La Barge area. Mapping by De Chadenedes (1975) clearly shows the Frontier source area to have been the orogenic highlands of the Sevier thrust belt. Figure 10 shows the Paris-Willard thrust sheet should have been uplifted at the time of Frontier deposition.

Local Structure

The deep Paleozoic section of Tip Top field is relatively unstructured (Figures 6, 11–15). Because Tip Top field occupies a near-crestal position at the northern end of the doubly plunging Moxa arch (the La Barge platform, Figure 1), the Paleozoic section falls within the closure of the Moxa arch crest. All of seismic line Z-Z' (Figure 15) and all of seismic line Y-Y' east of shot point 240 (Figure 13) are over the Paleozoic gas field.

Above the Paleozoic section, structures in the Tip Top area are related to décollement faulting and backthrusting. Webel (1977, 1987), Dorr et al. (1977), and Johnson (1984) are some recent authors who have recognized backthrusting as an important structural element in décollement faulting. Multiple levels of detachment and backthrusting can be seen in the Tip Top area (Figures 6, 13, and 15).

The producing Triassic–Jurassic Nugget structure in the Tip Top unit can best be described as a "pop-up" structure formed by a forward-directed thrust (east-bounding fault, Figure 6) and an antithetic backthrust (west-bounding fault, Figure 6). Webel (1977, 1987) recognized the Nugget trap as a thrust-faulted structure along with the structural analogs at Dry Piney and Hogsback fields (Figures 1 and 4). Seismic line Y-Y' (Figures 11–13) shows the Dry Piney Nugget structure at shot point 280. The Tip Top Nugget structure is not seen on this line. Strike continuity of the low relief "pop-up" structure is not extensive. Seismic line Z-Z' does show a strike projection of the Tip Top Nugget structure (Figures 14 and 15) at shot point 1740. Cross section A-A' (Figure 6) shows the best-developed portion of the Tip Top Nugget structure. Seismic data and well control suggest the décollement controlling these structures is in the Triassic red beds. The relatively small area of Nugget production, about 1260 ac (510 ha), is thought to closely match the area of simple closure at the Nugget horizon.

Frontier level structure can be seen in Figures 5, 6, 13, and 15. Although the stratigraphically trapped Frontier sands are productive both on and off structure, Frontier wells within the horst block shown in Figure 5 are generally the most prolific. This increased productivity may be due to less severe diagenesis of authigenic clays and the superior reservoir quality which is thereby less damaged. Fault duplication of the Frontier reservoir contributes to increased productivity in some wells. Seismic line Z-Z' (Figure 15) shows the Nugget and Frontier structures to have separate décollements. Cross section A-A' (Figure 6) shows the Nugget and Frontier structures sharing the same décollement. To the south, seismic line Y-Y' (Figure 13) shows only a weak, forward-directed thrust at shot point 460 with no expression of a Frontier-involved backthrust. This suggests the "pop-up" features have variable displacement along strike, and their related décollements can change stratigraphic level fairly quickly.

The surface geology of the Tip Top unit is somewhat deceptive. Figure 7 shows the Tip Top anticline very clearly. Seismic line Z-Z' (Figure 15) shows the reason for the surface anticline at shot

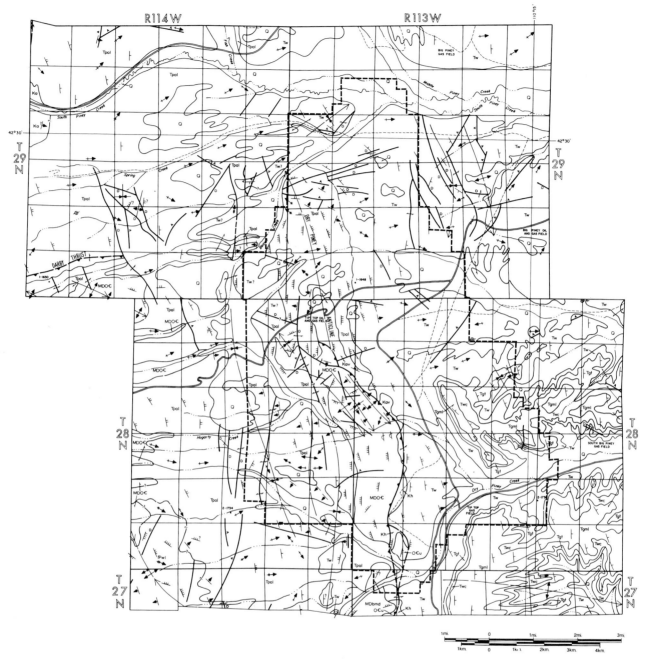

Figure 7. Surface geology of the Tip Top area. Note the pronounced Tip Top surface anticline (labeled here as "Dry Piney Anticline") and the trace of the Darby fault. (Mapping provided courtesy of Intrasearch, Inc. of Englewood, Colorado.) Key to symbols appears on page 9.

point 1720. A shallow décollement in the Hilliard Shale and an opposing backthrust create a structure that appears to be a simple anticline on the surface map. Drilling this surface feature would result in a Frontier gas discovery, but only by serendipity, because the Frontier and Nugget structures are independent.

Determining the time of formation for the Tip Top anticline and related structures is difficult. Webel (1977) states that the Darby thrust is folded by underlying thrusts, which requires the underlying thrusts have post-Darby movement. Seismic line Y–Y' (Figure 13) supports this observation. The Darby thrust is clearly deformed by the underlying Dry Piney structure at shot point 280. Because Dry Piney and Tip Top share a common Triassic décollement, they may both have a history of post-Darby movement. Figure 10 suggests this could be as young as Eocene (post-Wasatch) age. Surface mapping (Figure 7) clearly shows the Paleocene section is folded as part of the Tip Top structure. Dorr et al. (1977) have shown that movement on the Prospect thrust in the Hoback basin (Figure 1) ceased in early Eocene. However, these observations all point to an age of final

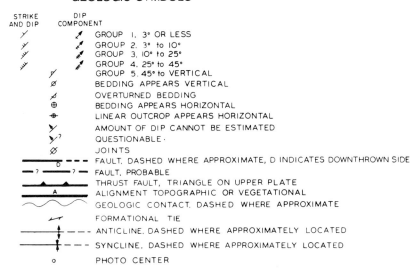

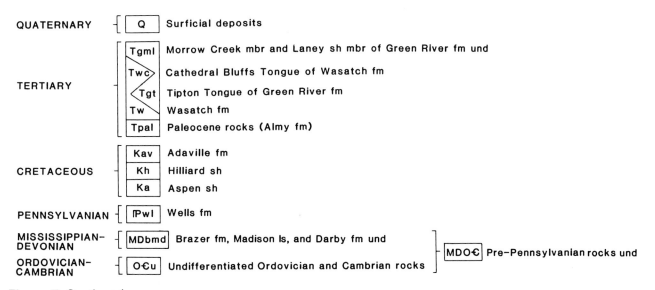

Figure 7. Continued

movement and do not define the time of initial movement. Kraig et al. (1987, 1988) have shown the eastern ramp of the Prospect thrust in Township 31 roots from a décollement in the Triassic clastic section, probably the same décollement that controls the Tip Top and Dry Piney structures. Kraig et al. (1988) demonstrated that the first movement of the Prospect fault, which they interpret to be older than the Darby thrust, was synchronous with initial movement of the Moxa arch fault in Late Cretaceous (Maastrichtian) time. Thus, the structural evolution of Tip Top field may range from Late Cretaceous (Maastrichtian) movement on the Triassic décollement for the Nugget structure to early Eocene movement on the shallower décollement that created the Tip Top surface anticline.

STRATIGRAPHY

Rocks penetrated at Tip Top range in age from Tertiary to Cambrian. A stratigraphic column showing the various units is given in Figure 3; Figure 16 shows gamma ray, resistivity, and velocity logs for the Cretaceous through Cambrian section and Figure 17 is a type log for the producing horizons. Marzolf (1965) has published detailed descriptions of these formations; therefore, only key reservoir and source intervals will be summarized here. Additional information on these units can be found in the reservoir and source sections, respectively.

Most Tip Top production is or has been from the Triassic-Jurassic Nugget and Cretaceous Frontier, although there is minor production from the

Figure 8. View to the south-southeast showing site of 4X-12G Nugget discovery well (mud pits) and rig on 26X-12G location. Ridge is the Death Canyon Limestone. (Photograph by Jack Rathbone.)

Cretaceous Muddy and Dakota. The Nugget is primarily an eolian deposit that is much more consistent lithologically than the Frontier. Typically the Nugget is a very fine to fine-grained subarkose or quartzarenite. Interbedded with the sandstone are siltstone, mudstone, and limestone that may represent interdune deposits.

In contrast, the Frontier comprises a diverse assemblage of siliceous clastics and coals deposited in a transition zone from a western fluviatile to an eastern marine environment. Incorporated within the Frontier are deltaic, point bar, channel fill, channel mouth bar, and marine bar deposits. Reservoirs within the Frontier are predominantly quartz sandstones that also contain chert, mica, glauconite, and clay.

The two major source intervals in the Wyoming-Utah-Idaho Overthrust belt are the Permian Phosphoria Formation and the Cretaceous (Lamerson, 1982; Warner, 1982). Maughan (1984) indicates the phosphatic shale members of the Phosphoria, which were probably deposited in an ancient coastal upwelling system, are especially organic rich. Within the Cretaceous, Rice and Gautier (1983) have suggested transgressive shales, progradational shales, and nonmarine coal swamps as the three settings most favorable for the production, accumulation, and preservation of organic matter.

TRAP

Tip Top field is a structural trap at the Triassic-Jurassic level and a stratigraphic-diagenetic trap at the Frontier level. Munsart (personal communication, 1988) has described the Paleozoic low BTU gas accumulation of the La Barge platform, of which Tip Top covers only a portion (Figures 1 and 4). The structural nature of the Nugget accumulation is seen in Figure 6. Although the Frontier structure map of Figure 5 shows about 500 ft (153 m) of simple closure at the Nugget level, Webel (1977) shows simple closure of about 200 ft (61 m). The Nugget

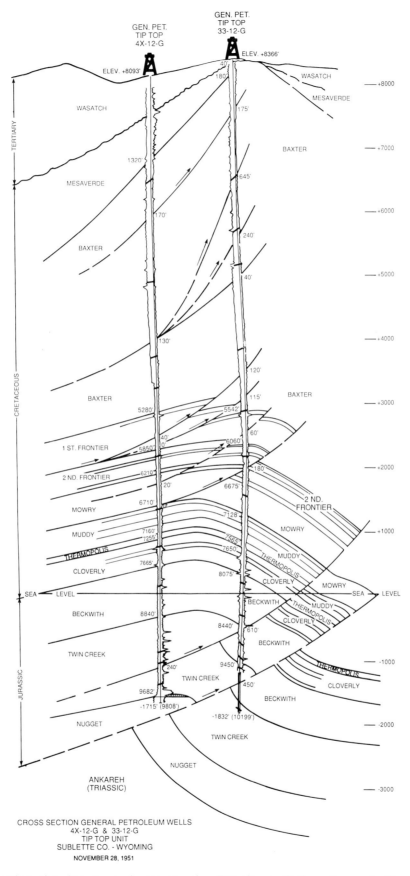

Figure 9. Cross section showing an early structural interpretation, as well as earlier stratigraphic terminology, of Tip Top made by geologists at General Petroleum. Both wells are in Sec. 12, T28N, R114W. Slanted lines indicating dip meter readings are shown between the electric log curves.

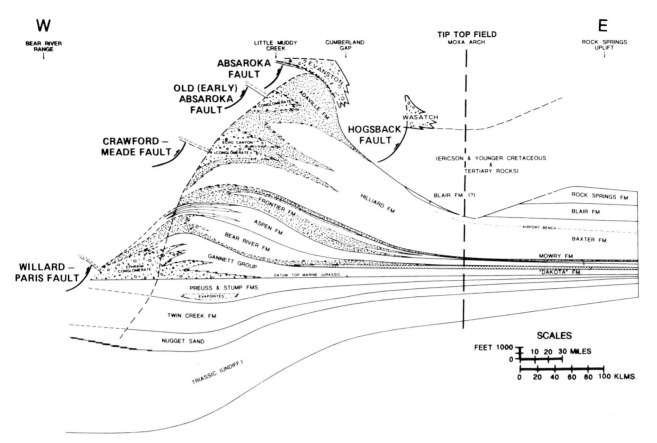

Figure 10. Stratigraphic diagram of restored Mesozoic and early Tertiary rocks in western Wyoming showing the relationship between thrust faulting and sedimentation. Hogsback fault is equivalent to the Darby fault. (Modified from Lamerson, RMAG, 1982.)

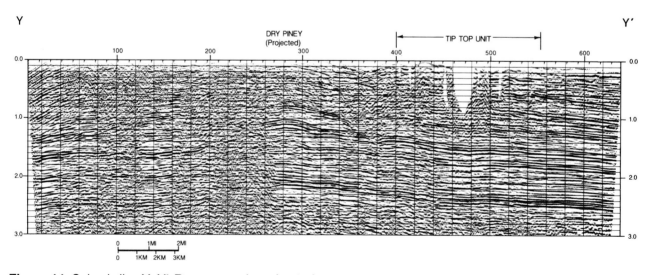

Figure 11. Seismic line Y-Y'. Reprocessed unmigrated data. See Figure 2 for location.

oil column is only 132 ft (40 m) (Webel, 1979), which means the closure is not filled to spill point for reasons not clear at this time. The anhydritic shales of the Gypsum Springs Member of the lower Twin Creek are probably the vertical seal for the Nugget reservoir.

The Frontier gas accumulation, although located near the crest of the Moxa arch, is a stratigraphic-diagenetic trap. The 500 ft (153 m) of closure in Figure 5 is insignificant in view of the 2000-plus ft (610 m) gas column in the Frontier. Downdip production is

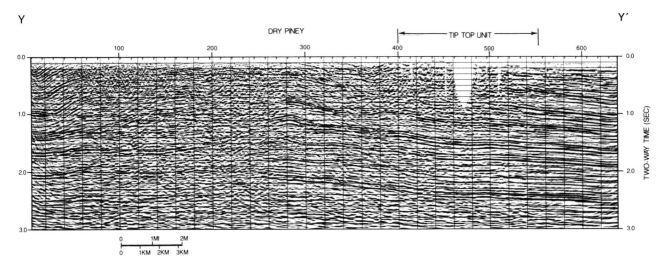

Figure 12. Seismic line Y-Y'. Reprocessed, migrated data. See Figure 2 for location.

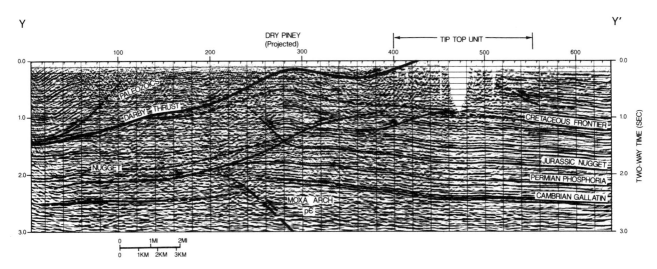

Figure 13. Seismic line Y-Y'. Reprocessed, migrated, and interpreted data. See Figure 2 for location. Key beds and major faults are marked. Stratigraphic control projected from the Mobil F22-19G well (Figure 16). Note the clear expression of the analogous Dry Piney "pop-up" structure at Nugget level.

limited by clay diagenesis within the sandstone units. Webel (1979) states there is no discrete gas-water contact in the Frontier. Log analysis and sample examination by SEM indicate the intensity of clay diagenesis and resultant water saturation values increase with depth. Thus the lower limits of production are controlled by reduced permeabilities and flow rates resulting in unsatisfactory reserves and poor economics. Many late development wells penetrated sands with virgin pressures, implying stratigraphic separation within the amalgamated sand bodies comprising the Frontier reservoir. The initial reservoir pressure of 3300 psi (22,754 kPa) represents a gradient of 0.60 psi/ft (13.56 kPa/m) at 5500 ft (1678 m), which means the Frontier is mildly overpressured. Thick marine shales of the overlying Hilliard (Figure 3) provide an excellent seal for the Frontier.

Reservoirs

The two major reservoirs at Tip Top are the Triassic-Jurassic Nugget and the Cretaceous Frontier. With the exception of Marzolf's (1965) description of cores from the Mobil 22-19G well, little detailed stratigraphic information specific to the Tip Top field has been published, and most of this comprises regional studies of these two reservoirs. Information on the Nugget, which has been shut-in since 1986 because of high water cut and low oil prices, is very limited because there are no recent production data.

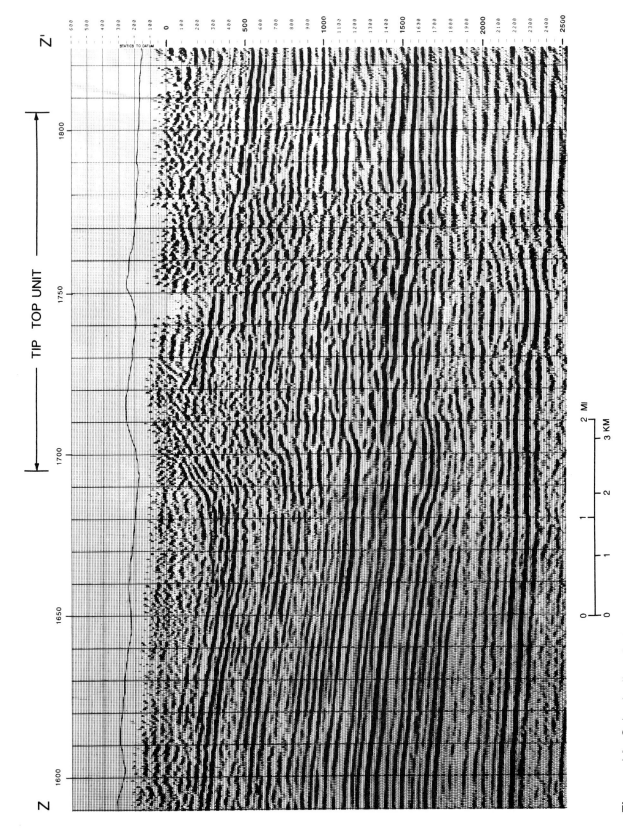

Figure 14. Seismic line Z-Z'. Original processing, migrated data from 1979 industry group shoot. See Figure 2 for location.

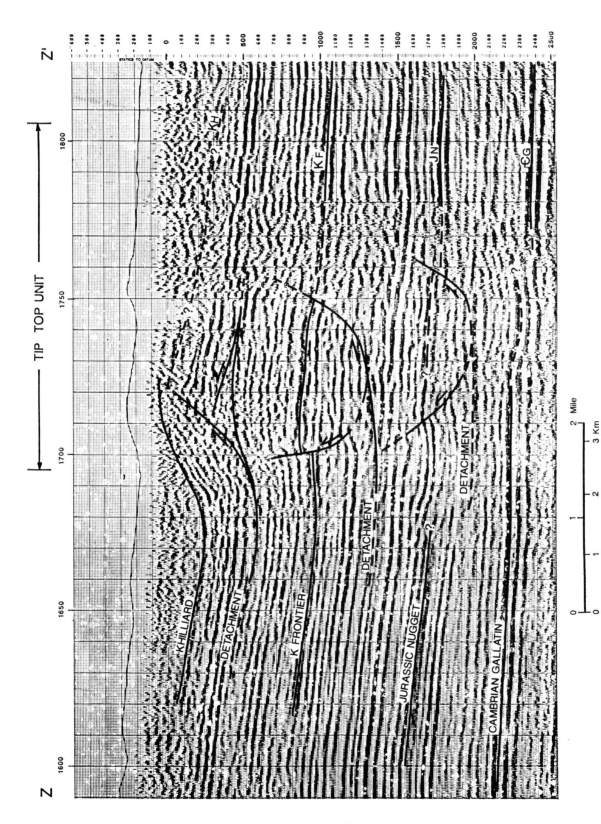

Figure 15. Seismic line Z-Z'. Original processing, migrated data from 1979 industry group shoot with interpretation. See Figure 2 for location. Stratigraphic control for key beds projected from the Mobil F22-19G well (Figure 16). Origin of the Tip Top surface anticline (shotpoint 1720) is clearly seen. Pop-up structures at the Frontier and Nugget levels appear to be derived from different décollements in this section.

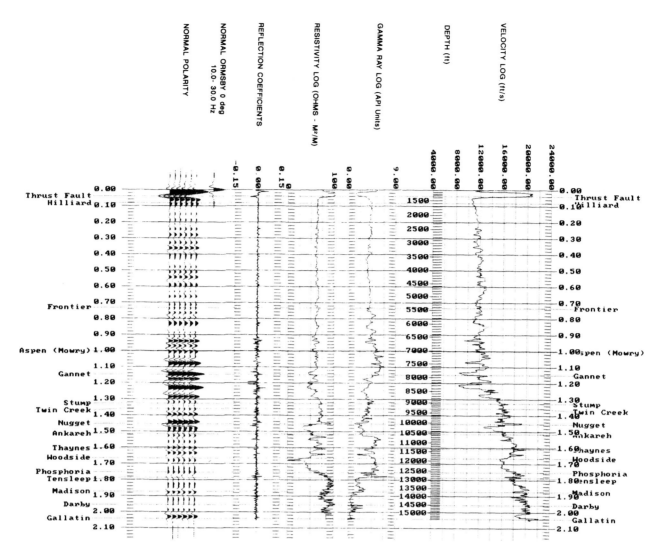

Figure 16. Deep stratigraphic control from the Mobil F22-19G well in NW NW Sec. 19, T28N, R113W. Gamma ray, resistivity, and sonic velocity curves are shown with synthetic seismogram.

Doelger (1987) provides a regional synthesis for the Nugget Sandstone. According to Doelger, the Nugget is generally assigned a Triassic–Jurassic age but may be entirely Jurassic.* Fossils are scarce in this predominantly eolian deposit, and as a result, most age assignments are determined by stratigraphic relationships with units of known age rather than by fossil evidence.

Doelger (1987) reports that the Nugget is usually lithologically consistent where it has been described and is "typically composed of very fine to fine grained, subangular to round, moderately to moderately well-sorted subarkose, quartz arenite, or both." Grains comprise 65–85% of the sandstones while matrix is less than 10%, and cement, which is mostly silica in the thrust belt, is usually less than 20%. Porosity varies from less than 1% to around 20%. Siltstone, mudstone, and limestone are interbedded with the sandstones. Photomicrographs of the Nugget Sandstone from Tip Top support these descriptions (Figure 18). These particular samples contain predominantly rounded quartz grains with apparently good intergranular porosity. Oversize pores suggest the presence of some secondary porosity as well.

In many areas, the Nugget sandstones are characterized by medium- to large-scale cross-stratification that is interpreted to be eolian in origin. Paleocurrent studies indicate the prevailing direction of transport for these eolian sands was to the south (Doelger, 1987). There are, however, water-laid sediments interbedded with the eolian units at most locations, and in some places the water-laid sediments are dominant.

The regional depositional environments of the Frontier Formation in southwestern Wyoming were interpreted by De Chadenedes (1975). He concluded the Frontier in western Wyoming contains siliceous

---

*For brevity on tables and figures, the Nugget is designated as either Jurassic Nugget or Jn.

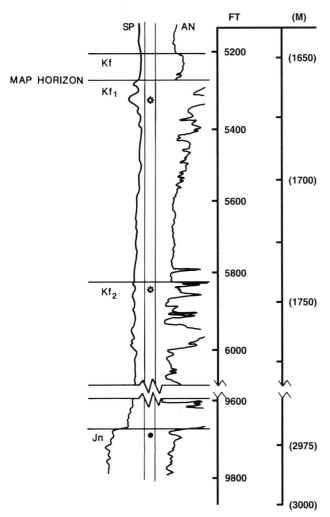

Figure 17. Tip Top Unit 4X-12 well of General Petroleum (SW SW NW Sec. 12, T28N, R114W) type log, spontaneous potential, and amplified normal resistivity for the two main producing horizons at Tip Top. KB, 8098 ft (2470 m). TD, 9808 ft (2991 m). (From Webel, 1979.) (See Figure 3 for formation abbreviations.)

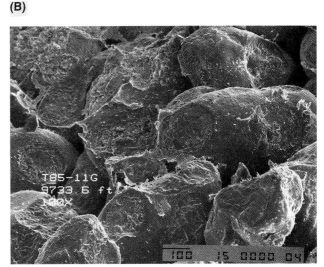

Figure 18. (A) Photomicrograph of Nugget Sandstone from 9733.6 ft (2966.8 m) in the T85-11G well. Scale bar, 0.5 mm. (B) SEM photomicrograph of Nugget Sandstone from 9733.6 ft (2966.8 m) in the T85-11G well. Scale bar on lower right is 100 microns.

clastics and coals deposited in a zone of transition from a western fluviatile to an eastern marine environment. Depending upon the relative balance between tectonism and absolute changes in sea level, marine shales transgressed to the west or coarse clastics prograded east. Within this regional framework, two large deltaic complexes developed: one to the south near Cumberland Gap (west and north of Church Buttes field, Figure 1) and the second to the north at La Barge. Localization of the La Barge delta may have been related to the presence of the structurally high La Barge platform. Furthermore, bar sands within the First and Second Frontier sandstone intervals tend to follow the trend of the Tip Top and Dry Piney anticlines, indicating that these features may also have been high during Frontier deposition (De Chadenedes, 1975).

In his cross section through Tip Top, De Chadenedes (1975) shows the presence of both the First and Second Frontier sandstones in the General Petroleum No. 33X well. Regionally, the First Frontier sandstone produces from two elongate sandstone bodies: an older, more widespread eastern bar and an overlying western bar. Both bars are quartz sandstones that also contain mica, glauconite, and some clay. The western bar is thicker and generally the better reservoir with 12-25% porosity while the eastern bar is commonly finer grained than the western bar and has about 10 to 18% porosity (De Chadenedes, 1975).

The Second Frontier sandstone produces over a larger area and contains most of the gas reserves of the greater La Barge platform. At Tip Top, the shallow estuarine and marine delta of the Second Frontier is overlapped by north–northwest-trending bar sandstones in the upper part of the interval. These marine bars were laid down as the sea transgressed westward across the point bars, channel fills, channel mouth bars, and marine bars that comprise the underlying deltaic deposits. According to De Chadenedes, the Second Frontier sandstones are usually fine to very fine grained and poorly to well sorted, with a variable silt content. They often contain chert pebbles up to 1 in. in diameter. These descriptions are consistent with photomicrographs of Frontier sandstones shown in Figure 19. The Third Frontier typically has the poorest reservoir quality of the three productive intervals.

In considering the Frontier within the larger context of the entire La Barge platform, Munsart (personal communication, 1988) described a complex, anisotropic, generally low-permeability reservoir where changes in petrophysical characteristics are frequently abrupt, occurring over relatively small distances. At Tip Top, however, porosity and permeability seem to be fairly uniform even though individual reservoir lenses are discontinuous (Lowell Martinson, personal communication, 1990). Porosity in the Frontier ranges from 12 to 15% and permeability varies from 0.4 md to as much as 8 md where the interval has been fractured during well treatment. There is no evidence of natural fractures in the few cores taken, but higher rates and drainage volumes along the axis of the Tip Top anticline suggest fractures may be present in this part of the field. Pay zones are stacked and total thickness varies from 80 to 230 ft (24 to 70 m), depending upon which intervals are productive in a given well. At Tip Top, the First Frontier is the most consistently productive while the Second Frontier is the most variable.

Gas produced from the Frontier is 86% methane with a gravity of 0.66; at the present time the market for it is variable. When the gas price is high enough, gas is sent to market via Northwest Pipeline. Condensate from the Frontier is 53° API with yields of 2 to 3 BC/MMCFG and about 1 bbl water/MMCFG. Initial flow rates after fracturing are sometimes as high as 5 MMCFGD; however, most wells in the field average between 300 and 600 MCFGD. Although there are no spacing requirements at Tip Top, development has largely been on 160 ac (65 ha) spacing. Infill locations are chosen by considerations of offset operator drainage, areas of high rates with high pressures, and to some extent by surface topography. Although development is not yet complete, overall ultimate recoveries should approach 90% with infill drilling and eventual gas compression.

Source

Possible Source Beds

Possible source intervals for the Wyoming-Utah-Idaho Overthrust belt fields, including Tip Top, are

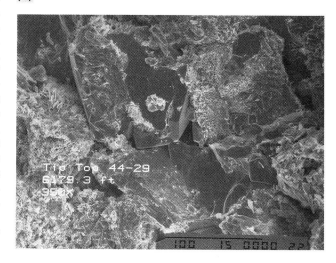

Figure 19. (A) Photomicrograph of Frontier Formation sandstone from 6179.3 ft (1883.4 m) in the 44-29 well. Scale bar, 0.25 mm. (B) SEM photomicrograph of Frontier sandstone from 6179.3 ft (1883.4 m) in the 44-29 well. Scale bar on lower right is 100 microns.

the Permian Phosphoria Formation and the Cretaceous (Lamerson, 1982; Warner, 1982). The Permian Phosphoria Formation has long been recognized as a possible source interval in the thrust belt (Claypool et al., 1978; Maughan, 1984). The phosphatic shale members of the Phosphoria are especially organic rich (Maughan, 1984) and are often cited as examples of deposition in an ancient coastal upwelling system (Parrish, 1982; Waples, 1982; Sheldon, 1963; and others). However, these units, while frequently exhibiting high residual total organic carbon contents, generally have no remaining generative potential within the thrust belt (Claypool et al., 1978; Maughan, 1984). It appears the Phosphoria was already in the stage of dry gas generation by the time most traps were formed by compressive stresses associated with thrusting (Warner, 1982; Valenti, 1987). When present, Phosphoria-sourced liquids can often be identified by a pristane/phytane ratio of <0.8 (Valenti, 1987), among other geochemical markers.

The depositional model for the Cretaceous is very similar to that for the Frontier, which was described earlier in the stratigraphy and reservoir sections. Basically, the Cretaceous contains interbedded transgressive-regressive deposits that represent environments ranging from fluviatile to marine, depending upon the relative balance between tectonism and absolute changes in sea level. Within this range of environments, Rice and Gautier (1983) have suggested transgressive shales, progradational shales, and nonmarine coal swamps as the three settings most favorable for the production, accumulation, and preservation of organic matter. Each of these environments has its own distinctive source characteristics. Transgressive shales probably contain the best quality, oil-prone kerogen, but progradational shales are volumetrically the most important. Both transgressive shales and progradational shales appear to contain a mixture of type II and III kerogen, but either preservation or original source quality (more abundant type II) was better in the transgressive shales. Coaly beds, although exhibiting high total organic carbon (TOC) contents, contain predominantly gas-prone kerogen (mostly type III). Thrust belt oils sourced from the Cretaceous can be easily identified by their pristane/phytane ratio, which is typically >1.8 (Valenti, 1987).

Geochemical Analyses at Tip Top and Maturation Modeling

Evidence available for determining the origin of liquid hydrocarbons produced at Tip Top comes from two types of geochemical data: geochemical analyses of the produced liquids themselves and analyses of possible source rocks.

Perhaps the most conclusive data showing that the liquids produced from the Frontier have a different origin than the condensate (60° API) produced out of the Nugget come from the C_{15+} gas chromatograms of these two liquids (Figures 20 and 21). The pristane/phytane peak ratio of the Frontier-produced oil is 3.50 (Figure 21) while the pristane/phytane ratio of the Nugget condensate is 1.26 (Figure 20). As discussed earlier, Valenti (1987) states that oils considered to have been generated from Cretaceous rocks have a pristane/phytane ratio >1.8 and Phosphoria-sourced oils usually have a pristane/phytane ratio <0.8. Consequently, liquids produced from the Frontier are believed to have a Cretaceous origin while the condensate from the Nugget appears to have been sourced from the Phosphoria. The slightly elevated pristane/phytane ratio in the condensate from the Nugget is probably due to the more rapid thermal destruction of phytane, which is preferentially destroyed over pristane at high temperatures. A Phosphoria source for the Tip Top Nugget condensate is consistent with the work of Seifert and Moldowan (1981), who used biomarker

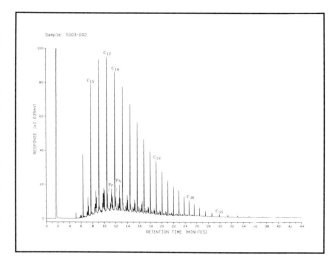

Figure 20. C_{15+} saturated hydrocarbons gas chromatogram for condensate produced from the Nugget Sandstone at Tip Top. Note pristane (Pr) and phytane (Ph) peaks.

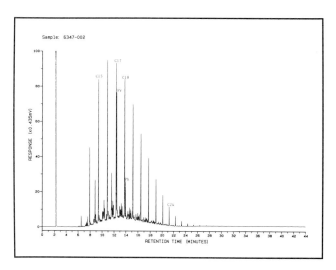

Figure 21. C_{15+} saturated hydrocarbons gas chromatogram for liquid produced from the Frontier Formation at Tip Top. Note pristane (Pr) and phytane (Ph) peaks.

data to determine that condensate produced from the Nugget at the nearby Dry Piney field also has a Phosphoria source.

Additional information on the origin of the Tip Top liquid hydrocarbons comes from source rock analyses performed on cuttings samples from the Mobil 22-19G well, NW NW Sec. 19, T28N, R113W (Tables 1-3). In this well, TOC (Table 1) was run on the thicker, predominantly shale intervals, and in turn pyrolysis was run on those samples with TOC >0.5% (Table 2). Samples for vitrinite reflectance were submitted at approximately 1000 ft (305 m) intervals.

Source rock analyses from the Phosphoria at Tip Top show this formation is presently too mature at $\%R_o = 2.59$ (Edman, 1982) to yield reliable information

Table 1. TOC analyses for Cretaceous* shaly intervals in the Mobil 22-19G well, NW NW Sec. 19, T28N, R113W.

Depth (ft)	Formation	TOC
1410-1460	Mesaverde	0.13
1510-1590	Mesaverde	0.14
1860-1910	Hilliard	0.19
2210-2250	Hilliard	0.41
2550-2600	Hilliard	0.27
2940-2990	Hilliard	0.20
3360-3410	Hilliard	0.17
3750-3800	Hilliard	0.18
4140-4190	Hilliard	0.42
4550-4600	Hilliard	0.35
4960-5000	Hilliard	0.27
5340-5390	Hilliard	0.28
5740-5780	Frontier	0.46
5980-6030	Frontier	1.17
6340-6360	Frontier	0.70
6580-6620	Frontier	0.60
7220-7250	Aspen	0.68
7590-7630	Bear River	0.77
8010-8030	Bear River	1.02
8350	Gannett	0.97
8750-8800	Gannett	0.91

*Gannett is Jurassic/Cretaceous in age.

regarding the initial generative potential of these facies. High residual TOC values (Edman, 1982; Claypool et al., 1978) indicate that if the original kerogen quality was good, significant volumes of hydrocarbons could have been generated from this formation as it matured.

Due to the difficulties in Phosphoria source rock analyses, other lines of evidence must be used to derive the relationship between the Phosphoria and hydrocarbons reservoired in the Nugget at Tip Top. As mentioned above, the pristane/phytane ratio suggests a Phosphoria source for the Nugget condensate. Furthermore, finite difference maturation modeling (Furlong and Edman, 1989) using measured %R_o values from Warner and Royse (1987) and an activation energy for type II kerogen indicates generation of Phosphoria oil occurred both during and immediately after Darby thrusting (63 Ma), as shown in Figure 22. The Phosphoria, therefore, was still generating oil after Maastrichtian movement on the Triassic décollement (Figure 6) formed the Nugget trap. The presence of dead oil that fills fractures in brecciated Phosphoria, Tensleep, and Madison core samples from Tip Top (Edman, 1982) suggests that fracturing accompanying thrusting probably provided migration pathways for Phosphoria liquids into the overlying Nugget.

To summarize, data and modeling indicate thrusting accelerated Phosphoria maturation and Phosphoria-sourced hydrocarbons migrated vertically upward along fractures into an overlying thrust-generated structural trap where they were preserved in the shallower, cooler Nugget reservoir.

In contrast to the now overmature Phosphoria, most of the subthrust Cretaceous shales in the Mobil 22-19G well have not yet entered the zone of active oil generation (Table 3), nor do they exhibit organic parameters generally associated with source intervals (Tables 1 and 2). Low TOC contents, many <0.5% (Table 1), combined with pyrolysis S2 yields typically <0.5 mg HC/g rock and hydrogen indices <90 (Table 2), show the immature to marginally mature (Table 3) Cretaceous shales in the immediate vicinity of Tip Top field have little potential to generate oil, even at higher maturity levels. Thus it seems unlikely that these horizons or their immediate lateral equivalents were the source of the liquids produced from the Frontier.

The seeming absence of a nearby source for liquid hydrocarbons produced from the Frontier creates a problem in a field that is a combination stratigraphic/diagenetic trap where the obvious first impression is of indigenous hydrocarbons and primary entrapment. Two different solutions to this problem, neither of which is entirely satisfactory, are offered below.

First, hydrocarbons produced from the Frontier may have migrated updip from more deeply buried and more mature Frontier horizons to the west. The Frontier to the west is downdip from Tip Top and would have been overridden and buried beneath a thicker Darby plate before an erosionally thinned thrust sheet advanced further to the east over Tip Top. In general, accelerated maturation of hydrocarbons in the subthrust Cretaceous occurs simultaneously with tectonic burial beneath an advancing thrust sheet (Edman and Furlong, 1987; Furlong and Edman, 1989), and this scenario fulfills the requirement for more mature source beds than exist in the subthrust Cretaceous at Tip Top. Furthermore, Oliver (1986) postulated that a thrust sheet acts as a giant squeegee, driving fluids ahead of it, which may account for west to east migration of Cretaceous hydrocarbons into the Frontier at Tip Top. Such a migration event undoubtedly occurred before Frontier reservoirs were diagenetically sealed, and there may even have been some fracturing associated with movement of the tectonic fluids.

Alternatively, the Frontier hydrocarbons could be hypothesized to represent immature condensates and dry gas generated from terrestrial kerogen at low maturity levels (0.35-0.60 %R_o) as described by Connan and Cassou (1980). At present, no evidence or data support this second hypothesis. The work presented by Connan and Cassou was primarily for Tertiary deltas containing abundant resinite (Snowdon and Powell, 1982) in which conditions were very different from those occurring in the Cretaceous of the western United States. Pristane/n-C_{17} ratios <1.0 in the Frontier liquids and the poor quality shales further argue against this hypothesis. Finally, because the Cretaceous shales in the 22-19G well are immature and no other measured Cretaceous maturity values are available in the area, no maturation model was constructed for these intervals.

Table 2. Rock-Eval data from Mobil 22-19G well, NW NW Sec. 19, T28N, R113W. (Formation abbreviations shown on Figure 3.)

Depth (ft)	Fm	TOC (wt. %)	S1 (mg/g)	S2 (mg/g)	T_{max} (°C)	HI
5980-6030	Kf	1.17	0.22	1.99	436	170
6340-6360	Kf	0.70	<0.10	0.44	432	64
6580-6620	Kf	0.60	<0.10	0.41	430	68
7220-7250	Ka	0.68	<0.10	0.27	431	39
7590-7630	Kbr	0.77	0.15	0.46	436	60
8010-8030	Kbr	1.02	0.18	0.88	437	86
8350	JKg	0.97	0.15	0.51	434	52
8750-8800	JKg	0.91	0.30	0.80	432	88

Table 3. Vitrinite reflectance data for Cretaceous from Mobil 22-19G well, NW NW Sec. 19, T28N, R113W.

Depth (ft)	Formation	%R_o
1410-1460	Mesaverde	0.49
1860-1910	Hilliard	0.44
2940-2990	Hilliard	0.47
3750-3800	Hilliard	0.51
4550-4600	Hilliard	0.49
5340-5390	Hilliard	0.48
6340-6360	Frontier	0.52
7220-7250	Aspen	0.56

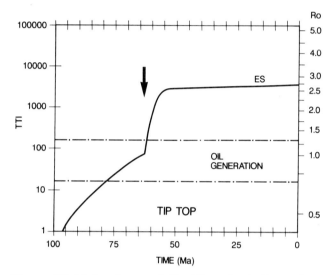

Figure 22. Maturation model for Phosphoria Formation at Tip Top field. Vertical arrow indicates time of thrusting (63 Ma). The Phosphoria had already entered the oil window prior to thrusting but emplacement of the Darby thrust accelerated maturation. (Modified from Furlong and Edman, 1989; ES indicates Edman-Surdam burial history used in this publication.)

EXPLORATION CONCEPTS

Exploration in the Tip Top area was initiated by the presence of seeps near a surface anticline. Fortuitously, the surface anticline at Tip Top is coincident with the deeper Nugget structure and Tip Top field was discovered. In a regional tectonic sense, the structural setting at Tip Top, in which a thrust belt impinges on a major arch, is unusual, and no productive analogs other than the nearby Hogsback and Dry Piney fields are immediately obvious. Considering the trap in more detail, the geologic concept of low-magnitude structures created by forward splays or blind thrusts has long been recognized and actively used to explore in overthrust belts. In addition to the fields already mentioned in Wyoming, Knowlton field in the Montana Disturbed belt and several fields in Alberta, Canada (Bally et al., 1966), represent structural analogs. Beyond these direct analogs, however, there are several more exploration concepts suggested by Tip Top that may be applicable in both the Wyoming-Utah-Idaho Overthrust belt and other overthrust systems.

First, in a local sense, the preservation of Phosphoria liquids at Tip Top suggests the possibility of producing Phosphoria-sourced liquids from other traps located in the frontal thrusts of the Wyoming-Utah-Idaho Overthrust belt. That is, even though the Phosphoria had been deeply buried and was already overmature by the time it was uplifted during the older Paris-Willard, Crawford, and Absaroka thrusting events, the Phosphoria in the footwall of the Darby thrust at Tip Top did not leave the oil window until after it was overridden by the Darby thrust. Phosphoria liquids generated in the Darby footwall were subsequently preserved in the overlying "cooler" Nuggett structure. The timing of events at Tip Top implies that Phosphoria liquids might be preserved elsewhere in the frontal thrust systems.

Second, extending the timing and maturation observations made at Tip Top to other thrust belts may result in development of additional exploration plays in those areas as well. Recognizing that source rocks which matured prior to trap formation in the older, internal sides of thrust belts may still have been generating hydrocarbons at the time traps were formed in the younger, frontal thrusts of a regional system could open additional plays in an area. This may be particularly applicable to areas that had previously been condemned on the basis of source rock or maturity considerations.

ACKNOWLEDGMENTS

This paper would not have been possible without the contributions and assistance of numerous individuals. Al Koch of Mobil was instrumental in obtaining cuttings, core, and oil samples. Lowell Martinson, an engineer working Tip Top for Mobil, helped with the data tables and production information. Union Pacific Resources did most of the drafting and also contributed the seismic sections. In particular, Barry Fish spent many hours reprocessing the seismic. Intrasearch generously donated the surface geology map. Many thanks are also due to Jack Rathbone who provided historical photographs and much insight on the discovery of Tip Top. Finally, we are greatly indebted to our reviewers, Donald S. Stone, Ted Beaumont, Bill Brownfield, Judy Mitchell, and Glenn Cole, whose comments and suggestions helped us improve the manuscript significantly.

REFERENCES CITED

Armstrong, F. C., and S. S. Oriel, 1965, Tectonic development of the Idaho-Wyoming thrust belt: AAPG Bulletin, v. 49, p. 1847-1866.

Bally, A. W., P. L. Gordy, and G. A. Stewart, 1966, Structure, seismic data, and orogenic evolution of the Southern Canadian Rocky Mountains: Bulletin of Canadian Petroleum Geology, v. 14, n. 3, p. 337-381.

Brown, W. G., 1988, Deformational style of Laramide uplifts in the Wyoming foreland, in C. J. Schmidt and W. J. Perry, Jr., eds., Interaction of the Rocky Mountain foreland and the Cordilleran thrust belt: GSA Memoir 171, p. 1-25.

Chapin, C. E., and S. M. Cather, 1981, Eocene tectonics and sedimentation in the Colorado Plateau-Rocky Mountain area, in W. R. Dickinson and W. D. Payne, eds., Relations of tectonics to ore deposits in the southern Cordillera: Arizona Geological Society Digest, v. 14, p. 173-198.

Claypool, G. E., A. H. Love, and E. K. Maughan, 1978, Organic geochemistry, incipient metamorphism, and oil generation in black shale members of Phosphoria Formation, Western Interior United States: AAPG Bulletin, v. 62, p. 98-120.

Connan, J., and A. M. Cassou, 1980, Properties of gases and petroleum liquids derived from terrestrial kerogen at various maturation levels: Geochimica et Cosmochimica Acta, v. 44, p. 1-23.

De Chadenedes, J. F., 1975, Frontier deltas of the western Green River basin: Rocky Mountain Association of Geologists Symposium, Deep drilling frontiers of the central Rocky Mountains, p. 149-157.

Doelger, N. M., 1987, The stratigraphy of the Nugget Sandstone: Wyoming Geological Association, 38th Annual Field Conference Guidebook, p. 163-178.

Dorr, J. A., D. R. Spearing, and J. R. Steidtmann, 1977, Deformation and deposition between a foreland uplift and an impinging thrust belt, Hoback Basin, Wyoming: GSA Special Paper 177, 82 p.

Edman, J. D., 1982, Diagenetic history of the Phosphoria, Tensleep, and Madison formations, Tip Top field, Wyoming: PhD dissertation, University of Wyoming, Laramie, Wyoming, 229 p.

Edman, J. D., and K. P. Furlong, 1987, Thrust faulting and hydrocarbon generation: reply: AAPG Bulletin, v. 71, p. 890-896.

Furlong, K. P., and J. D. Edman, 1989, Hydrocarbon maturation in thrust belts: thermal considerations, in R. A. Price, ed., Origin and evolution of sedimentary basins and their energy and mineral resources: International Union of Geodesy and Geophysics and American Geophysical Union, p. 137-144.

Gries, R., 1983, North-south compression of Rocky Mountain foreland structures, in J. D. Lowell, ed., Rocky Mountain foreland basins and uplifts: Rocky Mountain Association of Geologists, p. 9-32.

Hodgden, H. J., and R. E. McDonald, 1977, History of oil and gas exploration in the overthrust belt of Wyoming, Idaho and Utah: Wyoming Geological Association, 29th Annual Field Conference Guidebook, p. 37-69.

Hunter, J. K., and L. A. Bryan, 1987, LaBarge project: availability of CO_2 for Tertiary projects: Journal of Petroleum Technology, November, p. 1407-1410.

Johnson, E. H., 1984, Blackleaf Canyon field and Knowlton field, Teton County, Montana: Montana Geological Society, 1984 Field Conference and Symposium Guidebook, p. 325-330.

Jordan, T. E., 1981, Thrust loads and foreland basin evolution, Cretaceous, western United States: AAPG Bulletin, v. 65, p. 2506-2520.

Kraig, D. H., D. V. Wiltschko, and J. H. Spang, 1987, Interaction of basement uplift and thin-skinned thrusting, Moxa arch and the Western Overthrust belt, Wyoming: a hypothesis: GSA Bulletin, v. 99, p. 654-662.

Kraig, D. H., D. V. Wiltschko, and J. H. Spang, 1988, Interaction of the Moxa arch with the Cordilleran thrust belt, south of Snider basin, southwestern Wyoming, in C. J. Schmidt and W. J. Perry, Jr., eds., Interaction of the Rocky Mountain foreland and the Cordilleran thrust belt: GSA Memoir 171, p. 395-410.

Lamerson, P. R., 1982, The Fossil basin area and its relationship to the Absaroka Fault system, in R. B. Powers, ed., Geologic studies of the Cordilleran thrust belt: Rocky Mountain Association of Geologists, p. 279-340.

Marzolf, J. E., 1965, Description of cores and cuttings of rocks drilled by the Mobil Oil Company test well 22-19G in the Tip Top field, Sublette County, Wyoming: USGS Open-File Report, 37 p.

Maughan, E. K., 1984, Geological setting and some geochemistry of petroleum source rocks in the Permian Phosphoria Formation, in J. Woodward, F. F. Meissner, and J. L. Clayton, eds., Hydrocarbon source rocks of the greater Rocky Mountain region: Rocky Mountain Association of Geologists, p. 281-294.

Oliver, J., 1986, Fluids expelled tectonically from orogenic belts: their role in hydrocarbon migration and other geologic phenomena: Geology, v. 14, p. 99-102.

Oriel, S. S., and F. C. Armstrong, 1966, Times of thrusting in Idaho-Wyoming thrust belts: reply: AAPG Bulletin, v. 50, p. 2614-2621.

Parrish, J. T., 1982, Upwelling and petroleum source beds, with reference to Paleozoic: AAPG Bulletin, v. 66, p. 750-774.

Peterson, J. A., 1977, Paleozoic shelf-margins and marginal basins, western Rocky Mountains-Great Basin, United States: Wyoming Geological Association, 29th Annual Field Conference Guidebook, p. 135-153.

Rice, D. D., and D. L. Gautier, 1983, Patterns of sedimentation, diagenesis and hydrocarbon accumulation in Cretaceous rocks of the Rocky Mountains: SEPM Short Course No. 11 Lecture Notes.

Royse, F., Jr., M. A. Warner, and D. L. Reese, 1975, Thrust belt structural geometry and related stratigraphic problems, Wyoming-Idaho-northern Utah: Rocky Mountain Association of Geologists Guidebook, Deep Drilling Frontiers in the Central Rocky Mountains, p. 41-54.

Schultz, A. R., 1914, Geology and geography of a portion of Lincoln County, Wyoming: USGS Bulletin 543, 140 p.

Seifert, W. K., and J. M. Moldowan, 1981, Paleoreconstruction by biological markers: Geochimica et Cosmochimica Acta, v. 45, p. 783-794.

Sheldon, R. P., 1963, Physical stratigraphy and mineral resources of Permian rocks in western Wyoming: USGS Professional Paper 313-B, 273 p.

Snowdon, L. R., and T. G. Powell, 1982, Immature oil and condensate, modification of hydrocarbon generation model for terrestrial organic matter: AAPG Bulletin, v. 66, p. 775-788.

Valenti, G. L., 1987, Review of hydrocarbon potential of the Crawford thrust plate: Wyoming Geological Association, 38th Annual Field Conference Guidebook, p. 257-266.

Waples, D. W., 1982, Phosphate-rich sedimentary rocks: significance for organic facies and petroleum exploration: Journal of Geochemical Exploration, v. 16, p. 135-160.

Warner, M. A., 1982, Source and time of generation of hydrocarbons in the Fossil basin, western Wyoming thrust belt, *in* R. B. Powers, ed., Geologic studies of the Cordilleran thrust belt: Rocky Mountain Association of Geologists, p. 805-815.

Warner, M. A., and F. Royse, 1987, Thrust faulting and hydrocarbon generation: discussion: AAPG Bulletin, v. 71, p. 882-889.

Webel, S., 1977, Some new perspectives on the old Nugget oil fields of the LaBarge Platform: Wyoming Geological Association Guidebook, 29th Annual Field Conference Guidebook, p. 665-671.

Webel, S., 1979, Tip Top field: Wyoming Geological Association, 1979 Oil and Gas Fields Symposium, Greater Green River Basin, p. 388-389.

Webel, S., 1987, Significance of backthrusting in the Rocky Mountain thrust belt: Wyoming Geological Association, 38th Annual Field Conference Guidebook, p. 37-53.

Appendix 1. Field Description

Field name .... *Tip Top field*

Ultimate recoverable reserves .... *584 BCFG, 6.1 MMBO and NGL*

Field location:
- **Country** .... *U.S.A.*
- **State** .... *Wyoming*
- **Basin/Province** .... *Wyoming-Utah-Idaho Overthrust belt/western Green River basin*

Field discovery:
- Year first pay discovered .... *Cretaceous Frontier Sandstone (Kf)   1951*
- Year second pay discovered .... *Triassic-Jurassic Nugget Sandstone (Jn)   1951*
- Year third pay discovered .... *Cretaceous Muddy Sandstone   1962*

Discovery well name and general location:
- First pay .... *General Petroleum 32-1 G, SE SW Sec. 1, T28N, R114W (Cretaceous Frontier)*
- Second pay   *General Petroleum 4X-12G, SW SW NW Sec. 12, T28N, R114W (Triassic-Jurassic Nugget)*
- Third pay .... *Mobil Oil 22-19 Tip Top NW NW Sec. 19, T28N, R113W (Cretaceous Muddy)*

Discovery well operator .... *General Petroleum*
- Second pay .... *General Petroleum*
- Third pay .... *Mobil*

IP:
- First pay .... *1800 MCFGD (Cretaceous Frontier)*
- Second pay .... *266 BOPD (Triassic-Jurassic Nugget)*
- Third pay .... *913 MCFGD (Cretaceous Muddy)*

All other zones with shows of oil and gas in the field:

Age	Formation	Type of Show
Cretaceous	*Mesa Verde*	*Production*
	Hilliard	*Production*
	Dakota	*Production*
Paleozoic	*Madison*	*Production test*
	Darby	*Production test*

Geologic concept leading to discovery and method or methods used to delineate prospect
Mapped as a surface anticline with oil seeps by Schultz (1914). Seismic data shot and interpreted by General Petroleum in the 1940s and 1950s led to drilling of the discovery well in 1951.

Structure:
- **Province/basin type** .... *Western Overthrust: Bally 41, Klemme IIA;*
  Green River basin: Bally 222, Klemme IIA

Tectonic history
The Paleozoic tectonic and depositional history was dominated by a persistent miogeosyncline. This miogeosyncline was destroyed by thrusting that began during the Late Jurassic-Cretaceous and extended into the Cenozoic. Formation of Laramide, basement-involved structures overlapped in time with some of the younger thrusting events, and these Laramide features also exert an influence in the Tip Top area.

Regional structure
Crestal position on the Moxa arch in the boundary between the Wyoming-Utah-Idaho Overthrust belt and the western Green River basin.

Local structure
North-south-trending anticline created by a forward splay of a thrust fault (east-bounding fault) and an antithetic backthrust (west-bounding fault).

Trap:

    Trap type(s) .... *One stratigraphic/diagenetic trap with multiple pays; one anticlinal trap with one pay*

Basin stratigraphy (major stratigraphic intervals from surface to deepest penetration in field):
(From F22-19G well, Sec. 19, T28N, R113W.)

Chronostratigraphy	Formation	Depth to Top in ft (m)
Cretaceous	Frontier	5353 (1633)
Jurassic	Nugget	10,090 (3077)
Triassic	Ankareh	10,604 (3234)
Pennsylvanian	Tensleep	12,878 (3928)
Mississippian	Madison	13,720 (4185)
Cambrian	Open Door Limestone	15,368 (4687)

Reservoir characteristics:

    Number of reservoirs .... *2 primary reservoirs*
    Formations .... *Cretaceous Frontier (Kf) and Triassic–Jurassic Nugget (Jn);*
    (also Cretaceous Mesa Verde, Muddy, and Dakota)
    Ages .... *Frontier, Cretaceous; Nugget, Triassic–Jurassic*
    Depths to tops of reservoirs .... *Kf, 7000 ft (1680 m); Jn, 9850 ft (3000 m)*
    Gross thickness (top to bottom of producing interval) .... *Kf, 1100 ft (336 m); Jn, 520 ft (159 m)*
    Net thickness—total thickness of producing zones
        Average .... *Kf, 90 ft (27 m); Jn, 83 ft (25 m)*
        Maximum .... *Kf, 150 ft (46 m); Jn, 120 ft (37 m)*
    Lithology
Frontier: quartz sandstones that also contain chert, mica, glauconite, and clay
Nugget: very fine to fine-grained, subangular to round, moderate to moderately well sorted subarkose, quartzarenite, or both

    Porosity type .... *Intergranular and secondary porosity*
    Average porosity .... *Kf, 12–15%; Jn, 14%*
    Average permeability .... *Kf, 0.4 md; Jn, 8 md (fractured)*

Seals:

    Upper
        Formation, fault, or other feature .... *Kf, Cretaceous Hilliard; Jn, Jurassic Twin Creek*
        Lithology .... *Hilliard, marine shale; Twin Creek, anhydritic shale*
    Lateral
        Formation, fault, or other feature .... *Kf, diagenetic facies; Jn, OWC*
        Lithology .... *Kf, quartz sandstone with chert, mica, glauconite, and clay;*
        Jn, subarkose and/or quartzarenite

Source:

    Formation and age .... *Cretaceous and Permian Phosphoria*
    Lithology .... *Cretaceous: shales, coaly shales (possibly some coals);*
    Phosphoria: phosphatic shales
    Average total organic carbon (TOC)
Cretaceous shale source beds probably have 2–3% TOC with the coaly intervals much higher; Phosphoria phosphatic shales range 2–3% to greater than 30% TOC

    Maximum TOC .... *30+% (Phosphoria)*
    Kerogen type (I, II, or III) .... *Cretaceous, II and III; Phosphoria, II*
    Vitrinite reflectance (maturation) .... *Phosphoria $R_o = 2.59\%$*
    Time of oil maturation .... *Phosphoria, 78–62 MA*
    Present depth to top of source .... *Phosphoria, 12,600 ft (3843 m)*
    Thickness .... *Total Phosphoria, 280 ft (85 m), but max. phosphatic shale thickness near Tip Top is probably 60–90 ft (18–28 m)*
    Potential yield .... *NA*

Appendix 2. Production Data

Field name .... Tip Top field

Field size:
    Proved acres .... Kf, 31,800 ac (12,880 ha); Jn, 1260 ac (510 ha)
    Number of wells all years .... 103 (including 10 dry holes)
    Current number of wells .... 89 (a number of wells are dually completed)
    Well spacing .... Unspaced although drilled largely on 160 and 320 ac spacing in Kf
    Ultimate recoverable .... Kf, 565 BCFG, 2.2 MMBC (gas liquids); Jn, 3.9 MMBO, 19 BCFG
    Cumulative production .... 375.3 BCFG + 2.703 MMBO (combined reservoirs)
    Annual production .... Kf, 10.1 BCFG; Jn, shut in since 1986 (noneconomic)
    Present decline rate .... Kf, 5%
        Initial decline rate .... Kf, 5%
        Overall decline rate .... Kf, 5%
    Annual water production .... 1 BW/MMCFG
    In place, total reserves .... 20 MMBO and 740 BCFG
    In place, per acre foot .... Kf, 570 mcf/ac-ft; Jn, 631 bbl/ac-ft
    Primary recovery .... 4 MMBO and 590 BCFG (Kf and Jn combined)
    Secondary recovery .... NA
    Enhanced recovery .... NA
    Cumulative water production .... 5,427,658 bbl (mostly Jn) through 1988

Drilling and casing practices:
    Amount of surface casing set .... 500–1200 ft (153–366 m)
    Casing program .... Most casing is a single string from surface to TD; typically the casing is 4½-in. or 5-in. with some 9-in. production casing
    Drilling mud .... NA
    Bit program .... Varies, but generally use 8¾-in. bit size
    High pressure zones .... Third Frontier (Kf_3) is overpressured compared to First Frontier (Kf_1)

Completion practices:
    Interval(s) perforated .... Cretaceous Frontier, Muddy, and Dakota; Triassic–Jurassic Nugget
    Well treatment .... Frontier: 15,000–300,000 lb sand-water or sand-oil or sand-CO_2 foam fracs; Nugget: 10,000 lb sand-water fracs

Formation evaluation:
    Logging suites .... FDC-CNL, gamma, SP, DIL, CAL, and sonic
    Testing practices .... DST used in the past but presently the practice is to perforate then frac and flow test; repeat formation tester is also used
    Mud logging techniques
Mud logging is done from about 1000 ft (305 m) above the Frontier to TD; coring will be undertaken in 1990 to provide more detailed information

Oil characteristics:
    Type .... Frontier, paraffinic; Nugget, liquid chromatography not done because of high % volatiles
    API gravity .... Kf, 53° NGL; Jn, 60° API
    Base .... NA
    Initial GOR .... Kf, 250,000 scf gas/bbl oil; Jn, 141 scf gas/bbl oil
    Sulfur, wt% .... Jn, 13.0%
    Viscosity, SUS .... NA
    Pour point .... Jn, 5°F (−15°C)
    Gas-oil distillate .... NA

Field characteristics:

Average elevation	7600 ft (2135 m) approx.
Initial pressure	Kf, 3300 psi (22,750 kPa); Jn, 3750 psi (25,500 kPa)
Present pressure	NA
Pressure gradient	Kf, 0.61 psi/ft (13.56 kPa/m); Jn, 0.38 psi/ft (8.5 kPa/m)
Temperature	Kf, 170°F (77°C); Jn, 210°F (99°C)
Geothermal gradient	1.73°F/100 ft (3.15°C/100 m)
Drive	Frontier and Muddy, gas expansion; Nugget and Dakota, active water
Oil column thickness	±2200/132 ft
Oil-water contact	Variable/-1655 ft
Connate water	NA
Water salinity, TDS	Jn, 111,000 ppm
Resistivity of water	Kf, 0.38 ohm-m at 68°F; Jn, 0.085 ohm-m at 68°F
Bulk volume water (%)	NA

Transportation method and market for oil and gas:

Gas to market via Northwest pipeline; Nugget shut in since 1986

Rangely Field—U.S.A.
Uinta/Piceance Basins, Colorado

T. A. HEFNER
Chevron U.S.A., Inc.
Houston, Texas

K. T. BARROW
Bureau of Economic Geology
University of Texas at Austin
Austin, Texas

FIELD CLASSIFICATION

BASIN: Uinta/Piceance
BASIN TYPE: Foredeep
RESERVOIR ROCK TYPE: Sandstone
RESERVOIR ENVIRONMENT OF DEPOSITION: Eolian and Marine
TRAP DESCRIPTION: Asymmetric anticline in the hanging wall of a thrust fault

RESERVOIR AGE: Permian/Pennsylvanian
PETROLEUM TYPE: Oil and Gas
TRAP TYPE: Anticline

LOCATION

Rangely field is in Rio Blanco County in northwestern Colorado, U.S.A., about 5 mi (8 km) east of the Utah-Colorado state line. The southeastern end of the field underlies the town of Rangely, Colorado. The Rangely field lies within the Rocky Mountain province between the Douglas Creek arch to the south and the eastern plunge end of the Uinta Mountains to the north (Figure 1). The Douglas Creek arch separates the Piceance basin to the east from the Uinta basin to the west.

The Rangely field is on the Raven Park anticline, which is a large asymmetrical breached fold approximately 20 mi (32 km) long and 8 mi (13 km) wide. The highest structural point is in Sec. 31, T2N, R102W (Figure 2). The anticline is almost entirely surrounded by an escarpment of Castlegate Sandstone of the Upper Cretaceous Mancos formation that makes the flanks of the fold readily visible. Eroded shale of the Mancos formation forms the low center of the breached anticline. The area is now commonly known as the Rangely basin.

Thirty-four fields in the Uinta Mountains and northern Uinta-Piceance basins have produced approximately 1.2 billion bbl of oil and 1.1 tcf of gas (Osmund, 1986). About two-thirds of the oil has been produced from the Weber Sandstone from eight fields (Figure 1). Most of the remainder has been produced from lacustrine sandstones of the Eocene Green River and Uinta formations and fluvial sandstones of the Eocene Wasatch Formation. Red Wash and Bluebell-Altamont fields are the two most important oil fields producing from these Tertiary formations (Figure 1).

Other producing intervals are in the Jurassic Entrada Sandstone and Morrison Formation, the Triassic Gartra Member of the Chinle Formation, and the Cretaceous Niobrara limestone and Mancos Shale (Osmund, 1986).

Original oil in place at Rangely field is estimated to be 1.6 billion stock tank barrels (STB) or 250 million m³. Upon completion of the current carbon dioxide flood, 939 million STB (149 million m³) ultimately will be recovered. The gas to oil ratio of 300 ft³ per STB (52.8 Sm³/m³) is very low, and the field does not produce commercial quantities of gas.

HISTORY

History before the Weber Discovery

The history of the Rangely field is well documented and colorful. The large surface anticline was first recognized by C. A. White of the 1875 Hayden Survey (Figure 3). The earliest detailed geological description of the area was by H. S. Gale (1908). Gale did geological field work in August 1907 and made the following commentary on White's use of the name *Raven Park* for the Rangely basin (Gale, 1908, p. 9):

> According to popular usage in the Rocky Mountain region . . . the term (park) perhaps implies somewhat of the picturesque quality, a park being an open glade or valley surrounded or partly inclosed by timbered hills. With such a meaning the term would hardly be appropriately applied to the Rangely Basin. The basin

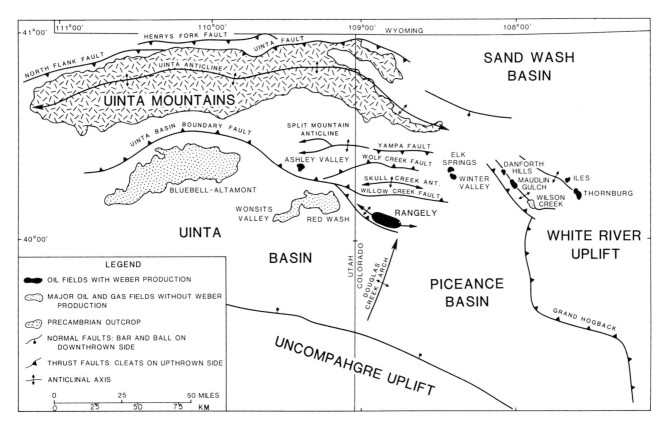

Figure 1. Regional tectonic map of Uinta basin and Piceance basin areas and fields with oil or gas production from the Weber Sandstone.

is in truth a desolate waste of dry washes and almost barren clay ridges . . . .

Petroleum exploration in the Rangely area dates to about 1890. The *Meeker Herald* (19 December 1890) reported that Rangely attorney William H. Clark and "Denver interests" (the Denver Gas and Petroleum Co.) had found an oil field at Rangely that would be explored the next summer. This group filed location certificates on 16,000 ac (6480 ha) of petroleum land in Raven Park. A group with greater financial backing formed the Raven Oil Co. in 1891. These earliest explorationists located prospects near numerous oil springs and seeps in the Mancos Formation, some of which are still active. Reports were made on intended drilling, but no drilling was done for ten years (Pollard, 1957).

Oil excitement in the Rangely area was rekindled in 1901. The *Meeker Herald* (15 June 1901) reported that a California company had examined the region and succeeded in getting an expert, George Demeron, from California "to go over the ground carefully." The Requena Co. of San Francisco was later reported to be the California company that was active in the area (*Meeker Herald*, 18 October 1902).

The first well spudded on 7 November 1901 and was known as the "Pool well" because combined interests backed the well. The Pool well was the discovery well for the Mancos Formation at Rangely field. The exact ownership was unknown, but the Colorado & Utah Co., Raven Oil Co., and The Meeker Co. were active in the area. It drilled from November 1901 to May 1902, when drilling was halted because of a severely deviated wellbore. Total depth was approximately 2100 ft (640 m). The well produced between 1 and 3 BOPD from a depth of 750 ft (225 m). The *Meeker Herald* (24 May 1902) reported: "when work was stopped they were still in the shale and a good deal of high grade oil was flowing in."

The next well, the Requena No. 0, was drilled in 1902 to 600 ft (183 m) and bailed about 3 BOPD. This well also became crooked and was abandoned. A redrill well, the Requena No. 1, reported stronger flows than the No. 0 (*White River Review*, 6 September 1902). By late 1902, three wells had been drilled to depths of 2080, 1000, and 600 ft (634, 305, and 183 m). Each had oil shows, but the anticipated gushers were not encountered. Total expenditures on the three wells were $35,000 (*White River Review*, 22 November 1902). Five more wells were spudded by the end of 1903, for a total of eight. The Meeker Co. drilled two wells to depths of 736 and 1002 ft (224 and 305 m). The Requena Co. spudded a third well, the No. 2, which was drilled intermittently until 1906. Drilling was abandoned at 2560 ft (780 m) because of junk in the hole (Gale, 1908).

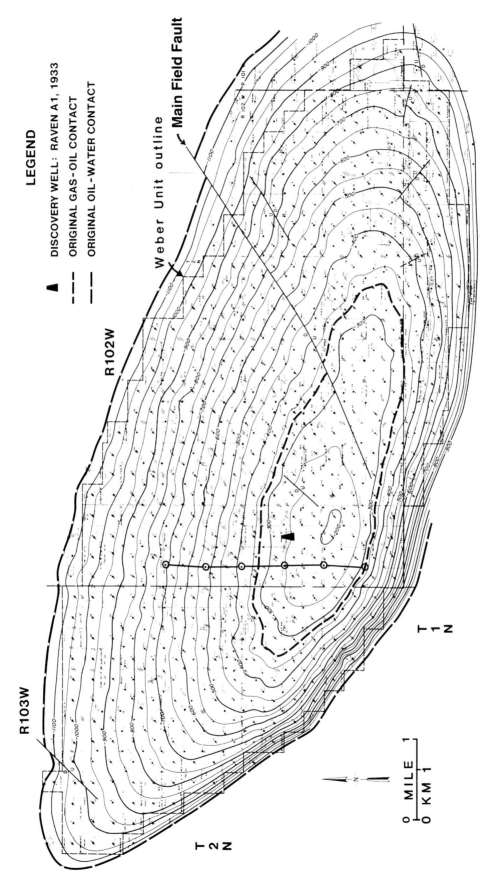

Figure 2. Rangely field structure map on top of the Weber Sandstone showing gas-oil and oil-water contacts and location of the field discovery well (Raven A-1). Indicated cross section is shown in Figure 13. Original gas-oil contact, −330 ft (−100.6 m) MSL; original oil-water contact, −1150 ft (−350.8 m) MSL. Contour interval, 500 ft (15.25 m).

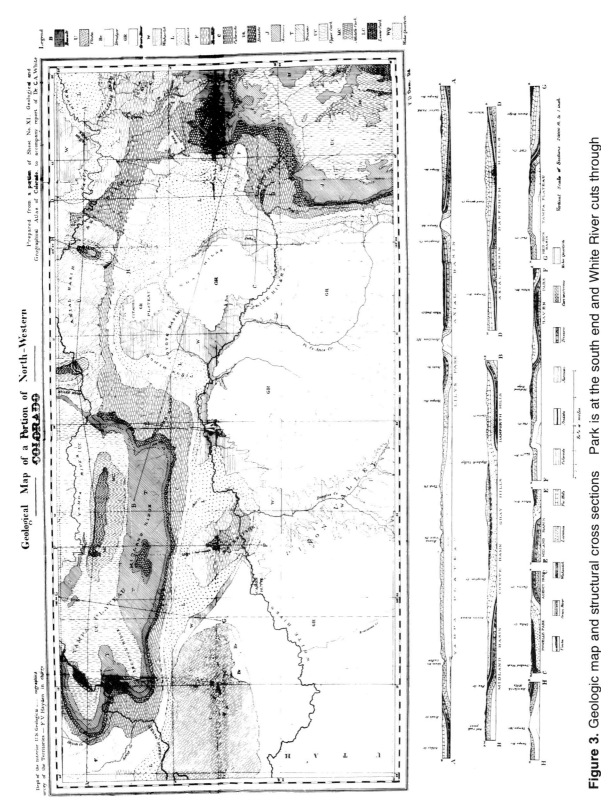

Figure 3. Geologic map and structural cross sections of northwestern Colorado. The Rangely structure is called Raven Park (cross section F) and is in the western-central part of the map. (Find the line of N-S section F, about 1½-in. west of map center; Raven Park is at the south end and White River cuts through the south end of Raven Park.) The stream labeled Ungatoo-roosh flows south into the White River at Raven Park. This creek is now known as Stinking Water Creek. From White (1878), plate II.

These early wells had the Lower Cretaceous Dakota Sandstone as their objective. However, the overlying Mancos Formation was thicker than anticipated (Gale, 1908), and the Dakota was not penetrated until later. Wells were drilled between 1901 and 1914 to obtain government patents on 160 ac (65 ha) oil placer claims (Pickering and Dorn, 1948, p. 134). By 1907 the area was covered by approximately 70 mi^2 (180 km^2) of unpatented locations for oil placer claims.

The Union Oil and Gas Co. operated a well with pooled interests of The Meeker Co., Requena Co., and Colorado-Utah Oil Co. This Union well was located 3 mi (5 km) north of the structural crest to encounter the Dakota objective *below* (it was hoped) a postulated gas cap (Gale, 1908). Their first well was drilled to 1300 ft (400 m) and encountered oil in the Mancos formation. It was abandoned after the casing collapsed and the tool string was lost. An offset was drilled to 3655 ft (1114 m) intermittently from 1903 to 1907. At this depth, the well was about 150 ft (50 m) above the Dakota objective. The rig burned (probably because of a boiler fire), but the well was said to be in good condition (Gale, 1908).

In 1907 the Colorado Pacific Co. of San Francisco drilled new wells and tested two old wells for Mancos potential. This work was part of an extensive exploration campaign on an 8640 ac (3500 ha) block (Owen, 1975, p. 240). The Colorado Pacific Co. and the Kern Trading and Oil Co., which bought and patented the acreage, were subsidiaries of the Southern Pacific Railroad (White, 1962, p. 417). Mr. J. H. Hunt, one of the company's directors and their local representative at Rangely, wrote to Gale that:

> Oil is found in all the wells except two at depths from 400 to 700 feet below the surface . . . The oil is of a uniform character and in all wells rises to the same level, approximately 360 feet below the surface. By continued pumping the production increases. (Gale, 1908, p. 42).

In 1909, the Colorado Pacific Co. attempted another Dakota test; however, their rig could not drill deep enough to reach the objective. Its 8640 ac (3500 ha) position was later purchased in 1917 by A. C. McLaughlin, an executive of the Southern Pacific Railroad. He resold 5000 ac (2025 ha) to the Standard Oil Company of California's subsidiary, the Richmond Oil Co., for $200,000 (Owen, 1975, p. 697). "It is impossible to determine how many geologists had their hands in this prime prospect before its long-delayed success" (Owen, 1975, p. 240).

Rangely provides an interesting footnote to the formative years of the Standard Oil Company of California, now Chevron Corporation. World War I made heavy demands on oil in storage. Although development of its California properties was accelerated, 1918 was the first time that the company drilled outside the state. Fourteen Mancos wells were drilled at Rangely in little more than a year's time, but the results were discouraging. Their total production for 1919 was less than 2100 bbl. By the end of 1919 these wells were shut in. The total cost of this venture, including land acquisition, exceeded $600,000. Standard retained the property, however, and its expectations were fulfilled more than a quarter of a century later (White, 1962, p. 417–418).

In 1919, the Emerald Oil Co. drilled a well to 485 ft (148 m) that was the most successful to date; it produced 150 BOPD from the Mancos at a depth of 521 ft (Pickering and Dorn, 1948, p. 134; Thomas, 1944). By 1924 The Mancos was producing between 1500 and 2000 bbl/month (Sears, 1924).

As of 1955 all Mancos wells were cable-tool drilled, with minimal surface casing set, and completed open hole (no casing through the reservoir) with retrievable production tubing (Peterson, 1955). These wells were drilled to intersect oil-bearing fractures, which are expressed at the surface as calcite-filled veins. The strike and dip of the veins were frequently mapped by drilling many shallow holes. Typically, good wells made only a few tens of thousands of barrels, although a few exceptional wells exceeded 100,000 bbl. By 1954 cumulative Mancos production was 4.7 million bbl, and monthly production was approximately 25,000 bbl. A more thorough discussion of Mancos fractures and oil production is given by Peterson (1955).

The Dakota Sandstone, the original exploratory objective at Rangely, never lived up to expectations. High initial gas flows and blowouts were common, but the Dakota wells watered out. There are no records of Dakota gas being exploited commercially or of any significant Dakota oil production. Heaton (1929, p. 108) reported that four Dakota gas wells had been completed on the crest of the structure. One well was gauged at 74 million ft^3/day, but a downdip test produced water. Holmes (1926) described the steps taken to shut in a well that blew out at a measured rate of 45 to 50 million ft^3 of gas/day. McMinn and Patton (1962, p. 104) noted that from 1924 to 1928 Midwest and Texas Production were active in the Dakota play and that the wells encountered strong flows of gas and water. The Emerald No. 1 drilled by Texas Production blew out at an estimated 90 million ft^3/day. The spewing gas was ignited by lightning and burned spectacularly for 20 days before going to water. Newton (1945, p. 63) states that the fire burned with a 1000 ft (300 m) flame, lighting the entire Uinta basin.

The first rotary well was spudded on 9 June 1926 by the Amazon Drilling Co. The well, Gray #4, was also known as "the Associated." The location was less than a mile from the structural crest. Total depth of 4772 ft (1455 m) was reached on 13 May 1927 in the Navajo Sandstone when the drillstring became stuck. The well later tested the Morrison Formation, which was not commercial, and the Dakota Formation, which was "wet" (Bench, 1946). This well indicated that original gas accumulation in the Dakota must have been quite limited, because the well produced water less than 200 ft (60 m) downdip from the Dakota structural crest.

Discovery and Development of the Weber Sandstone Reservoir

The discovery well for the Weber Sandstone reservoir, the giant oil accumulation at Rangely, was spudded on 18 May 1931 by the California Co., then a subsidiary of the Standard Oil Co. of California. The Raven A-1 was located in the NW 1/4 SE 1/4 Sec. 30, T2N, R102W (Figure 2). On 2 March 1933, almost two years after spudding, the well reached a total depth of 7173 ft (2186 m) in the lower Pennsylvanian Morgan Formation, about 800 ft (250 m) above its original deep objective, the Mississippian Madison Limestone. The well was plugged back to 6335 ft (1931 m) in the Weber Sandstone. The top of the Weber was encountered at 5704 ft (1739 m). Initial production was 229 bbl/day of 39° API oil. There has been speculation (Newton, 1945) that the economic potential of the Raven A-1 was underestimated, but field records show the potential was corrrectly reported. After producing approximately 8000 bbl, the Raven A-1 was shut in until September 1943.

Factors that kept Rangely crude off the market include the high cost of drilling, the distance to the nearest refinery (240 mi or 400 km at Salt Lake City), and depressed oil prices following discovery of the East Texas field.

Field development did not begin until World War II created a tremendous demand for petroleum. In 1943 the Raven A-1 was again tested, and the next deep well, the McLaughlin #1-32, was spudded by Sharples. In May 1944 three deep wells were drilling (Thomas, 1944). The California Co. Emerald #1 was completed as the second producer in the field on 26 August 1944. By June 1945, 18 wells had reached the Weber (Thomas, 1945). In May 1946 there were 45 Weber producers and many more drilling (Thomas et al., 1946).

In early 1945 the United States made plans to increase war efforts in the Pacific. The federal government requested producers to increase Rocky Mountain production by 20,000 bbl/day. Regional operators considered this order impossible to meet unless Rangely could supply most of the increase (Bench, 1946). Although the war ended shortly afterward, this appeal, and the operators' desire to increase production, spurred the Rangely boom.

From 1944 to 1949, 473 wells were drilled on 40 ac (16 ha) spacing, completing initial development of the field. At one time there were 63 rigs active in the field. The Rangely boom strained the resources of the adjacent communities from Vernal, Utah, to Craig, Colorado, but the spirit of cooperation overcame the "shortages of everything except red tape" (Ritzma, 1955).

One interesting aspect of this era was the development and perfection of diamond core bits and coring techniques (Williams, 1947; Christensen 1948; Stuart, 1947). Diamond coring was considerably less expensive than conventional drilling in the well-lithified Weber Sandstone, because faster penetration rates and longer bit life saved time and money despite frequent trips to retrieve core. By 1949, 326 wells were cored, and 93% of those were cored with diamond-studded bits (De Mohrenschildt, 1949, p. 124). Complete core recovery was typical. Continuous coring was attempted by reversing circulation and pumping core pieces to the surface inside the drill string. This method was soon abandoned because of mechanical complications. During the reservoir's initial development, 108,741 ft (33,136 m) of core were cut. De Mohrenschildt (1949) correctly predicted that, "this information will be of particular practical value during the later stages of production, when the remedial measures may become necessary and its value will be immense in the secondary-recovery projects."

On 19 September 1945, a 10-in. pipeline was completed to Wamsutter, Wyoming, where it joined the trunk line between Salt Lake City, Utah, and Laramie, Wyoming. The initial capacity of approximately 12,000 bbl/day was boosted to over 25,000 bbl/day by 1946. After a 10-in. pipeline from Rangely to Salt Lake City was added in November 1948, the daily production rate was raised to approximately 55,000 bbl/day (Cupps et al., 1951). Gas-oil ratios remained remarkably steady. Remedial work was done only on a few wells with high gas-oil ratios. By 1949 the field had produced 40 million bbl.

The Engineering Committee effectively managed the field before unitization in 1957. Their functions included data exchange and production allocations. Initially all wells had the same allowable, which was determined by the chairman of the committee (Williams, 1945). Production from individual wells was increased as pipeline capacity was added and decreased as new wells came on line. Later, the chairman allocated production in an effort to balance pipeline capacity with acreage drained, bottom-hole pressure, and the gas-oil ratio for each well (De Mohrenschildt, 1949). The need for pressure maintenance was recognized early. High oil production by several operators reduced reservoir pressure, expanded the gas cap, and increased gas-oil ratios. The maximum production reached 82,000 BOPD in mid-1956 (Figure 4). Flaring excess gas was a common practice, and 45 to 50 million ft^3/day were flared until compression and injection plants were put into operation.

Secondary Recovery from the Weber Sandstone

On 1 October 1957, Rangely was unitized with 45 operating interests and 450 royalty interests. It was the second largest unit in the United States. The California Co. with 45.6% interest was designated operator. Unitization was achieved on the eve of a hearing by the Colorado Oil and Gas Conservation Commission to consider measures for gas conservation. The Commission's chairman, Warwick M. Downing, played an important role in the unitization

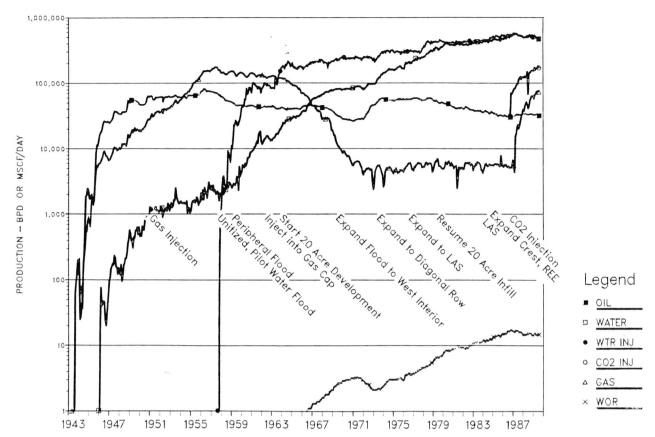

Figure 4. Rangely field history, production statistics and events. Initial sustained production was in 1943. By 1949 the field was completely drilled on 40 ac (16 ha) spacing. REE and LAS refer to the field engineering areas, *Remaining East End* and *Last Area Stimulated.* WOR, water to oil ratio.

drive. He threatened to reduce oil production by 20% to conserve gas (*Petroleum Week*, 1957). The Rangely Engineering Subcommittee estimated a primary recovery of 350 million bbl oil (current cumulative was 225 million bbl). Predicted incremental waterflood recovery was 385 to 437 million bbl.

A peripheral waterflood and increased gas reinjection into the gas cap were planned to boost recovery (*Oil and Gas Journal*, 1957). By late 1958, gas flaring was reduced from a high of about 50 million to approximately 5 million ft^3/day, but production had also fallen to approximately 53,000 bbl/day (Figure 4). Two pilot floods were begun while designs were made for the full-scale peripheral waterflood (*Oil and Gas Journal*, 1958).

A good description of the waterflood is provided by Bleakley (1973). The peripheral row project met with modest success. However, production engineers soon feared that the increased pressure in the periphery would drive oil into the lower pressure gas cap, thus reducing ultimate recovery. In response, water was injected into the crestal gas cap starting in 1962.

Wells immediately updip from the peripheral injectors had a good initial response, but by 1962 many of these wells had a 60 to 70% water cut. The second row was converted to injection, and 20 ac (8 ha) infill drilling was also begun in 1963. The third and fourth rows were converted by the early 1970s.

A major peripheral flood was never attempted in the east end of the field. This area had long been known to have lower permeability and less continuous reservoir sandstones than the rest of the field. In 1964 a staggered linedrive waterflood was begun in the east end and was expanded in 1966 and again in 1973.

The most successful waterflood program was begun in 1969 in the remaining north-central part of the field. Five-spot injection patterns in this area included 86 wells covering 3500 ac (1400 ha). Production increased by 14,000 BOPD. The expanded waterflood required a major enlargement of water handling facilities, including treatment plants and pump sizes.

In the early 1970s Rangely became a testing ground for experiments by the USGS (U.S. Geological Survey) on earthquake causes and control. Numerous small earthquakes in the Rangely vicinity had been observed since 1962, when the Uinta Basin Seismological Observatory began recording. In 1969, over 1000 small earthquakes were observed. This seismic activity was related to water injection on the south flank of the field along the extension of a fault mapped in the Weber Sandstone. By inducing hydraulic fractures, measuring injection pressures, and monitoring seismic activity, theories were confirmed that crustal movements can be caused by elevated pore pressures (Raleigh et al., 1972, 1976; Haimson, 1973; Gibbs et al., 1973). Parted casings and some other damage to production equipment have resulted from seismic activity over the years.

The waterflood was expanded in 1973 and again in 1976. The 1973 project expanded the linedrive injection pattern by drilling new dual injectors and infill producers. Other producers were also converted to injectors. The expanded patterns were referred to as modified 9-spots. The 1976 project included drilling 54 infill wells and converting former 5-spot patterns to a linedrive pattern.

After a break of several years, infill drilling resumed in 1979 and continued intermittently until 1988. Few undrilled 20 ac (8 ha) locations remain except in the watered-out periphery and along the main field fault. The remaining locations are not considered economic at this time. Between 1983 and 1986, 23 infill wells were drilled on 10 ac (4 ha) spacing. The 10 ac program has met with mixed economic success. Only a few of those wells are clearly profitable.

Tertiary Recovery from the Weber Sandstone

In 1978 a laboratory study was conducted to evaluate the feasibility of a miscible CO_2 flood (Graue and Zana, 1978). This study predicted a favorable incremental recovery. In this miscible flood, CO_2 lowers the viscosity of oil and causes it to swell. A later study specified cumulative CO_2 injection of 30% of hydrocarbon pore volume (HPV) on a 1-to-1 ratio of water alternating with gas, followed by a waterflood. Each of 20 water-alternating-gas cycles should consist of 1.5% HPV of CO_2 followed by an equal volume of water.

The CO_2 project was approved by the unit in 1984 and was certified as a tertiary recovery project in 1985. Flooding began in October 1986 with the injection of approximately 75 million ft^3/day into the prolific central portion of the field. The minimum miscibility pressure for CO_2 in Weber crude oil at Rangely is approximately 2500 psi at 160°F (17.24 Pa at 71°C). The average reservoir pressure in the project area has been raised to approximately 3100 psi (21.37 Pa). Injection is controlled by selective injection equipment that allows reservoir zones to be selectively opened or blanked off. In mid-1988 CO_2 injection rates totaled approximately 230 million ft^3/day (150 million ft^3 purchased and 80 million ft^3 recycled). The purchased CO_2 is produced by Exxon at the LaBarge field and processed at their Shute Creek plant near Kemmerer, Wyoming. It is transported by pipeline to Rock Springs, Wyoming, and then to Rangely via a 129 mi (208 km) 16-in. pipeline over the Uinta Mountains. The recycled CO_2 is produced with oil and water, and it is then dehydrated, recompressed, and mixed with purchased CO_2 for reinjection (Larsen, 1987).

The initial response to the tertiary project has been promising. Incremental production attributed to the tertiary flood is approximately 8000 BOPD, but three years is not yet enough to gauge the economic merit of this project.

DISCOVERY METHOD

The discovery of Rangely field can be traced to the identification of the Rangely anticline by C. A. White (1878) (Figure 3). Surface oil seeps confirmed the presence of oil in the giant structure. The first discoveries in the Mancos formation resulted from drilling near these oil shows. The best Mancos wells were drilled by mapping fractures, which are expressed at the surface as calcite-filled veins, and projecting them to the target depths.

Discovery of the giant Weber Sandstone oil reservoir resulted from the correct interpretation that the porous and permeable Paleozoic strata which outcrop in the Uinta Mountains would also underlie the Rangely basin. When deep drilling technology evolved, the Weber reservoir was discovered. Modern explorationists would readily recognize the potential of the Rangely area, even without geophysical technology.

STRUCTURE

The Rangely field is a large, asymmetrical, doubly plunging anticline (Figure 2). The axis of the structure trends approximately north 60° west at its western end and nearly due east at its eastern end. Dips on the gentle northeastern flank average 4°. The southwestern flank is steeper, dipping greater than 15° (Figure 5). The eastern and western ends plunge out gently at 2½°. The anticline is approximately 20 mi (30 km) long and 8 mi (13 km) wide. Closure exceeds 2100 ft (640 m) on the top of the Weber and includes an area of over 100 mi^2 (260 km^2). Only the top 950 ft (300 m), covering approximately 31 mi^2 (80 km^2), are hydrocarbon saturated. The Redwash syncline bounds Rangely on the north. The plunge ends of the structure flatten into regional dip and merge into flexures of adjoining areas. (Seismic profiles across Rangely are shown in Harding and Lowell, 1979, and in Stone, 1986a.)

The structure has a strong surface expression and was recognized by geologists as early as the Hayden Survey (White, 1878) (Figure 3). At the surface, the anticline is breached with a topographically low interior of eroded Upper Cretaceous Mancos formation. This surface basin is surrounded by an escarpment capped by the resistant Castlegate Sandstone Member of the upper Mancos formation (Cullins, 1971a, 1971b). The escarpment typically exceeds 200 ft (60 m), although the floor of the basin

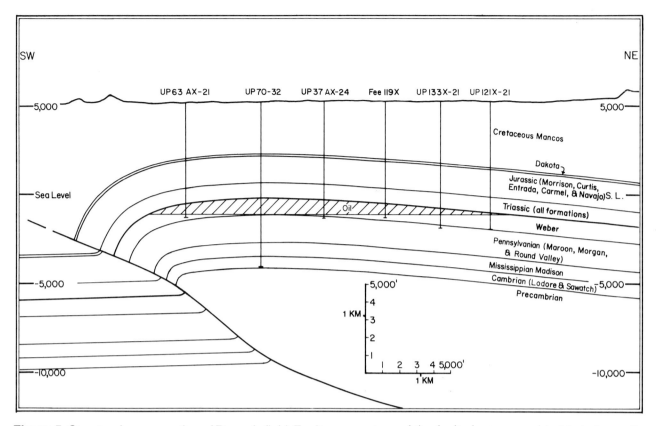

Figure 5. Structural cross section of Rangely field. Fault dips range from approximately 20° to 40°. Double curvature of the fault plane was added to balance the section without severe deformation of the footwall.

has several hundred feet of topographic relief and is dissected by steep-sided drainages.

Rangely is on the north-trending Douglas Creek arch, an intermediate sized uplift that separates the Uinta basin to the west from the Piceance basin to the east (Figure 1). The Douglas Creek arch developed as a north–south-trending anticline during the Late Cretaceous. Rangely is one of several anticlines on the arch showing similar trends. To the south is the broader Douglas Creek flexure, with Mesaverde rocks exposed at the surface. To the north are the larger and sharper Blue Mountain and Skull Creek anticlines, with exposed Carboniferous strata. Rangely and the other east–west-trending structures of the Uinta Mountain area are Eocene in age (Gries, 1983).

The Rangely anticline is located in the hanging wall of a splay of the Uinta basin bounding fault. This fault splits as it crosses the Douglas Creek arch forming the Rangely structure and the Skull Creek anticline to the north (Hansen, 1986). The fault splay has not been penetrated by wells in the Rangely field, but it is evident on seismic lines (Harding and Lowell, 1979; Stone, 1986a). Vertical separation of 7000 ft (2100 m) has been estimated at the base of the Paleozoic section (Stone, 1986a). Throw decreases upward, and displacement has not been observed at the surface (Cullins, 1971a, 1971b).

Several geologists have identified surface normal faults, which are made obvious by offsets in the escarpment (among others Thomas, 1944, 1945; Bench, 1945; and Newton, 1946). Most of these faults trend northeast, oblique to the north-northwest trend of the structural axis, and are down thrown to the southeast. Vertical displacement at the surface is usually less than 100 ft (30 m); however, the fault in Sec. 11 T1N, R102W has a throw of approximately 300 ft (100 m) (Bench, 1945).

There is only one major fault within Rangely field (Figure 2). It is informally known as the main field fault. The vertical fault plane trends east-northeast, roughly in line with several of the faults observed in the surface escarpment. However, the throw of no more than 70 ft (20 m) is down to the northwest, which is opposite to the surface faults and also disagrees with first motion studies reported by Raleigh et al. (1976). The main field fault acts as a barrier to fluid flow during production. The fault had not been a barrier to hydrocarbon accumulation in the reservoir, since the original oil-water contact was not affected by the fault. A well-developed fracture system on the northwest (down-thrown) side controls the flow and injection behavior of the wells within several hundred feet of the fault. A similar fracture system is not observed on the southeast side.

Calcite-filled fractures are found in a few areas at the surface, especially near the crest of the structure. The most numerous and important fractures trend northeast and dip to the southeast. This set of fractures is relatively persistent to a depth

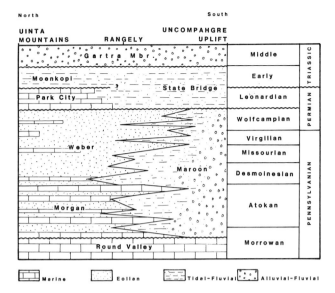

Figure 6. Pennsylvanian to Lower Triassic stratigraphic diagram showing complex relationships of marine and nonmarine strata related to the Uncompahgre uplift.

of 2500 ft (750 m) and has contained most of the oil produced from the Mancos formation. Calcite fillings near the surface have provided a barrier to flow, but oil seeps along these fractures indicate that the calcite has been only a partial seal to oil accumulation in the Mancos fractures. A second, less important, set of fractures lies south of and trends parallel to the structural axis. These fractures generally dip to the northeast and die out toward the core of the structure. Both sets of fractures are associated with the primary anticlinal flexure at Rangely, with the most dense fracturing occurring in areas that experienced the least compressive stresses during folding (Peterson, 1955).

STRATIGRAPHY

Rangely field lies about 50 mi (80 km) north of the Ancestral Uncompahgre uplift (Figure 1), which was active from Pennsylvanian Atokan to Permian Leonardian time (Rascoe and Baars, 1972; Kluth and Coney, 1981). During Pennsylvanian time active tectonism and possible glacially influenced eustatic sea level changes produced a complex interfingering of marine and nonmarine depositional environments (Crowell, 1978) (Figure 6). Tectonic activity waned during the Permian, and the Uncompahgre highland was gradually eroded during the Permian and Early Triassic.

The Pennsylvanian Morrowan Round Valley Limestone (Figure 7) can be traced from outcrop in the Uinta Mountains to the subsurface at Rangely, where it represents the final episode of pretectonic open marine sedimentation. The Round Valley is unconformably overlain by the Atokan and Desmoinesian Morgan Formation, which is composed of marginal marine sandstones and limestones and eolian sandstones. The Morgan Formation is more marine in character in the Uinta Mountains (Driese and Dott, 1984) than at Rangely, where it is interbedded with the lower portion of the Maroon Formation. The red beds of the Atokan to Permian Wolfcampian Maroon Formation were shed northward and eastward into the Eagle basin and deposited in fluvial, alluvial, and fan-delta environments. The upper portion of the Maroon Formation interfingers with the eolian Desmoinesian to Wolfcampian Weber Sandstone at Rangely. Eight miles (13 km) north of Rangely at Skull Creek anticline, the Maroon facies is missing; the eolian and marine Weber Formation lies above marine and eolian Morgan Formation. Marine facies of the Permian Park City Formation unconformably overlie the Weber Sandstone along the flanks of the Uinta Mountains. At Rangely, nonmarine to tidal red beds of uncertain age lie unconformably above the Weber. These strata are assigned to the Permian–Triassic State Bridge Formation (Maughan, 1980). The State Bridge Formation is unconformably overlain by the Middle Triassic Gartra Member of the Chinle Formation. The Gartra Member probably correlates to the Shinarump conglomerate of the Colorado plateau.

There is an abundance of actual and potential (unproven) source rocks in the Piceance and Uinta basins. These are discussed in *Sources* and include Eocene Green River and Uinta formations, Cretaceous Mancos Shale, Jurassic Curtis Formation, Permian Phosphoria Formation, and Pennsylvanian Belden Shale.

TRAP

Hydrocarbons at Rangely field are structurally trapped in a doubly plunging, asymmetric to the south, hanging wall anticline above a south-verging thrust fault (Figure 5). This Laramide structure contains, however, only localized hydrocarbons that were probably stratigraphically trapped at the facies transition between the eolian Weber Sandstone and fluvial Maroon Formation (Campbell, 1955; Hoffman, 1957; Larson, 1975; Fryberger, 1979; Koelmel, 1986; Fryberger and Koelmel, 1986) (Figure 7). Alternatively, a pre-Laramide paleostructural trap has been hypothesized based on Weber isopach patterns (Whiteker, 1975) and regional seismic data (Waechter and Johnson, 1986).

The current anticlinal trap is sealed by impermeable siltstones, shales, and sandstones of the Permian–Triassic State Bridge Formation (Figure 8A). These red beds were deposited in tidal and fluvial environments adjacent to the exposed Uncompahgre uplift in Late Permian and Early Triassic time. The coarse clastics are well cemented with calcite and anhydrite and have porosities less than 3% and permeabilities less than 0.1 md. The oil-water contact (OWC) was placed above a broad, approximately 200

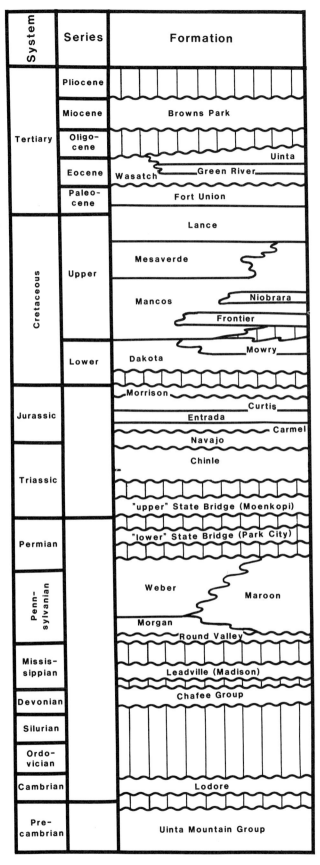

Figure 7. Generalized stratigraphic column of the Rangely area.

ft (65 m) transition zone at a depth of -1150 ft (-351 m) by the Rangely Engineering Committee. This level was thought to be the average upper limit of water production prior to waterflood.

Oil migration from the Phosphoria Formation began in the Early Jurassic and continued until at least the Late Cretaceous. The Eocene structure at Rangely developed where a hydrocarbon accumulation was already in place. The abrupt facies transition from porous and permeable eolian Weber Sandstone to impermeable, fluvial and alluvial Maroon Formation, from northwest to southeast, provided the original stratigraphic trap. The fluvial and alluvial Maroon is composed of arkosic sandstones and siltstones, with minor amounts of shale and conglomerate (Figure 9). The sandstones are micaceous, illitic, and well cemented with calcite. Maroon-facies porosity ranges from 0 to 8%, and permeability is usually less than 0.01 md. No oil saturation exists in the fluvial facies. The lack of mixing between the eolian and fluvial facies was important in forming the stratigraphic trap (Fryberger, 1979).

Paleostructures may have been important in trapping oil at some smaller fields having Weber production east of Rangely (Figure 1). Paleostructures have been identified based on Weber Sandstone isopach thins (Whiteker, 1975; Stone, 1986b). Paleostructural entrapment at Rangely was suggested by Waechter and Johnson (1986) from seismic evidence. However, structural relief present on the Mississippian Leadville (Madison) Limestone reflector is not observed at the top of the Weber, perhaps because of post-Morgan, pre-Weber erosion (Waechter and Johnson, 1986, their figure 10). No penecontemporaneous faults have been observed in the Weber at Rangely.

Reservoir Lithology

The eolian Weber Sandstone reservoir at Rangely field has been described by Fryberger (1979) and Bowker and Jackson (1989). Dune, interdune, and sand sheet facies can be identified from internal stratification seen in cores.

The dune facies contains the best reservoir rock. This facies is characterized by preserved high-angle (greater than 15°) cross-bedding commonly composed of thinly laminated (1-5 mm), inversely graded, translatent wind-ripple strata (Figure 10). Evenly textured, 1 to 2 cm thick, ungraded grain-flow strata are rare (Figure 8C). Both simple and complex dunes are preserved in the Weber. Simple dunes are generally less than 3 ft (1 m) thick. Complex dunes are composed of multiple beds and can be up to 80 ft (25 m) thick, averaging 10 ft (3 m). Both simple and complex dunes have flat to low-angle basal contacts above fluvial or interdune deposits. Bedding steepens progressively upward and is often overturned or contorted in the centers of complex dunes (Figure 8F). The tops of simple and complex dunes are moderately to extensively bioturbated (Figure 11).

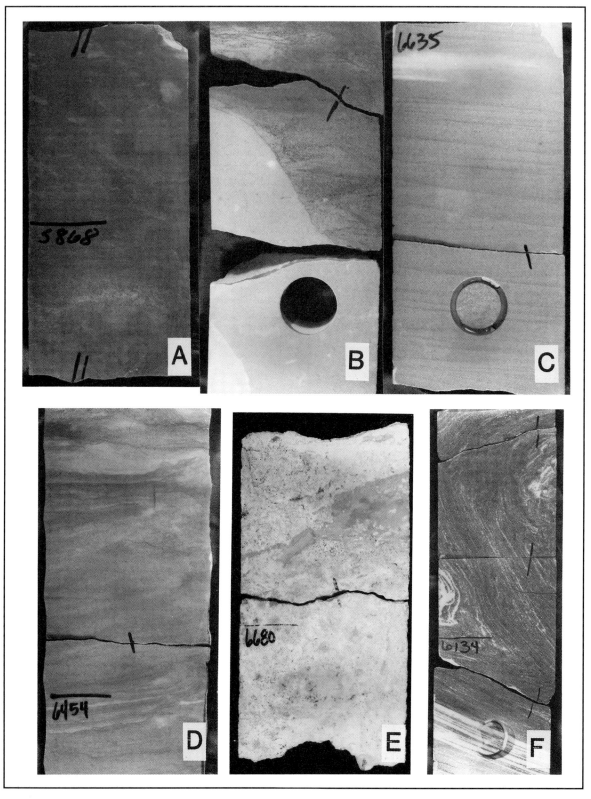

Figure 8. Core photographs of sealing and producing facies. (A) State Bridge Formation (Permian–Triassic): red mudstone deposited in tidal environment, 3 ft (1 m) above the Weber Sandstone. (B) Weber Sandstone (Pennsylvanian–Permian): angular contact between overlying fluvial arkosic sandstone of Maroon facies and underlying bioturbated eolian quartzose sandstone of dune facies. (C) Weber Sandstone dune facies: wind-ripple and possible grainflow strata. (D) Weber Sandstone wet-interdune facies: wavy bedded and bioturbated sandstone. (E) Weber Sandstone wet-interdune facies: mottled and brecciated arenaceous limestone. (F) Weber Sandstone dune facies: contorted wind-ripple strata. Dark-colored lamina are impregnated with residual bitumen (dead oil). Width of cores is about 3 in (7.5 cm).

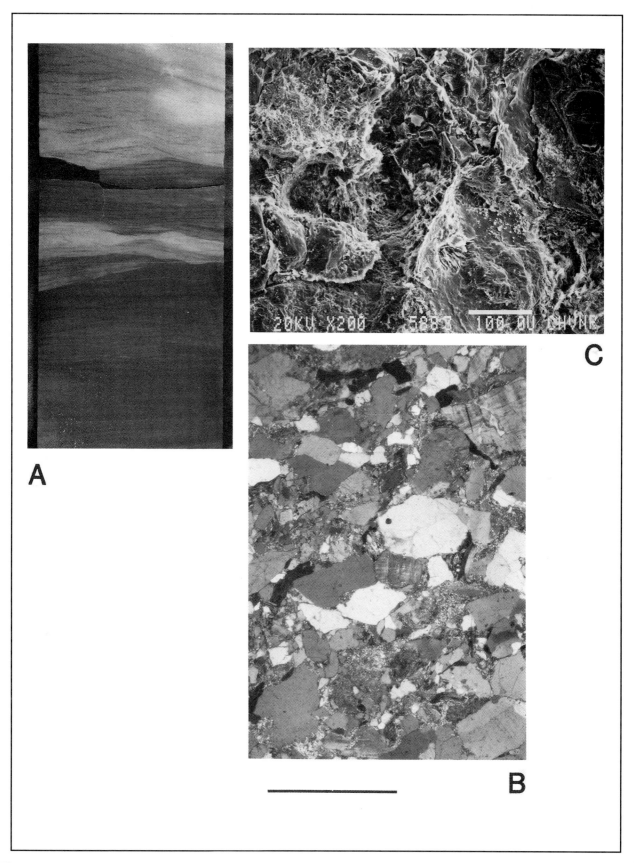

Figure 9. Fluvial and alluvial facies of the Weber Sandstone. (A) Core photograph of fluvial siltstone showing wavy bedding and climbing ripples. Width of core is about 3 in (7.5 cm). (B) Photomicrograph of fluvial sandstone. Bar, 1 mm. (C) SEM photograph of fluvial sandstone. The rock is tightly cemented. Permeability is 0.01 md, porosity is nearly 0. Bar, 0.1 mm.

Figure 10. Wind ripple strata of Weber Sandstone. (A) Core photograph of high-angle climbing translatent wind-ripple strata. Dark-colored lamina are coarser-grained tops of coarsening-upward lamina impregnated with residual bitumen (dead oil). Width of the core is about 3 in (7.5 cm). (B) Photomicrograph of wind ripple. Bar, 1 mm. Arrows show sharp contact of next coarsening-upward lamina. (C) SEM photograph of wind-ripple strata. Bar, 0.1 mm. Quartz grains are coated with druzy quartz. Euhedral quartz overgrowths are visible in the lower right. The amorphous substance is residual bitumen (dead oil).

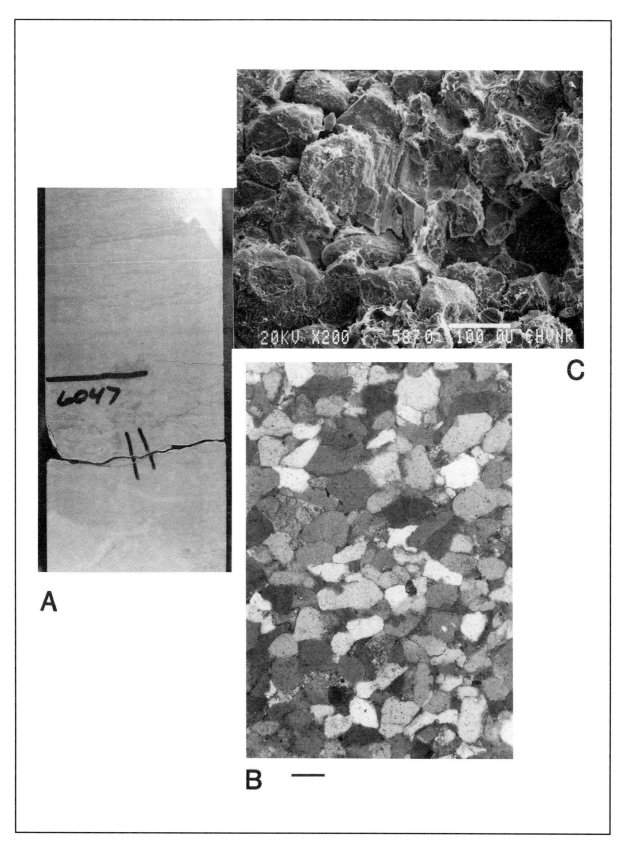

Figure 11. Bioturbated dune facies of Weber Sandstone. (A) Core photograph of heavily bioturbated dune top overlain by wet interdune facies. Core width is about 3 in (7.5 cm). (B) Photomicrograph of bioturbated quartzose eolian sandstone. Bar, 0.1 mm. Long and sutured quartz grain contacts result from quartz overgrowths. (C) SEM photograph of bioturbated wind-ripple strata. Bar, 0.1 mm. Quartz dolomite, and illite cements are visible. The hole at right is from a grain plucked during sample preparation.

Several bioturbated zones may be present within a single complex dune (Bowker and Jackson, 1989). Dune facies are commonly truncated by overlying interdune or fluvial deposits (Figure 8B).

Sand sheet and dry-interdune facies are characterized by abundant low-angle (less than 15°) wind-ripple strata (Kocurek and Nielson, 1986; Fryberger et al., 1979) (Figure 10). At Rangely, sand sheets are distinguished from dry interdunes by an absence of associated dune facies. Sand sheet facies can be comprised of either simple or complex beds. Flat upper and lower contacts with adjacent alluvial deposits are common. Complex sand sheets may be 30 ft (10 m) thick but are commonly less than 10 ft (3 m) thick. Dry interdunes usually unconformably overlie dunes and are overlain by dunes or fluvial deposits. Dry interdunes are generally thinner than sand sheets, rarely exceeding 3 ft (1 m). Both interdunes and sand sheets are moderately to extensively bioturbated.

Wet-interdune facies contain sandstones and siltstones with characteristic subaqueous structures and, more rarely, laterally discontinuous thin carbonates (Figure 8D, E). Sandstones and siltstones from a wet interdune environment are identified by low-index oscillation ripples, wavy bedding, associated finer-grained deposits, and rare fluid escape structures. Carbonates are arenaceous and have massive or mottled textures. Extensive bioturbation is common. Wet interdune deposits are rarely more than 2 ft (0.7 m) thick. No identifiable fossils have been recovered from these laterally restricted limestones and dolomites.

Deposits typical of the Maroon Formation form a distinct fluvial and alluvial facies interbedded with the eolian Weber Sandstone (Figure 9). Arkosic sandstones and siltstones comprise at least 80% of the Maroon facies; mudstones form the remaining 20%. The fluvial and alluvial strata are reddish brown or reddish gray and contrast sharply with the buff eolian Weber strata. Internal structures include planar and trough cross-bedding, current and climbing ripples, mudstone drapes, and mudstone rip-up clasts. The fluvial and alluvial facies are complexly and variably constructed. Grossly fining-upward sequences are most common, but coarsening-upward and massive packages are also present. The thickness of fluvial and alluvial deposits range from less than 1 ft (0.3 m) to over 40 ft (12 m).

Depositional Environment

Eolian strata at Rangely were deposited in an erg (sand sea) margin (c.f. Kocurek and Nielson, 1986). Coastal eolian dunes of the upper Morgan Formation were present intermittently during Morrowan and Atokan time (Driese and Dott, 1984). A Desmoinesian erg system succeeded the coastal dune environment throughout much of the northern and central Rocky Mountains. This erg was preserved in the eolian deposits of the Weber, Tensleep, Casper, and Quadrant sandstones in parts of Montana, Wyoming, Idaho, Utah, and Colorado (Blakey et al., 1988). At Rangely, the lower portions of the Weber Sandstone (below 6177 ft [1884 m], Figure 12; zones 9–11, Figure 13) were deposited in a sand sheet environment between the Maroon Formation alluvial plain to the south and the active dune field to the north. The sand sheet facies is interbedded with widely correlatable, subparallel alluvial facies. Later, the southern erg margin migrated across the Rangely area leaving dune and interdune deposits characteristic of the most productive portion of the reservoir (5717–6135 ft [1744–1871 m], Figure 12; zones 3–7, Figure 13). Ephemeral streams crossing the dune field produced characteristically channel-form and laterally discontinuous fluvial deposits. The sand sheet environment returned to the Rangely area at the end of Weber deposition (5612–5717 ft [1712–1744 m], Figure 12; zone 1, Figure 13). Widespread fluvial and alluvial deposits may reflect climatic effects, tectonic movement on the Uncompahgre uplift, or eustatic sea level changes related to Gondwanaland glaciation (Crowell, 1978).

Diagenesis

A complete discussion of the diagenesis of the Weber Sandstone at Rangely is given by Koelmel (1986). The Weber Sandstone contains residual, intergranular primary porosity (Figure 10) and minor secondary porosity (Figure 14). First stage diagenesis of the Weber Sandstone included deposition of illuviated clay and precipitation of calcite. Second stage quartz overgrowths were unaccompanied by significant pressure solution. In the Uinta basin and locally in the Rangely field, silica cementation completely filled porosity. Stage three diagenesis was characterized by the precipitation of calcite, dolomite, ferroan calcite, ferroan dolomite, anhydrite, and illite. Stage four dissolution of feldspar framework grains and stage three cements produced locally important secondary porosity. Hydrocarbon migration occurred sometime after stage four diagenesis.

Porosity and Permeability

The porosity and permeability characteristics of the Weber Sandstone are well known from extensive core analyses. Producible reservoir rock has been found to have 1 md or greater core permeability at surface conditions. Transform equations derived from crossplots are used to identify pay zones on modern porosity logs (Figure 15). A normalized 10% log porosity approximates an effective sandstone with 1 md permeability (Figure 12). A gamma-ray cutoff of 50 API units screens out tight alluvial and fluvial Maroon facies in which potassium-rich feldspars, micas, and clays cause a high gamma-ray response. Wet-interdune facies and heavily bioturbated tops of dune deposits have low gamma-ray values, but also low porosity. These strata are

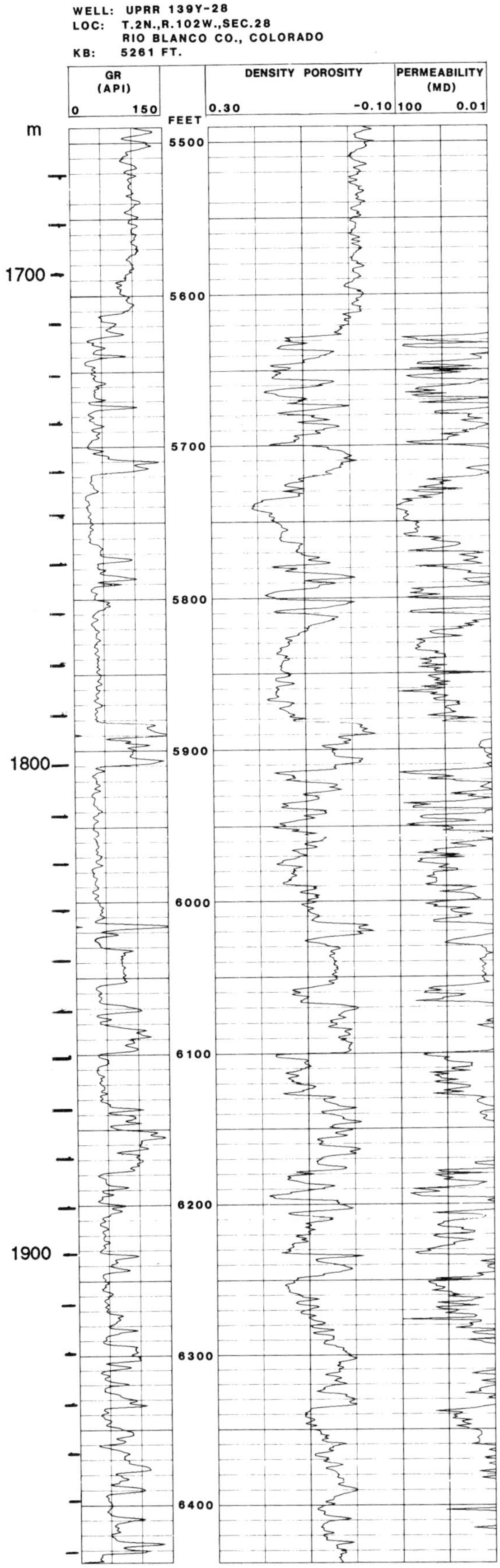

Figure 12. Type log of Weber Sandstone.

generally nonproductive. The best porosities and permeabilities are developed in the wind ripple strata of the dune and sand sheet facies. The field's average porosity is about 13% and the median permeability is about 10 md. The highest porosity in the field is about 25%, and the highest permeability is less than 400 md.

Pay zone thickness and production characteristics vary both vertically and laterally across the field (Figure 13). The upper 400 to 500 ft (120–150 m) of the Weber Formation contain the thickest complex dunes and the highest quality pay. These dunes are up to 80 ft (25 m) thick and average 15 to 20 ft (5–6 m). The lower several hundred feet of the reservoir contain pay zones in sand sheet deposits that average 5 ft (1.5 m) thick and are rarely greater than 15 ft (3 m) thick. The proportion of productive eolian facies decreases from 75% (of the Weber-Maroon section above the oil-water contact) at the northwest end of the field to less than 20% at the southeast end. Porosity and permeability also follow the depositional trend, decreasing from northwest to southeast. This reduction in reservoir quality is caused by increased calcite and authigenic illite cementation toward the southeast. Thus hydrocarbon production and recovery is lowest in the southeast end of the field.

Fluid Flow

Fluid flow at Rangely is controlled by an interplay of structural, depositional, and diagenetic factors. Horizontal permeability within pay zones is nearly isotropic because of the low bedding angle of the wind-ripple strata. Linear dune patterns probably control fluid flow within some sandstones. For example, a strong northwest-southeast trend exists in zone 3 in Figure 13. The dominant water flood pattern in the field is a northwest-southeast line drive that takes advantage of this lineation. However, dune trends shifted during Weber Sandstone deposition. Summed over the 820 ft (250 m) thick oil column, the net effect of depositional trends on fluid flow is variable and difficult to predict. Natural and induced fracturing locally control fluid flow, especially on the east end of the field. There, the fluvial influence was greatest and diagenetic processes were most effective in reducing primary porosity and permeability. Pulse tests in the eastern portion of the field have shown that fluid flows west to northwest, parallel to natural fractures.

Vertical fluid flow is controlled by depositional and structural factors. Vertical permeability is reduced by the alternating finer and coarser-grained layers of the wind-ripple strata. Further permeability reduction is caused by the residual bitumen (dead oil) that is frequently found in the coarser-grained layers of the wind-ripple strata (Figures 8F and 10). Bioturbated zones and the fluvial and alluvial facies are limited barriers to vertical fluid flow. Five relatively thick and laterally extensive alluvial intervals have been used to subdivide the reservoir's productive zones (Figure 13). Selective perforation

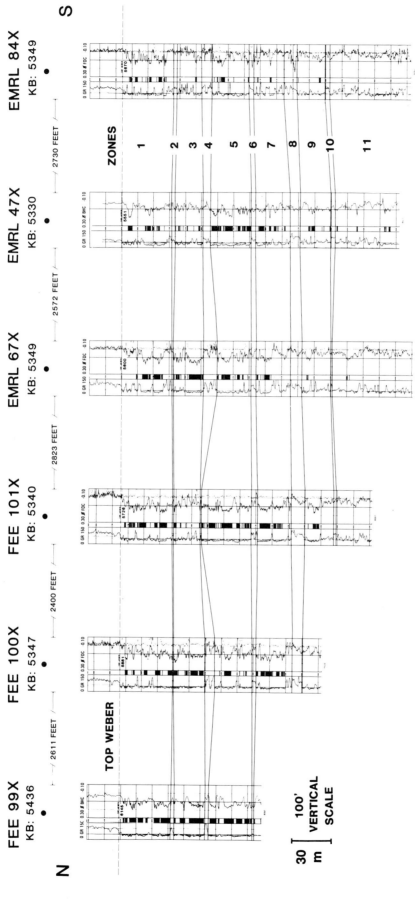

Figure 13. Rangely field north-south stratigraphic cross section. The amount of effective pay decreases to the south. Even numbered zones are fluvial and alluvial facies that divide the reservoir into six productive zones. Location of the cross section can be seen on Figure 2.

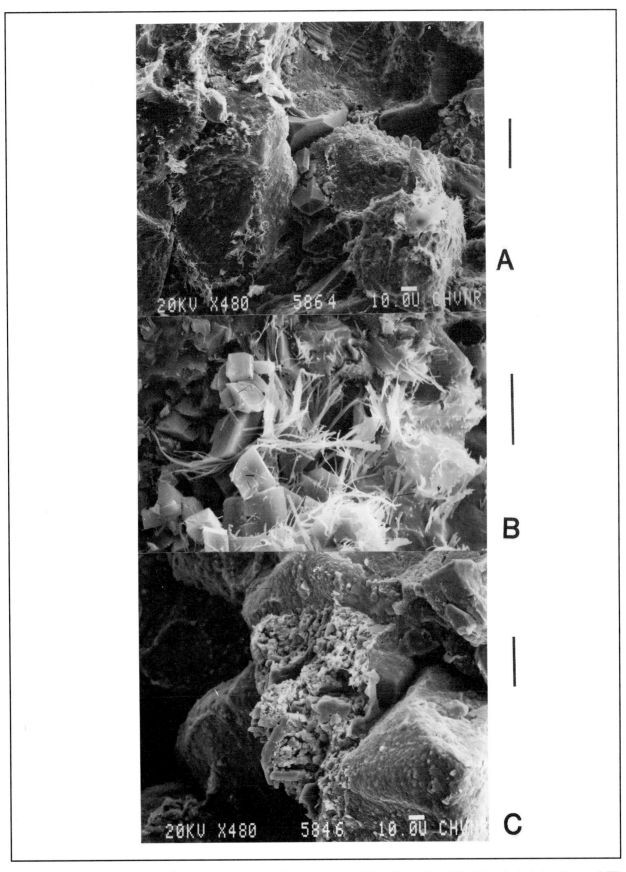

Figure 14. SEM photographs of diagenetic products in the Weber Sandstone. (A) Early druzy quartz coating, second stage euhedral quartz over-growths, and third stage illite. Bar, 30µ. (B) Euhedral dolomite and illite. Bar, 10µ. (C) Secondary porosity from dissolution of feldspar grain. Bar, 30µ.

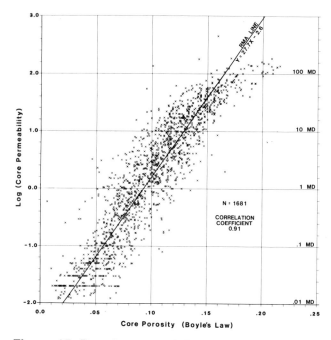

Figure 15. Porosity-permeability cross plot for Weber Sandstone. Horizontal core permeability vs. Boyle's law (helium injection) porosity for data from five recently cored wells. Log-derived porosity is closely correlatable to summation of fluids care porosity, which includes residual bitumen (dead oil) in the porosity calculation: 0% summation of fluids porosity is approximately equal to 9% Boyle's law porosity.

and injection equipment are used to enhance secondary and tertiary flooding of the productive zones. However, fluids move vertically around lateral pinch-outs of tight interdune and fluvial facies. In addition, the relatively competent arkosic sandstones and siltstones of the Maroon facies have fracture gradients similar to the well-lithified eolian Weber Sandstone. Natural and artificial fractures may cross both the Weber and Maroon facies and may have caused vertical communication between pay zones.

SOURCES

There are several known and several hypothesized source rocks for the Uinta and Piceance basins. Oil in the Weber Sandstone in the Piceance basin fields (Figure 1), including Rangely, was derived from the black phosphatic shale facies of the Permian Phosphoria Formation. Oil and associated gas from Jurassic- and Cretaceous-aged reservoirs are sourced from the Cretaceous Mancos formation. The Weber Sandstone production at Ashley Valley field is also sourced from the Mancos. Eocene oil and associated gas in the Uinta basin are sourced from lacustrine shales of the Eocene Green River and Uinta formations. Hypothetical but as yet unproven source rocks are contained in the overmature black marine shales of the Pennsylvanian Belden Shale of central Colorado (Nuccio and Schenk, 1986, 1987) and the Jurassic Curtis Formation marine shales in northwestern Colorado (Stone, 1986b).

Two different types of oil are produced from the Rangely anticline. Mancos fracture production is sourced indigenously from the Mancos formation, yielding a high-gravity, low-sulfur, paraffinic oil typical of Cretaceous sourced oils in the region. Weber oil is medium gravity, medium sulfur content, asphaltic, and resembles oil produced from other Permian and Pennsylvanian reservoirs in the Big Horn basin of Wyoming. An indigenous source for the Weber oil has not been found in the Rangely area. However, two sources requiring long-distance migration have been proposed.

Black phosphatic shales contained in the Phosphoria Formation in north-central Utah, eastern Idaho, and western Wyoming were proposed as source rocks for Permian Park City Formation and Pennsylvanian Tensleep Sandstone reservoirs in the Big Horn and Wind River basins of Wyoming (Sheldon, 1967; Stone, 1967). The Phosphoria Formation has also been proposed as the source for Rangely Weber oil by Fryberger (1979) and Koelmel (1986). Pyrolysis of Phosphoria shale sample extracts provides a good match to gas chromatographs of the Weber oil (Figure 16). Very high quality (greater than 1.5% TOC) source rocks producing type II kerogen are present in the Meade Peak and Retort Shale members of the Phosphoria Formation from southwestern Montana to central Utah. Maughan (1984) calculated a hydrocarbon yield of 30.7 billion metric tons (230 billion bbl) of oil for the Phosphoria Formation.

A burial history diagram of the Phosphoria Formation in western Wasatch County, Utah, is shown in Figure 17. The oil generation window shown was constructed by using a constant 1.3 heat flow units (HFU) and an activation energy model. Potential source rock in the area has TOC content of 7 wt. %. Initial oil generation occurred in the early Jurassic at this location. This is the earliest onset of oil generation in the Phosphoria, which has probably generated hydrocarbons continuously since that time. Migration from this site was controlled by paleodip and the availability of conduit beds having adequate permeability. An isopach map of the Triassic system (Figure 18) (MacLachlan, 1972) indicates that the paleodip on the top of the Permian Phosphoria–Park City in Early Jurassic time would have allowed migration of oil updip to the east and east-southeast. Exposures of the Weber Sandstone along the south flank of the Uinta Mountains (Bissell and Childs, 1958; Bissell, 1964) show the permeable Weber Sandstone in contact with organic-rich facies of the overlying Phosphoria Formation. The Weber Sandstone is the most likely conduit for hydrocarbon migration eastward into northwestern Colorado.

Recently, geologists working in the Pennsylvanian–Permian Eagle basin have proposed that the Lower Pennsylvanian Belden Formation of central

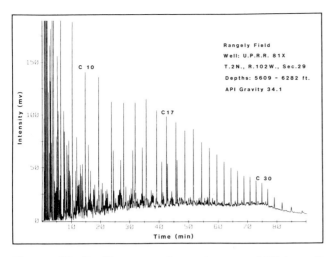

Figure 16. Capillary gas chromatogram of Weber oil from Rangely field. The oil is asphaltine based, unlike the paraffinic crude oils sourced by the Cretaceous.

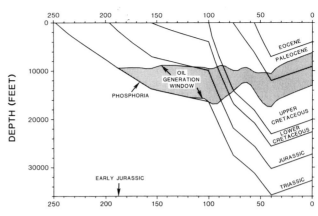

Figure 17. Geohistory diagram of the Phosphoria Formation, proposed source rock for the Weber Sandstone at Rangely. The diagram was constructed with an activation energy model and constant heat flow of 1.3 heat flow units. The sample was taken from outcrop in Wasatch County, Utah.

Colorado might be the source rock for oil in Weber Sandstone and Minturn Formation reservoirs of the Piceance basin (Nuccio and Schenk, 1986, 1987; Waechter and Johnson, 1986). The Belden Formation contains black shales with TOC content as high as 3.6 wt. % (Nuccio and Schenk, 1987). Unfortunately, all samples that have been collected from the Belden black shales have been supermature (vitrinite reflectance is 2.4% to 4.1%) and no "fingerprints" of Belden oil are available (Nuccio and Schenk, 1986, 1987). A residual bitumen (dead oil) phase is present both at Rangely and in Pennsylvanian and Permian sandstones in the Piceance and Eagle basins and may be the product of oil generated from the Belden Formation.

EXPLORATION CONCEPTS

Rangely field is an excellent example of a compressive, basement-folded foreland structure in the Rocky Mountain province. Perhaps the best known compressive foreland structures are in the Big Horn basin of Wyoming where large anticlines such as Elk Mountain are also exposed at the surface.

An important aspect of the Rangely field is the concept of superimposed traps through time. Evidence suggests that an oil accumulation was present at Rangely prior to formation of the Eocene age structure present today.

Continuing advances in geochemistry enable explorationists to time the onset of oil generation and migration more accurately than ever before. The period of hydrocarbon migration may, as at Rangely, predate the obvious trapping mechanism. Because of the surface anticline, the stratigraphically trapped oil in the Weber Sandstone at Rangely was discovered.

ACKNOWLEDGMENTS

Our thanks to Chevron U.S.A., Inc. for permitting publication of this paper. M. F. Mendeck, R. H. Elliott, R. E. Ladd, K. T. Bowker, and numerous other Chevron personnel supported our efforts and provided ideas. C. F. Dodge graciously provided numerous references and an outline for the appendix. R. H. Elliott and R. E. Ladd reviewed and improved the manuscript.

REFERENCES CITED

Bench, B. M., 1945, The Rangely oil and gas field and Uinta basin: Mines Magazine, v. 35, n. 10, p. 516-524, 532, 594, 596.

Bench, B. M., 1946, Rangely field geology and development: Oil Weekly, v. 123, n. 1 (Sept. 2), p. 18-25.

Bissell, H. J., 1964, Lithology and petrography of the Weber Formation in Utah and Colorado, in E. F. Sabatka, ed., Guidebook to the geology and mineral resources of the Uinta basin: Intermountain Association of Petroleum Geologists Thirteenth Annual Field Conference, p. 67-91.

Bissell, H. J., and O. E. Childs, 1958, The Weber Formation of Utah and Colorado, in Symposium of Pennsylvanian rocks of Colorado and adjacent areas: Rocky Mountain Association of Geologists, p. 26-30.

Blakely, R. C., F. Peterson, and G. Kocurek, 1988, Synthesis of late Paleozoic and Mesozoic eolian deposits of the Western interior of the United States: Sedimentary Geology, v. 56, p. 3-125.

Bleakley, W. B., 1973, Rangely flood looking good; production near 47,000 BO/D: Oil and Gas Journal, v. 71, n. 8 (Feb. 19), p. 90-92.

Bowker, K. A., and W. D. Jackson, 1989, The Weber Sandstone at Rangely Field, Colorado, in E. B. Coalson, ed., Petrogenesis and petrophysics of selected sandstone reservoirs of the Rocky Mountain region: Rocky Mountain Association of Geologists Annual Guide Book, p. 65-80.

Campbell, G. S., 1955, Weber pool of Rangely field, Colorado, in H. R. Ritzma and S. S. Oriel, eds., Guidebook to the Geology of Northwest Colorado: Intermountain Association of Petroleum Geologists—Rocky Mountain Association of Geologists, Guidebook, Sixth Annual Field Conference, 1955, p. 99-100.

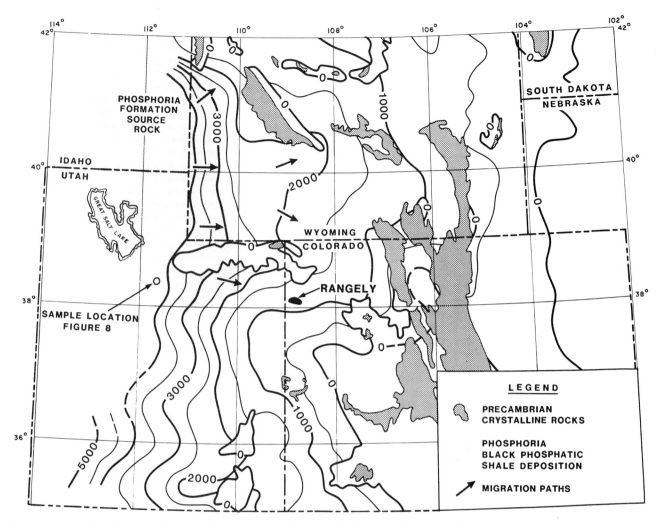

Figure 18. Triassic isopach map showing paleodip on top of the Permian Phosphoria Formation in Early Jurassic time. Contour interval, 500 ft. The burial history diagram of the sample located in Wasatch County, Utah, indicated oil generation and subsequent migration westward beginning in Early Jurassic. Modified from MacLachlan (1972, their figure 3, p. 169).

Christensen, C. J., 1948, Diamond coring in the Rangely field, Colorado: American Institute of Mining Engineering Transactions, v. 174, p. 206-218; American Institute of Mining Engineering Technical Paper no. 2301, Petroleum Technology, v. 11, n. 1 (Jan).

Crowell, J. C., 1978, Gondwanan glaciation, cyclothems, continental positioning, and climatic change: American Journal of Science, v. 278, p. 1345-1372.

Cullins, H. L., 1971a, Geological map of the Banty Point quadrangle, Rio Blanco County, Colorado: USGS Quadrangle, scale 1:24000.

Cullins, H. L., 1971b, Geological map of the Rangely quadrangle, Rio Blanco County, Colorado: USGS Quadrangle, scale 1:24000.

Cupps, C. Q., P. H. Lipstate, Jr., and J. Fry, 1951, Variance in characteristics of the oil in the Weber Sandstone reservoir, Rangely field, Colorado: U.S. Bureau of Mines Report of Investigation 4761, 68 pages.

De Mohrenschildt, G., 1949, Development of Rangely field, the Rocky Mountains' largest producer: Oil and Gas Journal, v. 48, n. 7 (June 23), p. 124, 126-128, 130.

Driese, S. G., and R. H. Dott, 1984, Model for sandstone carbonate "cyclothems" based on upper member of Morgan Formation (Middle Pennsylvanian) of northern Utah and Colorado: AAPG Bulletin, v. 68, p. 574-597.

Fryberger, S. G., 1979, Eolian-fluviatile (continental) origin of ancient stratigraphic trap for petroleum in Weber Formation, Rangely field, Colorado: Mountain Geologist, v. 16, n. 1 (Jan.), p. 1-36.

Fryberger, S. G., and M. H. Koelmel, 1986, Rangely field: eolian system-boundary trap in the Permo-Pennsylvanian Weber Sandstone of northwest Colorado, in D. S. Stone, ed., 1986, New interpretations of northwest Colorado geology: Rocky Mountain Association of Geologists, Denver, p. 129-149.

Fryberger, S. G., T. S. Ahlbrandt, and S. Andrews, 1979, Origin, sedimentary features, and significance of low-angle eolian "sand sheet" deposits, Great Sand Dunes National Monument and vicinity, Colorado: Journal of Sedimentary Petrology, v. 49, p. 733-746.

Gale, H. S., 1908, Geology of the Rangely oil district, Rio Blanco County, Colorado: USGS Bulletin 350, 61 p.

Gibbs, J. F., J. H. Healy, C. B. Raleigh, J. Coakley, 1973, Seismicity in the Rangely, Colorado area: Seismological Society of America Bulletin, v. 63, n. 5, p. 1557-1570.

Graue, D. J., and E. Zana, 1978, Study of a possible CO_2 flood in the Rangely field, Colorado: Proceedings Fifth Symposium on Improved Methods for Oil Recovery, Society of Petroleum Engineers of AIME, paper no. 7060, p. 253-260.

Gries, R., 1983, North-south compression of Rocky Mountain foreland structures, in J. D. Lowell, ed., Rocky Mountain foreland basins and uplifts: Rocky Mountain Association of Geologists, Denver, Colorado, p. 9-32.

Haimson, B. C., 1973, Earthquake related stresses at Rangely, Colorado, in New horizons in rock mechanics; earthquake and other dynamic phenomena: Symp. Rock Mech. Proc., n. 14, p. 689-708.

Hansen, W. R., 1986, History of faulting in the eastern Uinta Mountains, Colorado and Utah, *in* D. S. Stone, ed., 1986, New interpretations of northwest Colorado geology: Rocky Mountain Association of Geologists, Denver, p. 5-17.

Harding, T. P., and J. D. Lowell, 1979, Structural styles, their plate-tectonic habitats, and hydrocarbon traps in petroleum provinces: AAPG Bulletin, v. 63, n. 7 (July), p. 1016-1058.

Heaton, R. L., 1929, Relation of accumulation to structure in northwestern Colorado, *in* Structure of typical American oil fields, v. II: AAPG Symposium, AAPG, Tulsa, p. 93-114.

Hoffman, F. H., 1957, Possibilities of Weber stratigraphic traps, Rangely area, northwest Colorado: AAPG Rocky Mountain Section, Symposium on Stratigraphic Type Oil Accumulations in the Rocky Mountains, AAPG Bulletin, v. 41, n. 5 (May), p. 894-905.

Holmes, J. A., 1926, Shutting in Rangely gas well: AIME Petroleum and Technology in 1925, p. 179-182.

Kocurek, G., and J. Nielson, 1986, Conditions favorable for the formation of warm-climate aeolian sand sheets: Sedimentology, v. 33, p. 795-816.

Koelmel, M. H., 1986, Post-Mississippian paleotectonic, stratigraphic, and diagenetic history of the Weber Sandstone in the Rangely field area, Colorado, *in* J. A. Peterson, ed., Paleotectonics and sedimentation in the Rocky Mountain region, United States: AAPG Memoir 41, p. 371-396.

Kluth, C. F., and P. J. Coney, 1981, Plate tectonics of the ancestral Rocky Mountains: Geology, v. 9, p. 10-15.

Larsen, W. K., 1987, Rangely Weber Sand Unit: enhanced oil recovery field reports: Society of Petroleum Engineers, v. 12, n. 3, p. 2381-2391.

Larson, T. C., 1975, Geological considerations of Weber Sandstone reservoir, Rangely field, Colorado, *in* D. W. Bolyard, ed., Deep drilling frontiers of the central Rocky Mountains: Rocky Mountain Association of Geologists, p. 275-279.

MacLachlan, M. M., 1972, Triassic system, *in* W. W. Mallory, ed., Geologic atlas of the Rocky Mountain region: Rocky Mountain Association of Geologists, p. 166-176.

Maughan, E. K., 1980, Permian and lower Triassic geology of Colorado, *in* H. C. Kent and K. W. Porter, eds., Colorado geology: Rocky Mountain Association of Geologists, p. 103-110.

Maughan, E. K., 1984, Geological setting and some geochemistry of petroleum source rocks in the Permian Phosphoria Fm., *in* J. Woodward, F. F. Meisner, and J. L. Clayton, eds., Hydrocarbon source rocks of the greater Rocky Mountain region: Rocky Mountain Association of Geologists 1984 Symposium, p. 281-294.

McMinn, P. M., and H. L. Patton, 1962, Rangely field, Rio Blanco county, Colorado, *in* Exploration for oil and gas in northwestern Colorado: Rocky Mountain Association of Geologists, p. 104-107.

Meeker Herald, 12 December 1890; 15 June 1901; 18 October 1901; 24 June 1902, Meeker, Colorado.

Newton, S. M., 1945 (1946), The Rangely oil field, its early rise, its coma, and its resurrection: Rocky Mountain Petroleum Yearbook 1945, Golden, Colorado, p. 63-69.

Newton, S. M., 1946, Rangely faults—good and bad: Oil Reporter, v. 3, n. 18 (Oct. 22), p. 8-11, 27. Also printed in Rocky Mountain Petroleum Review 1946-47, p. 32-36.

Nuccio, V. F., and C. J. Schenk, 1986, Thermal maturity and hydrocarbon source-rock potential of the Eagle basin, northwestern Colorado, *in* D. S. Stone, ed., 1986, New interpretations of northwest Colorado geology: Rocky Mountain Association of Geologists, p. 259-264.

Nuccio, V. F., and C. J. Schenk, 1987, Burial reconstruction of the early and middle Pennsylvanian Belden Formation, Gilman area, Eagle basin, northwest Colorado; USGS Bulletin 1787-C.

Oil and Gas Journal, 1957, Rangely is unitized: v. 55, n. 34 (Aug. 26), p. 70.

Oil and Gas Journal, 1958, Rangely field ready for big flood; v. 56, n. 44 (Nov. 3), p. 50-51.

Osmund, J. C., 1986, Petroleum geology of the Uinta Mountains-White River Uplift, Colorado and Utah, *in* D. S. Stone, ed., 1986, New interpretations of northwest Colorado geology: Rocky Mountain Association of Geologists, p. 213-221.

Owen, E. W., 1975, Trek of the oil finders: a history of exploration for petroleum: AAPG Memoir 6, 1647 p.

Peterson, V. E., 1955, Fracture production from the Mancos Shale, Rangely field, Rio Blanco county, Colorado, *in* Intermountain Association of Petroleum Geologists—Rocky Mountain Association of Geologists Guidebook, 6th Annual Field Conference 1955, p. 101-105.

Petroleum Week, 1957, New era opens for unitized Rangely, v. 5, n. 9 (Aug. 30), p. 19-21.

Pickering, W. Y., and C. I. Dorn, 1948, Rangely oil field, Rio Blanco county, Colorado, *in* J. V. Howell, ed., Structure of typical American oil fields: AAPG Symposium v. III, p. 132-152.

Pollard, W. L., 1957, Development of the western frontier: Rangely, Colorado: Unpublished M.A. Thesis, University of Denver, Denver, Colorado.

Raleigh, C. B., J. H. Healy, J. D. Bredehoeft, 1972, Faulting and crustal stress at Rangely, Colorado, *in* Flow and fracture of rocks: American Geophysical Union, Monograph no. 16, p. 275-284, 337-352.

Raleigh, C. B., J. H. Healy, J. D. Bredehoeft, 1976, An experiment in earthquake control at Rangely, Colorado: Science, v. 191, p. 4223.

Rascoe, B. Jr., and D. L. Baars, 1972, Permian system, *in* W. W. Mallory, ed., Geologic atlas of the Rocky Mountain region: Rocky Mountain Association of Geologists, p. 143-165.

Ritzma, H. R., 1955, The Rangely boom, *in* H. R. Ritzma and S. S. Oriel, eds., Guidebook to the Geology of Northwest Colorado: Intermountain Association of Petroleum Geologists—Rocky Mountain Association of Geologists, Guidebook, Sixth Annual Field Conference, 1955, p. 100.

Sears, J. D., 1924, Geology and oil and gas prospects of part of Moffat county, Colorado and southern Sweetwater county, Wyoming: USGS Bulletin 751-G, from Contributions to Economic Geology, 1923-1924, Part II, p. 269-319.

Sheldon, R. P., 1967, Long-distance migration of oil in Wyoming: The Mountain Geologist, v. 4, p. 53-65.

Stone, D. S., 1967, Theory of Paleozoic oil and gas accumulation in Big Horn basin, Wyoming: AAPG Bulletin, v. 51, p. 2056-2114.

Stone, D. S., 1986a, Rangely field summary: 2. Seismic profile, structural cross section and geochemical comparisons, *in* D. S. Stone, ed., 1986, New interpretations of northwest Colorado geology: Rocky Mountain Association of Geologists, p. 226-228.

Stone, D. S., 1986b, Geology of the Wilson Creek field, Rio Blanco county, Colorado, *in* D. S. Stone, ed., 1986, New interpretations of northwest Colorado geology: Rocky Mountain Association of Geologists, p. 229-246.

Stuart, R. W., 1947, Diamond coring at Rangely Colorado: Petroleum Engineer, v. 18, n. 4, p. 43-48.

Thomas, C. R., 1944, Structure contour map of the exposed rocks in the Rangely anticline, Rio Blanco and Moffat Counties, Colorado: USGS Oil and Gas investigations Preliminary Map 7.

Thomas, C. R., 1945, Structure contour maps of the Rangely anticline, Rio Blanco and Moffat Counties, Colorado: USGS Oil and Gas Investigations Preliminary Map 41.

Thomas, C. R., J. W. Huddle, and N. W. Bass, 1946, Rangely oil field, Rio Blanco county, Colorado: AAPG Bulletin, v. 30, n. 5 (May), p. 749-750, abstract.

Waechter, N. B., and W. E. Johnson, 1986, Pennsylvanian-Permian paleostructure and stratigraphy as interpreted from seismic data in the Piceance basin, northwest Colorado: *in* D. S. Stone, ed., 1986, New interpretations of northwest Colorado geology: Rocky Mountain Association of Geologists, p. 51-64.

White, C. A., 1878, A report on the geology of a portion of northwest Colorado: 10th Annual Report, U.S. Geological and Geographical Survey of the Territories.

White, G. T., 1962, Formative years in the far west: New York, Appleton-Century Croft, 694 p.

White River Review, 6 September 1902; 11 November 1902, Meeker, Colorado.

Whiteker, R. M., 1975, Upper Pennsylvanian and Permian strata of northeast Utah and northwest Colorado, *in* D. W. Bolyard, ed., Deep drilling frontiers of the central Rocky Mountains: Rocky Mountain Association of Geologists, p. 75-84.

Williams, N., 1945, Difficult drilling and high costs attend development of Rangely field: Oil and Gas Journal, v. 44, n. 33 (Dec. 22), p. 44, 45, 53, 54.

Williams, N., 1947, Diamond-bit coring in the Rangely, Colorado field: Oil and Gas Journal, v. 45, n. 35 (Jan. 4), p. 48-50, 52.

SUGGESTED READING

Allen, D. D., 1963, High volume pumping in the Rangely Weber Sand Unit: Journal of Petroleum Technology, August, 1963. Discusses pumping techniques and hardware.

Bagzis, J. M., 1989, Refracturing pays off in Rangely field: World Oil, v. 209, n. 5 (November), p. 39-42, 44. A summary of fracturing techniques and their results.

Barb, C. F., 1946, Intensive drilling at Rangely: World Petroleum, v. 17, n. 4 (April), p. 62. Compact summary of field facts, less than a half page.

Bass, N. W., 1946a, Subsurface maps of the Rangely anticline, Rio Blanco county, Colorado: USGS Oil and Gas Investigations Preliminary Map No. 67. First published detailed Weber structure map. Includes history, stratigraphy, structure, and justification for subsurface faults.

Bass, N. W., 1946b, Geological survey completes new structure map for Rangely Field: Oil and Gas Journal, v. 45, n. 30 (Nov. 30), p. 86, 87, 93. Excerpts from USGS Oil and Gas Investigations Preliminary Map 67, see Bass 1946a.

Bench, B. M., 1946, Rangely oil field, Rangely, Colorado: Mines Magazine, v. 36, n. 11 (Nov. 2), p. 513-517. Current summary of activities, including discussion on Shinarump (Gartra) production, and comments on production curtailments.

Berman, A. E., D. Pooleschook, Jr., and T. E. Dimalow, 1980, Jurassic and Cretaceous systems of Colorado, *in* H. C. Kent, and K. W. Porter, eds., Colorado geology: Rocky Mountain Association of Geologists, p. 111-128. Regional stratigraphic summary.

Bleakley, W. B., 1960, Rangely flood moves ahead—gasoline plant gains satellites: Oil and Gas Journal, v. 58, n. 50 (Dec. 12), p. 110-111. Use of satellite processing plants to boost liquids recovery and current field statistics.

Bleakley, W. B., 1964, Colorado's biggest field engineered for maximum recovery: Oil and Gas Journal, v. 62, n. 9 (Mar. 2), p. 108-116. Thorough summary of production and engineering details after unitization and start of the water flood.

Bosco, F. N., and J. D. Brawner, 1946. Rangely field, largest addition to reserves in 1945: Oil Weekly, v. 120, n. 13 (Feb. 25), p. 18-20, 22, 24, 26, 28. Technical review of the field, engineering practices and problems, commentary on the place of petroleum in the post-war society.

De Voto, R. H., 1980a, Mississippian stratigraphy and history of Colorado, *in* H. C. Kent and K. W. Porter, eds., Colorado geology: Rocky Mountain Association of Geologists, p. 57-70. Regional stratigraphic summary.

De Voto, R. H., 1980b, Pennsylvanian stratigraphy and history of Colorado, *in* H. C. Kent and K. W. Porter, eds., Colorado geology: Rocky Mountain Association of Geologists, p. 71-101. Regional stratigraphic summary.

De Voto, R. H., and others, 1986, Late Paleozoic stratigraphy and syndepositional tectonism, northwestern Colorado, *in* D. S. Stone, ed., 1986, New interpretations of northwest Colorado geology: Rocky Mountain Association of Geologists, p. 37-49. Most complete summary of regional Pennsylvanian and Permian history.

Estergren, E. F., 1945, Deep drilling revives Rangely field, Colorado: Petroleum Engineer, v. 17, n. 3 (Dec.), p. 79, 82, 84. Brief technical summary of geology and activities.

Hendon, J., 1984, Chevron takes the risk; Rangely field due rejuvenation; AAPG Explorer, v. 5, n. 8, p. 10. Explains Chevron's plans for the CO_2 flood.

Johnson, R. C., 1985, Early Cenozoic history of the Uinta and Piceance Creek basins, Utah and Colorado, with special reference to the development of Eocene Lake Uinta, *in* R. M. Flores and S. S. Kaplan, eds., Cenozoic paleogeography of the west-central United States: Rocky Mountain Paleogeography Symposium 3: Rocky Mountain Section SEPM, p. 247-275. Regional stratigraphic summary.

Larson, T. C., 1974, Geological considerations of the Weber sand reservoir, Rangely field, Colorado: 49th Annual Symposium SPE of American Institute of Mining Engineering Fall Meeting Preprint No. SPE-5023, 8 p. Essentially the same paper as Larson, 1975.

Larson, T. C., 1977, Geologic considerations of the Weber Sandstone reservoir, Rangely field, Colorado: Bulletin of Canadian Petroleum Geology, v. 25, n. 3 (May), p. 410-418. Essentially the same paper as Larson, 1975.

Mallory, W. W., 1972, ed., Geological atlas of the Rocky Mountain region: Rocky Mountain Association of Geologists, 331 p. The "Big Red Book."

Mendeck, M. F., 1986, Rangely field summary: 1. development history and engineering data, *in* D. S. Stone, ed., 1986, New interpretations of northwest Colorado geology: Rocky Mountain Association of Geologists, p. 223-225. 1986 summary including engineering data.

Mining World, 1908, Rangely oil field: v. 29, p. 314. Brief note of oil development interest in the Rangely area.

Newton, S. M., 1945, Porosity and permeability as related to the Weber Sandstone, Rangely oil field, Rio Blanco County, Colorado: Rocky Mountain Petroleum Yearbook 1945, p. 69, Golden, Colorado, 1946. Follows other Newton article, and provides his justification for statements on reserves estimation. Questionable technical merit.

Newton, S. M., 1946, Looking at Rangely from the bottom up: Oil Reporter, v. 3, n. 20 (Nov. 19), p. 8, 9, 18, 19, 22, 24. Also printed in Rocky Mountain Petroleum Review 1946-47, p. 37-41. Nontechnical description of the stratigraphy and structure in the Rangely Area, including perceived potential of various horizons.

Newton, S. M., 1947, Rangely, yesterday, today, tomorrow: Mines Magazine, v. 37, n. 11 (Nov.), p. 53-56. A colorful account of Rangely's development, more interesting for its social and historical commentary than for technical merits.

Oil and Gas Journal, 1958, Two Rangely floods, v. 56, n. 7 (Feb. 17), p. 67. Brief note that two pilot waterfloods have begun.

Oil and Gas Journal, 1959, Rangely flood moving in high gear: v. 57, n. 50 (Dec. 7), p. 93-94. Gives status of waterflood and current field statistics.

Patton, H. L., 1950, Rangely oil field, *in* Utah Geological Society Guidebook to the Geology of Utah, n. 5, p. 127-133. Draws from Pickering and Dorn, 1948. Good picture of operations during this era of active primary development.

Ross, R. J., Jr., 1986, Lower Paleozoic of northwest Colorado: a summary, *in* D. S. Stone, ed., 1986, New interpretations of northwest Colorado geology: Rocky Mountain Association of Geologists, p. 99-102. Regional stratigraphic summary.

Ross, R. J., Jr., and O. Tweto, 1980, Lower Paleozoic sediments and tectonics in Colorado, *in* H. C. Kent and K. W. Porter, eds., Colorado geology: Rocky Mountain Association of Geologists, p. 47-56. Regional stratigraphic summary.

Thomas, C. R., 1945a, Rangely, one-time shallow field, now Rocky Mountains' most active area: Oil and Gas Journal, v. 44, n. 29 (Nov. 24), p. 90-96. Reprint of Preliminary Map 41 text, without surface map, and with very few minor changes.

Thomas, C. R., 1945b, Rangely, one-time shallow field, now Rocky Mountains' most active area: Petroleum Engineering, v. 17, n. 3 (Dec.), p. 79, 82, 84.

Williams, N., 1946a, Drilling operations in Rangely field—operations group themselves into three stages, representing formational divisions: Oil and Gas Journal, v. 44, n. 37 (Jan. 19), p. 70-73. Describes drilling procedures during the first phase of field development. Only minor geological content.

Williams, N., 1946b, Completions in Rangely field are complicated by tight formations: Oil and Gas Journal, v. 44, n. 40 (Feb. 9), p. 77-78, 81. Describes engineering practices used to assure good completions.

Appendix 1. Field Description

Field name .... *Rangely field*
Ultimate recoverable reserves .... *939 million bbl (149 million m³)*
Field location:
    Country .... *U.S.A.*
    State .... *Colorado*
    Basin/Province .... *Uinta/Piceance basins; Douglas Creek arch*
Field discovery:
    Year first pay discovered .... *Mancos formation 1902*
    Year second pay discovered .... *Weber Sandstone 1933*
Discovery well name and general location:
    First pay .... *Pool well, SE1/4 NE1/4 Sec. 33, T2N, R102W*
    Second pay .... *Raven A-1 well, NW1/4 SE1/4 Sec. 30, T2N, R102W*
Discovery well operator .... *California Company*
    Second pay .... *California Company*
IP:
    First pay .... *1–2 BOPD*
    Second pay .... *229 BOPD*

All other zones with shows of oil and gas in the field:

Age	Formation	Type of Show
Cretaceous	Dakota Sandstone	Early wells blew out at high rates; watered out by 1930s
Jurassic	Morrison Formation	High pressure gas in isolated sandstone lenses; limited commercial production
Triassic	Shinarump conglomerate (Gartra Member, Chinle Fm.)	Minor oil production in one well

Geologic concept leading to discovery and method or methods used to delineate prospect
Surface geology is a large breached anticline, with oil seeps at the surface. The prospective formations outcropping several miles away suggested to early explorers that rocks with good reservoir quality could be expected at depth.

Structure:
    Province/basin type .... *Bally 222; Klemme IIB (Piceance)/IIA (Uinta)*
    Tectonic history
The Rangely anticline and other regional structures are Laramide features. Earlier tectonic influences on depositional environment include a stable shelf during the Mississippian and a paleoarch developing on this paleotectonic platform during the Pennsylvanian. The arch produced a system boundary in the Weber Sandstone between eolian and alluvial fan environments.

    Regional structure
The Rangely field lies between the Douglas Creek arch to the south and the Blue Mountain uplift (of the Uinta Mountain front) to the north. This structural axis separates the Piceance basin to the east from the Uinta basin to the west.

    Local structure
A strongly defined asymmetrical anticline trending N60°W. The steep south flank dips up to approximately 30°, and the gentle north flank dips at approximately 4°. Plunge is approximately 2 to 3° on both ends. Surface closure exceeds 2000 ft (600 m).

Trap:

Trap type(s)

Structural, with probable stratigraphic trapping in the area before development of the structure. The stratigraphic trap was a facies change from nonporous alluvial sandstones and siltstones to eolian sandstones. Oil in the Mancos formation is trapped structurally in the Rangely anticline. Oil-filled fractures in the Mancos are partially sealed by calcite fillings near the surface. Oil seeps also occur from these fractures.

Basin stratigraphy (major stratigraphic intervals from surface to deepest penetration in field):

Chronostratigraphy	Formation	Meas.	Depth to Top in ft (m) M.S.L.
Cretaceous	Mancos formation	Surface	+5254 (+1601)
	Dakota Sandstone	2988	+2279 (+695)
Jurassic	Morrison Formation	3108	+2159 (+658)
	Curtis Formation	3728	+1539 (+469)
	Entrada Sandstone	3833	+1434 (+437)
	Carmel Formation	3980	+1287 (+392)
	Navajo Sandstone	4042	+1225 (+373)
Triassic	Chinle Formation	4648	+619 (+189)
	Gartra Member	4764	+503 (+153)
Triassic-Permian	State Bridge Formation	4842	+425 (+130)
Permian-Pennsylvanian	Weber Sandstone	5495	-228 (-69)
Pennsylvanian	Maroon Formation	6426	-1159 (-353)
	Morgan Formation	6650	-1383 (-422)
	Round Valley Limestone	7876	-2609 (-795)
Mississippian	Madison Limestone	8094	-2827 (-862)
Cambrian	Lodore Formation	8880	-3613 (-1101)
	Sawatch Quartzite	9300	-4033 (-2835)

Reservoir characteristics:

Number of reservoirs .... 1 primary (Weber Sandstone), 1 secondary (Mancos formation); others with minor or noncommercial production (Morrison, Shinarump or Gartra, Dakota)

Formations .... Weber Sandstone, Mancos formation

Ages .... Permian-Pennsylvanian, Cretaceous

Depths to tops of reservoirs .... Weber, 5500-6500 ft (1700-2000 m); Mancos, a few hundred feet to approx. 1600 ft (500 m)

Gross thickness (top to bottom of producing interval) .... Weber, average 700 ft; best porosity in upper 400 ft

Net thickness—total thickness of producing zones

    Average .... Weber, 250-300 ft (80-100 m)

    Maximum .... Approx. 600 ft (200 m)

Lithology

The Weber sandstone consists of fine- to very fine grained, well-sorted eolian sandstones with interbedded fluvial sandstone, siltstones, and shales; the effective reservoir lithology is usually limited to the eolian facies.

Porosity type .... Intergranular (Weber)

Average porosity .... 13% (Weber)

Average permeability .... 10 md (Weber)

Seals (Weber):

Upper

    Formation, fault, or other feature .... State Bridge Formation

    Lithology .... Red shale and anhydrite

Lateral
    Formation, fault, or other feature .... Maroon Formation
    Lithology .... Red sandstones and siltstones

Source (Weber):
    Formation and age .... Meade Peak and/or Retort Shale Members of Phosphoria Fm.; Permian
    Lithology .... Black shales
    Average total organic carbon (TOC) 3–5%
    Maximum TOC .... Approx. 8%
    Kerogen type (I, II, or III) II
    Vitrinite reflectance (maturation) .... Now supermature in central Utah
    Time of hydrocarbon expulsion .... Jurassic and later
    Present depth to top of source .... 150–200 mi (240–320 km)
    Thickness .... 30 ft (10 m)

Appendix 2. Production Data

Field name .... Rangely field

Field size:
    Proved acres .... Approx. 20,000 (8100 ha)
    Number of wells all years 891 (1989)
    Well spacing .... 20 ac
    Ultimate recoverable .... 939 million bbl (149 million m^3)
    Cumulative production .... 734 million bbl (117 m^3) est. year end 1989
    Annual production .... 12 million bbl (1.9 million m^3) est. for 1989
    Present decline rate .... Flat due to CO_2 flood
        Initial decline rate .... 7.8% (waterflood)
    Annual water production .... 175 million bbl (28 million m^3) est. 1989
    In place, total reserves .... 1578 million bbl (251 million m^3)
    In place, per acre foot .... Approx. 700 bbl (900 m^3)
    Primary recovery .... 821 million bbl (131 m^3)
    Enhanced recovery .... 118 million bbl (19 million m^3)
    Cumulative water production .... 2.4 billion bbl (380 million m^3)

Drilling and casing practices:
    Casing program
    Most early wells set 10-in. surface casing to cover Mancos fractures (800–2900 ft); most modern wells set casing to cover most of the Mancos (approx. 2600 ft); early wells set 7-in. production casing at the Weber top or approx. 50 ft below the gas-oil contact if the top was encountered in the gas cap; most modern wells have 7-in. casing to total depth; all wells are cemented above the top of the Dakota.

    Drilling mud .... Early wells, oil-based mud in Weber; modern wells, non-oil-based dispersed muds
    Bit program .... 12-in. for surface casing; 9-in. to total depth
    High pressure zones .... Isolated sandstone bodies in the top few hundred feet of Morrison Fm.

Completion practices:
    Interval(s) perforated ... Effective Weber sandstones from gas-oil contact to near top of oil-water contact
    Well treatment .... Early wells stimulated by nitroglycerine fracturing; modern wells hydraulically fractured to improve deliverability

Formation evaluation:

Logging suites
Varied through life of field; earliest logs were electrical surveys; neutron logs run when available; current evaluation based on lithodensity-neutron and high-frequency dielectric tools

Testing practices .... *Typically only production testing after drilling; few drillstem tests have been run*

Mud logging techniques .... *Mudlogs used intermittently through life of field, primarily to monitor for high-pressure gas in Morrison Fm.*

Oil characteristics:

API gravity	*34°*
Base	*Asphaltine*
Initial GOR	*300 ft³/STB*
Sulfur, wt%	*0.7*
Viscosity	*1.7 cp at 160°F (71°C)*
Viscosity, SUS	*44 at 100°F (38°C)*
Pour point	*10°F (-12°C)*

Gas-oil distillate
Total gasoline and naphtha, 26.1%; kerosene distillate, 10.3%; gas oil, 15.3%; nonviscous lubricating distillate, 12.7%; medium lubricating distillate, 7.6%; residuum, 26.5%; distillation loss, 1.5%

Field characteristics:

Average elevation	*5300 ft (1615 m)*
Initial pressure	*2750 psi (19,000 kPa)*
Present pressure	*3100 psi (21,000 kPa)*
Pressure gradient	*Approx. 0.45 psi/ft (10 kPa/m)*
Temperature	*160°F (71°C)*
Geothermal gradient	*Approx. 0.019°F/ft (0.034°C/m)*
Drive	*Primarily gas solution before water and CO_2 floods*
Gas column thickness	*135 ft (41 m)*
Gas-water contact	*-330 ft (-100 m) msl*
Oil column thickness	*820 ft (250 m)*
Oil-water contact	*-1150 ft (-350 m) msl*
Connate water	*27.5%*
Water salinity, TDS	*TDS variable averaging 110,000 ppm before waterflood*
Resistivity of water	*NA; originally 0.09 at 68°F (20°C)*
Bulk volume water (%)	*0.04*

Transportation method and market for oil and gas:
Pipeline to Salt Lake City, Utah, and Wamsutter, Wyoming

Matzen Field—Austria
Vienna Basin

NORBERT KREUTZER
ÖMV Aktiengesellschaft
Vienna, Austria

FIELD CLASSIFICATION

BASIN: Vienna
BASIN TYPE: Backarc
RESERVOIR ROCK TYPE: Sand and Dolomite
RESERVOIR ENVIRONMENT OF DEPOSITION: Deltaic, Nearshore Shallow Marine, and Carbonate Platform
RESERVOIR AGE: Miocene-Cretaceous-Triassic
PETROLEUM TYPE: Oil and Gas
TRAP TYPE: Complex of Anticline, Truncation, and Fault Traps
TRAP DESCRIPTION: Complex of traps that include steeply dipping anticline and thrust slices; anticlines combined with pinch-outs, truncation unconformity traps, and fault traps

LOCATION

Matzen, a giant field, is located in Lower Austria (Figures 1 and 2) about 30 km northeast of Vienna. It is the largest multipool oil field of the Vienna basin as well as of onshore Central and Western Europe. Total areal extent of Matzen field is about 100 km², entirely operated by ÖMV (Österreichische Mineralölverwaltung). Cumulative production from 1949 through 1987 is more than 62 million metric tons (MT) (440 million bbl) of oil (about two-thirds of the production of the Austrian part of the Vienna basin) and 23 billion m³ (812 bcf) of natural gas (Figure 6).

The intermediate structurally high Matzen area in the central part of the Vienna basin is a part of the downthrown block east of the synsedimentary Steinberg fault that separates the main part of the Vienna basin from the upthrown Mistelbach block in the west (Figure 7).

The Vienna basin is a Tertiary, rhomboidal pull-apart basin in the Alpine-Carpathian overthrust belt, which is superimposed on the allochthonous Flysch zone and Calcareous Alpine zone, and which, in turn, are superimposed on the Molasse and Autochthonous Mesozoic cover of the Bohemian massif of the European plate (Figure 3). A larger part of the basin is in Austria than is in Czechoslovakia. The basin, about 200 km long and up to 60 km wide, is the most important oil province of Middle Europe, with a cumulative production (Austrian part) since 1930 of about 89 million MT (650 million bbl) of oil and 45 billion m³ (1.6 tcf) of natural gas (Figure 6). The production is from both Miocene basin fill and from older sedimentary basement, mostly Paleogene sandstones of the Flysch zone in the northwest and Upper Triassic dolomites of the Calcareous Alps in the southeast.

Austria's total cumulative production to the end of 1987 is 96 million MT (700 million bbl) of oil and 58 billion m³ (2.0 tcf) of natural gas (Figure 6).

HISTORY

Discoveries and Discovery Methods

A farmer, Jan Medlen, accidentally found gas seepage when draining his land near Gbely (Egbell) in the Czechoslovakian part of the northern Vienna basin. He gathered the gas for heating purposes. In 1913, a gas explosion in his house alerted others to the presence of the gas seep. As a result of the interest aroused, the first oil reservoir of the Vienna basin was discovered in rocks of the Miocene Sarmatian Stage. The total depth of the first well was 136 m. In the years 1915–1917, during World War I, a military unit drilled a well designated as Raggendorf 1 in the northwestern margin of the yet undiscovered Matzen field to a total depth of only 1070 m, with weak gas and oil shows in the Sarmatian rocks.

In the years following World War I, the Vienna basin was subjected to various evaluations. After 1919, Eurogasco performed gravimetric measurements. Reflection seismic operations began in 1938, and micropaleontological methods were applied following their introduction in the mid-1930s. After geological surface mapping of the whole Vienna basin

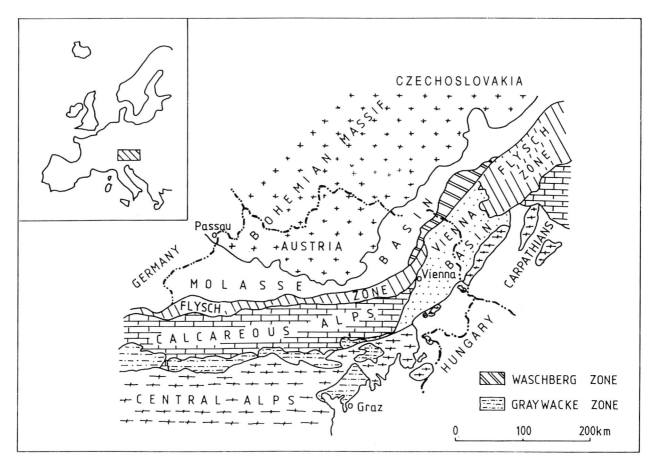

Figure 1. Schematic map showing the geographical location and general geological situation of the Vienna basin in the region of the Alps and Carpathians.

by K. Friedl, 1924–1926 (Vacuum Oil Company), several companies concentrated on further exploration along the Steinberg fault zone in the northern part of the Vienna basin. This led to trend drilling that began in 1934 and resulted in the discovery of many smaller and two larger oil fields (Gaiselberg, Mühlberg) in Sarmatian and Badenian rocks and in a larger discovery in the Paleogene Flysch zone (St. Ulrich-Hauskirchen) (Figure 2). Also discovered were small gas reservoirs in the Aderklaa structure (Karpatian) of the central Vienna basin.

The Matzen area came to be recognized between 1937 and 1940 as a regional high in Pontian Stage rocks (uppermost Miocene) through geological surface mapping and subsurface data from counterflush drilling down to 150 m (with gas blowouts) by RAG (an affiliate of Mobil and Shell). The lithologic correlation method that had been used successfully in the northern part of the Vienna basin could not be used to determine exact structural positions because the Pontian is almost nonfossiliferous in the central Vienna basin. Despite this complication, RAG selected a location for a deep well in 1939, but it could not be drilled because of the beginning of World War II. Also, after the occupation of Austria, the legal basis of exploration for hydrocarbons was altered by a new law that changed the apportionment of leases and concessions (Strauch, 1986).

After World War II, beginning in 1947, there was new counterflush drilling (again with blowouts), but the exact position of several structural highs in the Matzen area already had been determined by deeper structural drilling (to about 700 m) to the fossiliferous and correlatable middle and lower Pannonian Stage rocks. Drilling operations were conducted by the SMV (Sowjetische Mineralölverwaltung [Soviet-Mineraloil Management]) under Russian command after takeover of German property. One shallow structural well (Matzen K3) had a gas blowout in the 4th lower Pannonian (554 m). The first well, Matzen 1 (Figure 39), drilled in 1948 on one of RAG's local Pontian structural highs, resulted in the discovery of the first oil reservoir in the lower Miocene (1955 to 2046 m), but the drilling operation had a breakdown and the well was abandoned. The second well, Matzen 3 (1949), is considered the discovery well of Central Europe's largest oil field and of its largest reservoir in the Matzen sandstone,

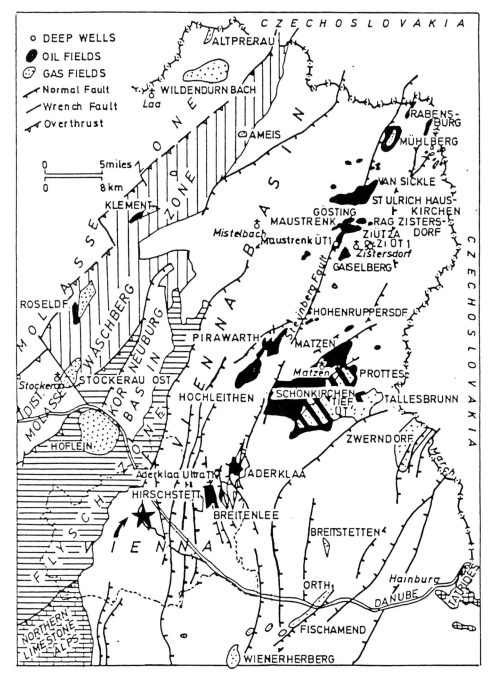

Figure 2. Oil and gas fields in the Austrian part of the Vienna basin and surrounding regions. (According to Wessely, 1988.)

at a depth of about 1650 m. This reservoir is the basal transgressive sandstone of the middle Miocene Badenian Stage (Figures 12, 13, 21, 23, and 39).

Before 1955, during the first exploration and development phase of the Miocene basin fill (referred to as the "first floor," Figure 3), numerous oil reservoirs in the Badenian (mostly with gas caps) and Ottnangian (or lower Karpatian) rocks and gas reservoirs in the lower Pannonian, Sarmatian, and Badenian in the Matzen field were discovered and developed (Figures 6, 11, 12, 13, 15, 21, and 26 to 39). In the central Vienna basin, the large gas reservoir, Zwerndorf, and smaller oil fields (including Aderklaa) were discovered in Badenian sandstones (Figure 2). Austria's peace treaty was signed in 1955, and the oil and gas fields were handed over to the Federal Republic of Austria. ÖMV was established for the operation of the fields. Between 1955 and 1971, three more very important discoveries within the multipool Matzen field were made.

By the end of the 1940s, K. Friedl (internal report) had considered the occurrence of possible oil

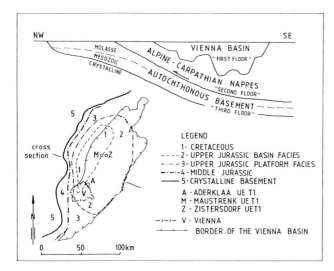

Figure 3. Diagrammatic cross section through the Vienna basin, illustrating that the basin is superimposed on allochthonous nappes of the Alpine-Carpathian thrust belt. The thrust belt, in turn, overlies the autochthonous basement of the European platform. (See Figure 4.) The distribution of the Mesozoic sediments and facies is shown. (According to Wessely, 1988.)

reservoirs in dolomites of the sedimentary basement of the Alpine-Carpathian thrust units ("second floor," Figure 3) beneath the Matzen field Miocene basin fill. The 1959 discovery in the Aderklaa field of a gas reservoir in a buried hill of Triassic dolomite in the Alpine nappe system triggered an extensive exploration phase for similar targets in the Matzen field (Figure 17). Because of difficulties in the application of seismic reflection methods for mapping basement relief, gravity data and "travel-time logging" were used. (The travel times from the individual shot points around a hole that already penetrates the basement—for instance Schönkirchen T1 [Figure 41]—in water-bearing dolomite to a geophone lowered in the well become shorter in the case of an ascending gradient of the [high velocity] basement [Kapounek et al., 1963].) A "high" in the southern part of the Matzen field was mapped by both methods. A second deep well, Schönkirchen T2 (Figure 41), drilled in 1962, discovered a large oil reservoir in a buried hill of fractured Upper Triassic dolomite ("Hauptdolomit") (Figures 8, 16, and 21) at a depth between 2700 and 2900 m. Two oil reservoirs, "Schönkirchen Tief" and "Prottes Tief," and a gas reservoir in buried hills of the dolomite on the top of the flat-lying Ötscher nappe, in Upper Cretaceous and Miocene dolomite debris, conglomerates, and sandstones, form a common energy unit (used throughout this paper to designate disparate reservoirs that are, in fact, in pressure and fluid communication). These were discovered and developed during this second exploration phase in the southern and northeastern part of the field (Figures 6, 12, 13, 40, and 41). Additionally, separate oil reservoirs in sandstones of the Bockfliess and Gänserndorf beds have been found.

In the third exploration phase, as the result of geological concepts developed by Kapounek regarding the occurrence of dolomite bodies in the internal Calcareous Alps, the first ultra-deep well, Schönkirchen T32, was drilled in 1968 to a depth of 6000 m in the southern part of Matzen field. This well led to the discovery of the large gas reservoir, "Schönkirchen Übertief," between 4800 and 5900 m (Figures 8, 21, and 42). The reservoir is in a very steeply dipping and, therefore, narrow zone of the Upper Triassic fractured dolomite ("Hauptdolomit") and Upper Cretaceous dolomite debris and conglomerate in the southeastern part of the Frankenfels-Lunz imbrication system (Kapounek and Horvath, 1968). The last important find came in 1971 with the discovery of the Reyersdorf gas (and oil) reservoir, which is a rather irregularly bedded, partly calcareous dolomite. The reservoir is steeply dipping, and on the southern flank it is overturned in the northwestern part of the Frankenfels-Lunz system (Figures 8 and 42).

The significant successes in the Vienna basin (Figure 2) in the past were based on the fact that the structural highs, containing multiple pools in Miocene beds (as in many Tertiary basins), as well as those of the sedimentary basement, are stacked and are only slightly shifted (Braumüller and Kröll, 1980). Most of these structural highs in the Miocene sequence are of synsedimentary origin; in other cases, postsedimentary structures or combinations exist. Therefore, the accumulations of hydrocarbons are concentrated in distinct, regional structural highs, whereas the great synclinal areas are almost devoid of any oil or gas pools. As a result, deeper drilling as well as stepout trend drilling was always of great importance in the whole Vienna basin.

Development

The different exploration phases were followed by stepout and infill drilling. This development resulted in the delineation of the other fault blocks and of the numerous reservoirs in the main block of the field. The first to be defined were the oil reservoirs (mostly with gas caps) in the middle Miocene Badenian over the whole field and in the lower Miocene Gänserndorf/Bockfliess beds in the northern part; second were the oil reservoirs in the buried hills of Triassic dolomites and in the Gänserndorf/Bockfliess beds of the southern and northeastern part of the field; and third were internal (within the nappes) dolomite gas reservoirs (Figures 6, 12, 13, and 26 to 42).

Two further drilling activities have started: Pressure maintenance waterflooding projects (mostly contour flooding) began in 1959 in the 8th, 9th, 10th, 12th, 13th, and 16th Badenian and in the Schönkirchen Tief reservoirs. Gas storage began in 1970 in the 3rd and 4th Pannonian, 5th Sarmatian, and 5th and 6th Badenian reservoirs. Some develop-

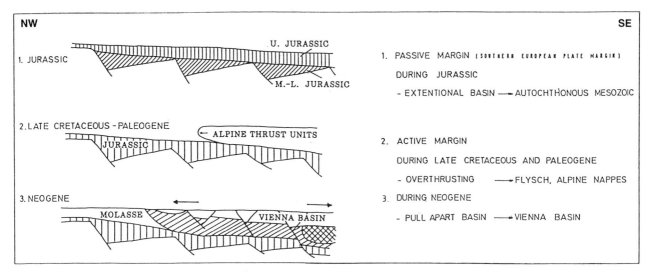

Figure 4. Schematic evolution of the Vienna basin area in three major tectonic phases. (According to Ladwein, 1988.)

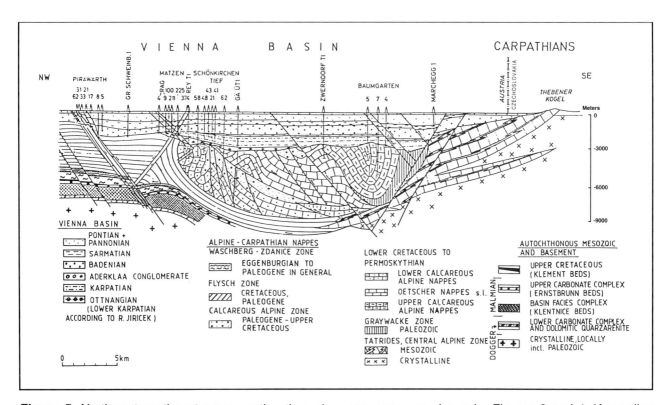

Figure 5. Northwest-southeast cross section through the Vienna basin in the area of the Matzen field. Explanation of the causal tectonism and the tectonic sequence are shown by Figures 3 and 4. (According to Wessely, 1988.)

ment has occurred since 1975 in the Gänserndorf/ Bockfliess beds, again in the northern margin of the field.

A total of about 1340 wells have been drilled in the Matzen area, of which 940 oil wells, 60 gas wells, 140 gas storage wells, and 60 water injection wells currently are active. The wells generally have been completed using casing perforations. Single or multiple intervals have been perforated and sometimes acidized.

In-place total reserves for the field are estimated to be 185 million MT (1.31 billion bbl) of oil and 26.7 billion Sm^3 (standard m^3) (936 bcf) of nonassociated gas, of which approximately 72.4 million MT (515 million bbl) of oil and 17.3 billion Sm^3 (611 bcf) of gas are recoverable. Cumulative production has been

Figure 6A. Annual oil production (millions metric tons [MT]) of Matzen field, Vienna basin, and Austria from before 1950 through 1987. Solid line, Matzen field; dashed line, entire Vienna basin; dot-dash line, Austria total.

62.4 million MT (444 million bbl) of oil and 11.9 billion Sm^3 (420 bcf) of nonassociated gas; present annual production (1987) is 640 thousand MT (4.5 million bbl) of oil and 240 million Sm^3 (8.5 bcf) of nonassociated gas (Figures 6, 12, and 13).

STRUCTURE

Tectonic History and Regional Structure

The Vienna basin is a post-alpine grabenlike downwarp, superimposed on the allochthonous units, Flysch and Calcareous Alps, which in turn are superimposed on the Molasse and the autochthonous Mesozoic cover of the Bohemian massif (Ladwein, 1988; Wessely, 1988; Ladwein et al., 1991) (Figure 3).

The Neogene basin fill consists predominantly of terrigenous sediments. Its thickness can exceed 5000 m in the depressions (Figure 9). The allochthonous Flysch and Calcareous Alpine zones have been overthrust from the south. The Calcareous Alps consist predominantly of carbonates (minor sandstones) of Triassic, Jurassic, and Cretaceous age; locally Paleogene sediments are found (Figure 10). Flysch sandstones and shales are of middle Cretaceous to Eocene age. The Molasse sandstones, shales, and conglomerates are of Paleogene to Neogene age. The autochthonous Mesozoic sediments (drilled to a maximum depth of 8553 m) are comprised of Lower to Middle Jurassic sandstones, shales, and coal, Upper Jurassic carbonates and marls, and Upper Cretaceous glauconitic sandstones (Figures 3 and 5).

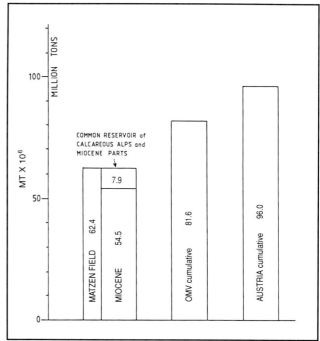

Figure 6B. Cumulative oil production of Matzen field, OMV, and Austria, through 1987.

The main evolutionary stages of the Vienna basin are summarized by Ladwein (1988), Wessely (1988), and Ladwein et al. (1991) (Figure 4), incorporating the information and concepts from the fundamental literature of recent years.

1. The autochthonous Jurassic sediments under the Vienna basin and under the Calcareous Alps were deposited on the southern passive margin of the European plate. The allochthonous Calcareous Alps were deposited on the other side of the intervening Penninic ocean (a rifted part of the Tethys with an ophiolitic suite of rocks).
2. The allochthonous units of Flysch and Alps overrode this passive margin during the Alpine orogeny (Cretaceous to Paleogene), and Neogene sediments transgressively covered these alpine thrust units.
3. Thrusting ended during Neogene in the Alps but was still active (as Carpathian orogeny), so the Vienna basin formed as a pull-apart basin.

The Vienna basin is classified as 321 under the modified scheme of Bally and Snelson (1980) (i.e., an epistructural basin located and mostly contained in compressional megasuture; backarc basin, associated with continental collision and on concave side of A-subduction arc; on continental crust of Pannonian-type basins) and is classified as III Bc using the scheme of Klemme (1971) (i.e., continental rifted basins; rifted convergent margin-oceanic consumption; median).

The Vienna basin contains a system of horsts and troughs in a distinctive arrangement (Wessely, 1988;

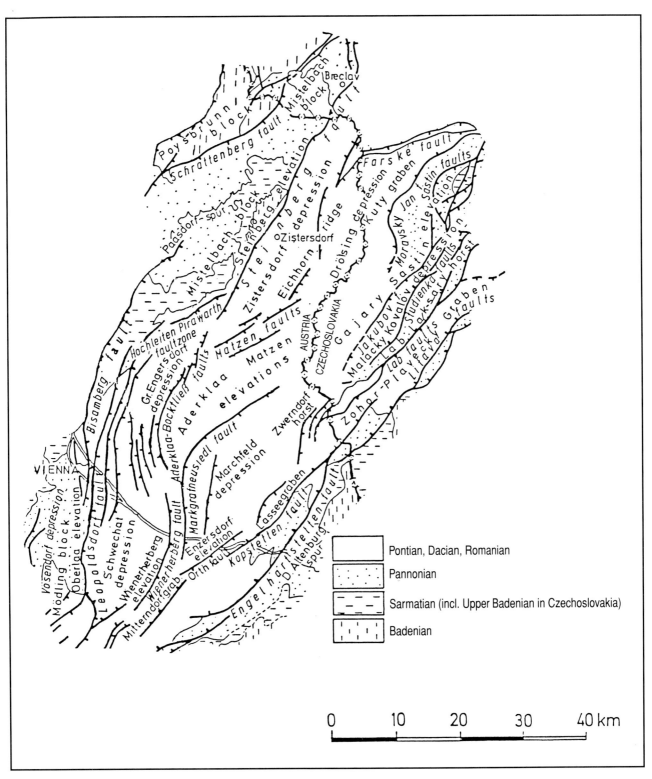

Figure 7. A structural synopsis of the central and northern part of the Vienna basin showing the major faults, depressions, and elevated blocks. (According to T. Buday, H. Unterwelz, K. Friedl, R. Grill, and R. Janoschek; compiled by Wessely, 1988.)

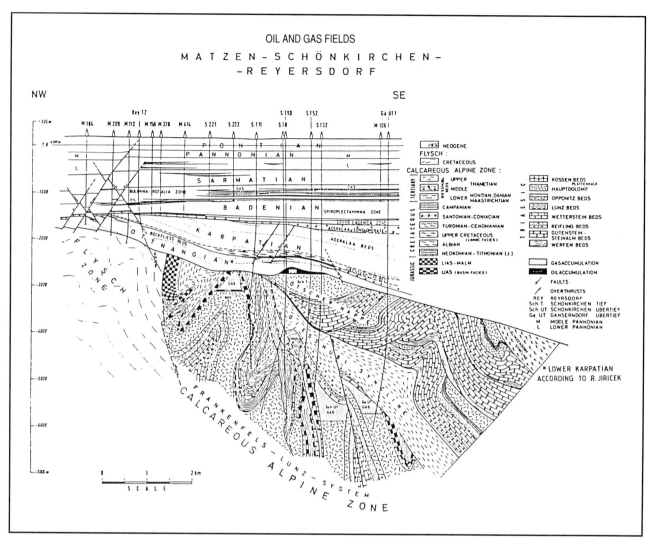

Figure 8. Cross section through the Matzen oil field. Production is from Neogene basin fill and from the Calcareous Alps, with its typical buried hill reservoir and dolomite reservoirs in anticlines. (According to Wessely, 1988, reprinted with permission.) (Wells ST 52 and REY T2 do not fall on the trace of the cross section; they are projected onto this section.) See Figures 41 and 42 for location of the section.

Ladwein et al., 1991) (Figure 7). In the northern part, the large, elevated, marginal Mistelbach basement block is bordered toward the basin interior by the important synsedimentary Steinberg fault; the highest part of the upthrown block (Steinberg) of the footwall is situated opposite the deepest part (oval depocenter) of the Zistersdorf depression on the downthrown block of the hanging wall. The Matzen and Aderklaa blocks, which are elevated horsts along the axis of the central part of Vienna basin arranged in a slightly sigmoidal fashion, are generally separated from adjacent depressions by curved faults. The synsedimentary Bockfliess-Aderklaa fault system, consisting of en echelon fault segments with a vertical throw of up to 500 m, separates both of these elevated blocks from the Gross-Engersdorf depression to the west. The median zone of elevated blocks is bounded against the Marchfeld depression to the southeast by the Markgrafneusiedl fault and against the Zistersdorf depression in the north by the Spannberg basement ridge together with the young Matzen fault system (see also Figures 9 and 15). The Matzen elevated block seems to continue to the northeast as the Eichhorn ridge.

Local Structure

Although the contours of the elevated Matzen block cannot be closed completely, the structural closures between the structural highs in the Matzen field and a structural low area about 10 km to the northeast are about 40 m at the top of the lower Pannonian, 100 m at the top of the Sarmatian, and 150 to 250 m within the Badenian. To the southwest, a structural closure of about 150 m can be mapped at the top of the Sarmatian.

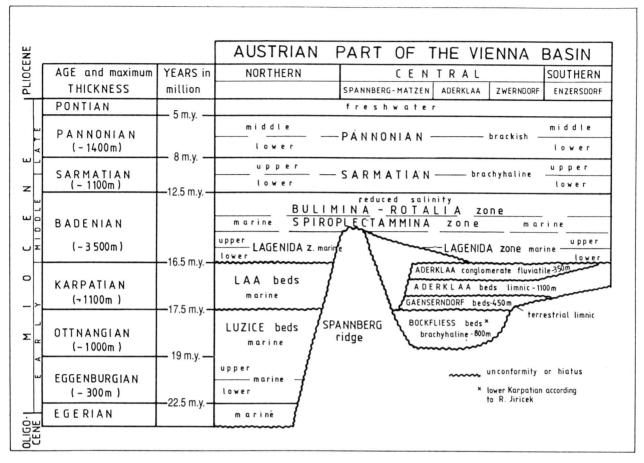

Figure 9. Schedule of the Neogene stratigraphy of the Vienna basin in Austria. (According to Wessely, 1988.) (Thicknesses shown are "to a maximum of," not negative values—Ed.)

STRATIGRAPHY AND STRATIGRAPHIC-STRUCTURAL RELATIONSHIPS

Alpine-Carpathian Units

Only three wells outside of the Matzen field have been drilled below depths of 6000 m into the autochthonous Mesozoic sedimentary cover of the basement ("third floor") beneath the nappes and the floor of the Vienna basin (Figures 2 and 3): Zistersdorf ÜT2A (total depth 8553 m) and Maustrenk ÜT1 to the north of the Matzen field, and Aderklaa ÜT1 to the southwest. All of these wells penetrated Malm platform carbonate and a basin facies of Malmian marl up to 1000 m thick having good source conditions and a very high pore volume of gas. Maustrenk ÜT1 produced oil until depletion from a limestone klippe; in Aderklaa ÜT1, drilling was terminated in crystalline schist of the Bohemian massif.

The Alpine-Carpathian thrust sheets in the Matzen field (Figures 5 and 8) consist of the structurally lower Flysch zone in the northwest and the structurally higher Calcareous Alps in the southeast (Figures 17 and 41). Neogene sediments unconformably cover these allochthonous units. The Flysch zone, overthrust by the Calcareous Alps in late Eocene time, comprises tight sandstones, shales, and limy marls from Early Cretaceous to Eocene in age. Two of the three tectonic units of the Calcareous Alpine zone exist within the fields: the strongly folded and fault-sliced Frankenfels-Lunz nappe in the northwest, overthrust in Paleocene time by the Ötscher nappe with a syncline in the southeast (Figure 8) (Kröll and Wessely, 1973; Wessely, 1984). The thrust plane below the Calcareous Alps is steep or overturned at this frontal part. The two nappes are separated by the Upper Cretaceous and Paleogene beds that discordantly overlap folded alpine structures of the underlying Frankenfels-Lunz nappe, forming an asymmetric syncline, the Giesshübl furrow (Wessely, 1988) (Figure 8). Cretaceous and Paleogene sediments also cover the front of the Ötscher nappe in a wide area between Schönkirchen Tief and Prottes Tief (Figures 16 and 41), resting unconformably on Upper Triassic beds, mostly dolomites.

The Calcareous Alpine sediments (Figure 10) include rocks of Permian Scythian to Paleocene age (Kröll and Wessely, 1973; Wessely, 1984, 1988). The

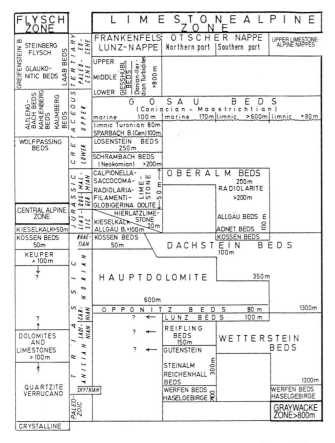

Figure 10. Stratigraphy of the Alpine-Carpathian basement of the Vienna basin in Austria (simplified). (According to Wessely, 1988.)

Permian Scythian comprises the evaporitic-dolomitic Haselgebirge complex and the pelitic-clastic Werfen complex. Middle Triassic strata consist partly of a basin facies of Gutenstein and Reifling limestone and partly of a platform facies of Wetterstein limestone and dolomite. Shale and sandstone of the Lunz beds form a distinct marker horizon between the carbonates of Middle and Late Triassic age. The upper Carnian Opponitz beds, a marly facies, underlie the Norian fractured Hauptdolomit, which was deposited in a lagoonal environment and has a thickness of up to 600 m. The porous and permeable Hauptdolomit is conformably overlain by the Norian-Rhaetian Dachstein limestone. The Rhaetian is represented by the dark limestone and marl of the Kössen strata. Liassic Allgäu beds and Kieselkalk represent the basin facies of the Hierlatz limestone, a shallower water facies. Limestone of Dogger and Malm ages consists of thin-bedded, condensed, red and brownish colored layers. Neocomian limy marls are restricted to the Frankenfels-Lunz nappe system. During Albian–early Cenomanian time, sandstones, conglomerates, and marls were deposited (Figure 8). Sediments of the Giesshübl syncline include Cenomanian marl and Turonian to Maastrichtian conglomerate, breccia, sandstone, and marl. Turbidites of upper Maastrichtian–Paleocene Giesshübl beds up to 1 km thick form the main sedimentary fill in the syncline and the caprock of the internal gas-bearing dolomites of the Frankenfels-Lunz nappe system.

Structure and Sedimentation of the Miocene Basin Fill

Terrigenous sediments were deposited in the Vienna basin during the early Miocene (Eggenburgian, Ottnangian, and Karpatian), coming predominantly from the south. During the middle and late Miocene (Badenian, Sarmatian, Pannonian, and Pontian) they are derived from the Molasse zone source to the northwest. The sediments were brought by an ancient Danube river through the Zaya furrow northwest of Mistelbach (Figures 2 and 45 to 48) and by an ancient March (Morava) river from the northeast. Submarine or subaqueous parts of several deltas were developed throughout all of these times in both the Austrian and Czechoslovakian parts of the central Vienna basin (Jiricek, 1978). The Miocene basin fill consists generally of terrigenous sands, sandstones, clay, and shales, with generally minor gravel beds. However, the Aderklaa conglomerate and the Gänserndorf conglomerate reach very great thicknesses. These were shed directly from southwest, south, and southeast. Additionally, some red algae beds were deposited on submarine swells in the Badenian. During the Miocene the water depth was always shallower than about 200 m.

A significant erosional angular unconformity exists between the older sedimentary basement—the Flysch zone and the Calcareous Alps—and the Molasse basin fill. Later erosion, especially prior to Badenian time, removed parts of the older Miocene beds, particularly in the region of the Spannberg ridge, a morphological structure that strongly influenced sedimentation (Wessely, 1988) (Figures 8, 9, and 15). The older Miocene sediments of the Eggenburgian, Ottnangian, and Karpatian, better preserved in the northern part of the Vienna basin, are considered to be Molasse overlapping the Alpine-Carpathian nappes (Janoschek and Matura, 1980). They were deposited on top of moving thrust sheets ("piggyback basins").

The older Miocene sediments south of the Spannberg ridge, in the Matzen field (Figures 8, 9, 11, and 15), are the brachyhaline (a term referring to sea water with somewhat reduced salinity) Bockfliess beds of the Ottnangian or lower Karpatian (according to Jiricek), resting unconformably (D1) on the Flysch zone in the north and on the Calcareous Alps in the south. They are overlain—separated by an erosional unconformity (D2)—by the limnic-lacustrine Gänserndorf beds of the lower or middle Karpatian, containing gray and variegated shale, sandstones, some anhydrite, and a basal conglomerate (alpine carbonate components) (Hladecek et al., 1971). The Aderklaa beds (upper Karpatian), a limnic (deltaic) facies of gray sandstones and shales, have a large areal extent south of the Spannberg ridge, nearly as great as that of the later basin sediments (Wessely, 1988).

The tectonic uplift of the Spannberg ridge (late Styrian phase) resulted in erosion that truncated the Aderklaa, Gänserndorf, and Bockfliess beds in the crest area (Figures 8, 9, and 15) down to the Flysch zone, creating a significant erosional and angular unconformity (D3). There was an inversion over the entire basin (Ladwein et al., 1991); the areas of maximum subsidence (depocenters) were shifted by tectonism from the north during the Karpatian to the south during the Badenian and later. In the southern part of the field, the Schönkirchen faults were active after the deposition of the Aderklaa beds and the pre-Badenian (Hladecek et al., 1971; Köves, 1971) (Figures 8 and 11).

The unfossiliferous and possibly fluviatile Aderklaa conglomerate, unconformably (D3) overlying the Aderklaa beds, consists of alpine carbonate and crystalline components. It pinches out on the flanks of the Spannberg ridge and is onlapped unconformably by marine transgressive Badenian beds of the lower *Lagenida* zone far south of the field and by beds of the upper *Lagenida* zone within the field, including the flysch-conglomerate bearing Auersthal beds (Figures 8 and 15). The Spannberg ridge, an island and an important facies divide, became completely inundated during the time of the deposition of the marine-brachyhaline *Spiroplectammina* and *Bulimina-Rotalia* zones (Figures 8, 9, 11, and 23). As later in Sarmatian and Pannonian times, the thickness of the sediments on the western margin of Matzen field depends on the position on the down- or upthrown blocks of the synsedimentary Bockfliess fault system, which became active during the deposition of the *Bulimina-Rotalia* zone (Figure 11). A general brachyhaline influence occurred during deposition of sandy beds of the latest Badenian as well as in the earliest Sarmatian sediments. The boundary, itself in local synsedimentary highs in the southern part of the field, is frequently marked by alpine carbonate gravel beds, indicating an erosional unconformity (Kreutzer, 1986).

During the time of Sarmatian deposition, the Vienna basin became the westernmost bay of the Paratethys (remnant of the Tethys). In the higher part of the lower Sarmatian as well as in the upper Sarmatian, interlayered sandstones and shales can be correlated over large distances. At the beginning of the Pannonian, the freshening of the water continued, and gradually an inland lake developed. Fresh-water conditions prevailed by Pontian time, when deposition of sands and gravels became predominant. The Matzen faults were active. During the late Pontian, the uplifting of the Vienna basin began. Quaternary sediments, mostly gravels, were deposited from the paleo-Danube river and its tributaries.

No volcanic activities took place in the Vienna basin, but distinct horizons contain volcanic tuffs and their altered products (bentonite) in the Badenian (the Matzen and Aderklaa markers) and Karpatian, used as stratigraphic markers for regional or local correlation (Kreutzer, 1986; Wessely, 1988) (Figures 11, 23, and 38).

Based on the paleoecology of Foraminifera and on geology, investigations by Jiricek (1978, 1985) in the Czechoslovakian (and Austrian) part of the central Vienna basin resulted in a determination of the predominantly deltaic nature of the gas-bearing upper Sarmatian and lower to middle Pannonian sand (gravel)-shale (clay) succession. Jiricek compared fingerlike distribution of several distinct northwest–southeast- north–south-oriented sandstones and gravel beds of the Pannonian, attaining a maximum thickness on the ends, with the modern bird-foot delta of the Mississippi River. These Pannonian fluctuating deltaic sand lobes of an ancient Danube river mouth are characterized by a fauna pointing to a lower salinity—or even fresh water—in the sands than in the surrounding lagoonal shales or clays. The Pannonian channels sometimes were eroded into deeper shale layers, for instance, through the Pannonian–Sarmatian boundary. On the other hand, other sandbars of different orientation have the same faunal content as the surrounding shales and are not of deltaic origin. Loading of the shales during burial resulted in structural highs by compaction over and around the sandstone lobes. These sandstone bodies became separate gas reservoirs (now used for gas storage) with individual gas-water contacts (Jiricek, 1978, 1985).

Kreutzer (1974) has indicated a possible deltaic origin of the brachyhaline Sarmatian in the Matzen area. Three conspicuous channel-like trends of sandstones with basal extra-formational alpine carbonate gravels, laterally and basally sharp-bounded, exist in the 3rd/4th, 6th, and 7th Sarmatian reservoirs (Figures 18 and 21). They are characterized by transgressive SP-log shapes (Kölbl, 1953; Wieseneder, 1959; Kreutzer, 1974). These partly meandering trends show a "pendulum effect" that points to a sediment transport from the northwestern Molasse zone to the south into the central Vienna basin. Such pendulum effect is characteristic of postulated deltaic environment, resulting from the lateral shifting, back and forth, of the depocenters. The lower Pannonian sand lobes in the Matzen area (Kreutzer, 1990) show a similar pendulum effect of the several northwest–southeast-oriented, lensoid sand and gravel beds, which are separated by shale or clay layers (Figures 19, 20, and 21). The gross and net thicknesses of sandstones and shales vary considerably toward the edges or flanks of the 5th, 4th, and 3rd lower Pannonian. These variations increase as the result of channel scouring and fill downward into older layers by the 5th and 4th Pannonian deposits; such scour and fill diminishes upward in the 2nd and 1st lower Pannonian. The transition between Pannonian and Sarmatian, also very abrupt—within 1 m, as shown by faunal contents—is generally a conformity within shale (clay) beds containing numerous and regionally

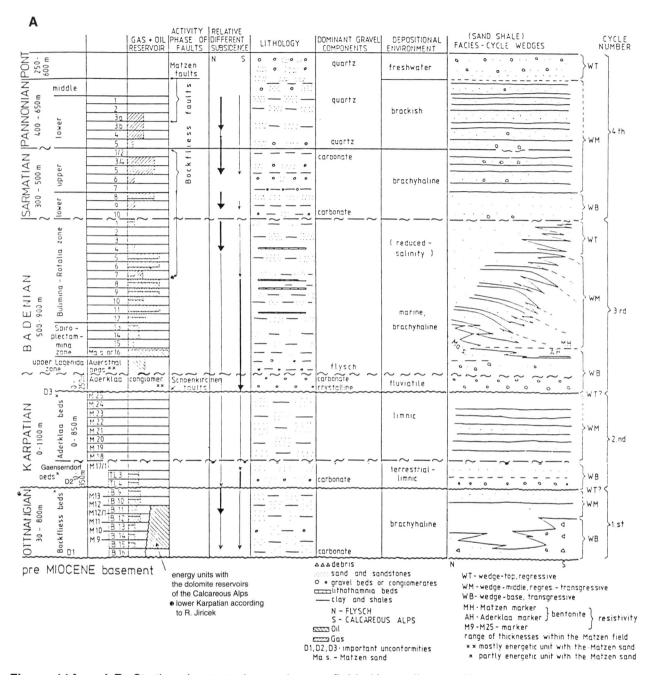

Figure 11A and B. Stratigraphy, tectonics, and gas and oil reservoirs of the Miocene basin fill in the Matzen field. (According to Kreutzer and Hlavaty, 1990, reprinted with permission.)

correlatable E-log resistivity markers. Where sandstone lobes of the 5th and the 4th lower Pannonian are very thick and have sharply defined bases, they have been eroded into the underlying Sarmatian beds. This kind of local sedimentary relationship occurs in many places in the Matzen area, similar to those observed in Czechoslovakia (Figure 20). The 4th Pannonian also was deposited in channels that sometimes eroded down into the Sarmatian; however, this occurred only on the flanks where the 5th Pannonian was shaly. In other places, resedimented shales seem to have been deposited below the 4th and 5th Pannonian. Erosional effects on the 3rd, 2nd, or 1st lower Pannonian, however, could not be detected (Kreutzer, 1990).

Important gas reservoirs, now used for gas storage, are contained in the 3rd and 4th lower Pannonian of the Matzen field (Figures 19, 21, 26, and 27). Investigations by Jiricek indicated that these sandstone lobes in the Matzen field should also be of deltaic origin, constituting parts of a subaqueous bird-foot delta of the ancient Danube river in the Austrian part of the central Vienna basin. The fluviatile sediments of the lower Pannonian Danube river are represented by the well-known Hollabrunn-Mistelbach gravel cone, seen in outcrops from the area

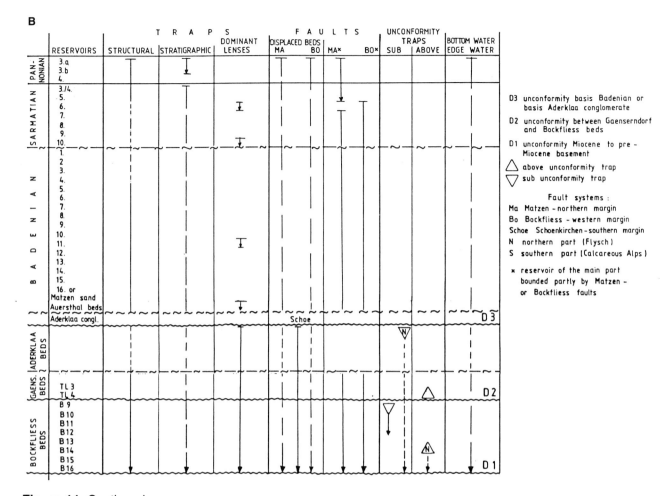

Figure 11. Continued

west of the paleocoast at Zistersdorf, westward through the Zaya furrow northwest of Mistelbach, and into the Molasse zone at Hollabrunn (Figure 48).

Transgressive-Regressive Facies-Cycle Wedges

The 2000 to 2800 m thick Miocene basin fill of the Matzen field can be divided on the basis of well data and some seismic data into different geological units, bounded mostly by erosional and angular unconformities, or at least by some paleontological or lithological criteria (Figures 8, 9, and 11). These geological units mostly correspond to depositional sequences (Mitchum et al., 1977) in a lower scale, as well as to transgressive-regressive facies-cycle wedges (Figures 11, 22, and 23) (White, 1980) in a higher scale. The latter model will be treated first.

Although at places bounded by regional unconformities, the cycle wedges, considered as fundamentally facies-defined bodies, are bounded by nonmarine tongues. The study by White (1980) was based on stratigraphic cross sections through the main producing areas of 80 basins of the western world (including the Vienna basin, but without a published cross section), as well as on Walther's law of facies (1893–1894) (the vertical succession of facies commonly is the same as the lateral order of their depositional environments, shifted by transgression and regression of the coast) (Middleton, 1973). The ideal wedge represents a transgressive-regressive cycle of deposition including, from base to top, the vertical succession of facies from nonmarine to coarse- and fine- to coarse-textured marine and back to nonmarine (Figure 22).

The Miocene basin fill of the Vienna basin can be divided into various numbers of sand-shale cycle wedges, depending on the occurrence of the pre-Karpatian sediments. In the Miocene basin fill of the Matzen field, four sand-shale cycle wedges can be recognized more or less distinctly (Figure 11): (1) Lower parts of a wedge-base are seen in the terrestrial limnic and conglomeratic Gänserndorf beds, the fluviatile Aderklaa conglomerate, the deltaic conglomeratic Auersthal beds, and the gravel beds at the Badenian–Sarmatian boundary. (2) Upper parts of a wedge-base occur in the transgressive sands of the lower Bockfliess beds, the transgressive Matzen sandstone of the lower Badenian, and the upper part of the lower Sarmatian (8th). (3) Wedge-middle occurs in the sand-shale succession in the middle part of the brachyhaline Bockfliess beds, of the limnic Aderklaa beds, of the brachyhaline to marine

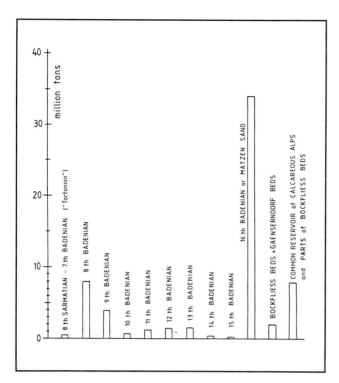

Figure 12. Annual production of oil from reservoirs in the Matzen field in millions MT.

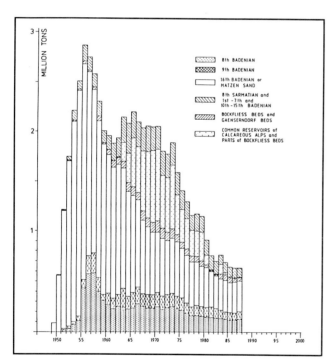

Figure 13. Cumulative production in millions MT through 1987 from the oil reservoirs in the Matzen field.

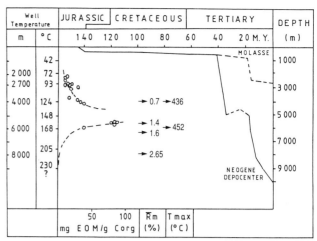

Figure 14. Burial histories of autochthonous Upper Jurassic source rocks in the Vienna basin depocenter. Extract curve and maturity parameter (R_o and T_{max}) show the position of the oil window between 4000 and 6000 m. (According to Ladwein, 1988.) EOM/g C_{org}, extractable organic matter/gram organic carbon.

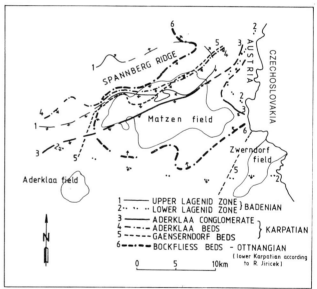

Figure 15. Areal extent of the Bockfliess, Gänserndorf and Aderklaa beds, Aderklaa conglomerate, and lower and upper *Lagenida* zones in the area of Matzen.

Badenian, of the brachyhaline Sarmatian, and of the brackish lower to middle Pannonian. (4) Wedge-tops are seen in the sand-rich beds of the uppermost Badenian (1st to 4th) and in the fresh-water gravel beds of the Pontian. Because of erosional and angular unconformities at the top of the Bockfliess (D2) and Aderklaa beds (D3), a wedge-top does not exist or is incomplete and is not represented by the overlying Gänserndorf (D2) or Aderklaa conglomerate (D3).

First Cycle—The oldest and first distinct transgressive-regressive sand-shale facies-cycle wedge is developed in the mostly lenticular sandstones of the brachyhaline Bockfliess beds. A thick transgressive wedge-base of coarse sandstones (B16) onlaps unconformably the flanks of the Spannberg Flysch ridge in the north, and dolomite debris and conglomerates (complexes B11 to B16) onlap the

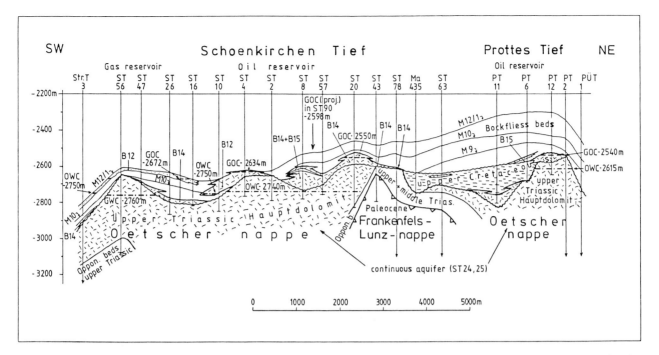

Figure 16. Geological cross section through the common reservoirs of Upper Triassic dolomites and Upper Cretaceous and Miocene dolomite debris, conglomerates, and sandstones. Location of cross section on Figures 40 and 41. Continuous aquifer (ST 24, 25) notation refers to the fact that communication between Prottes Tief and Schönkirchen Tief reservoirs is by the water-bearing Hauptdolomit as a common aquifer in the area of wells ST 24 and 25, southeast of the cross section.

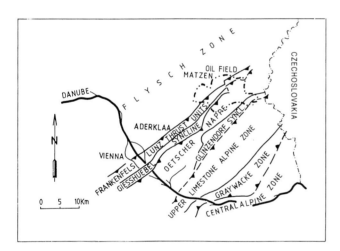

Figure 17. Simplified map of the tectonic zones of the sedimentary basement in the central Vienna basin.

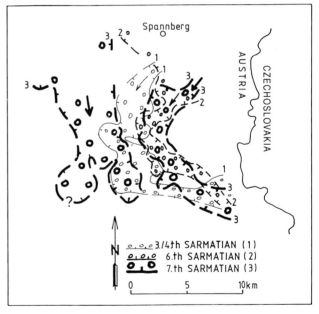

Figure 18. Areal distribution of extraformational carbonate gravel beds of the 3rd/4th, 6th, and 7th Sarmatian in the Matzen area. (According to Kreutzer, 1974, reprinted with permission.)

buried hills of the Calcareous Alps in the south, respectively (Figure 8). The transgressive-regressive wedge-middle comprises sandstones and thicker shale intervals, and perhaps also the overlying sand-rich beds (B9 to B11 in the north, and B9 and B10 in the south), although truncated by a regional unconformity (D2) (Figures 11 and 21).

Second Cycle—From the analysis of the first cycle, it follows that the overlying terrestrial limnic Gänserndorf beds (TL3, TL4) (Figure 21) with a basal conglomerate correspond to the nonmarine part of the wedge-base of the second cycle and not to the wedge-top of the first cycle. The limnic Aderklaa beds (marker M18-25) belong to the wedge-middle, although truncated on the top by the regional erosional and angular unconformity D3 (Figures 8 and 11). Both

71

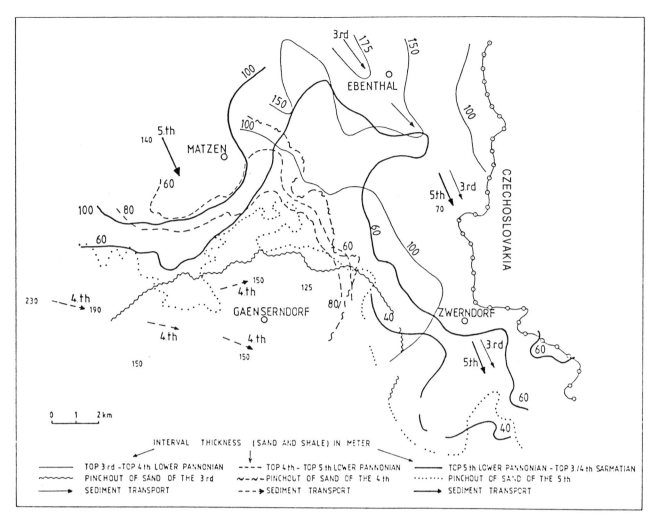

Figure 19. Areal trends of the 3rd, 4th, and 5th lower Pannonian in the Matzen area (deltaic lobes). (According to Kreutzer, 1990, reprinted with permission.)

the Gänserndorf and the Aderklaa beds also consist of mostly lenticular sandstones.

Third Cycle—The overlying Aderklaa conglomerate again probably corresponds to the nonmarine part of the wedge-base of the Badenian (Kröll and Wieseneder, 1972; Wessely, 1988) and not to the wedge-top of the second cycle. The lower *Lagenida* zone (far in the south of the field, Figure 15), the upper *Lagenida* zone (Auersthal beds), and the Matzen sandstone belong to the marine part of the wedge-base of the third cycle. The transgressive-regressive sand-shale facies-cycle wedge of the Badenian in the Matzen field is almost completely developed and is especially clearly recognizable (Figure 11; compare Figures 22 and 23) (Kreutzer, 1986; Kreutzer and Hlavaty, 1990).

The Matzen sandstone (Figure 21) is a typical basal, time-transgressive sandstone of the Badenian, corresponding to the transgressive wedge-base of the third facies cycle. This sandstone onlaps various older beds with increasing erosional and angular unconformity from south to north: first the conglomeratic Auersthal beds of the upper *Lagenida* zone, then the Aderklaa and Gänserndorf beds of the Karpatian, the Bockfliess beds of the Ottnangian or lower Karpatian, and at last the Upper Cretaceous Flysch of the Spannberg ridge (Figures 8 and 23). The within the field, up to 80 m (maximum up to 140 m) thick, diachronous Matzen sandstone is a high-quality reservoir (average porosity 26%, average permeability 1000 md) with a cumulative oil production of about 34 million MT (240 million bbl) since 1949.

The facies change between the Matzen sandstone and the overlying marine shale wedge occurs progressively farther marginward toward the island of the Spannberg ridge (or toward the basin margin in the northwest, respectively) in successively younger strata, representing a transgressive or up-to-margin pattern (Kreutzer, 1986; Kreutzer and Hlavaty, 1990) (Figures 11 and 23). The inner neritic Matzen sandstone (Rupp, 1986) (Figure 44) fines upward—apart from the conglomeratic and deltaic Auersthal beds in channel-like topographic lows at the base—from coarse sands into fine-grained clastics

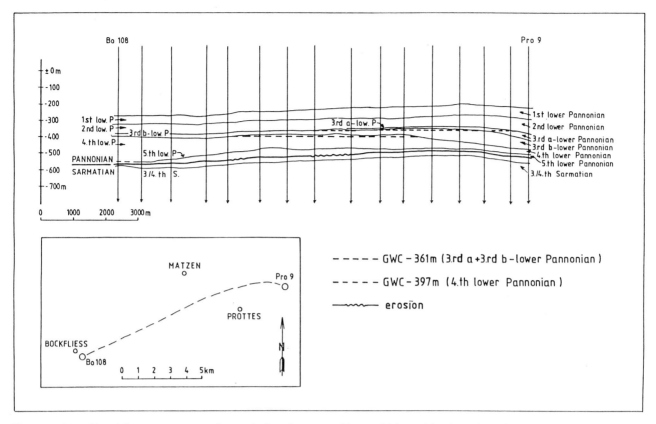

Figure 20. Simplified cross section of the lower Pannonian sandstone lobes in the Matzen field. (According to Kreutzer, 1990, reprinted with permission.) (A good idea of the location of the cross section shown by the inset can be gained by comparing the inset with other map figures, i.e., Figures 25 to 27.)

in response to a sea level rise and changes from a high- to a low-energy environment. The SP-log profile (Figure 21) is typically cylinder-shaped in the lower and middle part and more bell-shaped in the upper part as the sandstone grades into the overlying marine shales of the outer neritic environment (Rupp, 1986) (Figure 44).

The thickness of the basal transgressive sandstone is dependent on the gradient of the depositional surface (Abbott, 1985). Because of high sediment influx and the moderate to steep gradient of the flanks of the eroded Spannberg ridge (Figures 15 and 23) the transgression was slow, and unusually thick sands were deposited in the topographic lows, whereas over the topographic highs with a low to moderate surface gradient and a limited supply of source material, the transgression was relatively fast and thin sandstones have been preserved; in some places only shales have been deposited. This situation of the Matzen sandstone agrees well with the many examples of basal transgressive sandstones from North and South America and Australia, described and illustrated by Abbott (1985).

All the other sheet-like (and in some places lenticular) oil and gas sandstones in the Badenian (15th to 5th) above the Matzen sandstone belong to the wedge-middle plays (Figures 11, 21, 23, and 31 to 37). Numerous sandstone tongues alternating with shale interbeds, probably representing on the one side the transgressive fully marine parts, are on the other side the regressive marine parts of deltas (delta front and prodelta). These sandstone tongues extend first (15th, 14th) from the northeast *around*, and then also from the north and northwest *over* the subsided Spannberg ridge into the wedge-middle part—progressively farther south and basinward in successively younger strata. The lower and upper segments of such tongues commonly have up-to-center (regressive) and up-to-margin (transgressive) facies patterns and funnel-shaped or bell-shaped SP-log patterns, respectively (Figure 21). The oil and gas reservoirs of the Badenian (15th to 5th) follow this trend and occur in an up-to-center progression. The wedge-middle plays in the Matzen field are typically associated with marked depositional slopes and interval thickening (Figures 11, 23, and 24) (Kreutzer, 1986; Kreutzer and Hlavaty,1990).

The regressive wedge-top of the uppermost Badenian (4th to 1st) consists of sand-rich beds with a poor marine fauna and is truncated locally by an erosional unconformity in synsedimentary highs in the southern part of the field; these are marked by extra-formational alpine carbonate gravel beds at the Badenian-Sarmatian boundary (Figures 11, 23, and 24).

Fourth Cycle—Sand-rich beds of the lower Sarmatian (10th to 8th) (Figure 21) correspond to a new transgressive wedge-base, transitional to an upper

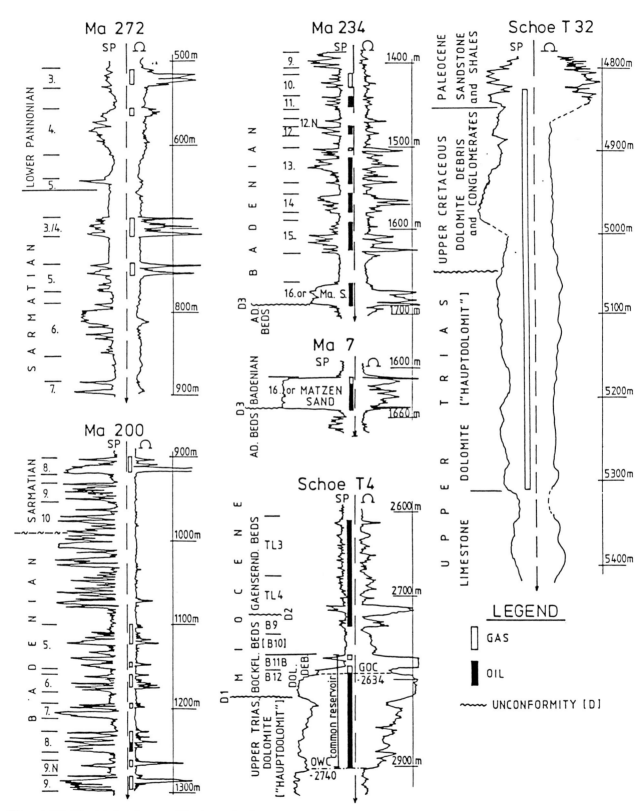

Figure 21. Electric log types of the important pay zones in the Matzen field. The locations of the wells are shown on the structure maps.

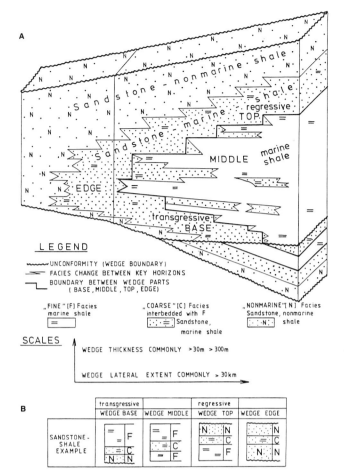

Figure 22. Sand-shale wedge. (Modified from White, 1980.) (A) Facies-cycle wedge. (B) Vertical facies successions.

Sarmatian thick shale interval (between 8th and 7th), followed again by sand-rich fining- or coarsening-upward beds (7th to 3rd/4th), including three channelized, abrupt-based sandstones with extra-formational alpine carbonate gravels (3rd/4th, 6th, and 7th Sarmatian) as well as several thicker shales (Figure 21). It seems that the transgressive-regressive wedge-middle part comprises not only the lower, thick shale interval but also the overlying sand (gravel) and shale beds of the deltaic upper Sarmatian, together even with the brackish and deltaic lower and middle Pannonian sand (gravel) and shale or clay succession (Figure 11). The lower Pannonian sandstones (1st to 5th) are characterized by an upward-coarsening facies (Kölbl, 1953; Wieseneder, 1959; Kreutzer, 1974) with regressive SP-log shapes, except in the cylinder-shaped SP-curve, channelized parts of very thick sandstone lobes. The wedge-top is represented by the sandstones and gravel beds of the fluviatile-lacustrine Pontian (Figure 11).

The Miocene basin fill has been subdivided into three sedimentary cycles by Kölbl, (1953 internal report, 1957, 1959). Because all his cycles begin with thick shale intervals and end with sand-rich beds, the wedge-bases of the 4th and 3rd cycle still belong to his 2nd and 1st cycle, respectively. The 2nd and 1st cycle correspond to his 1st cycle, but the sub-Badenian sediments have been only little known at this time and therefore could not have been differentiated by him.

Depositional Sequences

The Miocene geological units of the Matzen field generally also correspond to depositional sequences (Mitchum et al., 1977: "A depositional sequence is a stratigraphic unit composed of a relatively conformable succession of genetically related strata and bounded at its top and base by unconformities or their correlative conformities."). The Badenian of the central Vienna basin has been divided into two completely different geological units or depositional sequences: a lower trangressive unit or sequence of the lower Badenian and an upper alternating regressive-transgressive unit or sequence of the upper Badenian (Kreutzer, 1986; Kreutzer and Hlavaty, 1990). Using the criteria and definitions of Van Wagoner et al. (1990), the two geological units correspond to two parasequence sets, separated by a major marine-flooding surface or downlap surface or condensed section (parasequence set boundary) (Figures 23 and 24).

The lower transgressive unit of the Badenian in the Matzen field (Figure 23), showing an increasing base discordant onlap and hiatus toward north, comprises the Matzen sandstone as well as the overlying, parallel to slightly divergent-bedded, and rather homogeneous, bentonite-bearing shales, which have a facies change with the sandstone and increase in thickness southward. Using the criteria and definitions of Van Wagoner et al. (1990), the lower transgressive unit consists of a retrogradational parasequence set of the transgressive systems tract developed after a rapid rise in sea level. The conglomeritic part of the Auersthal beds below the Matzen sandstone, separated by a transgressive surface, represents an incised valley fill of a lowstand wedge of a lowstand systems tract. The underlying Aderklaa conglomerate, separated by a still deeper transgressive surface, represents a deeper lowstand systems tract after a rapid fall in sea level and a strong truncation of the Aderklaa beds (erosional unconformity D3) (Figure 11a).

The upper unit of the Badenian in the Matzen field above the regional Matzen bentonite marker, overlying the lower unit with increasing apparent base-discordant downlap and apparent hiatus toward south (Figures 23 and 24), is characterized by cyclic beds of lower regressive (coarsening-upward) and upper transgressive (fining-upward) sandstones and rather heterogeneous clays and shales. The upper and middle sandstones are thicker and laterally more extensive than the lower sandstones. Up to 1 m thick calcareous sandstones with a marine macrofauna are frequent on the top of such sandstone complexes and these, together with overlying shales, are transgressive sediments. These shales, 5 to 20 m thick and repeatedly intercalated, separate the sand complexes

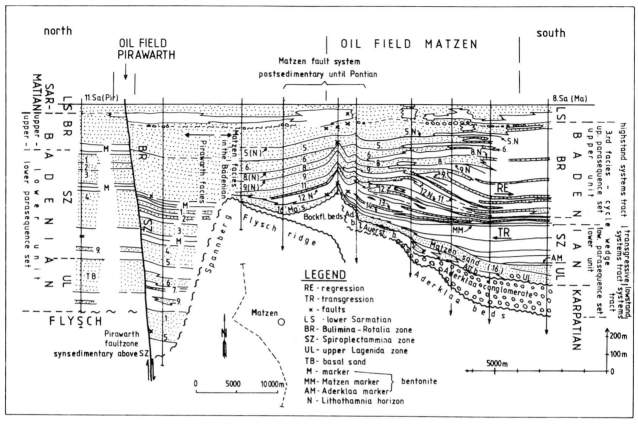

Figure 23. Stratigraphic cross section of the Badenian in the Matzen field with a facies-cycle wedge (south of the Spannberg ridge) and parasequence sets. (According to Kreutzer, 1986, modified and reprinted with permission.)

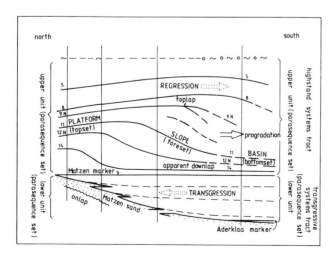

Figure 24. Schematic cross section of the depositional units or parasequence sets of the Badenian in the Matzen field. (According to Kreutzer and Hlavaty, 1990, reprinted with permission.)

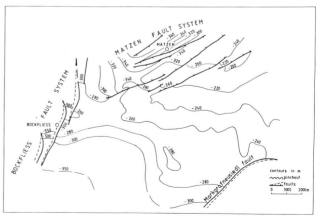

Figure 25. Structure map of 1st lower Pannonian.

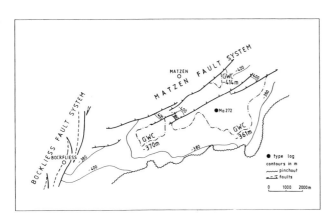

Figure 26. Structure map of 3rd lower Pannonian.

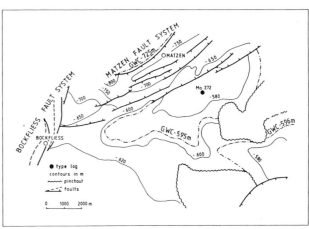

Figure 29. Structure map of 5th Sarmatian.

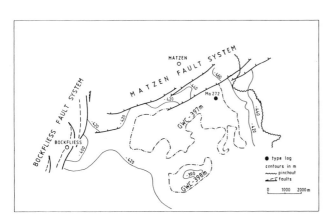

Figure 27. Structure map of 4th lower Pannonian.

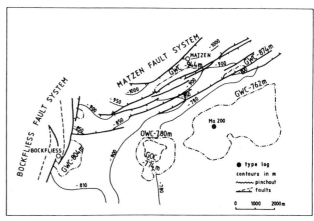

Figure 30. Structure map of 8th Sarmatian.

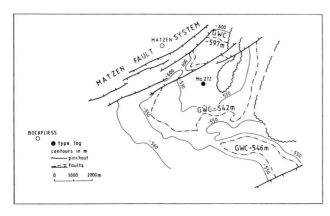

Figure 28. Structure map of 3rd/4th Sarmatian.

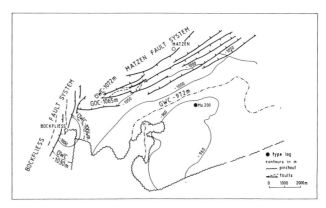

Figure 31. Structure map of 5th Badenian ("Tortonian").

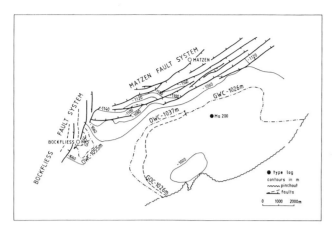

Figure 32. Structure map of 6th Badenian ("Tortonian").

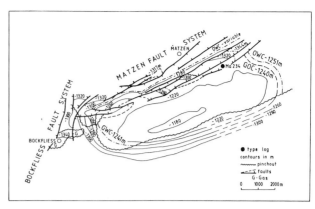

Figure 35. Structure map of 11th Badenian ("Tortonian").

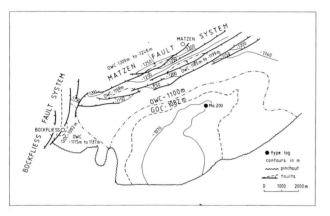

Figure 33. Structure map of 8th Badenian ("Tortonian").

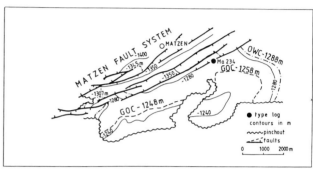

Figure 36. Structure map of 12th Badenian ("Tortonian").

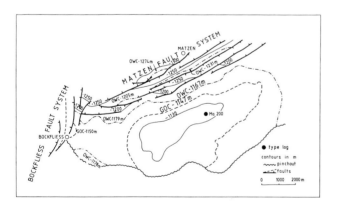

Figure 34. Structure map of 9th Badenian ("Tortonian").

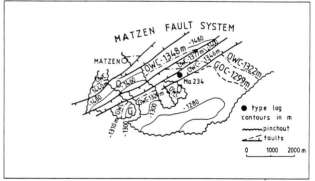

Figure 37A. Structure map of 13th Badenian ("Tortonian").

and reservoirs (Figure 21). Slightly increasing SP and resistivity values as well as silt or sand content of the shales indicate marine prodelta sediments, transitional to the marine delta front sands of such sand complexes. The upper unit (Figure 23) exhibits a sigmoid to oblique progradation or offlap, (1) with a paleotopographic differentiation into a northern upper platform (topset) zone, (2) with gently dipping and rather parallel-bedded segments of a sand-rich facies, a middle (foreset) zone, (3) with thicker more steeply dipping (up to 6°) segments locally terminating updip by toplap, and (4) a southern lower (bottomset) zone, again with gently dipping segments of a shale-rich facies, terminating downdip by apparent downlap on an apparent "nondepositional unconformity." The depositional environments generally have been shifted by regression southward, and therefore the platform facies of the younger beds, containing large oil and gas reservoirs, overlie the basin facies of the older beds ("Walther's law of

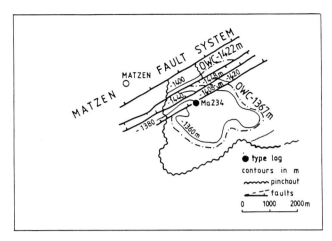

Figure 37B. Structure map of 14th Badenian ("Tortonian").

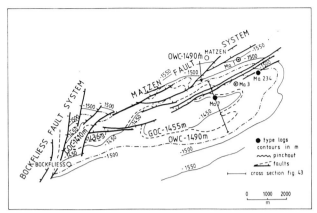

Figure 39. Structure map of Matzen sandstone or 16th Badenian ("Tortonian").

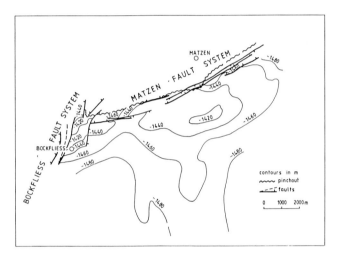

Figure 38. Structure map of Matzen bentonite marker.

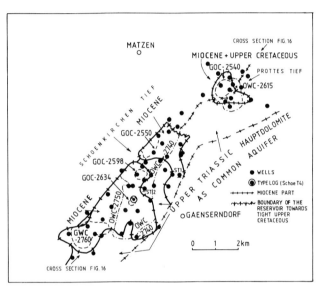

Figure 40. Structure map of the common reservoirs in Schönkirchen Tief and Prottes Tief, comprising Upper Triassic dolomites, Upper Cretaceous and Miocene dolomite debris, conglomerates, and sandstones (simplified). Notice that the location of the cross section of Figure 16 is indicated. Tight fossilliferous Upper Cretaceous marls separate the oil reservoirs of Schönkirchen Tief and Prottes Tief. These marls overlie Upper Triassic dolomites of the common aquifer of both reservoirs. Where their facies change into dolomite debris and conglomerates near the Triassic dolomites, they can be oil-producing (in Prottes Tief) or water-bearing (Figure 16).

facies") (Figure 23). Using the criteria and definitions of Van Wagoner et al. (1990), the upper unit consists of an aggradational to progradational parasequence set of the highstand systems tract developed after a slow rise, stillstand, and slow relative fall of sea level.

Several lines of evidence point to a subaqeous part of an ancient Danube delta of the upper unit of the Badenian in the Matzen field (Kreutzer, 1986; Kreutzer and Hlavaty, 1990): (1) cyclic succession (Figure 21), (2) geometrical configuration of the strata, and (3) alternating shale intervals of fully marine, "deeper water" Foraminifera and sand complexes of dominantly hyposaline shallow-water Foraminifera, as shown by Rupp (1986) (11th to 15th Badenian; Figure 44), indicating proximity to a river mouth system (as for instance, *Textularia earlandi*, the recent species typical of Mississippi mouth facies). The sediments consist generally of unbedded or evenly (parallel) to wavy bedded (flaser and lenticular bedding), sometimes bioturbated, sandstones and shales with common lignitic and plant debris. In the "deeper water" fully marine intervals, frequently up to several meters thick, layers of *Lithothamnion* nodules embedded in a matrix of calcareous clays and micritic limestone are repeatedly intercalated from the 5th to the 14th Badenian. Although the occurrence of the red algae indicates episodes of very shallow marine water over a submarine swell (50 to 150 m water depth), the composition of the matrix points to a low-energy environment. The extent of these *Lithothamnion* beds and the distribution of their thicknesses is for the

79

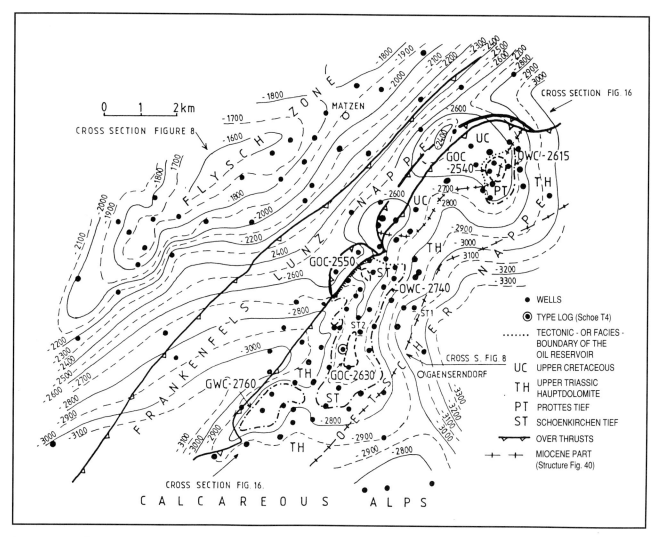

Figure 41. Structure map of sedimentary basement (Flysch zone and Calcareous Alps) in the Matzen field (simplified). (According to Kröll, 1984, reprinted with permission.) Notice that the locations of the cross sections shown by Figures 8 and 16 are indicated.

most part influenced by the synsedimentary structures (Kölbl, 1953; Wieseneder, 1956, 1964; Kreutzer, 1978). In cores from two wells, Müller and Schreiber (1991) have differentiated a subaquatic delta margin (in the 11th Badenian) with alternating and transitional zones of distributary mouth bars, distal bars, and prodelta to open marine subenvironments.

Seismic Facies

The appearance of facies seen in seismic sections around the Matzen field and the surrounding area is variable, depending on lithologies and stratal configurations. In the bedded sand-shale successions of the Pannonian Sarmatian and upper Badenian, the principal stratal configuration is parallel to subparallel or slightly divergent, sometimes hummocky, with continuous to discontinuous reflections and weak to strong amplitudes. The prograding and predominantly shaly part of the upper Badenian is almost reflection-free (seismically "transparent"), but sometimes distinct sigmoid and oblique clinoforms with apparent downlap and toplap are recognizable.

In the shale wedge of the lower Badenian, also nearly reflection-free, however, a parallel to slightly divergent seismic facies is indicated (although weakly expressed) and onlapping reflections over the strongly reflective Aderklaa conglomerate exist. The lenticular Aderklaa beds, truncated over the top of the structure, are characterized by rather indistinct and discontinuous reflections with weakly defined amplitudes and reflection-free zones. The conglomeratic Gänserndorf beds and the Bockfliess beds sometimes give stronger reflections, those of the latter beds onlapping the older sedimentary basement.

The Gänserndorf beds probably represent delta plain sediments, equivalent to beds in Czechoslovakia, and as shown by Jiricek (personal communication, 1988), indicate delta front and prodelta sediments, which distinctly prograde updip to the east in the seismic sections.

Based on the analysis of seismic sequences in the Vienna and Pannonian basins by Pogacsas and Seifert (1991), the sedimentary units of the Miocene in the Matzen area were attached to highstand, transgressive, and lowstand systems tracts as follows:

Lower Pannonian	highstand systems tract transgressive systems tract
Upper Sarmatian	highstand systems tract
Lower Sarmatian	transgressive systems tract lowstand systems tract
Upper Badenian (*Bulimina-Rotalia* zone)	highstand systems tract (progradational)
Middle Badenian (uppermost *Spiroplectammina* zone)	
Middle Badenian (*Spiroplectammina* zone)	
Lower Badenian (upper *Lagenida* zone) (lower *Lagenida* zone)	transgressive systems tract
Aderklaa conglomerate	lowstand systems tract
Aderklaa beds	highstand systems tract transgressive systems tract
Gänserndorf beds	lowstand
Bockfliess beds	highstand systems tract transgressive systems tract

Structural Development and Faults

The structural development that resulted in the Matzen field traps in the Miocene basin fill is governed predominantly by the differential subsidence of the sedimentary basement; these structures are the generally southwest–northeast-striking Spannberg Flysch ridge in the north and the north–south-striking ridge of the Calcareous Alps in the south (Figures 8, 11, and 41). This development has been investigated by use of isopach maps of the Pannonian, Sarmatian, and Badenian by Kreutzer (1971), and in the Aderklaa, Gänserndorf, Bockfliess beds by Köves (1971) and Hladecek et al. (1971). A southward convergence of electric log markers suggests less subsidence of the Calcareous Alps during the sedimentation of the upper Bockfliess beds (B11 to 9) as well as of the upper Badenian (above the 5th), Sarmatian, and Pannonian. (A short interruption of this tendency in the lowest Pannonian was caused apparently by the rapid deposition of some delta lobes.) A northward convergence suggests generally less subsidence of the Flysch ridge during the sedimentation of the Gänserndorf beds, the Aderklaa conglomerate, and from then until the deposition of the upper Badenian (8th). The almost parallel markers of the intervening intervals point to a balanced subsidence in the lower Bockfliess beds (B16 to 12), the Aderklaa beds, and the upper Badenian (between 8th and 5th) (Figure 11).

The structural development in the upper regressive-transgressive unit of the Badenian was influenced not only by differential subsidence but also by the progradation of the strata. The more steeply dipping segments of the individual foreset zones (5th to 15th Badenian) establish sedimentary structural flanks shifting laterally to southwest, south, and southeast, from the older to younger beds (Figures 23 and 24).

A final important change occurred in the Pontian owing to the origin of a new (postsedimentary) structure in the northern part of Matzen field in combination with the Matzen fault system. Numerous old synsedimentary structure elements are preserved, however.

In the Matzen field, three genetically unrelated fault systems can be recognized in the Miocene basin fill (Figures 11 and 25 to 39).

1. The oldest is the north–south-striking postsedimentary Schönkirchen fault system above the ridge of the Calcareous Alps in the southern part of the field. Activity began after the sedimentation of the Aderklaa beds and ended pre-Badenian. The vertical throw is up to about 50 m (Köves, 1971; Hladecek et al., 1971) (Figures 8 and 11).
2. The north–south-striking, west-dipping, and right-stepping (Ladwein et al., 1991) synsedimentary Bockfliess fault system, consisting of en echelon fault segments, on the western field margin, with a vertical throw of up to 400 m (Figures 25 to 39). Upthrown and downthrown blocks are directly connected by steeply dipping narrow strips between two faults (Figure 25). A large difference in the thickness of the Miocene sediments of the upthrown and downthrown blocks and increase in thickness within the narrow strips accentuate the synsedimentary character (Wessely, 1988). There were two activity phases, the one from the time of deposition of the 7th Badenian up to the top of Sarmatian, the other from the 3rd lower Pannonian up to the top of middle Pannonian, respectively (Kreutzer, 1971) (Figure 11).
3. The southwest–northeast-striking postsedimentary Matzen fault system in the northern part of the field consisting of northwest- and southeast-dipping rotational faults (Figures 8, 11, and 25 to 39). It is post-Pannonian in age (Kreutzer, 1971). The faults show a maximum vertical throw of 80 m and may be explained by tension above an updoming of the deeper (but still undrilled) autochthonous basement, which causes the Matzen-Spannberg elevation (Wessely, 1988). By displacing southeast-dipping faults, the northwest-dipping faults prove to be younger (Wessely, 1988).

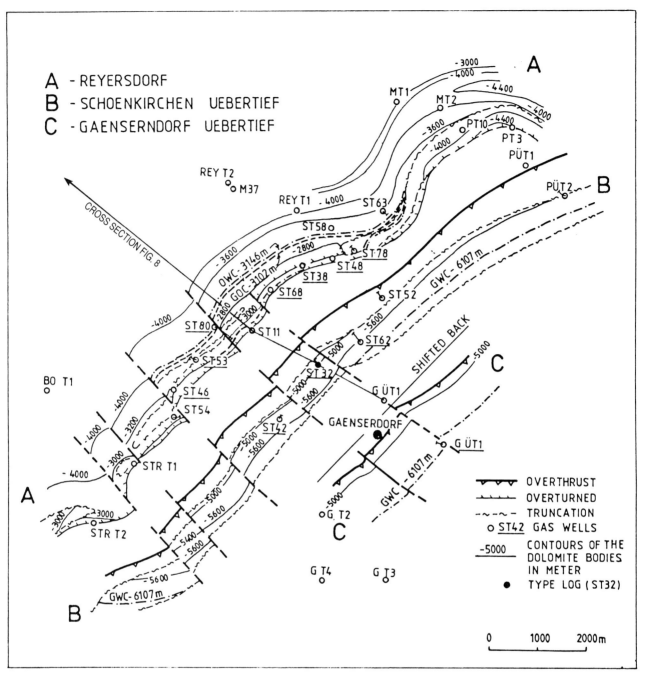

Figure 42. Structure map of internal dolomites of the reservoirs Reyersdorf, Schönkirchen Übertief, and Gänserndorf Übertief. (Well GüT 1 is shown in its real position in B—Schönkirchen Übertief. C—Gänserndorf Übertief—is shifted because of the actual overlap of Schönkirchen and Gänserndorf Übertief; GüT 1 locations indicate the amount of the shift.) (According to Wessely, 1984, reprinted with permission.)

Hydrocarbon Sources

Although in the past the source of the hydrocarbons was thought to be in the Tertiary fill, and later, also in underlying rocks of the alpine units, other explanations of their origin have been given by the Russian geologist Dolenko (1955) and by Kapounek et al. (1963, 1964), who believed that the autochthonous Mesozoic is the ultimate source of the hydrocarbons in the Vienna basin. (More discussion of hydrocarbon sources is given in the following section, *Traps and Reservoirs*.)

TRAPS AND RESERVOIRS

The oil and gas reservoirs of the Miocene basin fill in the Matzen field are predominantly combination structural-stratigraphic traps (Figure 11). Structural trap types are dominant, especially in the Pannonian, Sarmatian, and Badenian and in the buried hills of the Calcareous Alps, but these are mostly accompanied by a local or general pinch-out of the sands laterally or over the structural high (Figures 26 to 39). A typical structural trap is represented by the strongly folded anticlines in the Frankenfels-Lunz nappe (Figure 8). Stratigraphic or combined structural-stratigraphic traps are predominant in the lenticular Gänserndorf and Bockfliess beds (Figures 8 and 11). Additional trapping occurs against faults of the three fault systems discussed before; such fault traps occur in most reservoirs, at least in their marginal parts or in separate and smaller hydrocarbon-bearing areas of the same sand bodies (Figures 11 and 26 to 39). Traps above unconformity D2 occur frequently in the Gänserndorf beds and in traps below this unconformity in the upper Bockfliess beds B9 to 12 (Figures 8 and 11). In some wells, separate reservoirs exist in the Aderklaa, Gänserndorf, and Bockfliess beds below unconformity D3 and in the Bockfliess beds above unconformity D1.

On the other hand, common reservoirs (in communication) can exist on both sides of unconformity D1 in the case of the lower Bockfliess beds (B11 to 16) if they consist mostly of dolomite debris and conglomerates (and minor sandstones) separated by Miocene shales and if they are laterally adjacent to or covering directly the buried dolomite hills of the sedimentary basement (a boundary differentiated by sonic logs and cores) (Figures 11, 16, 40, and 41). Common reservoirs exist also between the Matzen sand and the Auersthal beds above unconformity D3 and the subjacent parts of the Aderklaa, Gänserndorf, and Bockfliess beds below it (Figure 43).

The reservoirs in the whole Miocene basin fill, as well as in the dolomites, are generally bounded by edge or bottom water (Figures 11 and 26 to 39). Common original gas-oil, oil-water, and gas-water contacts indicate probable communications between permeable zones within several sandstone complexes resulting from the discontinuity of shale layers that otherwise appear to segregate the sandstones. On the other hand, reservoirs distinctly separated by continuous shale layers show clearly different original fluid contacts (Kreutzer, 1985). Production and pressure data give evidence of the real connection of permeable beds. During production, fluid contacts generally are rising and individual sand layers are selectively invaded by water.

As a rule in the sand-shale successions of the Pannonian, Sarmatian, and Badenian, there are different original gas-oil, oil-water, and gas-water contacts in adjacent fault blocks, pointing to initial sealing fault planes (Kreutzer, 1985). Probable nonsealing faults exist where the sand thicknesses are greater than the vertical throw of the individual faults and where the faulted reservoir parts still remain in contact although displaced. Faults seeming

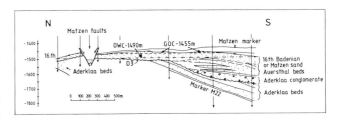

Figure 43. Schematic cross section through the common energy unit of the Matzen sandstone, including the Auersthal beds, Aderklaa conglomerate, and parts of the Aderklaa beds (Bockfliess beds not shown). Location on Figure 39. The vertical arrows are schematic representations of some deeper wells.

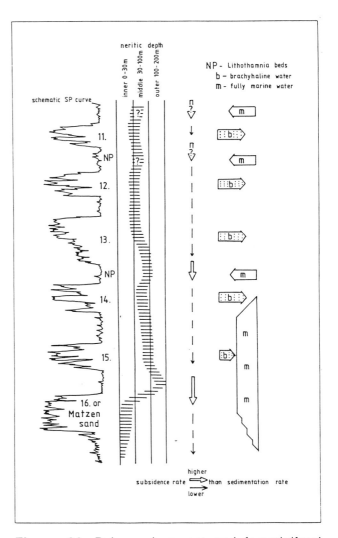

Figure 44. Paleoenvironment and foraminiferal paleoecology between the Matzen sandstone and the 11th Badenian, by Rupp (1986). (Modified and reprinted with permission.) Foraminiferal groupings in the sandstone complexes are made up of hyposaline shallow water foraminifers, and in the shale intervals of fully marine "deeper water" foraminifers.

to end within a reservoir (whether or not sealing) cannot be determined except by pressure data and fluid movements; even then, during production, initial sealing fault planes may rupture and become nonsealing. Generally, water drives are dominant in the reservoirs of the sheet-like beds of the Pannonian, Sarmatian, Badenian, and the common reservoirs of buried hills of dolomite and Bockfliess beds. Depletion drives are dominant in the lenticular sands of the Gänserndorf and Bockfliess beds, which are not in communication with the dolomite reservoirs.

The basal transgressive Matzen sandstone (Figure 21) overlies various older Miocene beds with increasing erosional and angular unconformity from south to north: first the conglomeratic Auersthal beds, then the Aderklaa and Gänserndorf beds, and last the Bockfliess beds (Figures 8 and 23). Nevertheless, the Matzen sandstone forms a common energy reservoir with Auersthal and other beds (Figures 11 and 43), as proven by production and/or pressure data and by a common original oil-water contact (Figure 39), both in the main block and in other fault blocks. The regional water-bearing Aderklaa conglomerate (Figure 15), sometimes over 200 m thick in the field, is part of the common aquifer not only of the Matzen sandstone but also of the lower Badenian reservoirs in the Aderklaa field and Zwerndorf field (Schröckenfuchs, 1975). Because of its great thickness, more than 2000 m^3 (12,600 bbl) of disposal water per well per day can be easily injected. Such water injection assists (in part indirectly) the natural water drive as well as the waterflood project in the Matzen sandstone. The evidence of probable communications between the layers of Matzen sandstone, the Auerthal beds, parts of the Aderklaa, Gänserndorf, and Bockfliess beds, and the Aderklaa conglomerate, even though these appear to be vertically separated by shale layers, indicates that as the result of unconformities, the shale layers pinch out and are discontinuous (Figure 43).

Several times between 1958 and the present, infill wells drilled into the Matzen sandstone clearly have had rising oil-water contacts in the central part of the field *although the original oil-water contact level still exists at its margins.* The natural water drive of the field is a bottom-water drive, and elevated oil-water contacts of new infill wells are distinctly higher than the perforations of the surrounding older oil-producing wells. This general phenomenon of the fluid withdrawal process from the Matzen sandstone, recognized by Horvath and Kreutzer (internal report by Krobot, 1960), is only possible if the perforations of the older wells are shielded from bottom water invasion by numerous but discontinuous vertical permeability barriers (Muskat, 1949). The Matzen sandstone contains numerous thin, interlayered, cemented sandstones and shales of limited areal extent with an average thickness of several decimeters. Based on analog methods described and illustrated by Muskat (1949), experiments with blotting paper electrolytic models have been performed by the ÖMV Laboratory to show the qualitative effects of permeability barriers on the nature of the injection fluid advance. The bottom water can be flowing only indirectly to the perforations, encroaching around and between the discontinuous barriers and invading (coning) to a higher position than some of the deeper perforations of the producing wells. Finally, the water flows laterally to the perforated intervals in some wells according to Kaufmann and Horvath (1961, internal report) and Horvath and Lachmayer (1964); these were the first papers dealing with such a phenomenon in an oil reservoir (subsequently it was described in relation to other oil reservoirs according to Horvath in a personal communication).

The Matzen sandstone (Figure 39) has a maximum gross oil thickness of 35 m in the main block and 65 m in the western Bockfliess block. Average porosity is 26% and average permeability is 1000 md, with a productive area of about 22 km^2. The cumulative production is more than 34 million MT (242 million bbl) of oil, of which the best well produced over a half million MT (3.6 million bbl) of oil (Figures 12 and 13).

A common aquifer (Kröll and Wessely, 1973; Schröckenfuchs, 1975) exists for three reservoirs in the Ötscher nappe of the Calcareous Alps (Figures 11, 16, 21, 40, and 41): the Schönkirchen Tief small gas reservoir in the southwest, the Schönkirchen Tief large oil reservoir of high quality in the middle part of the field, and the Prottes Tief small oil reservoir in the northeast. These reservoirs comprise not only buried hills of up to 650 m thick fractured Upper Triassic dolomite, but also—unconformably overlying—Upper Cretaceous and Miocene Bockfliess beds B11-16 of dolomite debris, conglomerates, and sandstones (Kapounek et al., 1964; Kröll and Wessely, 1973). The dolomites, underlain by Upper Triassic limestones and clastics, are in a syncline about 3 km wide and at least 15 to 20 km long, comprising part of the frontal Ötscher nappe (Figure 8). The boundaries of the oil pools are oil-water contacts as well as tight lithologies (marls) of Upper Cretaceous between Schönkirchen Tief and Prottes Tief. The three reservoirs have had clearly different original water contacts (Figures 16, 40, and 41), with elevations increasing from southwest to northeast: a gas-water contact at -2760 m (with some oil shows) in the gas reservoir Schönkirchen Tief and an oil-water contact at -2740 m; at least three different gas-oil contacts at -2634 m (Figures 16 and 21), -2598, and -2550 m in the Schönkirchen Tief oil reservoir; and an oil-water contact at -2615 m and a gas-oil contact at -2540 m in the Prottes Tief oil reservoir. A new evaluation of an early oil test at -2639 m has proved the reservoir as saturated and has indicated also primary gas-oil contacts at -2598, -2550, and -2590 m. The spillpoint between the gas and the oil reservoir is -2785 m in dolomite and dolomite debris; between the two oil reservoirs, it is about -2830 m in dolomite debris or around the Upper Cretaceous barrier. In accordance with the

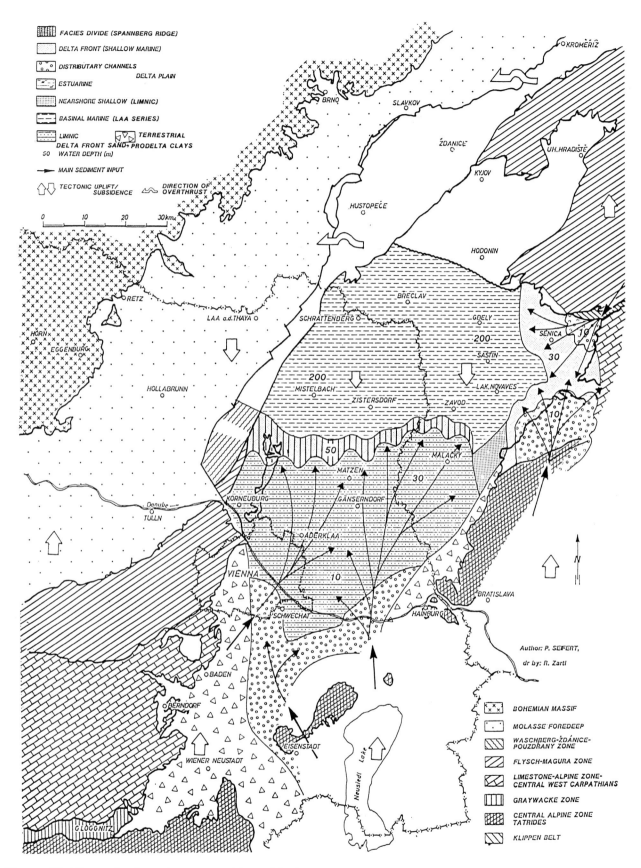

Figure 45. Vienna basin, paleogeographic and facies development in the Karpatian (17 Ma). (According to Seifert, 1989, unpublished, used with permission.)

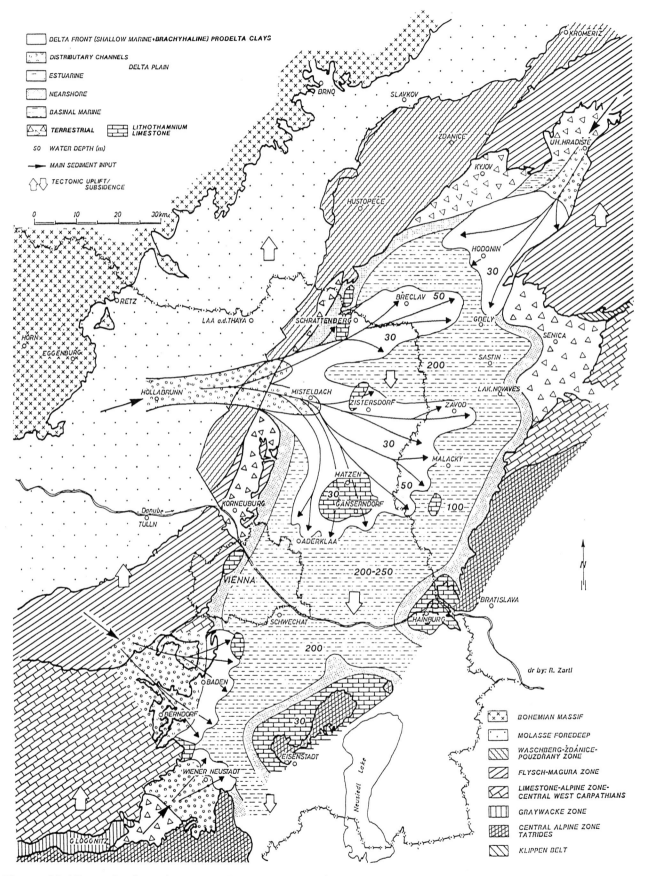

Figure 46. Vienna basin, paleogeographic and facies development in the middle Badenian (15 Ma). (According to Seifert, 1989, in Ladwein et al., 1991, their figure 22.14, reprinted with permission.)

theory of "differential trapping of hydrocarbons" by Gussow (1953), Kreutzer (internal report, 1966), Kröll and Wessely (1973), and Schröckenfuchs (in a comprehensive reservoir engineering study, 1975), a dominant flow direction of the migrating hydrocarbons from southwest to northeast resulting from successive spills from the closures was assumed, first of oil, then of gas, coming from the deepest gas reservoir and moving updip to the highest oil reservoir; this process resulted in selective trapping of oil and gas in the three reservoirs. Theoretically, a still-higher, water-bearing structure could exist (Gussow, 1953); however, none has been found although the water-bearing dolomites of the Ötscher nappe continue to the northeast.

The gas reservoir in the southwest has a maximum gross thickness of 119 m gas pay and an average net thickness of 55 m of gas pay. The Schönkirchen Tief oil reservoir has a maximum gross oil pay thickness of at least 106, 142, and 190 m in the three closures that had primary gas caps. In the Prottes Tief oil reservoir, the maximum gross oil pay thickness is 75 m where a primary gas cap exists. The average gross oil pay thickness of Schönkirchen Tief is 59 m, the average porosity 2.8 to 8.4% in dolomites and 11.3% in Miocene dolomite debris and sandstones. The average permeability is 1000 md. The cumulative oil production of these oil reservoirs is 7.9 million MT (56 million bbl), of which the best well produced over 1 million MT (7 million bbl) (Figures 12 and 13). Later infill wells have clearly shown elevated oil-water contacts resulting from the active water drive. The gas reservoir in the southwest produced 525 million m^3 (18.5 bcf) until its depletion.

The Reyersdorf gas and oil reservoir (Kröll and Wessely, 1973) is in a southwest-northeast-striking internal anticline in the northwestern part of the Frankenfels-Lunz imbrication system (Figures 8 and 42). The limbs of the narrow Reyersdorf anticline are steeply dipping and even overturned on the southeastern side. The core consists of a rather heterogeneous and bedded, fractured Upper Triassic Hauptdolomit, partly calcareous. The reservoir is sealed not only by Upper Triassic and Lower Jurassic tight limestones and Upper Cretaceous to Paleocene tight beds of the Giesshübl furrow, but also by Miocene shales of the basin fill. It is only about 0.7 km wide and 7 km long. Seven wells have been drilled into this reservoir, which is at a depth of about -2700 m at the crest of the structure. It has a gas-oil contact at -3102 m, an oil-water contact at -3146 m, a gross gas pay thickness of 400 m, an average net gas pay thickness of 77 m, and an average porosity of 2.8%. In the northeast, the reservoir is bounded by the structural dip of the anticline, and in the southwest, by one of the several northwest-southeast-trending wrench faults that dissect the anticline (Wessely, 1984). Because of the steeply dipping and bedded heterogeneous dolomites and of the thinness of the oil zone, a breakthrough of bottom water and gas occurred during oil production. Therefore, the oil-producing intervals had to be abandoned and the wells were recompleted for gas production. The cumulative oil production was only 33,700 MT (240,000 bbl) and cumulative gas production was 368 million m^3 (13 bcf).

The Schönkirchen Übertief gas reservoir (Kapounek and Horvath, 1968; Kröll and Wessely, 1973; Wessely, 1984) (Figures 8, 21, and 42) is in a southwest-northeast-striking internal thrust slice of Upper Triassic Hauptdolomit of the southeastern part of the Frankenfels-Lunz imbrication system as well as of the Upper Cretaceous dolomite debris of the discordantly overlapping Giesshübl beds. This narrow anticlinal thrust slice is very steeply dipping to southeast and has an overturned stratigraphic sequence of rather homogeneous fractured Upper Triassic dolomites and tight Upper Triassic and Lower Jurassic limestones, capped discordantly by Upper Cretaceous dolomite debris and sealing Paleocene sandstones and shales of the Giesshübl furrow. Thrust faults divided this slice from others in the northwest as well as from the dolomite slice of the Gänserndorf Übertief gas reservoir (Wessely, 1984) immediately to the southeast. Both gas reservoirs, however, belong to a common energy unit (as shown by pressure data) with a common gas-water contact of -6107 m. The thrust faults dissect the sediments of the Giesshübl furrow, but not the overlying Ötscher nappe (Figure 8) that contains the Schönkirchen Tief oil reservoir. The length of the gas reservoir Schönkirchen Übertief, extending to the northeast as well as to the southwest, is yet unknown and could be at least 13 km, with a width of about 0.8 km (Figure 42). The extent of the Gänserndorf Übertief gas reservoir, penetrated by only one well, the Gänserndorf Übertief 1, is completely unknown. Both reservoirs probably have been cut, as was the Reyersdorf reservoir, by northwest-southeast-trending wrench faults (Wessely, 1984). Three wells have been drilled into the Schönkirchen Übertief gas reservoir, which has a structural top at about 4890 m in the dolomites and 4700 m in the Upper Cretaceous dolomite debris. It has an apparent gross gas pay thickness of 900 m, an average net gas pay thickness of 300 m, an average porosity of 3.8%, and an average permeability of 2 md. The cumulative production is 5.6 billion m^3 (200 bcf) sour gas. The height of the gas column is about 1300 m, and the initial pressure gradient to the top of the reservoir is about 1.25 bar/10 m (approximately 590 bar at 4800 m).

Oil and Gas Sources

As stated before, hydrocarbon sources were originally thought to be in the Tertiary fill, and later also in the underlying alpine unit rocks. The autochthonous Mesozoic was believed by Dolenko (1955) and Kapounek et al. (1963, 1964) to be the ultimate Vienna basin source. Due to detailed geochemical investigations by Kratochvil and Ladwein (1984a), Ladwein (1988), and Ladwein et al. (1991), the most important source rocks for both oil and thermocatalytic gas were found to be in the autochthonous Jurassic sequence beneath the overthrust belt.

Much of the source potential of these Jurassic rocks was preserved during thrusting and most of the hydrocarbon generation took place post-thrusting during Miocene Vienna basin subsidence (Figures 3 and 4). The distribution of the hydrocarbon pools (vertical stacking) is indicative of predominantly vertical migration through major faults (Dolenko, 1955). Because of the activity of the Bockfliess fault system since late Badenian (Figure 11) and the occurrence of synsedimentary structures in the northern part (Flysch zone) as well as in the southern part (Calcareous Alps) of the Matzen field, accumulations of hydrocarbons were made possible.

The total organic carbon (TOC) of the Upper Jurassic marls of the restricted marine basin facies (Figure 14) ranges between 0.3% and 5% and the average lies between 1.5% and 2%. The kerogen of this facies is classified as types II and III.

The Upper Jurassic basin sediments are in the "oil window" between 4000 and 6000 m. The main stage of thermocatalytic gas generation occurs below 6000 m at vitrinite reflectance values higher than $R_o = 1.6\%$. Most of the crude oils of the Vienna basin are altered and belong to one major family. The sediments of the Neogene basin fill as well as of the Calcareous Alps and Flysch zone are generally too low in organic content to be classed as source, so only minor, noncommercial amounts of hydrocarbons could have generated from them (Ladwein, 1988; Ladwein et al., 1991).

Oil Properties

Differences between oils from different reservoirs in the Vienna basin are obvious; a general relationship was observed (Wieseneder, 1972) between increasing reservoir depth, increasing paraffinicity, and decreasing specific gravity of the crude oils. In the Matzen field, for instance, 8th Badenian oil sp.gr. is 0.933 at about 1200 m; 16th Badenian oil is 0.905 at about 1600 m; and Bockfliess beds and Triassic dolomites oils in the southern part of the field have a sp.gr. of 0.875 at 2600 to 2900 m. The naphthenic crude oils (Badenian) result from surface influences and are derived from paraffinic oils (Gänserndorf and Bockfliess beds and Triassic dolomites) (Welte et al., 1982). Despite the differences, all crude oils in the Vienna basin belong to one major family (Welte et al., 1982; Ladwein, 1988; Ladwein et al., 1991), and additionally have an advanced maturation level (dominance of n-alkanes relative to pristane and phytane in the saturated hydrocarbons, suggesting an origin from mature source rocks) (Welte et al., 1982).

Many of the crude oils of the Vienna basin are biodegraded (preferential removal of n-alkanes) and/or altered by water-washing (preferential removal of water-soluble aromatics). Biodegradation is observed only in oil from shallow reservoirs (above 1700 m, Badenian in Matzen) because of the critical survival temperature (65–70°C) for microbial organisms (Welte et al., 1982). The crude oils represent different stages of biodegradation. This may account for the depth dependence of specific gravity and paraffinicity and means that these variations do not necessarily correspond to genetic differences but are caused by various degrees of biodegradation. In spite of biodegradation, the crude oils show similar sterane patterns for altered and unaltered crudes, pointing to one family of high-maturity oils (Welte et al., 1982; Ladwein, 1988; Ladwein et al., 1991). Oils of this major oil family correlate well with extracts of basin marls from the autochthonous Jurassic.

Brines

The concentration of Miocene brines never exceeds that of sea water, but those of the Mesozoic Calcareous Alps of the basement are two or three times greater than sea water (Kröll and Wieseneder, 1972). In the Matzen field, the concentration of brines increases with increasing depth and reaches its maximum in the 5th to 9th Badenian (27,000 mg/L), but then decreases with depth down to the Matzen sand (18,000 mg/L), staying at this minimal value in the Aderklaa conglomerate as well as in the Aderklaa, Gänserndorf, and Bockfliess beds of the northern part of the field. Salinity rises sharply in the Gänserndorf and Bockfliess beds of the southern part of the field, reaching a maximum of 105,000 mg/L in the Calcareous Alps.

The abnormal decrease in concentration relative to increased depth in the Badenian and the uniform lower concentration in the deeper Miocene beds of the northern part of the field has been explained as the result of an upward migration of less concentrated waters, i.e., a migration toward an increase in concentration (Schröckenfuchs, 1975). The movement of cool near-surface waters from the edge of the basin in a large, very thick aquifer (Aderklaa conglomerate) displaced upward (up to the 6th Badenian) and laterally more concentrated and heated formation waters into regional high-structured areas (Schröckenfuchs, 1975). The concentration of Cl^- ions in the formation waters of the Badenian shows a common trend in decreasing from the structural highs to the edge water. Additionally there is a decrease of Cl^- concentration from west to east in the Matzen sand, approximating in the east the value of the Aderklaa conglomerate. On the other hand, CO_3 and SO_4 concentrations are higher in the east as well as in the edge-water

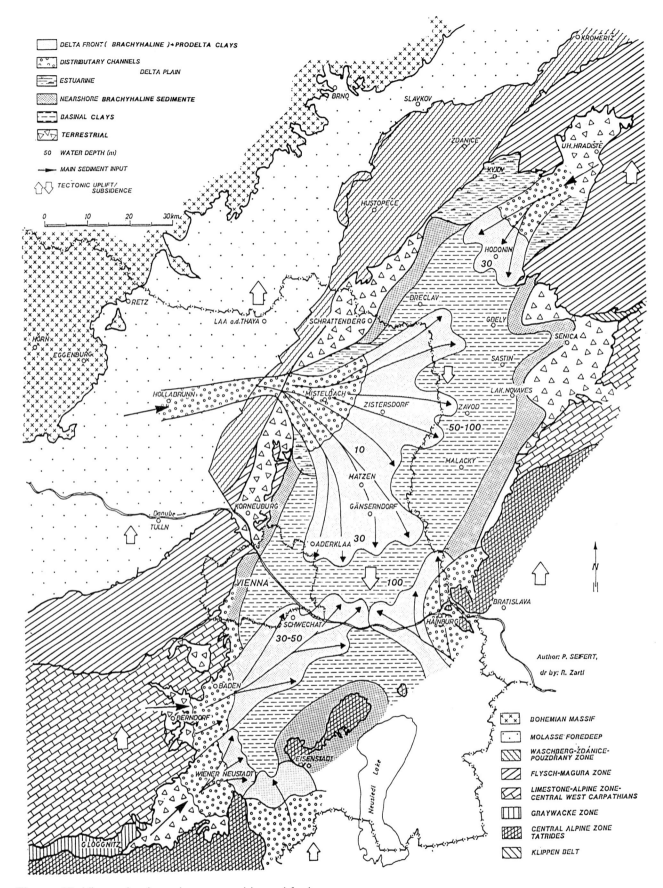

Figure 47. Vienna basin, paleogeographic and facies development in the lower Sarmatian (12 Ma). (According to Seifert, 1989, unpublished, used with permission.)

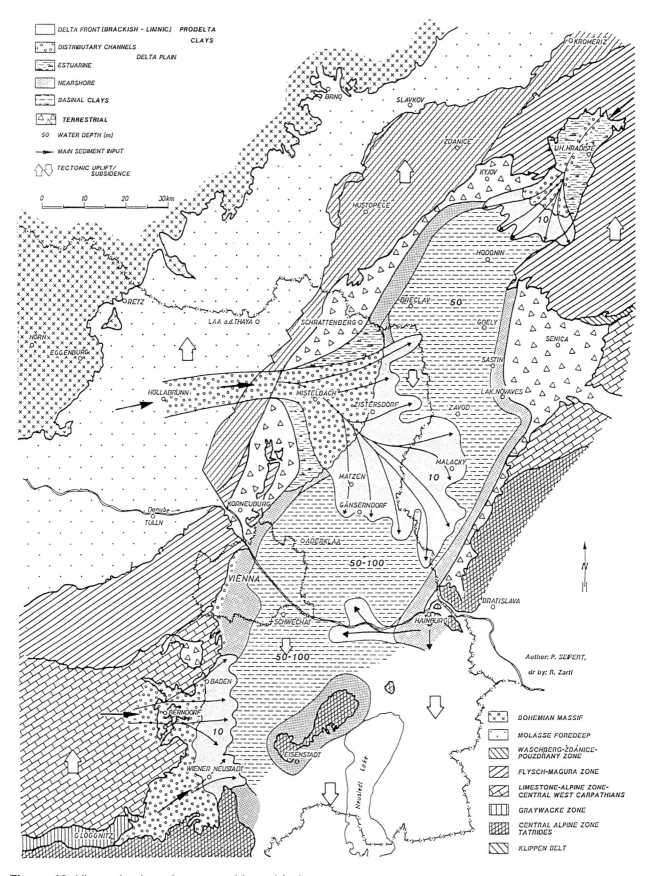

Figure 48. Vienna basin, paleogeographic and facies development in the lower Pannonian (10 Ma). (According to Seifert, 1989, unpublished, used with permission.)

area, corresponding again to those of the Aderklaa conglomerate.

In the Triassic dolomite, total dissolved solids decrease from the southwest (Schönkirchen Tief) to the northeast (Prottes Tief).

ACKNOWLEDGMENTS

G. Wessely and W. Grün from the Geological Department of ÖMV initiated the description of the Matzen field for the AAPG Atlas of Oil and Gas Fields. I am especially grateful to G. Wessely and W. Ladwein from ÖMV for the use of their manuscripts regarding the regional geology and hydrocarbon generation in the Vienna basin, respectively. I am also indebted to the AAPG editorial staff and William G. Brownfield for reviewing and editing. I express my special thanks to G. Schröckenfuchs and H. Murer from the ÖMV Reservoir Engineering Department for reviewing much of the reservoir data of the Matzen field, and as much to co-workers F. Sobotka, I. Widhalm, M. Taschl, and R. Grünner for reviews of drawings and manuscript drafts. I wish to acknowledge ÖMV for permission to publish this paper.

REFERENCES CITED

Abbott, W. O., 1985, The recognition and mapping of a basal transgressive sand from outcrop, subsurface, and seismic data *in* Orville Roger Berg and Donald G. Woolverton, eds., Seismic Stratigraphy II: AAPG Memoir 39, p. 157-167.

Braumüller, E., and A. Kröll, 1980, Erdölgeologie (Petroleum geology): Erdöl-Erdgas-Zeitschrift 96.Jg, H. 5, Hamburg-Wien, Urban Verlag, p. 153-160.

Dolenko, G. N., 1955, Conditions of the formation of crude oils in deposits of the Vienna basin. Contributions to the problems of the origin and migration of oil. (Original in Russian.) Izd., A k, USSR, Kyev, str, 238.

Friedl, K., 1950, Geologischer Bericht über die im Raume von Bockfliess im Jahre 1949 durchgeführten Bohrarbeiten (Geological report from the drilling activities in the area of Bockfliess): Internal ÖMV report of the Geological Department.

Friedl, K., 1959, The oilfields of the Vienna basin: Fifth World Petroleum Congress, Section I, Paper 48, p. 865-881.

Gussow, W. C., 1953, Differential trapping of hydrocarbons, cit. *in* A. I. Levorsen, Geology of Petroleum: San Francisco, W. H. Freeman and Company, p. 555-556.

Hladecek, K., S. Köves, and W. Krobot, 1971, Das tiefere Neogen im Raume Matzen-Süd (The lower Neogene in the southern area of Matzen): Internal ÖMV report of the Geological Department.

Horvath, S., and O. H. Lachmayer, 1964, Einige Besonderheiten beim Ausbeuten eines relativ homogenen, flachen Ölträgers mit Gaskappe und Wassertrieb (Specific phenomena during the recovery of a relatively homogeneous reservoir with water drive): Erdölzeitschrift, 80. Jg, H.9, Wien-Hamburg, Urban Verlag, p. 331-341.

Janoschek, W. R., and A. Matura, 1980, Outline of the Geology of Austria: Abh. Geol. B.A., 26e C.G.I., 34, Wien, p. 7-98.

Jiricek, R., 1978, Mikropaleontologie a jeji nove aplikace v geologickem pruzkumu (The micropaleontology and new applications in geological investigations): Zemni plyn a nafta—Rocnik XXIII—Cislo 4 a Seminar o mikropaleontologii, zari 1978, Hodonin.

Jiricek, R., 1985, Deltovy vyvoj spodniho panonu v jizni casti videnske panve (The development of deltas in the lower Pannonian in the southern part of Vienna basin) (Remark: southern part = Czechoslovakian part in the central part of the Vienna basin.): Zemni plyn a nafta—Rosnik XXXI (1985)—Cislo 2.

Kapounek, J., and S. Horvath, 1968, Die Bohrung Schönkirchen Tief 32 als Beispiel für den Aufschluss einer Lagerstätte im tiefen Anteil der Kalkalpen (The well Schönkirchen Tief 32 as an example of the exploration in the deep part of the Calcareous Alps): Erdöl-Erdgas Zeitschrift, 84. Jg, H. 11, Hamburg-Wien, Urban Verlag, p. 396-407.

Kapounek, J., L. Kölbl, and F. Weinberger, 1963, Results of new exploration in the basement of the Vienna basin: Sixth World Petroleum Congress, Section I, Paper 2.

Kapounek, J., A. Kaufmann, H. Kratochvil, and A. Kröll, 1964, Die Erdöllagerstätte Schönkirchen Tief im alpin-karpatischen Beckenuntergrund (The oil reservoir Schönkirchen Tief in the alpine-carpathian basement): Erdölzeitschrift H. 8, Hamburg-Wien, Urban Verlag, p. 305-317.

Kaufmann, A., 1984, Das Feld Matzen—Lagerstättentechnik (The Matzen field—reservoir engineering technique): Erdöl-Erdgas, 100. Jg., H. 5, Hamburg-Wien, Urban Verlag, p. 175-179.

Kölbl, L., 1953, Korrelation der Profile der Erdöllagerstätten des Wiener Beckens. Unveröffentlichter Bericht, Wien (The correlation of the profiles of the oil-reservoirs of the Vienna basin): ÖMV internal geological report.

Kölbl, L., 1957, Sedimentationsformen tortoner Sande im mittleren Teil des inneralpinen Wiener Beckens (Configuration of Badenian sands in the middle part of the Vienna basin): Jahrbuch der Geol. Bundesanstalt, Wien, 100.Bd, H.1.

Kölbl, L., 1959, Art und Verteilung der Sedimentkörper im Torton des Erdölfeldes Matzen (Wiener Becken) (Distribution of Badenian sediments in the Matzen oilfield): Eclogae Geologica Helvetiae, 51, 3, Basel, p. 999-1009.

Köves, S., 1971, Palaöstrukturen und tektonische Phasen im tieferen Neogen des Raumes Matzen-Süd (Paleostructures and tectonic phases in the lower Neogene of the southern Matzen area): Internal ÖMV report of the Geological Department.

Kratochvil, H., and H. W. Ladwein, 1984a, Die Muttergesteine der Kohlenwasserstofflagerstätten im Wiener Becken und ihre Bedeutung für die zukünftige Exploration (The Vienna basin hydrocarbon source rocks and their importance for future exploration): Erdöl-Erdgas, 100. Jg. H. 3, Hamburg-Wien, Urban Verlag, p. 107-115.

Kratochvil, H., and H. W. Ladwein, 1984b, Die Entwicklung der geochemischen Vorstellungen über das Wiener Becken und ihre Bedeutung als Explorationswerkzeug (The development of a geophysical conception for the Vienna basin and its importance as an exploration tool): Erdöl-Erdgas, 100 Jg. H. 10, p. 334-338, Urban Verlag, Hamburg-Wien.

Kreutzer, N., 1971, Mächtigkeitsuntersuchungen im Neogen des Ölfeldes Matzen, Niederösterreich (Distribution of thicknesses in the Neogene of Matzen field, lower Austria): Erdöl-Erdgas, 87. Jg., H. 2, Hamburg-Wien, Urban Verlag, p. 38-49.

Kreutzer, N., 1974, Lithofazielle Gliederung einiger Sand und Schotterkomplexe des Sarmatien und oberstens Badenien im Raume von Matzen und Umgebung (Wiener Becken) (Distribution of some sand and gravel beds of the Sarmatian and uppermost Badenian in the Matzen area, Vienna basin): Erdöl-Erdgas Zeitschrift, 90. Jg, H. 4, Hamburg-Wien, Urban Verlag, p. 114-127.

Kreutzer, N., 1978, Die Geologie der Nulliporen (Lithothamnien)—Horizonte der miozänen Badener Serie des Ölfeldes Matzen (Wiener Becken) (The geology of the Lithothamnia horizons of the Badenian, Miocene, in the Matzen oilfield, Vienna basin): Erdöl-Erdgas Zeitschrift, 94. Jg, H. 4, Hamburg-Wien, Urban Verlag, p. 129-145.

Kreutzer, N., 1985, Die Probleme der Lagengliederung von Öl- und Gashorizonten (Correlation problems in the subdivision of hydrocarbon bearing sand layers): Erdöl-Erdgas, 101. Jg, H. 1, Hamburg-Wien, Urban Verlag, p. 11-14.

Kreutzer, N., 1986, Die Ablagerungssequenzen der miozänen Badener Serie im Feld Matzen und im zentralen Wiener Becken (The depositional sequences of the Miocene Badenian in the Matzen field and in the central part of the Vienna basin): Erdöl, Erdgas, Kohle, 102. Jg, H. 11, Hamburg-Wien, Urban Verlag, p. 492-503.

Kreutzer, N., 1990, The lower Pannonian sand and the Pannonian-Sarmatian boundary in the Matzen area of the Vienna basin, *in* Dagmar Minarikova and Harald Lobitzer, eds., Thirty years of geological cooperation between Austria and Czechoslovakia (Festival Volume): Federal Geological Survey Vienna and Geological Survey Prague, p. 105-111.

Kreutzer, N., and V. Hlavaty, 1990, Sediments of the Miocene (mainly Badenian) in the Matzen area in Austria and in the Southern part of the Vienna basin in Czechoslovakia (Remark: Southern part = CSFR part in the central part of the Vienna basin), *in* Dagmar Minarikova and Harald Lobitzer, eds., Thirty years of geological cooperation between Austria and Czechoslovakia (Festival Volume): Federal Geological Survey Vienna and Geological Survey Prague, p. 112-118.

Kröll, A., 1984, Die Erdöl-Erdgasregion Matzen/Schönkirchen aus geologischer Sicht (The geology of the oil and gas region of Matzen/Schönkirchen): Erdöl-Erdgas, 100. Jg, H. 5, Hamburg-Wien, Urban Verlag, p. 185-195.

Kröll, A., and G. Wessely, 1973, Neue Ergebnisse beim Tiefenaufschluss im Wiener Becken (New results from deep drilling in the Vienna basin): Erdöl-Erdgas Zeitschrift, 89. Jg, H. 11, Hamburg-Wien, Urban Verlag, p. 400-413.

Kröll, A., and H. Wieseneder, 1972, The origin of oil and gas deposits in the Vienna basin (Austria): 24th International Geological Congress, Section 5, p. 153-160.

Ladwein, H. W., 1988, Organic geochemistry of the Vienna basin: model for hydrocarbon generation in overthrust belts: AAPG Bulletin, v. 72, n. 5, p. 586-599.

Ladwein, W., F. Schmidt, P. Seifert, and G. Wessely, 1991, Geodynamics and generation of hydrocarbons in the region of the Vienna basin, Austria, *in* A. Spencer, ed., Generation, accumulation and production of Europe's hydrocarbons: Special Publication of the European Association of Petroleum Geoscientists, n. 1, Oxford, Oxford University Press, p. 289-305.

Middleton, G. V., 1973, Johannes Walther's law of the correlation of facies: Geological Society of America Bulletin, v. 84, p. 979-987.

Mitchum, R. M., Jr., P. R. Vail, and S. Thompson III, 1977, Seismic stratigraphy and global changes of sea level, part 2: The depositional sequence as a basic unit for stratigraphic analysis, *in* Charles E. Payton, ed., Seismic stratigraphy—applications to hydrocarbon exploration: AAPG Memoir 26, p. 53-62.

Müller, A. H., and O. S. Schreiber, 1991, Faziesstudie an Bohrkernen der Bohrungen Prottes 200 and Prottes 201 (11th TH) (Facies study of the cores in the wells Prottes 200 and Prottes 201, 11th Badenian): Internal report of the Geological Laboratory.

Muskat, M., 1949, Physical principles of oil production: McGraw-Hill Book Company, Inc., p. 670-680.

Pogacsas, G., and P. Seifert, 1991, Vergleich der neogenen Meeresspiegelschwankungen im Wiener und Pannonischen Becken (Comparison of Neogene sea level rises and falls in the Vienna and Pannonian basins): Jubiläumsschrift 20 Jahre geologische Zusanmenarbeit Österreich-Ungarn, Teil 1., Geologische Bundesanstalt Wien und MAFI Budapest, Wien, p. 93-100.

Rupp, Cr., 1986, Paläoökologie der Foraminiferen in der Sandschalerzone (Badenien, Miozän) des Wiener Beckens (Paleoecology of Miocene [Middle Badenian] Foraminifera from the Vienna basin): Beiträge zur Paläontologie von Österreich, Institut für Paläontologie der Universität Wien, p. 1-97.

Schröckenfuchs, G., 1975, Hydrogeologie, Geochemie und Hydrodynamik der Formationswasser des Raumes Matzen-Schönkirchen Tief (Hydrogeology, geochemistry and hydrodynamics of the formation waters in the Matzen-Schönkirchen Tief area): Erdöl-Erdgas Zeitschrift, H. 9, Hamburg-Wien, Urban Verlag, p. 299-321.

St. John, B., A. W. Bally, and H. D. Klemme, 1984, Sedimentary provinces of the world, hydrocarbon productive and nonproductive: AAPG Map Series booklet, 35 p.

Strauch, E., 1986, Prospects for onshore petroleum exploration *in* Oil and gas prospects in ireland: Cork.

Van Wagoner, J. C., R. M. Mitchum, K. M. Campion, and V. D. Rahmanian, 1990, Siliciclastic sequence stratigraphy in well logs, cores, and outcrops: Concepts for high-resolution, correlation of time and facies: AAPG Methods in Exploration 7.

Walther, Johannes, 1893-1894, Einleitung in die Geologie als historische Wissenschaft; Beobachtungen über die Bildung der Gesteine und ihrer organischen Einschlüsse. Jena: G. Fischer. 1055 p.

Welte, D. H., H. Kratochvil, J. Rullkötter, H. Ladwein, and R. G. Schaefer, 1982, Organic geochemistry of crude oils from the Vienna basin and an assessment of their origin: Chemical Geology, 35: Amsterdam, Elsevier Scientific Publishing Company, p. 33-68.

Wessely, G., 1984, Der Aufschluss auf kalkalpine und subalpine Tiefenstrukturen im Untergrund des Wiener Becken (The exploration to limestone and subalpine deep structure in the basement of the Vienna basin): Erdöl-Erdgas Zeitschrift, 110. Jg, H. 9, Hamburg-Wien, Urban Verlag.

Wessely, G., 1988, Structure and development of the Vienna basin in Austria *in* L. Royden and F. Horvath, eds., The Pannonian basin, a study in basin evolution: AAPG Memoir 45, p. 333-347.

White, D. A., 1980, Assessing oil and gas plays in facies-cycle wedges: AAPG Bulletin, v. 64, p. 1158-1178.

Wieseneder, H., 1956, Zur Kenntnis der neuen Erdöl- und Erdgas-vorkommen im Wiener Becken (The new oil- and gas-reservoirs in the Vienna basin): Erdöl u. Kohle, 9. Jg.

Wieseneder, H., 1959, Ergebnisse sedimentologischer und sedimentpetrographischer Untersuchungen im Neogen Österreichs (Results of sedimentological and petrographical investigations in the Neogene of Austria): Mitt. Geol. Ges. Wien, 52.Bd.

Wieseneder, H., 1964, Die Erdölmuttergesteinsfrage im Wiener Becken (The problem of the hydrocarbon source rocks in the Vienna basin): Erdöl-Erdgas-Zeitschrift, 80.Jg., H.12.

Wieseneder, H., 1972, New Aspekte zur Geochemie der Erdöl- und Erdgaslagerstätten des Wiener Beckens (New aspects of the geochemistry of the oil and gas reserves of the Vienna basin): Mitt. Geol. Ges. Wien, 65. Bd., p. 159-170.

SUGGESTED READING

Braumüller, E., and A. Kröll, 1980, Erdölgeologie (Petroleum geology): Erdöl-Erdgas Zeitschrift, 96. Jg, H. 5, p. 153-160. A historical review of the oil fields in the Vienna basin.

Friedl, K., 1959, The oil fields of the Vienna basin: Fifth World Petroleum Congress, Section I, Paper 48, p. 865-881. Presents the geologic setting of the oil fields in the Miocene and Flysch zone until 1959.

Kaufmann, A., 1984, Das Feld Matzen-Lagerstättentechnik (The Matzen field-reservoir engineering technique): Erdöl-Erdgas, 100. Jg, H. 5, Hamburg-Wien, Urban Verlag, p. 175-179. A summary of production data.

Kreutzer, N., 1986, Die Ablagerungssequenzen der miozänen Badener Serie im Feld Matzen und im zentralen Wiener Becken (The depositional sequences of the Miocene Badenian in the Matzen field and in the central part of Vienna basin): Erdöl, Erdgas, Kohle, 102.JG, H.11, Hamburg-Wien, Urban Verlag, p. 492-503. Depositional sequences and facies-cycle wedges in the Badenian reservoirs.

Kröll, A., 1984, Die Erdöl- und Erdgas Region Matzen/Schönkirchen aus geologischer Sicht (The geology of the oil and gas region of Matzen/Schönkirchen): Erdöl-Erdgas, 100. Jg, H. 5, Hamburg-Wien, Urban Verlag, p. 185-195. A summary of the geology of the Matzen field.

Ladwein, H. W., 1988, Organic geochemistry of Vienna basin: Model for hydrocarbon generation in overthrust belts: AAPG Bulletin, v. 72, n. 5, p. 586-599. A summary of geochemistry.

Ladwein, H. W., F. Schmidt, P. Seifert, and G. Wessely, 1991, Geodynamics and generation of hydrocarbons in the region of the Vienna basin, Austria, *in* A. Spencer, ed., Generation, accumulation and production of Europe's hydrocarbons: Special Publication of the European Association of Petroleum Geoscientists, n. 1: Oxford, Oxford University Press, p. 289-305.

Schröckenfuchs, G., 1975, Hydrogeologie, Geochemie und Hydrodynamik der Formationswässer des Raumes Matzen/Schönkirchen Tief (Hydrogeology, Geochemistry and Hydrodynamics of the Formation Waters in the Matzen/Schönkirchen Tief Area): Erdöl-Erdgas Zeitschrift, H. 9, Hamburg-Wien, Urban Verlag, p. 299-321. A summary of combined considerations.

Welte, D. H., H. Kratochvil, J. Rullkötter, H. Ladwein, and R. G. Schaefer, 1982, Organic geochemistry of crude oils from the Vienna basin and an assessment of their origin: Chemical Geology, 35, Amsterdam, Elsevier Scientific Publishing Company, p. 33-68. A summary of the new investigations of the crude oils.

Wessely, G., 1988, Structure and development of the Vienna basin in Austria, *in* L. Royden and F. Horvath, eds., The Pannonian basin, a study in basin evolution: AAPG Memoir 45, p. 333-347. A geological summary of the Vienna basin.

Appendix 1. Field Description

Field name .... Matzen field
Ultimate recoverable reserves .... 72.4×10^6 MT (515 million bbl) oil;
17.3×10^9 m^3 (611 bcf) nonassociated gas

Field location:
    Country .... Austria
    State .... Lower Austria
    Basin/Province .... Vienna basin

Field discovery:
    Year first pay discovered .... Multipay zones in the Miocene basin fill,
particularly Matzen sandstone   1949
    Year second pay discovered .... Buried hills of Upper Triassic dolomites of Calcareous Alps   1962
    Year third pay discovered .... Internal dolomites of Calcareous Alps   1968

Discovery well name and general location:
    First pay .... Matzen sand, Matzen 3 well, oil (about 35 km northeast of Vienna)
    Second pay .... Buried hills dolomite, Schönkirchen Tief 2 well,
oil (about 6 km south-southwest of Matzen 3)
    Third pay .... Internal dolomite, Schönkirchen Tief 32 well, gas (near Schönkirchen Tief 2)

Discovery well operator
    First pay .... S.M.V.
    Second pay .... ÖMV
    Third pay .... ÖMV

IP
    First pay .... Matzen sand: Matzen 3 120 MT (850 bbl) oil per day
    Second pay .... Dolomite Schönkirchen T2 150 MT (1066 bbl) oil per day
    Third pay .... Dolomite Schönkirchen T32 1×10^6 m^3 (35 mmcf) gas

All other zones with shows of oil and gas in the field:

Age	Formation	Type of Show
Pannonian	Pannonian	Gas
Sarmatian	Sarmatian	Gas, some oil
Badenian	Badenian	Gas, oil
Karpatian	Aderklaa beds	Gas, oil
	Gänserndorf beds	Gas, oil
Ottnangian or lower Karpatian	Bockfliess beds	Gas, oil
Upper Triassic	Reyersdorf dolomite	Gas, oil

Geologic concept leading to discovery and method or methods used to delineate prospect

1. Geological surface mapping. 2. Shallow and deeper drilling on structures. 3. Geological considerations of the occurrence of Triassic dolomites of the buried nappes of the Calcareous Alps of the basement, assisted by gravity data and seismic reflection methods, led to deep and ultradeep drilling.

Structure:
    Province/basin type .... Bally 321; Klemme III Bc
    Tectonic history

1. During the Jurassic the autochthonous sediments beneath the Vienna basin were deposited on the southern edge of the European plate. 2. During the Alpine orogeny (Late Cretaceous-Paleogene) the allochthonous units of Flysch and Calcareous Alps overrode this passive margin. 3. During the Neogene,

when thrusting ceased in the Alps, the Carpathians were still active. In between, the Vienna basin formed as a pull-apart basin and the Neogene transgressively overlapped Alpine thrust units.

Regional structure

Differential subsidence created depression and elevated zones with numerous faults, some of which are growth faults. In the basin's median structurally high Matzen block, both the Spannberg flysch ridge and a ridge of Calcareous Alps governed the differential subsidence, creating synsedimentary structures. The Aderklaa-Bockfliess growth fault system is another regional structural element.

Local structure

The local structures in the Miocene basin fill have been created by the combined effect of syn- and postsedimentary movements of the basement, accompanied by the activity of three genetically unrelated normal fault systems: the Schönkirchen, Bockfliess, and Matzen faults.

Trap:

Trap type(s)

1. Anticlinal traps, mostly combined with local or general pinch-outs laterally or on structural highs in the Miocene. Buried hills in the Calcareous Alps.
2. Fault traps.
3. Truncation unconformity traps in the Miocene.
4. Traps in steep dipping folds and thrust slices of dolomites in the Calcareous Alps.

Basin stratigraphy (major stratigraphic intervals from surface to deepest penetration in field):

Chronostratigraphy	Formation	Depth to Top in m
Quaternary		Surface
Miocene Pontian	Pontian	Surface or some 10 m
Miocene Pannonian	Pannonian (Pa)	North, 250-400 m
		South, 250 m
		West, 300-600 m
Miocene Sarmatian	Sarmatian (Sa)	North, 700-900 m
		South, 650 m
		West, 700-1200 m
Miocene Badenian	Badenian (Ba)	North, 1000-1300 m
		South, 1000-1100 m
		West, 1050-1400 m
Miocene Karpatian	Aderklaa conglomerate	South, 1750-1900 m
	Aderklaa beds	North, 1600 m
		South, 2050 m
	Gänserndorf beds (Gä)	North, 1700 m
		South, 2100-2800 m
Miocene Ottnangian or lower Karpatian	Bockfliess beds (Bo)	North, 1700 m
		South, 2500-2900 m
Eocene-Cretaceous (Flysch)	Flysch zone	North, 1850-2900 m
Paleocene-Lower Triassic (Calcareous Alps)	Ötscher nappe	South, 2700-3200 m
	Frankenfels-Lunz nappe	South, 2500-4500 m

Reservoir characteristics:

  Number of reservoirs .... About 25 important oil and gas reservoirs
  Formations .... Pa, 2 gas pays; Sa, 3 gas; Ba, 10 oil (gas); Gä, 1 oil; Bo, 4 oil; buried dolomite (dol.) hills, 2 oil and 1 gas; internal dol., 3 gas
  Ages .... Miocene (basin fill), Upper Triassic dol. (Hauptdolomit) of sedimentary basement

Depths to tops of reservoirs
Important reservoirs: Pa, 500-550 m; Sa, 650-900 m; Ba, 1100-1600 m; Gä, 1900 m (N) and 2400 m (S); Bo, 1800 m (N) and 2500 m (S); buried dol. hills, 2700-2800 m; internal dol., Reyersdorf (Rey), 2800 m; Schönkirchen (Sch) Übertief (ÜT) and Gänserndorf Übertief, 4800 m

Gross thickness (top to bottom of producing interval)
Important pay zones (real or apparent): Pa, 57 m (gas); Sa, 82 m (gas); Ba, 225 m (gas); 234 m (oil); Gä+Bo not eval.; Sch Tief (T) dol., 190 m (oil); Rey dol., 400 m (oil); Sch ÜT, 900 m (gas)

Net thickness—total thickness of producing zones

Average

Pa, 12 m (gas); Sa, 12 m (gas); Ba, 34 m (gas), 44 m (oil); Gä, 6 m (oil); Bo, 10 m (gas), 17 m (oil); Sch T, 59 m (oil); Rey, 77 m (gas); Sch ÜT, 300 m (gas)

Maximum

Pa, 27 m; Sa, 52 m: Ba, 91 m (gas), 121 m (oil); Gä, 25 m (oil); Bo, 34 m (gas), 187 m (oil) in the case of common reservoirs with Schönkirchen Tief; Sch T, 190 m (oil)

Lithology
Miocene basin fill: (shaly) sands and sandstones
Miocene and Upper Cretaceous: dolomite debris and conglomerates surrounding buried hills
Upper Triassic: dolomites in the buried hills and internal reservoirs of the Calcareous Alps

Porosity type .... Intergranular in sandstones; fractures in dolomites
Average porosity .... Pa, Sa, Ba, 25%; Gä, Bo, 6-23%; Sch T, 2.9 to 8.4% and 11.8%; Sch ÜT, 3.8%
Average permeability .... Pa, Sa, Ba, 300 md; Gä, Bo, 20 md; Sch T, 1000 md; Sch ÜT, 2 md (Matzen, sand, Ba, 1000 md)

Seals:

Upper

Formation, fault, or other feature ..... Miocene (1) clay or shale layers, (2) unconformities, (3) faults
Lithology .... Calcareous Alps: (1) Miocene shales covering buried hills, (2) Paleocene sandstones and shales and Mesozoic limestones, sandstones, and shales

Lateral

Formation, fault, or other feature .... Miocene: (1) pinch-out of sands, (2) faults
Lithology .... Calcareous Alps: same as upper seals

Source:

Formation and age
The main source rock is a basinal marl of the autochthonous Upper Jurassic underlying the allochthonous Alpine-Carpathian nappes in the Vienna basin

Lithology .... Dark basinal marls
Average total organic carbon (TOC) 1.5-2%
Maximum TOC 5%
Kerogen type (I, II, or III) .... II-III
Vitrinite reflectance (maturation) ..... R_o = between 0.7 and 1.4 to 1.6 in the oil window (4000-6000 m)
Time of hydrocarbon expulsion .... Post-thrusting during Miocene subsidence and burial
Present depth to top of source .... About 6700-9000 m in the Matzen field*
Thickness .... Up to 1000 m
Potential yield .... 2-5 kg/MT

*Drilled between 7641 and 8553 m in Zistersdorf ÜT 2; drilled between 6050 and 6223 m in Aderklaa ÜT 1, outside of the field.

Appendix 2. Production Data

Field name .... Matzen field
Field size:
    Proved acres .... NA
    Number of wells all years .... Total drilled wells, 1340
    Current number of wells .... 944 oil, 60 gas, 140 gas storage, 60 water injection
    Well spacing .... About 220-600 m in the oil reservoirs
    Ultimate remaining recoverable .... 7.4×10^6 MT (53 million bbl) oil;
        17.3×10^9 m^3 (611 bcf) nonassociated gas
    Cumulative production .... 62.4×10^6 MT (444 million bbl) oil and 11.9×10^9 m^3
        (420 bcf) nonassociated gas*

*Matzen sand (or 16th Ba): 34.2 million MT (243 million bbl) oil
Upper Triassic buried hills of dolomite and Cretaceous and Miocene dolomite debris: 7.9 million MT (56 million bbl)
Upper Triassic internal dolomite: 5.6 billion m^3 (200 bcf) gas

    Annual production .... 640×10^3 MT (4.5 million bbl) oil
        and 240×10^6 m^3 (8.5 bcf) nonassociated gas
    Present decline rate 7%
        Initial decline rate .... NA
        Overall decline rate 7%
    Annual water production .... 7.6 million m^3 (47.8 million bbl)
    In place, total reserves .... 185 million MT (1.31×10^9 bbl) oil;
        26.7 billion m^3 (943 bcf) nonassociated gas
    In place, per acre foot .... NA
    Primary recovery .... 54.7 million MT (389 million bbl) oil
    Secondary recovery .... 7.7 million MT (55 million bbl) oil
    Enhanced recovery .... 9th Ba, 9600 MT (68 MM bbl) oil
    Cumulative water production .... 164.6×10^6 m^3 (1035 MM bbl)

Drilling and casing practices:
    Amount of surface casing set .... 150–250 m of 13 ⅜-in. or 9⅝-in. casing
    Casing program

7-in., 6⅝-in., or 5½-in. to T.D.; inside casing gravel packs more frequent than open-hole packs for sand stabilization

    Drilling mud .... Water + bentonite, lignite + lignosulfonate, KCl-polymer
    Bit program .... 17½-in., 12¼-in., 8½-in., 6-in.
    High pressure zones .... Generally none in Matzen field

Completion practices:
    Interval(s) perforated .... Multiple intervals in most reservoirs
    Well treatment .... Stimulation with HCl acid; inside and open-hole gravel packs

Formation evaluation:
    Logging suites .... SP, CE, IEL, ML-MLL, CNL-GR, LL3, Cal., IC, CDM, Temp.,
        (BHC), in normal bentonite muds; in KCl-polymer muds, also DLL, HDT, LDT
    Testing practices .... Drill-stem tests in open holes and casing tests
    Mud logging techniques .... In exploration and stepout wells: cuttings analysis, gas chromatograph; high-pressure prediction by a data unit in ultradeep wells

Oil characteristics:

Type	Ba, naphthenic; Gä, Bo, Sch T, paraffin basic
API gravity	Sp. gr. 0.944 to 0.826 kg/L (28.4° API)
Base	NA
Initial GOR	Matzen sand: 45 Sm^3/m^3 (253 ft^3/bbl); Sch T, 88 Sm^3/m^3 (382 ft^3/bbl)
Sulfur, wt%	Up to 2.1% H$_2$S in all internal dolomites
Viscosity, SUS	At 20°C: Ba, 4-20 cp; Gä, Bo, Sch T, 1 cp
Pour point	NA
Gas-oil distillate	NA

Field characteristics:

Average elevation	150-220 m
Initial pressure	Hydrostatic in most reservoirs; in Bo beds and in internal dol. reservoirs, partly 0.103 bar/m; top Sch ÜT, 0.125 bar/m
Present pressure	Matzen sand, 124 bar
Present gradient	Matzen sand, 0.1 bar/m
Temperature	175°C (at 6000 m)
Geothermal gradient	3.0°C/100m above 2600 m and 4.5°C/100 m below 2600 m
Drive	Water drive dominant in Pa, Sa, Ba, Sch T; depletion dominant in Gä, Bo beds
Oil column thickness	Multiple pay zones
Oil-water contact	Variable: between –780 m and –3146 m
Connate water	Pa, Sa, Ba, 11-49%; Gä, Bo, 16-50%; Sch T, 15-35%; Sch ÜT, 23%
Water salinity, TDS (mg/L)	Pa, 7000; Sa, 11,000 to 23,000; Ba, 16,000 to 28,000; Bo, 9000 to 105,000; Sch T, 105,000
Resistivity of water (ohms at 20°C)	Pa, 0.96, Sa, 0.56 to 0.3; Ba, 0.4 to 0.29; Bo, 0.7 to 0.07; Sch T, 0.07
Bulk volume water (%)	NA

Transportation method and market for oil and gas:

Major oil and gas pipelines, refinery

Cortemaggiore Field—Italy
Po Plain, Northern Apennines

MARCO PIERI
Consulting Geologist
Firenze, Italy

FIELD CLASSIFICATION

BASIN: Po
BASIN TYPE: Foredeep
RESERVOIR ROCK TYPE: Sandstones
RESERVOIR ENVIRONMENT OF
 DEPOSITION: Marine Shelf and Submarine Channel Complex
TRAP DESCRIPTION: Elongate anticline in the hanging wall of a thrust fault

RESERVOIR AGE: Miocene
PETROLEUM TYPE: Oil and Gas
TRAP TYPE: Anticline

LOCATION

The Cortemaggiore field is located in the Emilia Region (Northern Italy) in the Po River alluvial plain, approximately 80 km east-southeast of Milan (Figure 1). Ultimate recoveries are estimated to be 1.2 million m^3 (7.5 million bbl) of oil, 14.8 billion m^3 (523 bcf) of gas, and 26,000 m^3 (160,000 bbl) of condensate; production during the first productive years averaged 2.8 to 3.0 million m^3/day (99 to 110 MMCFG/day) of gas and 190 to 200 m^3/day (1200 to 1300 bbl/day) of oil (Figure 2).

The field is currently used for storage of imported gas. Production of oil virtually came to an end in 1978; one new well (Cortemaggiore 103) was afterward drilled and six old wells were worked over and recompleted. All of them proved to be minor oil producers.

Laboratory experiments for enhanced oil recovery (miscible displacement with carbon dioxide, micellar polymer flooding, and residual oil vaporization into cycled dry gas) have been carried out with the view of possible field testing after primary oil production (Causin et al., 1982).

Many other gas and oil fields are located in the Po Plain, northwest and east-southeast of the Cortemaggiore field. Most of them produce gas from upper Tertiary clastic reservoirs discovered during the 1950s and 1960s. A few discovered since 1973 produce oil and/or gas and condensates from Mesozoic carbonates.

HISTORY

Pre-Discovery

Petroleum exploration in Italy started during the second half of the nineteenth century in the Northern Apennines margin (Figure 1), where gas and oil seeps were known and exploited since the Middle Ages. Many of these seeps are from the "Argille Scagliose" unit, a sequence composed of argillites and calcareous and sandy flysch with a complex and, in some places, a chaotic structure. This unit overlies the Tertiary flysch of the Northern Apennine sequence and is overlain by Oligocene–Miocene and Pliocene clastics. Most nineteenth century Italian geologists assumed that the Argille Scagliose is itself the source of the petroleum seeps, while others favored a deeper origin. Whatever their origin, oil and gas accumulations in the Northern Apennines, as a rule, were small in size and low in productivity, and their discovery was mainly due to chance.

In 1921 C. Porro, an Italian petroleum geologist of wide international experience,* summarized the

---

*Cesare Porro (1865-1940) was one of the most eminent petroleum geologists of his generation. His experience spanned the world from the East Indies to the United States (where he discovered the Salt Creek field in Wyoming). Further references concerning his professional activity may be found in T. S. Harrison (1952) and in E. W. Owen, *Trek of the Oil Finders: A History of Exploration for Petroleum* (AAPG Semicentennial Commemorative Volume, 1975).

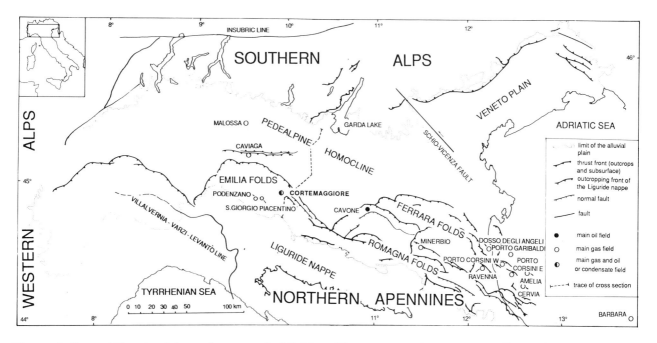

Figure 1. Po and Veneto plain; main gas and oil fields and structural elements (Pliocene and post-Pliocene deformations). (From Pieri and Groppi, 1981, modified.) The cross section corresponding to the trace is shown on Figure 7.

status of exploration in Northern Italy. In his opinion, the Argille Scagliose could be the source of petroleum, but accumulations in economic quantities require a reservoir rock sealed in a trap; reservoirs, seals, and traps are all lacking in the Northern Apennine area, but he thought these elements possibly could be present in the external part of the Apennine chain where tectonism was presumably less intense and where Pliocene clays could provide an effective cap rock. This "external Apennine" could be buried under the alluvial cover of the southern Po Plain where geophysical surveys could be used to identify possible traps.

Encouraged by this geologic model, regional gravimetric surveys were carried out by Agip, Italian State Oil Company, during the years 1927 to 1935 and later on by SPI (a subsidiary of Standard Oil Company of New Jersey). Several tests were drilled on positive anomalies, but results were disappointing.

The sedimentary sequence of the Po Plain subsurface was found to be different in thickness and facies from the sequence exposed in the Apennines; tectonism was found to have been greater than predicted and structural conditions could not be reliably interpreted from the gravimetric survey. Furthermore, gas and oil shows were scanty.

During the 1930s, geologic knowledge of the Northern Apennine chain greatly improved, stratigraphy was revised, and structures were reinterpreted. It was recognized that the Argille Scagliose unit was in fact an allochthonous nappe mainly composed of Cretaceous-Eocene formations over the Tertiary top of the autochthonous Apennine

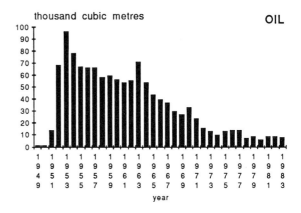

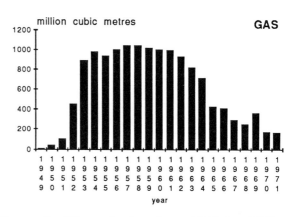

Figure 2. Oil production from 1949 to 1983; gas production from 1949 to 1971. Since 1972 the field has been used for storage of imported gas.

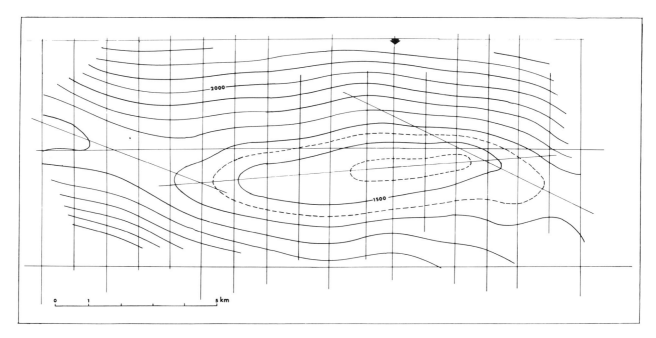

Figure 3. Contour map of the Miocene top of the Cortemaggiore anticline based on reflection seismic data. Contour interval, 100 m. The mapped horizon actually corresponds to the main unconformity of Messinian age. The arrow indicates the position of the seismic profile of Figure 4. This map is dated June 1949, but the seismic survey was carried out in previous years (from Agip's files, redrawn). (The approximate location of the anticline by latitude and longitude may be seen by comparison with Figure 5.)

sequence. This assumption opened new problems, in particular the reciprocal stratigraphic and structural relationships between the Tertiary sequences below and above the nappe and the feasibility of extrapolating the Apennine geology to the subsurface of the plain.

Meanwhile, exploration in the plain was at a standstill that lasted until Agip employed Western Geophysical Company to conduct the first reflection seismic survey in Italy, beginning in June 1940. Operations were hampered by wartime restrictions; nevertheless, the first wildcat of the new generation (S. Giorgio Piacentino 2) was spudded in 1942 with the objective of testing the upper Tertiary section on the anticline of Podenzano-S. Giorgio Piacentino (Figure 1). Surface petroleum seepages were known in the area, and previous wells had encountered promising gas and oil shows in basal Pliocene sands at 530 m depth. Small accumulations of wet gas and light oil were discovered in Pliocene sands and in the eroded underlying middle Miocene shaly and sandy rock section.

A second test was spudded in 1943, 35 km northnorthwest of Podenzano, on the structurally lower and better defined anticline of Caviaga (Figure 1), and gas was found in upper Miocene sands sealed by lower Pliocene clays at a depth of 1300–1400 m. Original reserves were estimated at 14 billion m³ (490 billion ft³). This discovery was the starting point of a new exploratory cycle aimed at testing the Miocene-Pliocene clastic section in several similar anticlinal traps that seismic surveys were gradually revealing in the central and eastern Po Plain.

Discovery

Agip resumed the seismic surveys after the end of World War II, and at the end of 1947, the Cortemaggiore prospect was adequately defined (Figures 3 and 4). A report by C. Contini and E. Di Napoli (respectively the chief geophysicist and chief geologist of Agip) summarized the structure and presumed stratigraphy of the anticline: "... the structure essentially consists of an intensely folded basal nucleus ... and an overlying less deformed complex. By analogy with the nearby S. Giorgio and Podenzano structures ... the nucleus could be Langhian, possibly Tortonian and Messinian. Strata above the erosional unconformity should be Pliocene or younger." The trend of the main axis of the fold was east-west, with a steeper north flank and an estimated closure area of about 6 by 3 km.

Three possible targets were defined: (1) basal Pliocene clastics, (2) Tortonian and Messinian sandstones, and (3) possible coarse clastics of lower Langhian age. The first two objectives could be reached at a depth of about 2000 m, while Langhian horizons were presumably 2500 to 3000 m deep. Two wells were spudded almost contemporaneously in March 1948: Cortemaggiore 1, with the main objective of testing the Miocene sequence, and Cortemaggiore 2, for the exploration of the section above the unconformity (Figure 5). In January 1949, Cortemaggiore 2 was the discovery well for gas sands penetrated at depths between 1464 and 1471 m (pool A) and between 1526 and 1541 m (pool C, IP 80,000 m³/day) [2.8 MMcfd] gas); Cortemaggiore 1 was the

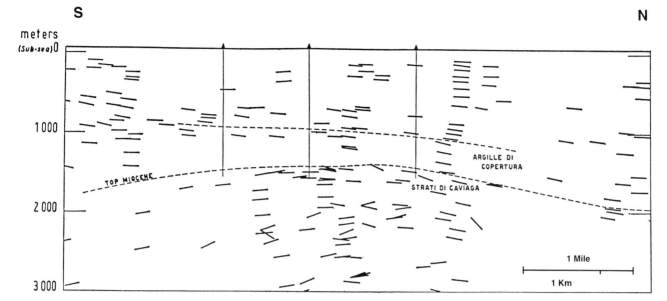

Figure 4. Seismic profile ("dip shooting," circa 1947) across the Cortemaggiore anticline (trace on Figure 3). "Argille di Copertura" and "Strati di Caviaga" are informal lithostratigraphic units corresponding to Santerno and Cortemaggiore sandstones, respectively. (From Facca, 1951.)

discovery well for a deeper gas and oil pool at depths between 1585 and 1610 m (pool D, IP 50,000 m³/day [1.8 MMCFGD] gas and 8.55 m³/day [53.8 bbl] oil). Productive levels, initially attributed to the basal Pliocene, are now dated upper Miocene (pools A and C) and Tortonian (pool D). Drilling of Cortemaggiore 1 had to be interrupted at a depth of 1950 m, and exploration for deeper Miocene pools was not resumed until 1950–1952. In February 1952, Cortemaggiore 29 drilled Tortonian oil sandstones at depths between 1925 and 1950 m, discovering the two deepest pools of the field (pools E and F; pool F, IP 28 m³/day [176 bbl] of oil).

Post-Discovery

A total of 85 exploratory and development wells were drilled in the field (Figure 5), 65 of them oil and/or gas producers. In 1964, pool C began to be used for storage of imported gas, and 15 more wells were drilled.

DISCOVERY METHOD

The geologic history of the Po Plain area, as interpreted at the time of the discovery of Cortemaggiore field, may be outlined as follows (Facca, 1951; Jaboli, 1951; Pieri, 1984).

At the end of the Miocene, a strongly compressive tectonic phase resulted in the Apenninic orogeny; the large Miocene sea was reduced on the external, northwestern side of the Apennines to the area of the present Po Plain. The crests of the anticlines partially emerged and were eroded. The basin was filled with brackish and fresh-water sediments during upper Miocene time. A lower Pliocene marine transgression following widespread regional subsidence resulted in deposition of a predominantly shaly section, with basal coarse clastics and sandy intercalations. A few of the highest Miocene anticlines were still emergent at the end of Pliocene and were later covered by the Quaternary sea. In this scheme, compressive tectonism was not active during the Pliocene. Folding and faulting of the Pliocene section were interpreted to be mainly the result of draping accompanied by extension.

In later years, cumulative evidence for Pliocene diastrophism led geologists to accept the idea of two main Pliocene folding phases of different intensities in the various sectors of the basin. However, persisting prejudice against the hypothesis of Pliocene compressive tectonism is evident in the geologic cross sections drawn in 1955 (Rocco) (Figure 6), 1958 (Rocco and Jaboli), and 1959 (Agip); the concept of reverse faulting was confined to the Miocene part of the stratigraphic section, whereas many of the major faults bordering the northern flanks of the Pliocene anticlines were still interpreted to be normal and extensional. Moreover, poor seismic data in critical zones did not help interpretation.

Nonetheless, the geologic model existing at the end of the 1940s and the quality of seismic surveys were sufficient to locate the Cortemaggiore anticline trap, as well as others of the same type. However, progressively improved models of the general structural and stratigraphic evolution of the Pliocene foredeep basin adopted during the past 20 years and the high-quality modern seismic methods greatly improved the ability to predict and locate Tertiary clastic reservoirs.

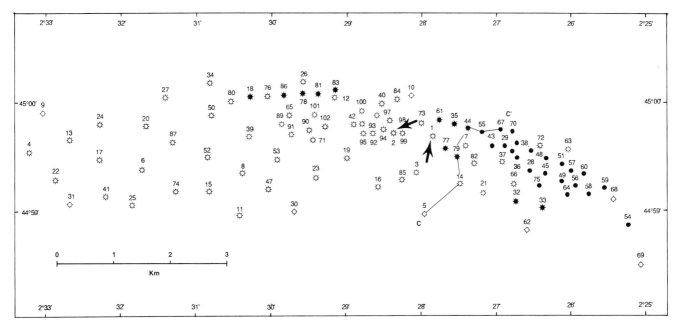

Figure 5. Well locations in the Cortemaggiore field. The two discovery wells, Cortemaggiore 1 and 2, are arrowed. The trace C–C' shows the location of the electric log cross section shown on Figure 18.

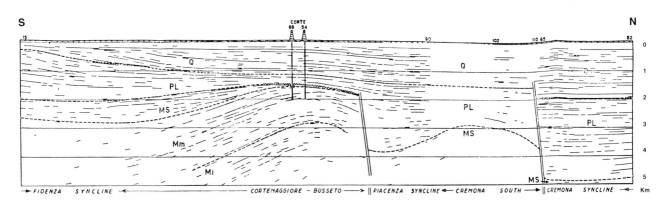

Figure 6. South to north seismic profile (circa 1952) across the Cortemaggiore anticline. The main fault on the northern flank of the anticline is interpreted to be the result of extensive movement. (From Rocco, 1955.) Q, Quaternary; PL, Pliocene; MS, Messinian; Mm, middle Miocene; Ml, lower Miocene. Horizontal scale = vertical scale.

STRUCTURE

Tectonic History

Tectonic history of the Northern Apennines, including its external elements, buried at present below the Po Plain post-tectonic sediments, may be divided into two main time intervals.

Upper Permian to Middle Oligocene

The area was part of the northwestern continental margin of the Adria (Apulian) plate and was subjected to extensional stresses that predated, accompanied, and followed rifting and oceanic spreading and that separated the European and Adria continental margins. With the possible exception of a few localized, stable highs, sedimentation was uninterrupted, starting with continental clastics and sabkha evaporites, which were followed during earliest Lias by carbonates deposited on a widespread shallow water shelf and which became dissected by a system of extensional, possibly listric faults. Beyond this shelf were deep-water carbonate basins. During the early Lias, the sea progressively encroached onto the shelf margins and continued to advance in Jurassic–Cretaceous time.

The oceanic crust that originated during Upper Jurassic–Lower Cretaceous spreading was progressively subducted under the European continental margin in Upper Cretaceous, Paleocene, and Eocene.

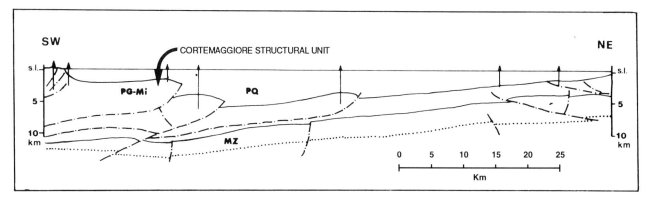

Figure 7. Regional cross section in the western Po Plain (trace on Figure 1), based on seismic, gravimetric, and magnetometric data. The Cortemaggiore structural unit (see arrow) consists of the Miocene clastic section, detached and overthrust onto lower structural units and the southwest-dipping foreland. In the eastern part of the section, the external structural units of the South Alpine chain are covered by the Pliocene foredeep sediments. PQ, Pliocene–Quaternary; PG-Mi, Paleogene–Miocene; MZ, Mesozoic; dotted line, top of magnetic basement. (From Cassano et al., 1986, redrawn and simplified.) No vertical exaggeration.

At the end of the Oligocene, the European and Adria continental margins collided, and shortening and/or subduction of the Adria continental margin caused the detachment of its sedimentary cover and the birth of the fold-and-thrust Northern Apenninic chain.

Upper Oligocene-Miocene to Quaternary

Progressive subduction of the Adria continental margin created a foredeep basin bordering the external (northern) side of the Apenninic chain; the deformation of the clastic sediments deposited in the foredeep caused the addition and growth of external tectonic elements to the Apenninic chain.

Relations between tectonism and sedimentation are complex. Orogenesis along the internal margin of the foredeep caused uplift and erosion and provided one of the sources for clastic material (the other main source being the Alpine chain along the external margin of the foredeep). Incipient deformation of the internal foredeep modified the depositional surface and eventually controlled distribution, thickness, and facies of deposits. Porosity distributions and trends follow the paths of turbidity currents flowing longitudinally along the axis of embryonic synclines, while on the tops of structural highs, shaly sedimentation generally prevails. The foredeep basin may therefore be subdivided into several subbasins corresponding to structurally lower areas. The most internal of these subbasins may actually be superimposed on moving thrust sheets or on active allochthonous nappes of internal origin, and consequently have been named "piggyback basins" (Ori and Friend, 1984).

Regional Structure

The Cortemaggiore anticline is located in the Emilia folds arc, one of the main structural units forming the external Apennines (Figure 1). It is worth noting that, according to interpretations of magnetometric and gravimetric data, the Emilia folds consist of the Tertiary clastic section, detached from the Mesozoic carbonates. The Mesozoic rocks and the underlying basement dip regionally southward below the chain (Figure 7) (Cassano et al., 1986). The wells drilled in the Emilia arc area did not reach the pre-Miocene section, and the depth and nature of the detachment level are unknown. It is even possible that the Miocene clastics of the Emilia folds were actually deposited on the allochthonous Liguride nappe and carried to their present position during the Pliocene tectonic phases (Pieri and Groppi, 1981).

Reinterpretation of recent seismic surveys demonstrates that the Cortemaggiore anticline is actually thrust over a lower structural unit (Figure 8); the thrust plane, 4000 to 5000 m deep, has not been reached by wells, and the age and nature of the lower unit is therefore unknown.

Local Structure

The Cortemaggiore anticline originated from the detachment of the Marnoso-arenacea section from its incompetent basal levels. Deformation started during upper Miocene time ("intra-Messinian phase"), when the upper Messinian Colombacci shales were deposited with onlap over the lower Messinian Verghereto shales (Figure 9). The main set of compressive faults on the northern flank of the fold are overthrust, and structural relief controlled the sedimentation of storm layers (Cortemaggiore sands) over the top of the anticline.

Episodes of submarine erosion occurred during Pliocene and Pleistocene times when compression was still active. In the crestal area, lower Pliocene Santerno clays directly overlie the eroded Miocene section.

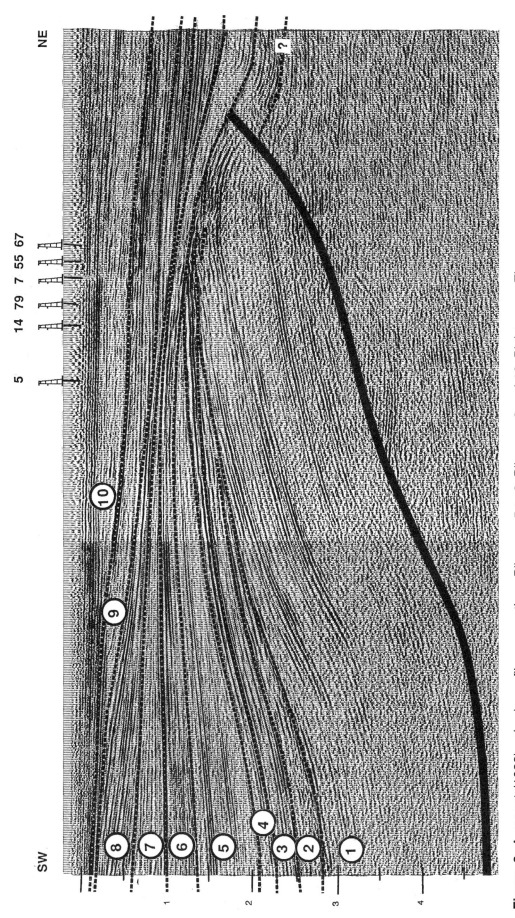

Figure 8. A recent (1983) seismic profile across the Cortemaggiore anticline (Agip). The dashed lines correspond to sequence boundaries: 1, Serravallian–early Messinian; 2, late Messinian (lower part); 3, late Messinian (upper part); 4, late Messinian–early Pliocene; 5 to 8, Pliocene; 9 and 10, Pleistocene. The seismic profile passes near the line of the log cross section C–C' shown by Figure 18, continuing to the northeast and southwest. Horizontal scale approximately 1:50,000.

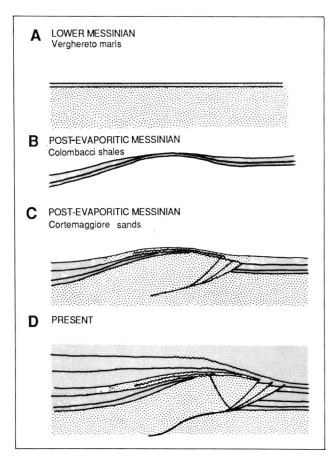

Figure 9. Structural evolution schematic of the Cortemaggiore anticline. (From Madeddu, 1988.) No scale.

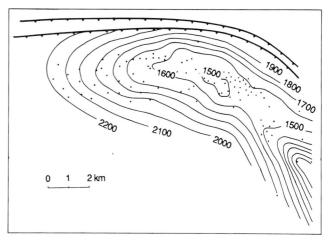

Figure 10. Structure at the top of the Marnoso-arenacea Formation. Contours in meters; datum, sea level. (From Madeddu, 1988.)

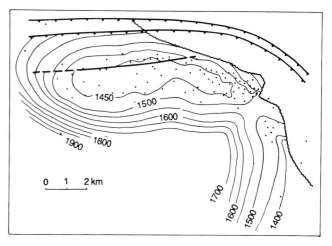

Figure 11. Structure at the top of the Cortemaggiore sandstones. Contours in meters; datum, sea level. (From Madeddu, 1988.)

The structural configuration shown by contours therefore differ when drawn at different stratigraphic levels (Figures 10 and 11), and isopach maps of the Cortemaggiore sandstones (Figure 12), where the main gas reservoirs are situated, show thicknesses controlled mainly by post-Miocene erosion.

STRATIGRAPHY

The stratigraphic log of the Cortemaggiore 6 well (Figure 13) is typical of the area. The sedimentary sequence includes the following formations (Figure 14).

Marnoso-Arenacea (Serravallian-Tortonian)

This formation consists of marls and sandy marls with sandstone, conglomerate, and pebbly mudstone intercalations. The formation has been drilled to 1783 m (Cortemaggiore 29 well), but its total thickness probably reaches 3 to 4 km. The type area of the Marnoso-arenacea is the Romagna Apennines, 150 km east-southeast from Cortemaggiore; here the formation, a classic turbiditic flysch, has extensive outcrops, and its stratigraphic and sedimentological features have been thoroughly studied (Ricci Lucchi, 1975a,b, 1987). However, correlation of the middle-upper Miocene clastic section in the western Po Plain subsurface with the type section of the Marnoso-arenacea Formation is still a matter of discussion as far as tectonic, environmental, and paleogeographic conditions are concerned.

Verghereto Marls (Lower Messinian)

This formation is made up of marls, with siltstone and fine sandstone intercalations. Thickness varies from 127 m in the western part of the field to none in the eastern part. The formation represents condensed basinal deposition in conditions of increasing salinity that in other areas eventually led to evaporitic sediments.

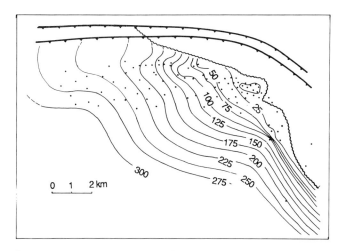

Figure 12. Isopach of the Cortemaggiore sandstones. Contours in meters. (From Madeddu, 1988.)

Colombacci Shales (Upper Messinian)

The Colombacci shales are characterized by marls and silty marls; thickness varies from 118 m in the western part of the field to none in the eastern part. In the lower parts of the basin, these deposits may be replaced by olistostromes.

Cortemaggiore Sandstones (Upper Messinian)

Figure 15 shows the location of the Cortemaggiore sandstones. This formation consists of sandstones and silty sandstones with shale intercalations. Thickness varies from 265 m on the anticlinal flank to none on the crest where the sands have been eroded. The formation was deposited in a fresh and relatively shallow water environment; the sandy levels are interpreted to be storm layers and distal turbidite deposits (Figure 16) (Madeddu, 1988).

Santerno Shales (Lower, Middle, and Late Pliocene)

This formation is made up of shales and silty shales, with thin, sandy intercalations. The average thickness is 250 to 450 m. The environment is open and relatively deep marine, following the Messinian "Mediterranean salinity crisis."

Asti Sandstones (Pleistocene)

These sandstones and shaly sandstones with shale intercalations average 500 to 800 m thick. Regressive sandstones were deposited in a progressively shallower marine environment.

Alluvium (Holocene)

The alluvium is characterized by gravels, conglomerates, sandstones, and shaly sandstones; average thickness is 300 m.

TRAP

Trap Type

Petroleum in the Cortemaggiore field is accumulated in seven pools (or groups of pools) in sandstone reservoirs (Figures 17 and 18):

Pool A

Pool A is an anticlinal trap. Lower Pliocene shales provide the vertical and lateral seal; lateral sealing is also due to the shale-out and to onlap termination of sandstone levels over the intra-Messinian unconformity on the eastern side of the pool. The original GWC was −1476 m.

Pool C

Pool C is an anticlinal trap. Overlying upper Miocene shales, 25 to 35 m thick, provide the seal; in the eastern part of the field the lateral seal is also due to the onlap of the sandstone levels on the intra-Messinian unconformity. The original GWC was −1545 to −1555 m.

Pool C_1

Pool C_1 is a stratigraphic trap with sandstone lenses sealed by upper Miocene shales. The GWC was −1550 m.

Pool D

Pool D is a stratigraphic trap. Truncated levels sealed by upper Miocene shales deposited on the main intra-Messinian unconformity. The OGC was −1533 m; and the OWC was −1547 m.

Pool $D_1 + D_2$

This pool group is a stratigraphic trap and includes several sandy levels below the D pool with the same trap characteristics.

Pool E

Pool E is an anticlinal trap, sealed by Tortonian shales 20 to 30 m thick. The OWC was −1895 m.

Pool $F_1 + F_2$

This pool group consists of anticlinal traps, sealed by Tortonian shales 20 to 30 m thick. The OWC was −1910 m.

The traps were mainly formed during the upper Miocene–lower Pliocene time interval; however, deformation leading to the present structural setting continued through Pliocene and Pleistocene.

Reservoirs

The seven Cortemaggiore reservoirs are discussed below. See Figure 19 and Tables 1–4 for further information.

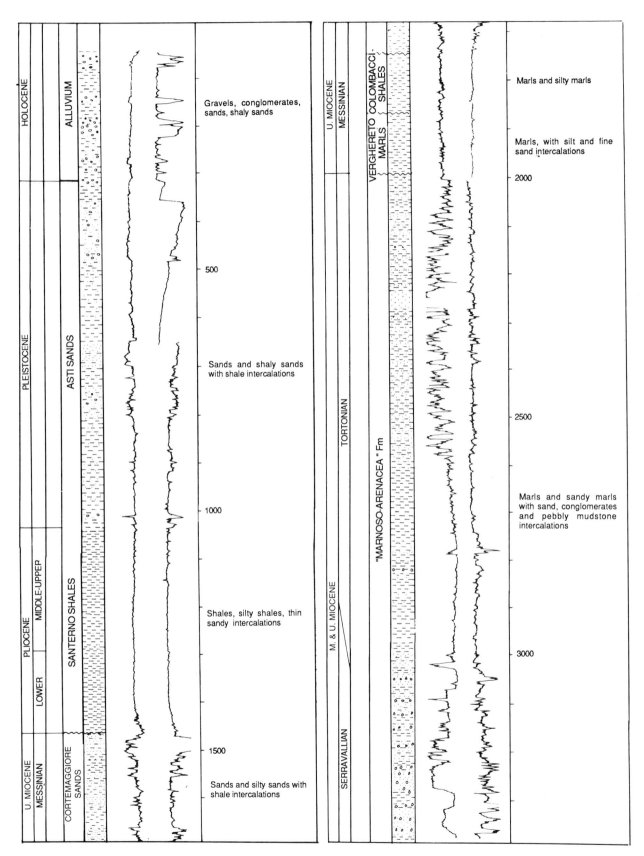

Figure 13. Stratigraphic log of Cortemaggiore 6 well. (Unquantified log curves show characteristic log signatures.)

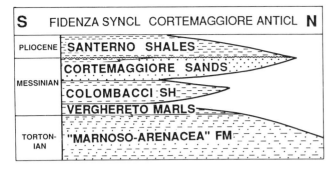

Figure 14. Stratigraphic diagram of the Cortemaggiore area. (From Madeddu, 1988.)

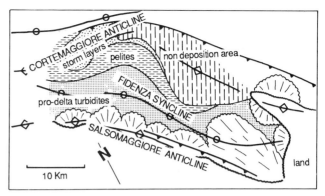

Figure 16. Paleogeographic sketch of the Cortemaggiore-Salsomaggiore area during post-evaporitic upper Messinian time; position of marginal fan deltas and facies distribution are shown (cf. Figure 6). (From Madeddu, 1988.)

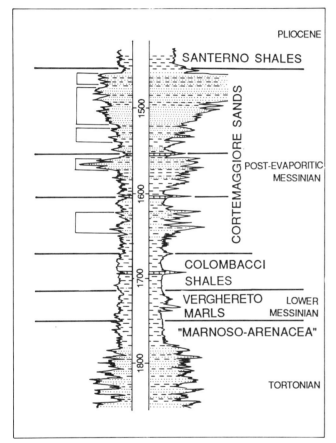

Figure 15. Typical log of the upper Miocene section. (From Madeddu, 1988.)

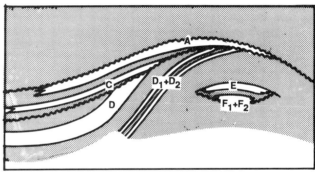

Figure 17. Pool and traps schematic diagram. Letters are pool designations described in the text.

Pool A

Pool A is located in upper Messinian clean, well-sorted sandstones grading downward to shaly sandstones alternating with thin, sandy shales. There are turbiditic distal deposits and storm layers in the upper part of the deposits. The pool area is 17.85 km². Maximum gross pay is 60 m. Average temperature is 46°C. Initial formation pressure was 170 kg/cm². Wet gas to gas condensate ratio was 31.37 cm³/Nm³. The original condensate content gradually decreased with pressure reduction and retrograde condensation. In the eastern part of the field the shaly sandstones in the lowest part of the pool contain a small quantity of oil. The pool produced by gas expansion with subordinate water drive.

Pool C

Pool C is located in upper Messinian sandstones, deposited as storm layers on relatively high areas. Pool area is 7.9 km². Maximum gross pay is 15.5 m. Pool C produced wet gas. In the northern part of the pool a few meters of oil pay underlie the gas-bearing sandstones. The pool produced by gas expansion and water drive. It is used at present for imported gas storage.

Pool C_1

Pool C_1 is located in upper Messinian lenticular sandstones. The pool area is 0.3 km²; average thickness of net pay is 1.5 m. This pool produced wet gas.

Pool D

Pool D is located in Tortonian sandstones on the southwestern end of the anticline, with an area of about 1 km²; average thickness is 28 m. Oil was produced by gas cap expansion and water drive.

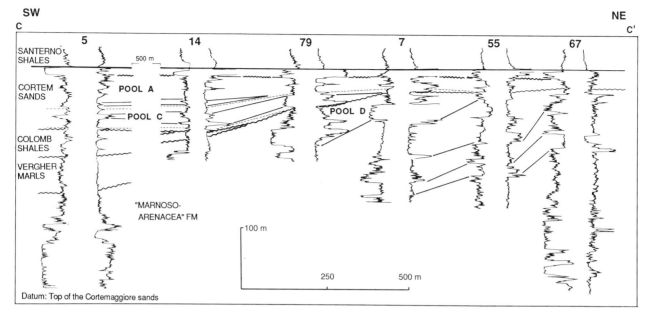

Figure 18. Log cross section (cf. Figure 8). (From Madeddu, 1988.) (Unquantified log curves for correlation only.) See Figure 5 for location of the section.

Pool $D_1 + D_2$

This group includes a number of shaly sandstone layers of Tortonian age at a depth of about 1700 m. The pool area is 0.03 km². Average net pay is 2 to 3 m; production of oil has been small.

Pool E

Pool E is located in fine-grained silty Tortonian sandstones in the eastern part of the field. Average thickness is 5 m. The pool produced oil by gas expansion and water drive.

Pool $F_1 + F_2$

This group is located in Tortonian conglomerates and sandstones in the eastern part of the field. The pool area is 0.96 km² with an average thickness of 31 m. The quantity of oil produced from the pool by gas expansion and water drive indicated an average 25% recovery factor.

According to a recent sedimentological interpretation (Rossi, 1988), the clastic reservoirs of the deeper pools of the field (E, $F_1 + F_2$) represent a complex channel body deposited on a submarine erosional surface. The channel complex, which has a general northwest-southeast trend, is over 2 km wide and was cut into the highstand pelites of the Marnoso-arenacea Formation during a relative fall of sea level.

Cumulative production data for the pools of the field for the years 1950–1956 are shown on Figure 20.

Faults

The main thrust plane underlying the Cortemaggiore anticline is accompanied by a set of reverse faults that are very evident on its northern limb (Figures 7 and 9). These faults, however, do not contribute to the trap closure. Faults affecting the crest of the anticline (Figure 11) probably had a role in the migration of petroleum from deeper levels in the Cortemaggiore sandstone reservoirs.

Source

On the basis of chemical and physical characteristics, the oils of Po Plain fields may be subdivided into different genetic groups and correlated with specific source rocks (Figures 21 and 22) (Riva et al., 1986). The "Cortemaggiore group" (including the oils of the Cortemaggiore pools and of nine other oils from wells of the southwestern Po Plain and northwestern Apennines) is characterized by a specific set of geochemical parameters. The presence of 18 α (H) oleanane suggests an origin from Tertiary sources. The most likely candidate for source rock is the Miocene clastic sequence (Marnoso-arenacea Formations), which has been drilled by several wells in the southwestern Po Plain and which forms the core of the Cortemaggiore anticline.

The Marnoso-arenacea Formation is immature in the Cortemaggiore wells, contrasting with the conditions necessary for the maturation of the oils and the thermogenic requirements for the generation of the gas accumulated in the pools (Mattavelli et al., 1983). Therefore, the origin of the oil and thermogenic gas should be sought at depth in the structurally lowest areas and in the structural units underlying the overthrust Cortemaggiore anticline where the "thermal blanket effect" of the emplacement of the thrust sheets caused an increase in temperature (Figure 23) (Mattavelli, 1987).

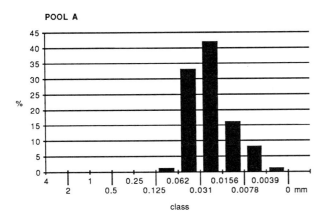
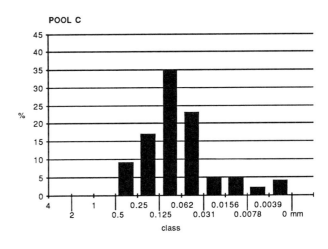
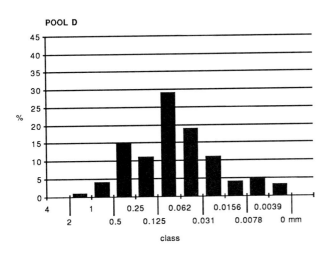
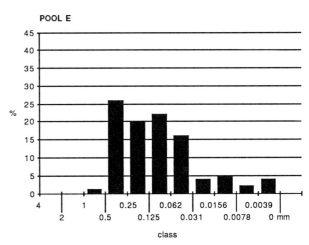
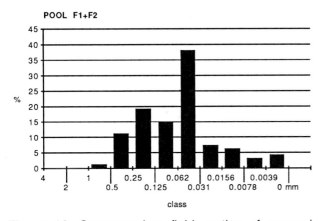

Figure 19. Cortemaggiore field: sorting of reservoir sands grain sizes. (Data from Agip, 1959.)

EXPLORATION CONCEPTS

The Cortemaggiore field produces from two groups of reservoirs: the Messinian Cortemaggiore sandstones and the Tortonian Marnoso-arenacea Formation. Wet gas and oil of both groups were sourced by the organic matter in the Tortonian and pre-Tortonian section at depths in the range of 5 to 6 km. This set of geological circumstances differentiates the Cortemaggiore fields from other Po Plain fields, where gas in Tertiary clastic reservoirs is predominantly of biogenic origin and oil in Mesozoic carbonate reservoirs is mostly derived from Mesozoic source rocks.

The reasons for this uniqueness are not completely clear. It seems likely that in the wide Miocene foredeep basin, where thick turbiditic sediments accumulated in a relatively short time, deposition

Table 1. Porosity and permeability of the Miocene sandstones of the Cortemaggiore field.

Pool	Porosity (%)			Permeability (md)		
	min	max	avg	min	max	avg
A	26	39	33	24	1100	300
C	27	39	30.2	8	150	50
C_1			5-6		low	
D	32	35	33.5	125	1130	500
E, $F_1 + F_2$	20	37	29	26	1500	300

Data from Agip, 1959.

Table 2. Analysis of the gas from the two stratigraphically highest pools.

Pool	C_1 (%)	C_2 (%)	C_3 (%)	C_4 (%)	C_5^+ (%)	N_2 (%)	S (gr/Nm³)
A	92.57	4.82	1.30	0.59	0.30	0.42	0.0063
C	91.69	4.97	1.34	0.94	0.56	0.50	0.0063

Data from Agip, 1959.

Table 3. Physical properties of the condensate produced with gas from upper pools.

Pool	Density (at 15°C)	API (°)	S (%)
A	0.776	50.7	0.034
C	0.772	51.8	0.03

Data from Agip, 1959.

Table 4. Properties of oil produced from the stratigraphically deeper pools.

Pool	Density (at 15°C)	Viscosity (at 20°C) cP
D	0.826	1.86
$D_1 + D_2$, E, $F_1 + F_2$	0.852	2.80

Data from Agip, 1959.

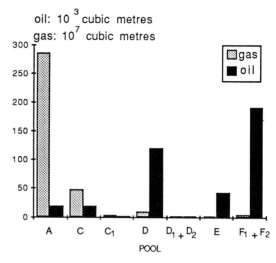

Figure 20. Cumulative oil and gas production from the pools of the field from 1950 to 1956.

and preservation of organic matter were not uniform. Frequency of oil and gas shows may indicate that the area now corresponding to the Emilia folds has been privileged both in regard to organic content ratio and to burial conditions favorable to its maturation.

A second determinant factor is porosity distribution. Messinian and basal Pliocene coarse clastics deposited in shallow-marine environment are related to unconformities consequent to sea level changes during the Messinian "salinity crisis" and the subsequent Pliocene marine ingression, but are also controlled by local structural setting and synsedimentary tectonism. On the other hand, Miocene pre-Messinian reservoirs appear to be related to depositional mechanisms and to patterns of turbiditic deposition in a deep water basin plain.

Further development of the Cortemaggiore plays therefore seems to depend mainly on the possibility of applying the geochemical and sedimentological data and the concepts that have been recently acquired.

ACKNOWLEDGMENTS

Agip S.p.A. has permitted publication of the paper. The author is particularly indebted to G. Paulucci and L. Mattavelli, who allowed free access to Agip's files and the use of the company facilities; to M. Tonna, M. Rossi, and G. Dalla Casa for their help with data and discussion; a special thanks to L. Madeddu, whose thesis paper on the Cortemaggiore field is the actual source of a large part of the data and illustrations.

REFERENCES

Agip, 1959, Descrizione dei giacimenti gassiferi padani (Description of the Po Plain gas fields), in Atti del Convegno sui Giacimenti Gassiferi dell'Europa Occidentale, v. 2, p. 131-497.
Cassano, E., L. Anelli, R. Fichera, and V. Cappelli, 1986, Pianura Padana: interpretazione integrata di dati geofisici e geologici (Po Plain: integrated interpretation of geological and geophysical data): 73° Congresso Società Geologica Italiana, Roma.
Causin, E., G. L. Chierici, M. Erba, G. Mirabelli, and C. Turriani, 1982, Study of EOR processes, Cortemaggiore oil field, Italy: Hungarian National Society for Mining and Metallurgy, 18th Congress, Siófok, Hungary, 30 p.
Contini, C., and E. Di Napoli, 1947, Breve relazione sul programma di esplorazione della struttura di Cortemaggiore (Short report on the exploration program of the Cortemaggiore structure): Agip company files.

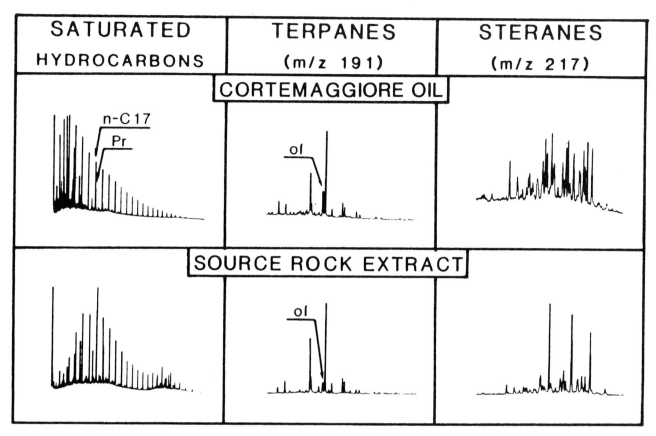

Figure 21. C_{15}^+, GC, and GC/MS traces of saturated hydrocarbons used for oil source-rock correlation between the Cortemaggiore group oil and the Marnoso-arenacea, a Tertiary source rock. Oleanane (ol) is indicated by arrow. (From Riva et al., 1986.)

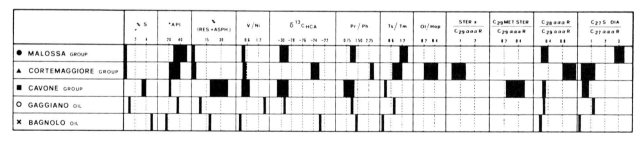

Figure 22. Summary of the bulk and analytical parameters characterizing the different groups of oils in the Po Plain. (From Riva et al., 1986.)

Facca, G., 1951, Three gas-bearing fields in the Po Valley (North Italy): 3rd World Petroleum Congress Proceedings, Section 1, The Hague, p. 256-265.

Harrison, T. S., 1952, Cesare Porro (Memorial): AAPG Bulletin, v. 36, p. 1681-1686.

Jaboli, D., 1951, Le gaz et le pétrole dans la plaine du Po (Gas and oil in the Po Plain): 3rd World Petroleum Congress Proceedings, Section 1, The Hague, p. 220-235.

Madeddu, L., 1988, Thesis, Department of Earth Sciences, University of Milan, Italy.

Mattavelli, L., 1987, Influence of Neogene geological events on the origin, accumulation and distribution of oil and gas in the Po basin: Ann Inst. Geol. Publ. Hung. (Proceedings of the VIII RCMNS Congress), Budapest, v. 70, p. 571-580.

Mattavelli, L., T. Ricchiuto, D. Grignani, and M. Schoell, 1983, Geochemistry and habitat of natural gases in Po basin, northern Italy: AAPG Bulletin, v. 67, n. 12, p. 2239-2254.

Ori, G. G., and P. F. Friend, 1984, Sedimentary basins formed and carried piggyback on active thrust sheets: Geology, v. 12, p. 475-478.

Pieri, M., 1983, Three seismic profiles through the Po Plain, *in* Seismic expression of structural styles: AAPG Studies in Geology 15, v. 3, p. 3.4.1-8 to 3.4.1-26.

Pieri, M., 1984, Storia delle ricerche nel sottosuolo padano fino alle ricostruzioni attuali (A history of exploration of the Po Plain subsurface up to the present model): Cento Anni di Geologia Italiana, Volume giubilare della Società Geologica Italiana, Roma, p. 155-176.

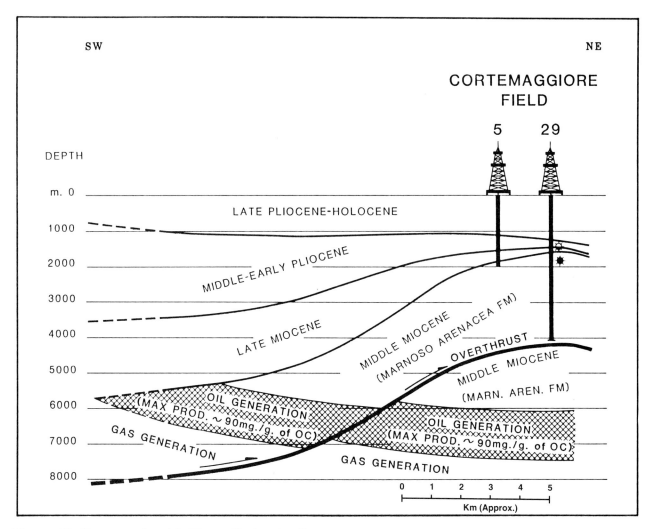

Figure 23. Cross section (cf. Figure 8) showing the oil window interval and the amounts of generated petroleum. (From Mattavelli, 1987.)

Pieri, M., and G. Groppi, 1981, Subsurface geological structure of the Po Plain, Italy: Progetto Finalizzato Geodinamica, sottoprogetto Modello Strutturale, Consiglio Nazionale delle Ricerche, Pubbl. n. 414, Roma, 13 p.

Porro, C., 1921, In tema di recerche petrolifere in Italia (On the exploration for petroleum in Italy): La Miniera Italiana, v. 12, Roma, p. 145-148.

Ricci Lucchi, F., 1975a, Miocene paleogeography and basin analysis in the Periadriatic Apennines, in C. H. Squyres, ed., Geology of Italy: Earth Sciences Society of the Libyan Arab Republic, 15th annual field conference, p. 129-236.

Ricci Lucchi, F., 1975b Depositional cycles in two turbidite formations of Northern Apennines: Journal of Sedimentary Petrology, v. 45, p. 3-43.

Ricci Lucchi, F., 1987, The foreland basin system of the Northern Apennines and related clastic wedges: a preliminary outline: Giornale di Geologia, v. 48, p. 165-185.

Riva, A., T. Salvatori, R. Cavaliere, T. Ricchiuto, and L. Novelli, 1986, Origin of oils in Po basin, northern Italy, in D. Leythaeuser et al., eds., Advances in Organic Geochemistry, 1985, pt. I. Petroleum Geochemistry, v. 10, p. 391-400.

Rocco, T., 1955, Comparative geological and geophysical study of the Po basin: 4th International Petroleum Congress, Rome, v. 1, p. 675-690.

Rocco, T., and D. Jaboli, 1958, Geology and hydrocarbons in the Po basin, in Lewis G. Weeks, ed., Habitat of oil: AAPG, p. 1153-1167.

Rossi, M., 1988, Campo di Cortemaggiore: sedimentologia e caratteristiche geometriche dei livelli E, F, G, H (Formazione Marnoso-arenacea) (Cortemaggiore field, sedimentology and geometric features of the E, F, G, H levels [Marnoso-arenacea Formation]): Agip company files.

SUGGESTED READING

Chiaramonte, M. A., and L. Novelli, 1986, Organic matter maturity in Northern Italy: some determining agents, in D. Leythaeuser et al., eds., Advances in organic geochemistry, 1986, pt. I, Petroleum geochemistry, v. 10, p. 281-290.

Dainelli, L., and M. Pieri, 1988, The evolution of petroleum exploration in Italy: Memorie Società Geologica Italiana, v. 31, p. 243-254.

ENI, 1969, Italia (geologia e ricerca petrolifera) (Italy: [geology and exploration for petroleum]), in Enciclopedia del Petrolio e del Gas Naturale, v. 6, Colombo, Roma, p. 336-571.

Pieri, M., and L. Mattavelli, 1986, Geologic framework of Italian petroleum resources: AAPG Bulletin, v. 70, p. 103-130.

Rizzini, A., and L. Dondi, 1978, Erosional surface of Messinian age in the subsurface of the Lombardian plain (Italy): Marine Geology, v. 27, p. 303-325.

Rizzini, A., and L. Dondi, 1979, Messinian evolution of the Po basin and its economic implications (hydrocarbons): Palaeogeography, Palaeoclimatology, Palaeoecology, v. 29, p. 41-74.

Appendix 1. Field Description

Field name .... *Cortemaggiore field*
Ultimate recoverable reserves .... *1,200,000 m³ (7.55 million bbl) oil;*
14.8 × 10⁹ m³ (523 bcf) gas
26,000 m³ (164,000 bbl) condensate

Field location:
  Country .... *Italy*
  Basin/Province .... *Po Plain*

Field discovery:
  Year first pay discovered .... *Pools A, C, C₁, D, D₁ + D₂   1949*
  Year second pay discovered .... *Pools E, F₁ +F₂   1952*

Discovery well name and general location:
  First pay .... *Cortemaggiore 2 (pools A, C, C₁)*
  Second pay .... *Cortemaggiore 1 (pool D)*
  Third pay .... *Cortemaggiore 29 (pools E, F₁ + F₂)*

Discovery well operator .... *Agip*

IP
  First pay .... *35,000 m³/day (1.24 MMcf) gas (pool A)*
  Second pay .... *50,000 m³/day (1.77 MMcf) gas (pool D); 80,000 m³/day (2.80 MMcf) gas (pool C)*
  Third pay .... *28 m³/day (176 bbl) oil (pool F)*

All other zones with shows of oil and gas in the field:

Age	Formation	Type of Show
Pliocene	Santerno Formation	Gas

Geologic concept leading to discovery and method or methods used to delineate prospect
After many decades of exploration in the Northern Apennines, exploration was extended northwards on the Po Plain, with the objective to test the Neogene reservoirs of the external Apennine folds that could possibly be present in the subsurface of the alluvial plain. Reflection seismic ("dip shooting" lines) was carried out in 1940-1943 and confirmed the folded structure of the Neogene section.

Structure:
  Province/basin type .... *Bally 41; Klemme II Cb/IV*
  Tectonic history
  The Northern Apennine foredeep basin originated from the subduction of the western continental margin of the Adria (or Apulian) plate during late Oligocene and Neogene time. The sediments deposited into the inner (i.e., southwestern) margin of the basin were progressively involved in the chain structure. The present external folds of the Apennine were mainly formed during Pliocene and Pleistocene times.

  Regional structure
  The fold belongs to the Emilia arc, i.e., one of the more external (and younger) structural units of the northwestern Apennines.

  Local structure
  East-west-trending anticline overthrust on a lower structural unit

Trap:
  Trap type(s)
  Anticlinal trap: lateral sealing due to the shale-out and/or onlap of sand levels (pools A, C); sand lenses (pool C₁); stratigraphic truncation trap (pools D, D₁ + D₂); anticlinal trap in channel fill sands and conglomerates (pool E, F₁ + F₂)

Basin stratigraphy (major stratigraphic intervals from surface to deepest penetration in field):

Chronostratigraphy	Formation	Depth to Top in m
Pleistocene	Asti sands formation	-700
Pliocene	Santerno clays formation	-1200
Upper Messinian	Cortemaggiore sandstones	-1500
	Colombacci shales formation	(wedged out)
Lower Messinian	Verghereto marls	(wedged out)
Tortonian-Serravallian	"Marnoso-arenacea" formation	-1600

Reservoir characteristics:

- **Number of reservoirs** 7
- **Formations** .... Cortemaggiore Formation (pools A, C, C_1)
  Marnoso-arenacea formation (pools D, $D_1 + D_2$, E, $F_1 + F_2$)
- **Ages** .... Upper Messinian to Tortonian
- **Depths to tops of reservoirs** .... Pools A, C, C_1, approx. -1450 m;
  pools D, $D_1 + D_2$, E, $F_1 + F_2$, approx. -1500 to -2000 m
- **Gross thickness (top to bottom of producing interval)**
  Pools A through C, approx. 100 to approx. 40 m (wedging out) (A max. gross pay 60 m; C max. gross 15.5 m); pools D, $D_1 + D_2$, thick > 120 m off flank wedging out on crest; pools E, $F_1 + F_2$, NA (> 40 m, crestal mudstones in eastern part of field)
- **Net thickness—total thickness of producing zones**
  - **Average** .... C, 1.5 m; D, 28 m; $D_1 + D_2$, 2 to 3 m; E, 5 m; $F_1 + F_2$, 31 m
  - **Maximum** .... NA
- **Lithology**
  Fine-grained sandstones and silty sandstones (pools A, C, D, $D_1 + D_2$); conglomerates and sandstones (pool $F_1 + F_2$)
- **Porosity type** .... Intergranular porosity
- **Average porosity** .... Pools A, C, C_1, 25%; pools D, $D_1 + D_2$, E, $F_1 + F_2$, 30%
- **Average permeability** .... Pools A, C, C_1, 180 md; pools D, $D_1 + D_2$, E, $F_1 + F_2$, 80 md

Seals:

- **Upper**
  - **Formation, fault, or other feature**
    Pool A, Santerno Formation shales; Pools C, C_1, D, $D_1 + D_2$, shale zones in the Cortemaggiore Formation; pools E, $F_1 + F_2$, shale zones in the Marnoso-arenacea formation
- **Lateral**
  - **Formation, fault, or other feature**
    Permeability loss due to lateral shaling of sands, onlap termination against unconformity with shale seal; anticlinal closure with downdip OWC

Source:

- **Formation and age** .... Marnoso-arenacea formation (middle-late Miocene)
- **Lithology** .... Clays and marls
- **Average total organic carbon (TOC)** 0.7%
- **Maximum TOC** 1.5%
- **Kerogen type (I, II, or III)** III
- **Vitrinite reflectance (maturation)** .... NA
- **Time of hydrocarbon expulsion** .... Pliocene-Pleistocene
- **Present depth to top of source** .... 5 to 6 km (estimated)
- **Thickness** .... >2000 m
- **Potential yield** .... 3000 cm^3/m^3 (1 kg/metric ton)

CORTEMAGGIORE

Appendix 2. Production Data

Field name .... *Cortemaggiore field*

Field size:
    Proved acres .... *2400 ha*
    Number of wells all years .... *113 (68 gas, 45 oil)*
    Current number of wells .... *35 (25 gas, 10 oil)*
    Well spacing .... *Approx. 150 m*
    Ultimate recoverable .... *14.8×10^9 m^3 (523 bcf) of gas;*
        1.2×10^6 m^3 (7.55 million bbl) of oil
    Cumulative production .... *14.145×10^9 m^3 (500 bcf) of gas;*
        1.2×10^6 m^3 (7.55 million bbl) oil;
        26,000 m^3 (164,000 bbl) condensate
    Annual production .... *NA*
    Present decline rate .... *NA*
        Initial decline rate .... *NA*
        Overall decline rate *7%*
    Annual water production .... *NA*
    In place, total reserves .... *21.0×10^9 m^3 (742 bcf) of gas; 4×10^6 m^3 (25 million bbl) oil*
    In place, per acre-foot .... *NA*
    Primary recovery .... *14.8×10^9 m^3 (523 bcf) of gas; 1.2×10^6 m^3 (7.55 million bbl) of oil*
    Secondary recovery .... *NA*
    Cumulative water production .... *18,400 m^3 (116,000 bbl) gas pools;*
        5,690,000 m^3 (36,000,000 bbl) oil pools

Drilling and casing practices:
    Casing program .... *13⅜-in. at 200 m; 9⅝-in. at 1400 m; 7-in. at 1700 m*
    Drilling mud .... *LS type; dens 1150 g/L*
    Bit program .... *17½-in. SDGH; 12¼-in. x3A; 8½-in. J1; 6⅛-in. DGJ*
    High pressure zones .... *NA*

Completion practices:
    Interval(s) perforated .... *NA (various reservoirs)*
    Well treatment .... *Gravel packing plus 5-in. tubing*

Formation evaluation:
    Logging suites .... *IND + sonic; Dens; Neutron; MSFL; HDT*
    Testing practices .... *Drill-stem test and long production tests*
    Mud logging techniques .... *Standard*

Oil characteristics:
    Type .... *Naphthenic*
    API gravity .... *34.7–43.7°*
    Base .... *Naphthenic-mixed naphthenic*
    Initial GOR .... *120 to 140*
    Sulfur, wt% .... *Traces to 0.09*
    Viscosity, SUS .... *1.55 to 2.80 cp*
    Pour point .... *NA*
    Gas-oil distillate .... *NA*

Field characteristics:
    Average elevation .... *50 m a.s.l.*
    Initial pressure .... *179.9 to 222.5 kg/cm^2*
    Present pressure .... *79.4 to 130 kg/cm^2*

Pressure gradient .... *0.12 to 0.11 kg/cm²/m*
Temperature .... *47–54°C*
Geothermal gradient .... *0.03°C/m*
Drive .... *Expansion; weak water drive*
Oil column thickness .... *Oil, 45 m; gas, 20 m*
Oil-water contact .... *OWC, -1900 m; GWC, -1500 m*
Connate water .... *Gas, 32%; oil, 25%*
Water salinity, TDS .... *113.69 to 233.21 g/L*
Resistivity of water .... *0.075 to 0.040 ohm*
Bulk volume water (%) .... *NA*

Transportation method and market for oil and gas:
Pipeline

Malossa Field—Italy
Po Basin

L. MATTAVELLI
V. MARGARUCCI
Agip S.p.A.
Milan, Italy

FIELD CLASSIFICATION

BASIN: Po
BASIN TYPE: Foredeep
RESERVOIR ROCK TYPE: Dolomite
RESERVOIR ENVIRONMENT OF DEPOSITION: Intertidal
TRAP DESCRIPTION: Basement involved thust anticline

RESERVOIR AGE: Triassic to Jurassic
PETROLEUM TYPE: Gas and Condensate
TRAP TYPE: Anticline

LOCATION

The Malossa gas-condensate field (Figure 1) is located near the Adda River in the northwestern sector of the Po Valley, approximately 25 km (16 mi) east of Milan, the most important Italian industrial city. This valley is a wide, flat region surrounded by the Alps to the north and west and by the Apennines to the south. The field area belongs to the external sector of the Southern Alps domain. The Caviaga gas field, 30 km (19 mi) south of Malossa, is the nearest gas accumulation and was also the first significant discovery of natural gas in Italy (1944).

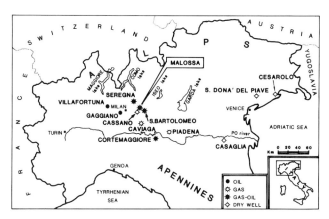

Figure 1. Location map of the fields and wells mentioned in the text.

HISTORY

Pre-Discovery

The hydrocarbon exploration of the area northeast of Milan, where the Malossa field lies, began in the late 1950s. However, the initial seismic mapping of this area was characterized by a lack of indications of folding phenomena in Pliocene–Pleistocene strata (pedealpine-homocline; Pieri and Groppi, 1981) and obtained good reflections only down to the unconformity between the Miocene and the Pliocene (practically, only down to a depth of 2000 m [6600 ft]). The main exploration target in this sector was the Sergnano gravel formation, which had been deposited and deeply eroded during the early Messinian regression (Mediterranean salinity crisis) and rapidly covered by the marine shales of the Pliocene ingression.

A few gas fields were found in stratigraphic traps with reservoirs formed by discontinuous and isolated alluvial plain bodies of the Sergnano gravel formation. Notwithstanding the poor seismic responses, a few wells were drilled on the tops of Pliocene anticlines (i.e., Piadena, Cassano D'Adda, etc.) in order to explore the petroleum potential of pre-Miocene sequences. These wells penetrated more than 1500 m (4920 ft) of impervious, thick, and sometimes strongly dipping Miocene sediments at depth below 5000 m (deepest well 5251 m; 17,228 ft).

The shallower Pliocene folds were not considered to be consistent with the tectonic trends of the deep Mesozoic carbonate sequences. Nevertheless, some wells were drilled during the 1950s in the stable area of the eastern Po Valley sector (S. Donà de Piave 1, Cesarolo 1) and in the external part of the Apennines (Casaglia 1; Ferrara folds). These wells penetrated Mesozoic carbonates (down to 4300 m; 14,107 ft) characterized by good reservoir properties but without any hydrocarbon shows.

Discovery

In 1967–1968, the introduction of digital seismic techniques provided better information at great depths, with improved definition of the reflecting horizons belonging to the Mesozoic carbonate substratum. As a consequence of this new development, Agip (State Oil Company of Italy) carried out another seismic survey of the entire Po basin. In 1969, 25 km (16 mi) east of Milan, this survey detected a good reflector at 3 seconds (two-way time), interpreted to be the top of the Mesozoic (Figures 2 and 3). A preliminary isochron contour map outlined a regular, gently folded anticline with a slight east-to-west elongation.

Well no. 1 (whose name is derived from the nearby Malossa farm) was drilled at a location selected on the axis of the anticline by the late Dr. Lucchetti (AGIP Domestic Exploration Manager) to evaluate potential petroleum accumulations contained in the deep Mesozoic carbonate sequences. The well was spudded in July 1972 and reached a total depth of 5545 m (19,193 ft) in February 1973. After drilling a thick Tertiary terrigenous sequence (4700 m; 15,420 ft), the well penetrated about 800 m (2625 ft) of Jurassic and Cretaceous pelagic limestones and marls and bottomed in Lower Jurassic calcareous dolomites (Zandobbio formation).

The pay zone was penetrated only 90 m (295 ft) due to serious overpressure problems from approximately 4000 m (13,124 ft) downward. The first drill-stem test carried out in the 5510 to 5545 m (18,078–18,193 ft) depth interval detected overpressured gas-condensate with a static pressure of 1050 kg/cm^2 (15,000 psi). The entire pay zone was eventually penetrated by well no. 3, which was drilled to the Ladinian sandstones and shales. The pay zone (from 5521 m to 5672 m; 16,802 to 18,610 ft) is represented, in the upper part by pink or reddish, fine- to medium-grained dolomite and calcareous dolomites associated with the Zandobbio formation (Liassic). However, the main reservoir (from 5672 m to 6131 m; 18,610 to 20,116 ft) consists of gray, fine- to medium-grained intraclastic and fossiliferous dolomites characterized by tidal flat deposits that are typical of the Dolomia Principale formation (Norian). A minor pay zone was found from 5088 to 5152 m (16,694–16,904 ft) in white, "tight" and cherty limestones belonging to the Maiolica formation (Upper Jurassic to Lower Cretaceous).

Because of the problems connected with the abnormal pressure, and the need to impose strict safety controls, it was impossible to carry out the first production test until the end of 1973 (about 10 months after completion of the discovery well). This test gave the following production results: gas, 509,000 m^3/d (17,975 MDFGD) and oil, 371 m^3/d (2334 bbl/day).

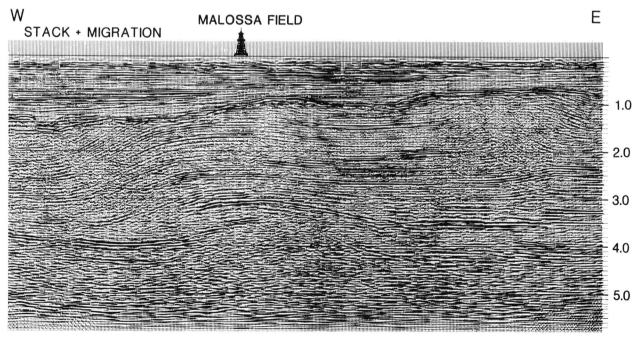

Figure 2. Seismic line of the Malossa structure after migration processing.

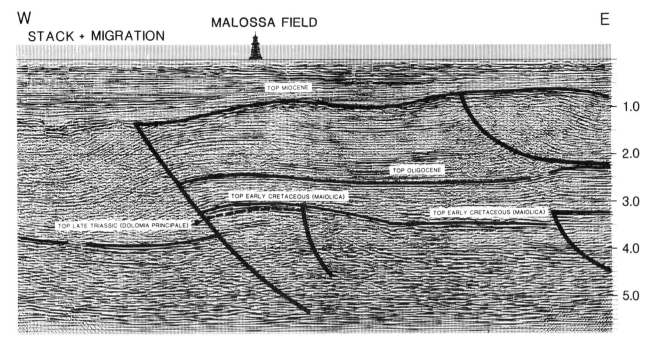

Figure 3. Interpreted seismic line of the Malossa structure.

Post-Discovery

Field development continued in early 1974 with well nos. 2 and 3. Well no. 2 was drilled to define the southward limit of the Malossa field. In fact, after having penetrated very thick middle to upper Liassic pelagic limestone sequences (Medolo formation > 550 m; 1800 ft), it encountered a fault and penetrated through the lower part of the main reservoir (Dolomia Principale formation) for 100 m (330 ft). This well, the deepest in the field (6471 m; 21,231 ft), bottomed in lower Ladinian sediments.

Well no. 3, located in the central sector of the structure and bottomed in upper Ladinian deposits, provided important information on the thickness and lithological characteristics of the two main pay zones (Zandobbio and Dolomia Principale) and on the depth of the oil-water contact (5800 m; 19,030 ft).

The development and delineation of the Malossa field (Figure 4) was complete in 1979 with the drilling of 13 wells, ten of which were completed for production. Ten wells have produced from the two main pay zones (Dolomia Principale and Zandobbio formations) and four wells from the upper pool (Maiolica formation). The field area is represented by a faulted and overthrust anticline about 4 km (2.5 mi) long and 2.5 km (1.6 mi) wide.

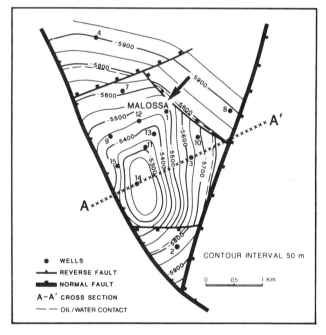

Figure 4. Structure map of the top of the main reservoir (Zandobbio formation). Depth in meters below sea level. (Redrafted from Errico et al., 1980.) Cross section A-A' is shown by Figure 5.

DISCOVERY METHOD

The Malossa structure was detected by the new digital seismic survey carried out in 1969. However, the interpretation of this structure from seismic data was a complex operation. In fact, after drilling the first well, the preliminary isochron contour map (based on a good reflection at 3 seconds, two-way time) was found to delineate the top of Maiolica formation (Late Jurassic–Early Cretaceous; Errico et al., 1980), while no other evident and continuous reflectors were observed in deeper strata. The lack of seismic data from sediments underlying the Maiolica formation is explained by the small acoustic impedance contrast within the carbonate sequence. In order to overcome this problem, new seismic lines were recorded in 1975 in the field using more sophisticated acquisition techniques. The improved quality of the new seismic cross sections (Figures 2 and 3) resulted in a correct recognition of the top of the reservoir (Figure 4).

Another negative factor that created serious problems with seismic interpretations was the presence of thick and irregular Quaternary alluvial plain conglomerates. These bodies deformed the paths of the seismic waves and complicated velocity studies and migrated-depth determinations. However, at the end of the 1970s a better knowledge of the complex Mesozoic tectonic setting, together with improved seismic techniques, both in data acquisition and processing methodologies, led to the discovery of two minor condensate gas fields (S. Bartolomeo and Seregna) in the Malossa area (Figure 1). To these can be added the recent (1984) important find of the Villafortuna oil field 50 km (31 mi) to the west. The latter, whose pay zone is located at a depth between 6118 m and 6212 m (20,073–20,382 ft), represents one of the deepest liquid hydrocarbon accumulations in the world.

STRUCTURE

Tectonic History

The Malossa field is located in the northwestern part of the Po basin, which has been defined according to Klemme's classification (St. John et al., 1984) as "a continental multicycle crustal collision basin (II cb/IV)." In fact, it can be considered a syntectonic and post-tectonic sedimentary basin filled by Tertiary sediments that replaced Mesozoic carbonates as a result of the Alpine orogeny.

The Po basin area was a passive continental margin during the Late Triassic and Early Cretaceous and became an active margin during the Late Cretaceous and Tertiary. In short, taking into account only the main tectonic and sedimentologic events, the Malossa area during the Late Triassic (Norian) was characterized by the deposition of thick carbonate sequences (Dolomia Principale formation) that outcrop all over the Southern Alps. At the end of the Norian, rifting activities that preceded the opening of the Jurassic ocean caused the segmentation of this carbonate platform. Synsedimentary movements along the listric faults created paleohighs (or seamounts) characterized by emersion, erosion, and/or nondeposition processes and troughs (or basins) with euxinic conditions.

During this period the Malossa area was a paleohigh as indicated by the lack of Rhaetian facies (Figures 5 and 6) which, to the contrary, were well developed in the contiguous north–south-trending anoxic basins. The latter were filled by black shales and argillaceous limestones (Argilliti di Riva di Solto and Calcari di Zu formations), locally more than 2000 m (6550 ft) thick (Iseo trough, toward the east, Figure 6). With the continuation of the rifting phase in the Early Jurassic, the sedimentation became more marine, but the thickness of the deposits varied considerably among the paleohighs and the troughs as a consequence of synsedimentary fault movements. In fact, while the thickness of the Medolo formation (Liassic) is around 100 m (330 ft) at the top of the Malossa structure, it exceeds 550 m (1800 ft) (Errico et al., 1980) only 2 km (1.2 mi) southward (well 2) and measures, in the depocenter of the Iseo trough, 1800 m (5900 ft).

In the Late Jurassic–Early Cretaceous, the rifting phenomenon reached the spreading phase and a new ocean (Tethys or Mesogea) was generated at the same time as the opening of the North Atlantic Ocean. In the Lombardy area, the deep water conditions led to the deposition of cherty limestones, radiolarites, and marls (Winterer and Bosellini, 1981). At the end of the Early Cretaceous, terrigenous turbidites (Flysch Lombardo) found near the Malossa area represented the first evidence of the Alpine compressional tectonic phase. The continuation of tectonic compression during the Eocene, Oligocene, and early Miocene and the subsequent emersion and erosion of larger areas of the Alpine domain caused the deposition of extremely thick (>3000 m; 9850 ft) sandstone and marl sequences in the rapidly subsiding Malossa area.

A strong compressional tectonic phase, which occurred during the middle and late Miocene, was mainly responsible for the formation of the Southern Alps, which are characterized by the presence of south-verging thrust sheets. As a consequence of the last tectonic phase, the northern part of Lombardy emerged and was deeply eroded during the Messinian salinity crisis. A new sedimentary cycle, which had begun in the late Messinian, ended with the deposition of Quaternary alluvial plain sediments. That the Pliocene-Quaternary sequences were not affected by compressional tectonism is indicated by their horizontal position shown on the seismic sections (Figures 2 and 3).

Regional Structures

The Malossa field is located in the external part of the Southern Alp domain, more specifically in its central sector—the Mesozoic carbonate reservoirs cropping out between Como and Iseo lakes, about 30 km (19 mi) to the north. The Southern Alps consist of south-verging tectonostratigraphic units that were thrust over Adria (African promontory), mainly during the Miocene.

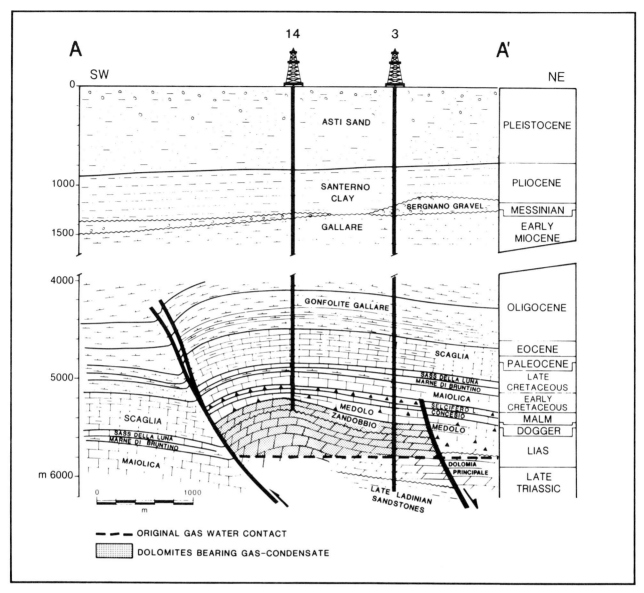

Figure 5. Structural cross section through Malossa field. Lateral and upper seals are represented by Lower to Upper Cretaceous marls and argillaceous limestones and by early Tertiary sandy marls. The location of the cross section is shown on Figure 4.

At the time of the Malossa discovery, the regional geologic setting was believed to consist of numerous overthrust folds whose south-southwest displacement was considered to range from a few kilometers to 10 km (6 mi). A recent study, based on seismic, magnetic, and gravimetric data (Cassano et al., 1986), seems to indicate that the crystalline basement was also involved in the shortening phenomena that affected the Southern Alp domain (Figure 6). Therefore, the overthrust folds present in the Malossa area should be interpreted as the outer part of a Southern Alp thrust sheet. Moreover, the complexity of this area is also related to the fact that the diversified paleogeographic features originating during the Jurassic rifting phase and passive margin regime were modified by the subsequent Neogene compressional tectonic events.

It should also be pointed out that the strongly reflecting horizon corresponding to the Maiolica formation (Figures 2 and 3) gives little information on the complex paleogeologic characteristics of the underlying Liassic and Upper Triassic sequences.

Local Structure

As shown in Figure 7, a simplified isochron map, the top of the Mesozoic in the Malossa structure belongs to a faulted anticline trending northwest-southeast for a distance of 30 km (19 mi). This slightly

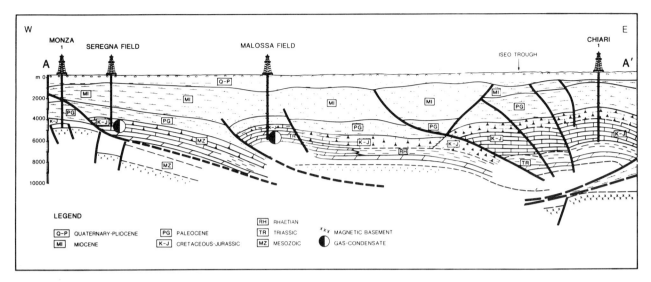

Figure 6. Geological cross section of the Malossa area based on seismic, magnetic, and gravimetric data. (Redrafted from Cassano et al., 1986.) Location of the cross section is shown on Figure 7.

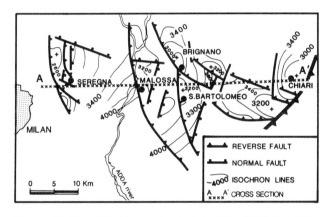

Figure 7. Simplified tectonic sketch of the Malossa area near the top of the Mesozoic. (Modified after Errico et al., 1980.) Cross section A-A' is shown by Figure 6.

arched tectonic element, overthrust toward the southwest, represents (according to the recent regional study previously discussed) the external part of a greater Southern Alp thrust sheet (Figure 6). Its formation was related mainly to a strong tectonic event of the late Miocene that intensely affected the outer sector of the Southern Alps. The characteristics of this folded anticline in which the Malossa structure is located probably were influenced also by the pre-existing paleotectonic pattern. The influence of the latter seems to be substantiated by the fact that faults defining the eastern and northern extension of the Malossa field probably represent the reactivation phase of those generated during the Jurassic passive margin regime (Figure 4).

Stratigraphy

The simplified stratigraphic section of the Malossa area shown in Figure 8 was reconstructed on the basis of information obtained from well nos. 2 and 3, which are the deepest ones in the field. The oldest unit in Malossa field is the Middle Triassic Esino formation, which consists of light gray, fine- to medium-grained dolomites with algal remains (Dasycladaceae). It probably is overlain unconformably by a thin sequence of varicolored sandy shales and feldspathic sandstones that have been attributed on the basis of recent palynological analyses to the late Ladinian.

Another unconformity separates the Ladinian sequences from Norian deposits. The latter are characterized by thick tidal-flat deposits consisting of gray and light brown, fine- to medium-grained dolomites (Dolomia Principale formation). This formation, which constitutes the main reservoir of the Malossa field, is characterized by sequences that are thinner (500 m; 1640 ft) than those found in the outcrops (1000–2000 m; 3280–6560 ft).

This fact, together with the lack of Rhaetian anoxic facies, clearly indicates that the Malossa area was affected by emersion and erosion phenomena during the late Norian and Rhaetian. The Rhaetian anoxic facies, present in the adjacent troughs and recently recognized as source rocks of the Malossa hydrocarbons (Riva et al., 1986), consist of black argillaceous limestones, marls, and shales (Calcari di Zu and Argilliti di Riva di Solto formations). Their thickness is extremely variable, ranging from more than 2000 m (6560 ft) to less than 200 m (660 ft) within a short distance of 10 km (6 mi) (Iseo lake).

The Norian sediments are unconformably overlain by 150 m (490 ft) of reddish, fine-grained, calcareous dolomites, sometimes with chert nodules and sponge

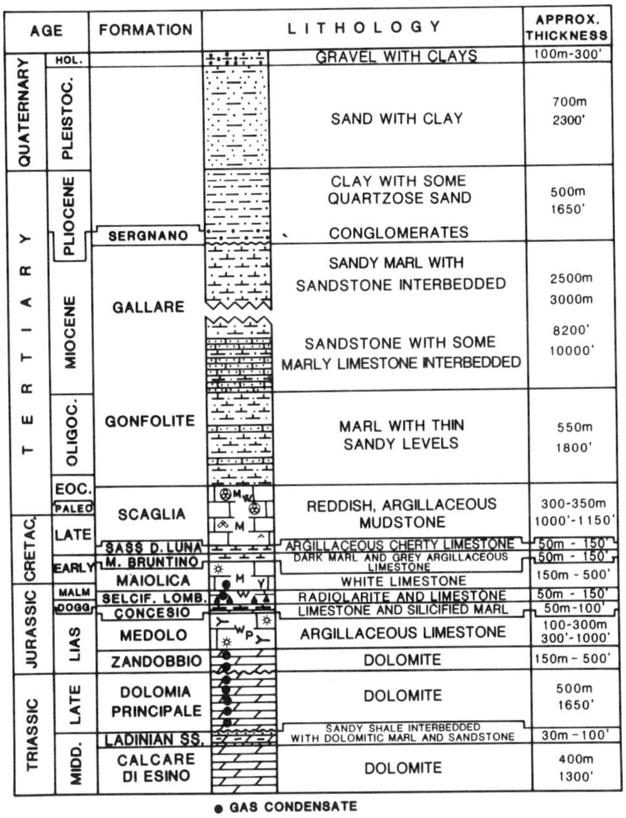

Figure 8. Simplified stratigraphic section of the Malossa area obtained from data from the deepest wells (nos. 2 and 3).

spicule remains from the Zandobbio formation. The age of this formation, which represents the upper part of the main pay zone, was assigned to the Early Jurassic on the basis of the lithological and sedimentological similarities with the Zandobbio formation cropping out 30 km (19 mi) to the north in the foothills of the Southern Alps at the Zandobbio quarry.

The middle-late Liassic, Dogger, and Malm are characterized by deep-water sediments consisting of gray, cherty, argillaceous limestones with Radiolaria and sponge spicules and radiolarites (Medolo, Concesio, and Selcifero Lombardo formations). The pelagic conditions continued with the sedimentation of Late Jurassic-Early Cretaceous cherty limestones that form the upper reservoir of the Malossa field (Maiolica formation). This formation is overlain by a thin sequence (30 m; 100 ft) of dark gray marls deposited in a euxinic environment (Marne di Bruntino formation). The latter, together with the argillaceous limestones of the Sass de la Luna and Scaglia formations (Late Cretaceous-early Eocene), represent the roof rock of the Malossa field.

The period of time from the middle Eocene until the early Miocene was characterized by the deposition of synorogenic sediments, consisting of thick terrigenous turbidites (Gallare and Gonfolite formations). The Langhian sediments are unconformably overlain by alluvial plain polygenic conglomerates attributed to the late Tortonian-early Messinian (Sergnano formation). The Pliocene-Quaternary cycle, which began with deposition of deep marine clay, ended with Holocene alluvial plain deposits.

TRAP

Trap Type

The Malossa field is a structural trap consisting of a northwest-southeast-trending anticline faulted and overthrust toward the southwest. The reservoir is bordered by faults to the southwest and southeast, respectively (Figure 4). The closure is determined principally by these two main faults and by other minor normal faults present in the northern and southern sector of the trap. The upper and lateral seals are formed by the impervious sediments of the Marne di Bruntino, Sass de la Luna, and Scaglia formations together with the lower terrigenous synorogenic deposits of the Gonfolite-Gallare formations (Figure 5).

The original oil-water contact was at a depth of 5800 m (19,030 ft) according to production tests carried out in well no. 3. The main unconformity, strictly related to the field's paleogeographic characteristics, is present between the Norian and early Liassic, as previously mentioned, by the lack of Rhaetian anoxic sediments. Though the Malossa area was already a paleohigh during late Norian and Liassic time, the actual shaping of the structure was determined mainly by the strong tectonic event of the middle-late Miocene. The reactivation of some synsedimentary Late Triassic-Early Jurassic faults during the Neogene cannot be ruled out. However, on the basis of the new regional reconstruction (Cassano et al., 1986), we disagree with the previous interpretation (Errico et al., 1980) which stated that the closure conditions present during the Jurassic were preserved even after subsequent tectonic movements.

Reservoir

Stratigraphy and Lithology

As previously stated, the main reservoirs are formed by Late Triassic-early Liassic dolomites of the Dolomia Principale and Zandobbio formations, while a minor upper pay zone is made up of Late Jurassic-Early Cretaceous cherty limestones of the Maiolica formation. The Dolomia Principale consists of white, gray, fine- to medium-grained, intraclastic, and fossiliferous dolomites with a few intercalations of dolomitized marl in the lower part. Desiccation structures (fenestra) and algal mat features (stromatolites) are also present. The Zandobbio formation is formed by gray, reddish, fine- to medium-grained, sandy, calcareous and argillaceous dolomites with some relics of intraclasts, oolites, sponge spicules, and ammonites. Thin chert nodule levels (well no. 4) and a few karst-cavity intervals filled with breccia (well no. 14) are also observed. The Maiolica formation is lithologically uniform, characterized by white, cherty mudstone and wackestones with thin, black argillaceous layers.

Abnormal Pressures

One of the most peculiar aspects of the Malossa field is the presence of abnormal pressures (about twice the normal hydrostatic pressures detected in the other sectors of the Po basin) in the reservoir (over 15,000 psi or over 1050 kg/cm^2), which created serious problems during drilling and production operations. The origin of the abnormal pressure was dealt with in a recent study (Novelli et al., 1987) in which hydraulic conditions were simulated using a finite element monodimensional mathematical model. Briefly, the hydrogeological framework of the northwestern Po basin is represented by a two-layered aquifer system with an interbedded aquitard formed by the impervious terrigenous sequences of the early Tertiary (Eocene-Aquitanian). As indicated by pressure gradient profiles (Figure 9), very abnormal high pressures found at a depth of about 4000 m (13,100 ft) characterize the deeper carbonate aquifer.

The origin of this overpressure regime in the Lombardy area can be ascribed to the concurrence of tectonic and sedimentological factors. In particular, compressive movements, related to the late-middle Miocene tectonic phase, interrupted the hydraulic continuity in the Mesozoic carbonate series in an area (on the basis of well data) extending at least for 100 km (60 mi) and 70 km (45 mi) in the east-west and north-south direction, respectively. In essence, this hydraulic closed system was created in the northern and southern sector of the Lombardy area by thrusting phenomena related to the formation of the Southern Alps and Apennines, respectively. On the other hand, the hydraulic discontinuity on the western and southern part of this area probably was caused by transpression faults that created impervious lateral boundaries (Cassano et al., 1986). The subsequent high sedimentation rate (>150 m/Ma; 500 ft/Ma) during the Pliocene-Pleistocene created an abnormal pressure regime inside the early Tertiary impervious sediments that overlay the Mesozoic carbonate sequences.

The abnormal pressures start at a depth generally greater than 3000 m (9850 ft), reaching maximum value (pressure gradient = 1.9 atm/10 m or 89 psi/100 ft depth) in the lower part of the Gallare-Gonfolite formation near the boundary with the Mesozoic carbonate (Figure 9). This overpressure regime is transmitted consequently to the underlying, closed carbonate rock aquifer. It is important to note that within the carbonate series, the pressure gradient then decreases downward, tending toward the hydrostatic one. For further detailed information on the origin of these abnormal high pressures in the

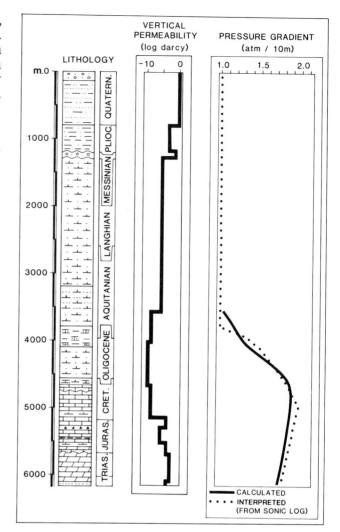

Figure 9. Pressure profile of the Malossa no. 3 well.

Lombardy area, the reader is referred to the paper of Novelli et al. (1987).

Depositional Setting

The Dolomia Principale formation originated in a shallow-water carbonate environment ranging from subtidal to supratidal. The small number of cores taken in the Malossa field does not permit a precise reconstruction of the original depositional setting. However, these sediments seem to be represented by typical tidal flat facies widespread in outcrops all over the Lombardy Southern Alps.

The depositional environment of the Zandobbio formation was characterized by an appreciable variation in water depth. In fact the presence of karst features on top of the Malossa paleohigh indicates emersion and subaerial exposure (well no. 15), while in the external zone marine conditions (open shallow platform) prevailed during the entire Liassic as indicated by the existence of slumping phenomena (well no. 4).

The depositional setting of the Maiolica formation was characterized, on the other hand, by rather uniform deep-water conditions prevalent in the entire Lombardy basin.

Porosity Types

The main types of pores present in the sequences of the Dolomia Principale and Zandobbio formations are: (1) vugs (Figure 10); (2) intercrystalline pores (Figures 11 and 12); (3) karstic cavities (Figure 13); and (4) vugs and pores in or connected by the fractures (Figure 14). Porosity values measured on the only three cores taken from the Dolomia Principale are rather low and their range is from 1% to 6% (Figure 15). On the other hand, values obtained from electrophysical log analyses vary from 1% to 7% with an average value of around 3.0%. Porosity values, measured on the two continuous corings (well nos. 4 and 14) carried out in the Zandobbio formation, range from 1% to 13% but have a low mean value of around 2.5% (Figure 16). Permeabilities determined on full-size plugs (6.4 cm; 2.5 in.) are characterized by an average value of 21 md.

Diagenesis strongly modified the characteristics of the original sediments in the two main reservoirs, although in different ways. In particular, in the Dolomia Principale, the considerable amount of early porosity created by penecontemporaneous dolomitization, desiccation, leaching, and solution of non-replaced calcite was almost destroyed during the subsequent diagenetic stages by the deposition of dolomite cement, pressure solution, etc. Nevertheless, the dolomitization homogenized the different original characteristics of the Zandobbio formation as indicated by the similar porosity values detected both in shallower water sediments (well no. 14) and in deeper ones (well no. 4). Therefore, dolomitization, which took place in different stages, overlapping in space and time in various and complex ways, created a rigid framework that determined two favorable conditions for preserving a small but definite amount of porosity at great depth. Dolomitization resulted in a better resistance to pressure solution phenomena in comparison to limestone and a greater tendency to fracturing during the tectonic movements.

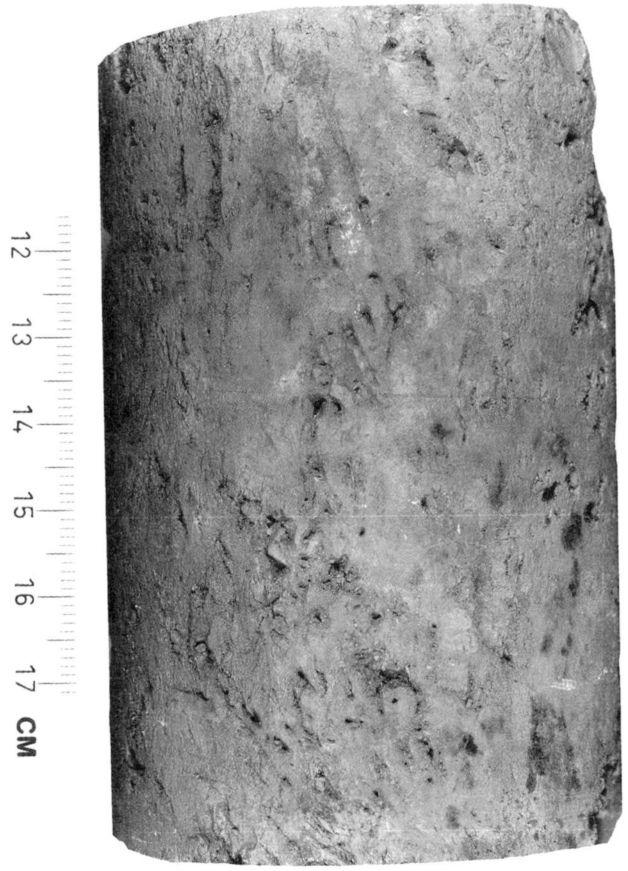

Fractures

In both the main reservoir facies (Dolomia Principale and Zandobbio formations), porosity was increased by fracturing caused by the various tectonic phases that affected the Southern Alp domain. In particular, north–south-trending fractures and listric faults (see *Tectonic History*) were generated in the Late Triassic–early Liassic carbonate platform by extensional tectonics related to the opening of the Jurassic Ocean. These fractures represented the migration avenues for the first generation of hydrocarbons during the Jurassic, as indicated by the presence of pyrobitumens in the Zandobbio formation (see *Source Rocks and Migration*). A second important system of fractures, trending mainly northwest-southeast, was later created by the strong compressional tectonics of the late-middle Miocene that probably partly reactivated the previous extensional fractures.

Figure 10. Vuggy porosity in the main reservoir (Dolomia Principale formation, Malossa no. 2 well, 5979 m; 19,617 ft).

In practice, fractures are the only porosity type present in the tight pelagic limestone of the Maiolica formation, since by SEM (scanning electron microscopy) analysis, only a few intercrystalline unconnected micropores were detected in this limestone in which any remains of the original nannofossils were obliterated. The fractures observed in the Zandobbio formation show a subvertical trend, with dips ranging from 60° to 80° (Figure 14). Furthermore, the zones affected by stronger tectonic movements are characterized by the presence of fault breccia intervals. The fractures, which are in general only partly cemented, constitute the vertical connection between the two main pay zones. The upper pool, located in the Maiolica formation, also probably communicates with the main reservoirs through an irregular system of micro- and macrofractures. On the whole, the fracture volume has been evaluated, on the basis of reservoir engineering considerations, on the order of 0.5%.

Figure 11. SEM photomicrograph of the Dolomia Principale formation showing intercrystalline micropores (Malossa no. 2 well, 5979 m; 19,617 ft).

In conclusion, even though fractures give only a minor contribution to total porosity, their presence nevertheless caused extremely variable permeability distribution in the reservoir and strongly influenced production. This seems to be demonstrated by the fact that the wells located in the crestal area of the structure where fractures are more frequent are characterized by higher production rates.

Pay-Zones Thicknesses

The maximum total thickness at the top of the structure of the two main reservoirs (Dolomia Principale and Zandobbio formations) is about 580 m (1900 ft), while the net pay averages 300 m (985 ft) thick. On the other hand, the upper pay zone found in the Maiolica formation shows a rather irregular distribution, reflecting the greater occurrence of fractures in the crestal area of the Malossa faulted anticline. Generally, the thickness of the Maiolica formation pay zone is on the order of 60 m (200 ft).

Hydrocarbon Characteristics

Owing to the abnormal pressure system in the deep Mesozoic carbonate aquifer, the reservoir hydrocarbon fluid consists of a gas-condensate characterized by a pressure of 1067 kg/cm^2 (104.6 M Pa = 15,179 psi) with a temperature of 155°C (311°F). The composition of this monophase fluid shown in Table 1 (well no. 3) is indicative of a gas-condensate rich in heavy hydrocarbons with a GOR of approximately 1000 sm^3/m^3 (approximately 5600 ft^3/bbl). Carbon dioxide and small amounts of hydrogen sulfide, which cause corrosion as a result of the combined effects of high pressure and temperature, are also present. The reservoir gas also contains small quantities of asphaltenes that cause some production problems by reducing the cross-sectional area of the tubing. Dew point pressure is 398 kg/cm^2 (39 M Pa = 5661 psi) according to laboratory experiments. The liquid hydrocarbon separated at the treatment plant is made up of light oil having a gravity of 53° API.

Figure 12. SEM photomicrograph of dolomite of the Zandobbio formation (Malossa no. 14 well, 5389 m; 17,681 ft).

Figure 13. Karst cavity in the dolomite sequences of the Zandobbio formation (Malossa no. 14 well, 5389 m; 17,681 ft).

FAULTS

At the end of 1973, after the first well had been successfully drilled and a velocity survey was run, a preliminary depth contour map of the structure was reconstructed. On this map, reflections corresponding to the Maiolica formation show two important faults converging toward the south. Further information obtained by drilling well nos. 2 and 3 and by a new 2D seismic survey allowed a more precise definition and location of the two main faults delimiting the Malossa structure to the west and east, respectively (Figure 4).

In particular, the southwest fault acts as a good seal because the reservoirs are juxtaposed against the younger impermeable sequences (Figure 5).

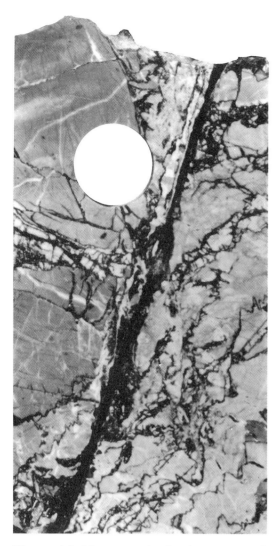

Figure 14. Macro- and microfractures and stylolites in the Zandobbio formation (Malossa well no. 14). Note the presence along the fractures (in black) of pyrobitumen indicating the first oil migration that occurred during the Liassic. (The white round spot shows where a 1-in. [2.54 cm] plug for petrophysical analyses was taken.)

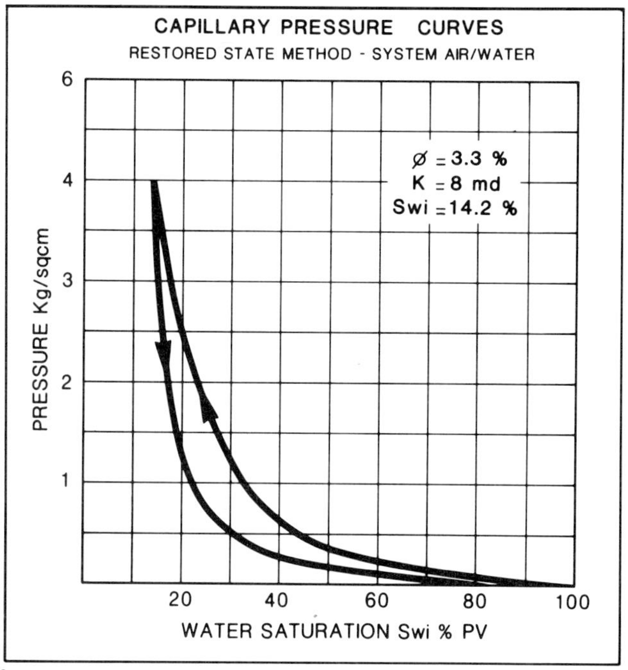

SOURCE ROCKS AND MIGRATION

Rhaetian black shales were considered the hydrocarbon source rocks at the time of the Malossa discovery (Argilliti di Riva di Solto formation, Errico et al., 1980). However, this assumption could not be proved at that time by geochemical data for two main reasons: (1) Rhaetian black shales were absent in the Malossa area; and (2) the organic matter of the classical Rhaetian sequences outcropping on the western side of the Iseo lake is overcooked ($R_o > 2$; metagenesis stage).

More geological surveys during the early 1980s in the western sector of the Lombardy Southern Alps

Figure 15. Capillary pressure curves relative to the main reservoir (Dolomia Principale formation, no. 2 well, 5971 m; 20,188 ft).

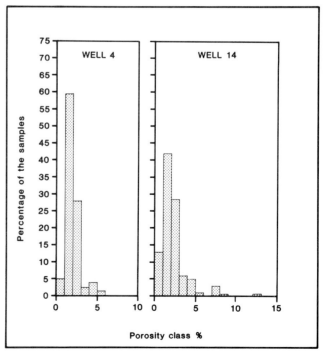

Figure 16. Histograms of porosity classes relative to continuous corings taken in the Zandobbio formation of the Malossa nos. 4 and 14 wells.

Table 1. Composition of the Malossa gas-condensate.

Component	Moles (%)
N_2	0.83
CO_2	0.59
H_2S	0.4 to 0.6 ppm
C_1	74.77
C_2	6.27
C_3	3.17
(i+n) C_4	2.62
(i+n) C_5	1.72
C_6	1.84
C_7	1.07
C_8	1.48
C_9	1.23
C_{10}	0.75
C_{11}	0.59
C_{12}	0.51
C_{13+}	2.56
Total	100.00

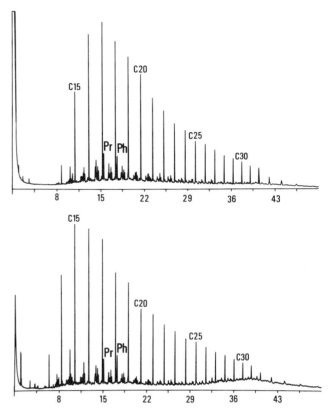

Figure 17. Gas chromatograms of the Malossa oil (above) and Argilliti di Riva di Solto formation source rock extract (below).

led to the discovery of the Rhaetian Argilliti di Riva di Solto formation, which has organic matter in the catagenesis stage. Consequently, a good correlation was established on the basis of molecular organic geochemistry analyses (GC, Figure 17, and GC-MS, Riva et al., 1986) between extracts from this formation and the Malossa oil.

The lack of source rocks in the Malossa area leads us to assume that hydrocarbons were generated in nearby basins, particularly in the eastern Iseo trough where Rhaetian black shales are well developed, and then migrated toward the Malossa paleohigh (Figure 6). A possible burial history of the source rocks, reconstructed on the basis of wells and seismic data, seems to indicate that hydrocarbon generation and migration took place at different times (Figure 18). The first generation, in fact, occurred during the rapid sedimentation of the Liassic. The direction of the migration paths was from east to west, therefore, from the Iseo trough toward the Malossa paleohigh, presumably through extensional synsedimentary faults. This seems to be substantiated by the presence of pyrobitumens with different maturity degrees in the Zandobbio formation (Figure 14, Malossa well no. 14, Novelli et al., 1987). These Early Jurassic oils dispersed toward the surface because of the lack of efficient seal rocks. Other possible hydrocarbon accumulations occurred during the passive margin regime, which lasted until the Early Cretaceous but which were probably destroyed by Tertiary compressive tectonic movements.

The formation of the Malossa gas-condensate took place probably during the strong subsidence of the Pliocene-Pleistocene when the shaping of the structure was no longer modified by compressive tectonics. The generation of the reservoired hydrocarbons occurred at great depth (over 7000 m; 22,970 ft) because, on the basis of different maturity parameters (R_o, TAI, T_{max}), organic matter looks immature in many wells down to 6000 m (19,690 ft) (Chiaramonte and Novelli, 1986). In particular, this assumption seems to be confirmed by isotopic values of the Malossa gas ($\delta^{13}C_{1-36}$ λ; δD - 153 λ), which suggest highly mature source rocks close to an anthracite coalification level (Mattavelli et al., 1983). The total organic carbon (TOC) in the outcrop sequences (Iseo lake) of the Argilliti di Riva di Solto is rather low, with a mean value of 0.8%. This value does not represent the original organic contents of these sediments because they were affected by a strong thermal alteration corresponding to the metagenesis stage. In the subsurface, the Argilliti di Riva di Solto sequences were seldom penetrated by the wells. Their TOC ranges from 0.7% to 1.63%, with average values around 1.25%. The kerogen, on the basis of Rock Eval and TAI data, is immature

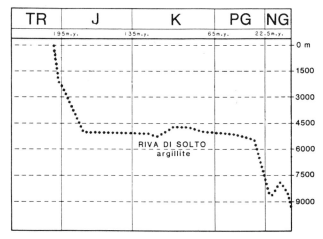

Figure 18. Simplified burial history of the Argilliti di Riva di Solto formation in the Iseo trough. Oil generation based on the Tissot and Espitalié method occurred also during the Liassic, but the lack of efficient seals prevented its accumulation.

and mainly composed of terrestrial organic matter (>80%; type III). Its potential yield is on the order of 3.5 kg of hydrocarbons per ton of rock.

EXPLORATION CONCEPTS

The new digital seismic survey was an essential technology in locating the Malossa structure, but other geological methods played an important role in the discovery of this gas-condensate field. In fact, the decision to drill a deep well in the Lombardy basin also was influenced by the favorable geological conditions for hydrocarbon accumulations existing in the Mesozoic sequences that outcrop about 30 km (19 mi) north of Malossa.

The presence of Cretaceous marls and argillaceous limestones (Bruntino and Scaglia formations) was considered a crucial factor in determining the existence of trapping conditions within the deep Mesozoic series. In fact, in addition to providing distinctive seismic reflectors, these impervious pelagic sediments could have acted as an essential seal rock for the underlying Upper Triassic dolomite reservoir. Furthermore, the Rhaetian black shales (Argilliti di Riva di Solto formation), deposited in intracarbonate platform basins (originated during the first phases of the Triassic continental rifting), were regarded as potential petroleum source rocks.

These specific features of the Mesozoic sequence in the Lombardy basin were the geological "key ideas" that, coupled with the new seismic techniques, led to the discovery of Malossa field.

Conversely, it was the absence of similar favorable conditions in the thick carbonate platform of the eastern Po basin that seems to account for the unsuccessful drilling results in the 1950s: (1) lack of a cap rock seal (San Donadi Piave no. 1 and Cesarolo no. 1 wells); (2) absence of source rock (Casaglia no. 1 well).

Critical factors that, in our opinion, may be important in developing prospects in the province of the Malossa field are the correct seismic interpretation and definition of the paleogeographic features of the complex Mesozoic paleohighs (resulting from extensional Jurassic tectonics) and the determination of the age of the trap formation in relation to the main timing of hydrocarbon migration.

ACKNOWLEDGMENTS

We are grateful to Agip management and, in particular, to the Senior Vice-President G. Paulucci.

We are also indebted to N. Erba and F. Benelli who supported us in the collection of data and to L. Anelli for his helpful suggestions.

REFERENCES

Bellotti, P., R. Deidda, and D. Giacca, 1979, SPE-AIME Deep Drilling and Production Symposium, Amarillo, TX, Paper SPE 7847, p. 69-83.

Bongiorni, D., V. Crico, and D. Fenati, 1977, Geophysical and drilling problems encountered in the exploration of deeper structures in the Po basin: 10th World Energy Conference, Istanbul, Turkey, p. 1-16.

Cassano, E., L. Anelli, R. Fichera, and V. Cappelli, 1986, "Pianura Padana. Interpretazione integrata di dati geofisici e geologici," 73° Congresso Società Geologica Italiana, 29 Settembre-4 Ottobre 1986, Roma, Agip Report, annex to Bollettino Società Geologica Italiana, v. CVI, fasc. 4, p. 1-27.

Chiaramonte, M. A., and L. Novelli, 1986, Organic matter maturity in Northern Italy: some determining agents, in D. Leythaeuser and J. Rullkötter, eds., "Advances in organic geochemistry," p. 281-290.

Chierici, G. L., G. Sclocchi, and L. Terzi, 1980, Pressure, temperature profiles calculated for gas flow: Oil and Gas Journal, Jan. 7, p. 65-72.

Errico, G., G. Groppi, S. Savelli, and G. C. Vaghi, 1979, AAPG/SEPM Convention on Giant Discoveries of the Past Decade, Houston, Texas, April 1979.

Errico, G., G. Groppi, S. Savelli, and G. C. Vaghi, 1980 Malossa field: a deep discovery in Po Valley, Italy, in M. T. Halbouty, ed., Giant oil and gas fields of the decade: 1968-1978: AAPG Memoir 30, p. 525-538.

Gardner, F. J., 1975, Italy's deep Po Valley play yields its first major field: Oil and Gas Journal, v. 73, n. 10, p. 44-45.

Groppi, G., A. Muzzin, and G. C. Vaghi, 1977, Erdoel Kohle, Erdgas, Petrochem Brennst-Chem., Ergänzungsband DGMK-Vortrage der 4. Gemeinschaftstagung OGEW/DGMK, Oktober 1976 in Salzburg, p. 50-74.

Mattavelli, L., G. V. Chilingarian, and D. Storer, 1969, Petrography and diagenesis of the Taormina Formation, Gela oil field, Sicily (Italy): Sedimentary Geology, v. 3, p. 59-86.

Mattavelli, L., T. Ricchiuto, D. Grignani, and M. Schoell, 1983, Geochemistry and habitat of natural gases in Po basin, Northern Italy: AAPG Bulletin, v. 67, p. 2239-2254.

Novelli, L., M. A. Chiaramonte, L. Mattavelli, G. Pizzi, L. Sartori, and P. Scotti, 1987, Oil habitat in the northwestern Po basin. International Meeting on Migration of Hydrocarbons in Sedimentary Basins, B. Doligez, ed., 2nd Expl. Res. Conference, Carcans, France, Edition Technip, p. 27-57.

Pieri, M., and G. Groppi, 1981, Subsurface geological structure of the Po plain, Italy: Consiglio Nazionale delle Ricerche,

Pubblicazione n. 414 del Progetto finalizzato Geodinamica, p. 1-13.

Pieri, M., and L. Mattavelli, 1986, Geologic framework of Italian petroleum resources: AAPG Bulletin, v. 70, p. 103-130.

Rizzini, A., and L. Dondi, 1979, Messinian evolution of the Po basin and its economic implications (hydrocarbons), *in* M. B. Cita et al., eds., Geodynamic and biodynamic effects of the Messinian salinity crisis in the Mediterranean: Palaeogeography, Palaeoclimatology, Palaeoecology, v. 29, p. 41-74.

Riva, A., T. Salvatori, R. Cavaliere, T. Ricchiuto, and L. Novelli, 1986, Origin of oils in Po basin, Northern Italy, *in* D. Leythaeuser and J. Rullkötter, eds., Advances in organic geochemistry 1985, p. 391-400.

Rocco, T., and O. D'Agostino, 1972, Sergnano gas field, Po basin, Italy, A typical stratigraphic trap, *in* R. E. King, ed., Stratigraphic oil and gas fields: AAPG Memoir 16, p. 271-285.

Sajgo, C., 1984, Organic geochemistry of crude oils from southeast Hungary, *in* P. A. Schenck, J. W. De Leew, and G. W. M. Lijmbach, eds., Advances in organic geochemistry 1983: Organic Geochemistry, n. 6, p. 569-578.

St. John, B., A. W. Bally, and H. D. Klemme, 1984, Sedimentary provinces of the world, hydrocarbon productive and nonproductive: Tulsa, Oklahoma, AAPG, p. 1-35.

Suau, J., C. Boyeldieu, R. Roccabianca, and M. Cigni, 1978, 19th Annual SPWLA Logging Symp., El Paso, CA, June, Paper No. W., p. 1-23.

Tricotti, M., 1978, Deep drilling in Italy's Po Valley: Petrol Engineer, v. 50, n. 6, p. 24-28.

Vaghi, G. C., L. Torricelli, M. Pulga, D. Giacca, G. L. Chierici, and D. Bilgeri, 1980, Production in the very deep Malossa field, Italy: 10th World Petrol. Congr. (Bucharest, 1979) proc. v. 3, p. 371-388.

Winterer, E. L., and A. Bosellini, 1981, Subsidence and sedimentation on Jurassic passive continental margin, Southern Alps, Italy: AAPG Bulletin, v. 65, p. 394-421.

Appendix 1. Field Description

Field name .... Malossa field
Ultimate recoverable reserves .... NA
Field location:
    Country .... Italy
    Basin/Province .... Po basin/Lombardy
Field discovery:
    Year first pay discovered .... Lower Jurassic Zandobbio-Dolomia Principale 1973
    Year second pay discovered .... Upper Jurassic-Lower Cretaceous Maiolica 1974
Discovery well name and general location:
    First pay .... Malossa 1, 24 km (15 mi) east of Milan, Lombardy
    Second pay .... Malossa 4, same location
Discovery well operator .... Agip
    Second pay .... Agip
IP in barrels per day and/or cubic feet or cubic meters per day:
    First pay .... 500,000 m³/d gas + 435 m³/d oil (17.66 MMCFG/d + 2740 BOPD)
    Second pay .... 100,000 m³/d gas + 87 m³/d oil (3.5 MMCFG/d + 547 BOPD)

All other zones with shows of oil and gas in the field:

Age	Formation	Type of Show
Middle-Late Jurassic	Selcifero Lombardo	Condensate gas
Early Jurassic	Ammonitico Rosso	Condensate gas
Early Jurassic	Medolo	Condensate gas

Geologic concept leading to discovery and method or methods used to delineate prospect
Hydrocarbon accumulations in the deep Mesozoic carbonate series of the Lombardy Plain were assumed on the basis of the favorable geologic characteristics found in the sequences outcropping at the Southern Alp foothills. The prospect was delineated by using the new digital seismic techniques in the late 1960s.

Structure:
    Province/basin type .... Bally 221, Klemme II Cb/IV
    Tectonic history
    The Malossa field area was affected by extensional tectonics (passive continental margin regime) during Late Triassic-Lower Cretaceous. This area was later involved in the compressional Miocene tectonic movements related to the formation of the Southern Alps.
    Regional structure
    The Malossa field is located in the external sector of the Southern Alps, consisting of anticlines and overthrust structures generally with reverse faults on the southern flank.
    Local structure
    Structural northwest-southeast-trending anticline faulted and overthrust toward the southwest.

Trap:
    Trap type(s)
    Anticlinal trap bordered to the southwest by reverse fault and delimited toward the southeast by a normal fault

Basin stratigraphy (major stratigraphic intervals from surface to deepest penetration in field):

Chronostratigraphy	Formation	Depth to Top in m (ft)
Quaternary	Asti Sand	113 (371)
Pliocene-Miocene	Santerno Clay-Sergnano Gravel	802 (2632)

Miocene-middle Eocene	Gonfolite-Gallare	1279 (4198)
Lower Eocene-Lower Cretaceous	Scaglia-Bruntino	4677 (15,350)
Lower Cretaceous-Malm	Maiolica	5146 (16,889)
Jurassic	Medolo	5280 (17,329)
Lower Jurassic	Zandobbio	5521 (18,120)
Upper Triassic	Dolomia Principale	5672 (18,616)
Middle Triassic	Calcare di Esino	6131 (20,122)

Reservoir characteristics:

    Number of reservoirs 3
    Formations .... Dolomia Principale, Zandobbio, Maiolica
    Ages .... Late Triassic, Early Jurassic, Late Jurassic-Early Cretaceous
    Depths to tops of reservoirs .... Dolomia Principale, 5670 m (18,600 ft); Zandobbio, 5520 m (18,120 ft); Maiolica, 5150 m (16,890 ft)
    Gross thickness (top to bottom of producing interval) .... 580 m (1900 ft)
    Net thickness—total thickness of producing zones
        Average .... 300 m (985 ft)
        Maximum .... 580 m (1900 ft)
    Lithology
    Maiolica: white, cherty mudstones and wackestones with radiolarians and Tintinnidae
    Zandobbio: fine- to medium-grained dolomite and calcareous dolomite
    Dolomia Principale: fine- to medium-grained intraclastic and fossiliferous dolomite
    Porosity type .... Fracture porosity with minor intercrystalline and moldic porosity
    Average porosity 3%
    Average permeability .... 50 md

Seals:

    Upper
        Formation, fault, or other feature .... Bruntino Formation
        Lithology .... Marls
    Lateral
        Formation, fault, or other feature .... Fault

Source:

    Formation and age .... Argilliti di Riva di Solto (Late Triassic)
    Lithology .... Shales and marls
    Average total organic carbon (TOC) 0.8%
    Maximum TOC 2.26%
    Kerogen type (I, II, or III) .... III and II
    Vitrinite reflectance (maturation) .... $R_o = 2\%$
    Time of hydrocarbon expulsion .... Late Neogene
    Present depth to top of source .... 7000 m (22,950 ft)
    Thickness .... 500 m (1640 ft)
    Potential yield .... 3.5 kg of hydrocarbon per ton of rock

Appendix 2. Production Data

    Field name .... Malossa field
    Field size:
        Proved acres .... 10 km^2
        Number of wells all years 13

MALOSSA

Current number of wells 4
Well spacing .... 500 m (1640 ft)
Ultimate recoverable .... NA
Cumulative production .... NA
Annual production .... NA
Present decline rate .... NA
    Initial decline rate .... NA
    Overall decline rate .... NA
Annual water production .... NA
In place, total reserves .... NA
In place, per acre-foot .... NA
Primary recovery .... NA
Cumulative water production .... NA

Drilling and casing practices:

Amount of surface casing set .... None
Casing program
20 in. set at 300–600 m (980–1970 ft); 13⅜ in. set at 2400–2900 m (7875–12,800 ft); 9⅝ in. set at 5000 m (16,400 ft); 7 in. set at 5500–5900 m (18,050–19,360 ft)
Drilling mud .... AR/BS 1200–1300 g/L (10.0–10.8 lb/gal);
                                                                                                                                                      CLO/El 1300–2000 g/L (10.8–16.7 lb/gal)
Bit program .... 26 in. OSC 3A; 17½ in. OSC3A, DST; 12¼ in. 16 3A; 8⅜ in. V, F4J
High pressure zones .... From 4000 m (13,120 ft) to TD

Completion practices:

Interval(s) perforated .... 5500–6000 m (18,050–19,690 ft)
Well treatment .... Single completion

Formation evaluation:

Logging suites .... FDC/CNL/GR-DLL-IES-SL-HDT
Testing practices .... DST—long production test
Mud logging techniques .... Standard

Oil characteristics:

Type .... N + iso-alkanes = 55.3; cyclo-alkanes = 40.4; aromatic + NSO = 4.3
API gravity .... 52–55°
Base .... Paraffinic/mixed paraffinic
Initial GOR .... 950 m³/m³ (5335 ft³/bbl)
Sulfur, wt% .... Nil
Viscosity, SUS .... 1.73 Cst 20°C
Pour point .... NA
Gas-oil distillate .... NA

Field characteristics:

Average elevation .... 126 m (413 ft) asl
Initial pressure .... 1067 kg/cm² (15,179 psi)
Present pressure .... 800 kg/cm² (11,380 psi)
Pressure gradient .... 0.18 kg/cm²/m (0.78 psi/ft)
Temperature .... 155°C (311°F)
Geothermal gradient .... 0.023°C/m (1.26°F/100 ft)
Drive .... Gas and water drive
Oil column thickness .... 5.0 m (16 ft)
Oil-water contact .... 5800 m (19,030 ft)
Connate water 45%
Water salinity, TDS .... 14 g/L

Resistivity of water .... 0.5 ohm
Bulk volume water (%) .... NA
Transportation method and market for oil and gas:
Pipeline

Penal/Barrackpore Field—West Indies
South Trinidad Basin, Trinidad

BRIAN L. DYER
PHILIP COSGROVE
Trinidad and Tobago Oil Company Limited (TRINTOC)
Pointe-a-Pierre, Trinidad, West Indies

FIELD CLASSIFICATION

BASIN: South Trinidad
BASIN TYPE: Foldbelt
RESERVOIR ROCK TYPE: Sandstone
RESERVOIR AGE: Miocene and Pliocene
PETROLEUM TYPE: Oil and Gas
TRAP TYPE: Thrust Anticlines
RESERVOIR ENVIRONMENT OF DEPOSITION: Deep Water Turbidites
TRAP DESCRIPTION: Multiple traps in complex of thrust anticlines and thrust sheets; local stratigraphic control of reservoirs

LOCATION

The Penal/Barrackpore oilfield lies in southern Trinidad, West Indies (Figure 1), 6.8 mi (11 km) southeast of the town of San Fernando and southeast of the giant Forest Reserve oil field on the northern perimeter of the Siparia syncline (Figures 2 and 4). The Trinidad and Tobago Oil Company Limited (TRINTOC) owns and operates the Penal/Barrackpore field. Ultimate recoverable reserves for this field are estimated at 127.9 MMBO and 628.8 BsCFG.

HISTORY

Pre-Discovery

Sporadic drilling in the Barrackpore area during 1911 by South Naparima Oilfields Ltd. established oil production from the Pliocene Forest sandstone (Figure 3). This early drilling was based on field mapping of the surface expression of the Penal/Barrackpore anticline (Figure 4) where numerous oil seeps are present at the outcrop of the Forest sandstones.

Discovery

United British Oilfields of Trinidad (U.B.O.T.) initiated their Penal/Barrackpore drilling program in 1936, and although it was not then realized, the first well Penal 1 (TD 5570 ft [1699 m]) penetrated the Cipero Formation at 5400 ft (1647 m) but narrowly missed the Herrera sandstone.

Significant oil seeps within the Cipero outcrop to the northeast prompted continued interest in locating a possible subsurface Cipero play. In 1938, well Penal 23 was drilled to 4656 ft (1420 m) and produced 500 BOPD of a lighter oil (29.0° API) than the Forest (Wilson) sandstones (14.0° API); the Herrera sandstones had been penetrated but were not so identified (Bitterli, 1958). Penal 55, completed in 1940, was the next well to produce this lighter crude, but, because it was adjacent to the Digity mudflow, paleontological analyses indicated a mixture of Miocene and younger fauna, and the producing sandstone was not conclusively identified as the Herrera.

Finally, with the completion of Penal 64 in 1941 (800 BOPD), the sandstone in this well was paleontologically established as part of the Herrera sandstone member of the Cipero Formation, and the suspicion that Penal 23 had already penetrated the Herrera was confirmed.

Post-Discovery

In 1944, a bold stepout well, Penal 77, downdip of the "wet" Penal 74 well, established substantial Herrera production of 1700 BOPD in the southwest area of the Penal lease.

Drilling results to 1946 indicated the possibility of a faulted structure with a different strike south of the surface anticline. This "intermediate limb," as it is now known, was confirmed in 1946 by a 3000 ft (915 m) deepening of Penal 92 because of the

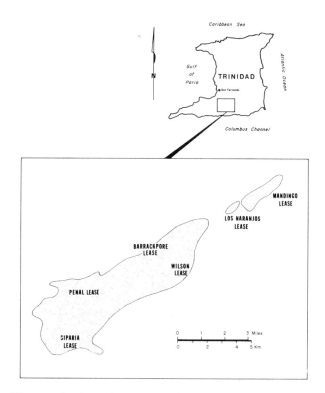

Figure 1. Location map showing area of Penal/Barrackpore field complex.

Limited (TRINTOC) and is called the Penal/Barrackpore field.

TRINTOC was formed when the government of the Republic of Trinidad and Tobago acquired the holdings of Shell Trinidad Ltd. in 1974 and, as such, assumed operatorship of the Penal lease. In 1985, the majority of the assets of Texaco Trinidad Incorporated (TEXTRIN) were vested in the government of Trinidad and Tobago and the assets of TEXTRIN were placed under the operations of TRINTOC. Since 1985, TRINTOC has drilled several successful extension tests in the former lease area. Extension drilling in this area continues to date for the shallow intermediate and underthrust structural levels.

(Note: The emphasis of this paper will be on the detailing of the middle Miocene accumulations and its associated structural and stratigraphic complexities. However, descriptions of the Pliocene play are necessarily undertaken in some detail because of important linkages with the middle Miocene play with respect to structure and the migration and trapping of hydrocarbon in the Penal/Barrackpore field.)

absence of the shallow Herrera sandstone. A good development of deeper Herrera sandstone was found below the Cipero marls. The initial production of this well was 600 BOPD. These sandstones were determined to be of the same age as the shallow Herrera but in an overturned position (Figure 5).

In 1949, Trinidad Leaseholds Limited drilled a deep test based on seismic (Barrackpore 336 well) to 10,087 ft (3076 m). The well flowed 790 BOPD initially and thus the deep Herrera sandstone was discovered. This sandstone is now recognized as the underthrust limb of the third structural level of the Penal/Barrackpore subsurface anticlinal feature.

In an attempt to extend the development of the shallow Herrera sandstones to the east of the Barrackpore field, wells were drilled in Los Naranjos (1953) and in Mandingo (1957) (Figure 1), and production was obtained from the shallow and intermediate limbs of the Penal/Barrackpore subsurface anticline (Figures 6 and 7). Extension tests to the west into the Siparia area to the east-southeast along the extrapolated anticlinal trend (1957) also proved successful.

The Barrackpore, Mandingo, and Los Naranjos leases were bought by Texaco International in 1956. The former Siparia, Penal, Barrackpore, Wilson, and Los Naranjos leases are now understood to belong to one large hydrocarbon field within a highly faulted anticlinal complex. This field is now owned and operated by the Trinidad and Tobago Oil Company

DISCOVERY METHOD

Methods employed included mapping of the surface geology and oil seeps in the Cipero and Forest formations. The Cipero and Forest formations outcrop to the northeast of the Penal/Barrackpore anticline (Figure 4). Numerous oil seeps and gas shows recorded in these outcrops prompted the drilling of exploration tests to the subsurface anticline, the position of which was extrapolated from surface geological data.

It should be noted, however, that the surface anticline, which was formed during Pliocene folding, is sometimes diapiric in nature and overprints folds formed during the earlier middle Miocene tectonic phase. Hydrocarbon accumulations within subsurface middle Miocene structures have been found in close proximity to the surface anticline but more normally are found in anticlinal substructures that are subparallel to but just south of the surface anticline (Figure 6).

Recommended exploration methods that should be employed today include (1) the acquisition of a 3D seismic data set at the existing field and to the possible eastern extension of the field, (2) the use of balanced cross sections (cross sections that account for stratal shortening as a result of deformation), and (3) accurate fault plane mapping of the numerous reverse faults. These methods and techniques can result in the drilling of exploration and development wells to exploit the extensions of known accumulations and to prove the existence of hydrocarbon in overthrust and subthrust structural traps.

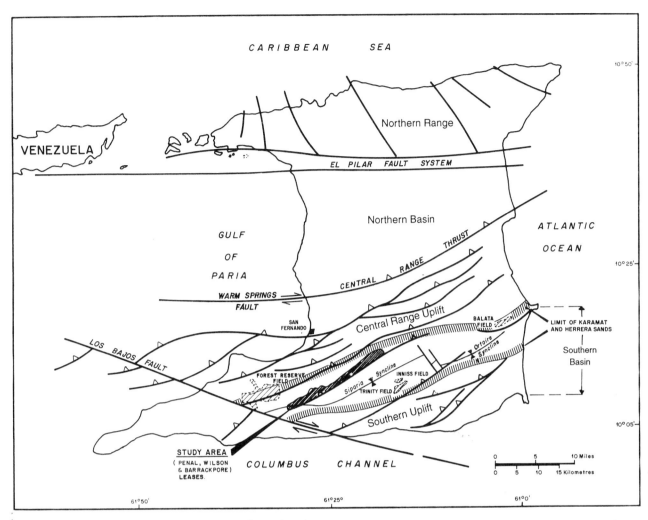

Figure 2. Index map of Trinidad, West Indies, showing Miocene lineaments of southern Trinidad (onshore) and other major tectonic elements.

STRUCTURE

Tectonic History

Detailed stratal analyses of geologic formations recognized at Trinidad's Southern basin have revealed five major tectonic episodes at middle Cretaceous, Late Cretaceous, late Eocene, middle Miocene, and late Pliocene (Tyson et al., 1989). The Tertiary tectonic episodes resulted from the interactions of the proto-Caribbean and South American plates. Cretaceous tectonics occurred with the rifting of the South American and African plates, the clockwise rotation of the South American plate, and the resultant compression and uplift at the proto-Caribbean and South American plate boundary. The tectonics of the Tertiary are of greater relevance to the Penal/Barrackpore structure.

The initiation of late Eocene rifting between South America and Antarctica, with the consequent northward movement of the former, resulted in the convergence of the South American and proto-Caribbean plates and tectonism in the Eastern Venezuelan and Trinidad basins. At Trinidad's Southern basin, a basement-involved thrust was formed at the present-day Central Range, with propagation southward as a series of imbricate thrusts with associated folding (Tyson et al., 1989).

A continuation of the northward movement of the South American plate with compressive interaction at the margin with the Caribbean plate climaxed in the middle Miocene tectonic event, which was a much stronger event than that of the late Eocene. This event produced southward-directed basement-involved thrusting that propagated southward as an imbricate thrust belt. Evidence of this tectonic phase

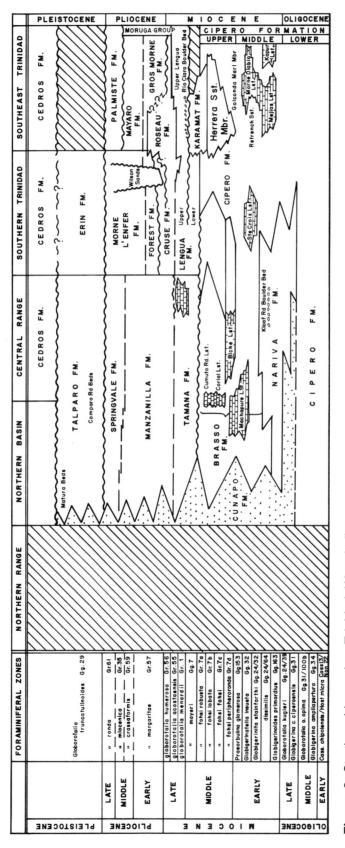

Figure 3. Stratigraphic section, Trinidad, West Indies. Cipero Formation rests unconformably on the Eocene Navet Formation.

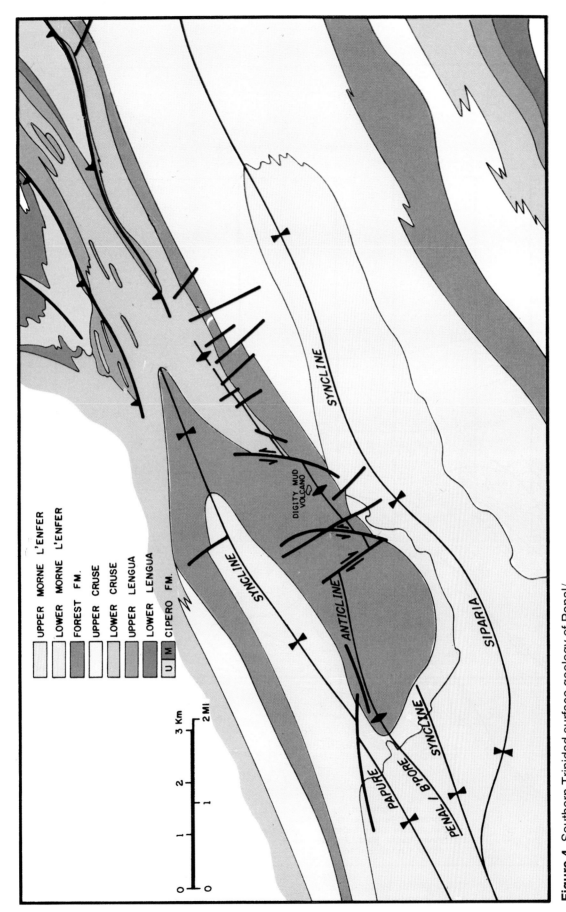

Figure 4. Southern Trinidad surface geology of Penal/Barrackpore anticline area.

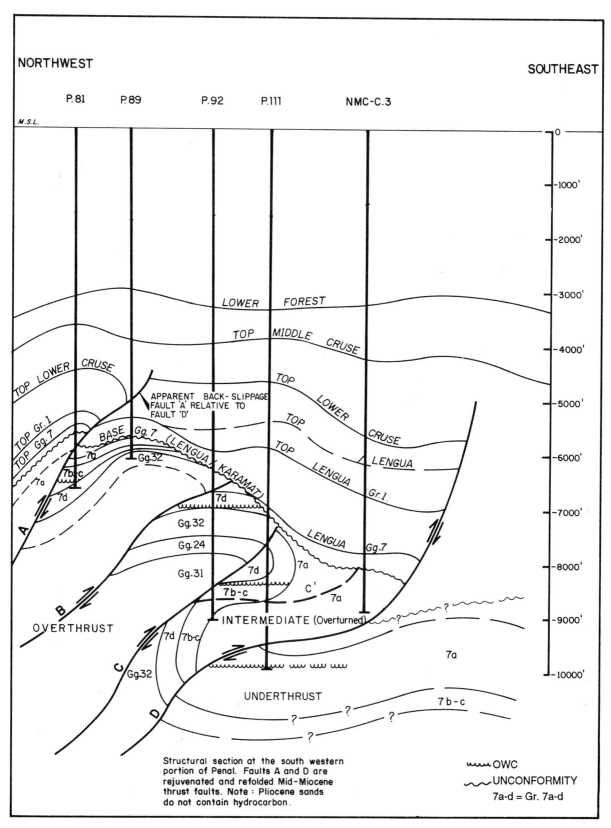

Figure 5. Schematic structural cross section, Penal field. Location shown on Figures 9 and 10.

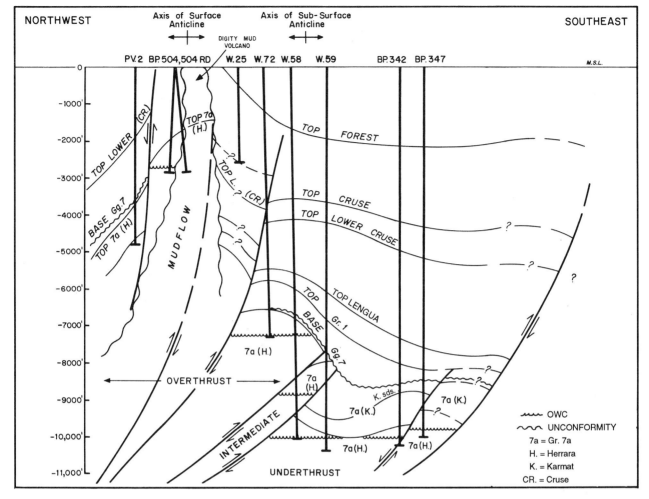

Figure 6. Schematic structural cross section, Barrackpore field. Location shown on Figures 9 and 10.

is recognized at the Penal/Barrackpore anticline where a distinct unconformity is mapped at the base of the *Globorotalia* 7 (Gg 7) biozone (Figures 5 and 6).

Heightened compressive interaction between the Caribbean and South American plates during the late Pliocene produced northwest–southeast-oriented stress release transcurrent faults, one of the most significant of which is the Los Bajos fault (Figure 2).

Other structures produced by the late Pliocene transpressional tectonics include basement thrust faults, listric normal faults, and detached simple folds that sometimes contain diapiric cores, e.g., the Penal/Barrackpore surface anticline (Figure 4).

Regional Structure

The Penal/Barrackpore field is located on one of a series of northeast-trending en echelon middle Miocene anticlinal structures found in the southern subbasin of onshore Trinidad (Figures 2 and 4).

The Herrera and Karamat sandstones of middle Miocene age (Figure 3) were sourced primarily from a major uplift to the north and were deposited with the deep water marls and clays of the Cipero Formation. The latter rests unconformably on previously deformed Eocene strata (Figure 8).

A major compressive event began near the end of the deposition of the Herrera sandstone. This middle Miocene compression occurred in response to the oblique collision of the Caribbean and South American plates and produced a series of en echelon folds and overthrusts (Figure 2), especially in areas close to the northern flanks of the present-day Siparia and Ortoire synclines where deformation is most complex (Figure 10).

At this complex uplifted area, there is an absence of Gr 7a and/or 7b biozones in some areas along the Miocene folds (Tyson et al., 1989). Isopach mapping of sand depositional trends within the Karamat further exemplifies these paleohighs (Figure 9) as reservoir sands within the Gr 7a biozones deposited around and along the flanks of the recognized Miocene paleohighs (Dyer, 1985).

Compression was accompanied by a period of extension that produced normal faults roughly perpendicular to the fold axes. A major unconformity (end of Gr 7a) (Figure 3) indicates the period of

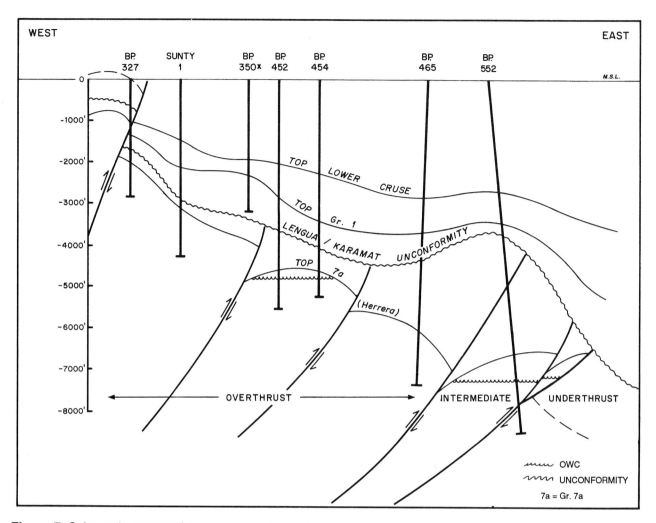

Figure 7. Schematic structural cross section, Barrackpore field. Location shown on Figures 9 and 10.

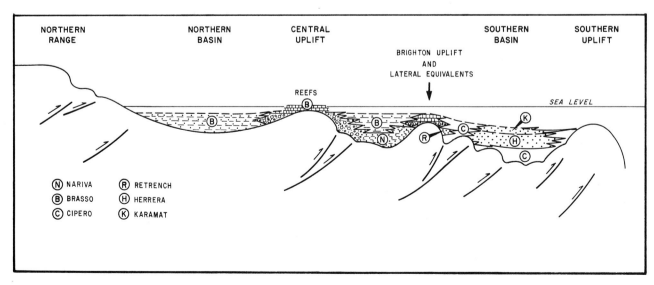

Figure 8. North-south diagrammatic cross section, Trinidad, just prior to the middle Miocene compression. See Figure 2.

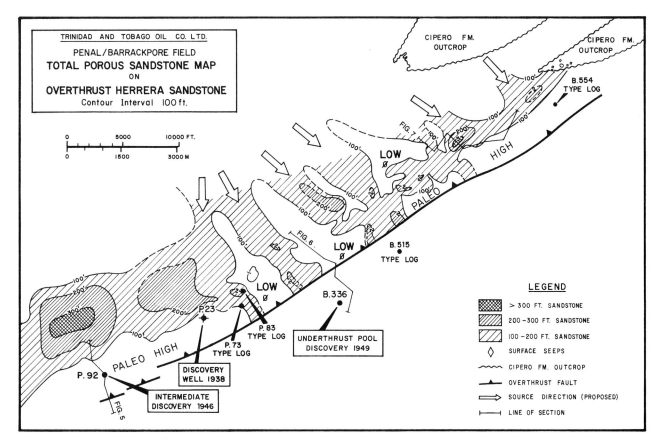

Figure 9. Total porous sandstone in the overthrust Herrera sandstone. Note oil seeps in Cipero Formation outcrop. LOW ϕ, areas of low porosity. Contour interval, 100 ft.

maximum compressive stress of the middle Miocene compression (Figure 5). Older rocks of various ages subcrop on this synorogenic surface, and in some areas the Herrera and Karamat sandstones have been either completely eroded or not deposited on the unconformity surface (Figure 5), as observed at wells Penal 92 and Penal 113.

Deep water clays, shales, and silts of Lengua (Gr 7 and Gr 1) (Figure 3) and Karamat (Gg 7) (Figure 3) age then covered the middle Miocene synorogenic surface. Boulder beds of Lengua age were deposited in a continuous band along the northern flank of the basin and generally to the north of the limit of the Herrera and Karamat sandstone deposits.

Karamat and Lengua deposition was superseded by the deposition of the Cruse Formation followed by the Pliocene Forest (Wilson) and Morne L'Enfer formations. These are primarily deltaic sandstones that onlap onto the previously formed middle Miocene structures. (Figures 5 and 6).

A transpressional episode occurred during late Pliocene producing: (1) shallow-rooted folds "piggybacking" older highs with associated high-angle reverse faults (Figures 5 and 6); (2) folds associated with low to high-angle reverse faults, generated along some of the pre-existing lineaments, formed during earlier compressional episodes (this resulted in the accentuation and overprinting of the smaller amplitude middle Miocene folds, Figures 5 and 6); (3) localized occurrences of back-faulting (Figure 5); (4) tear faults generally oriented northwest-southeast and sometimes northeast-southwest (Figure 4); (5) transcurrent faults as exemplified by the Los Bajos and the Warm Springs lineaments and their associated features (Figure 2); (6) listric fault development perpendicular to subperpendicular to the folds and on the flanks of pre-existing highs (Figure 10); and (7) mud diapirism associated with transcurrent faults, high-angle reverse faults (Figure 6), and some listric faults.

Local Structure

The southwest-plunging Penal/Barrackpore anticline is the main structural feature of the field and exhibits a regional tilt of 11° southwest. The internal detail of the structure (Figure 5) shows a highly folded and overthrusted middle Miocene structure below and asymmetrical to Pliocene folds, which are more open and of a larger amplitude. The reverse faults associated with these larger folds transect the middle Miocene structure (Figures 5 and 6). Bitterli (1958) broadly classified the middle Miocene thrust sheets into overthrust, intermediate (overturned), and underthrust structural levels, and this nomenclature has been successfully applied to a large extent to the Penal lease.

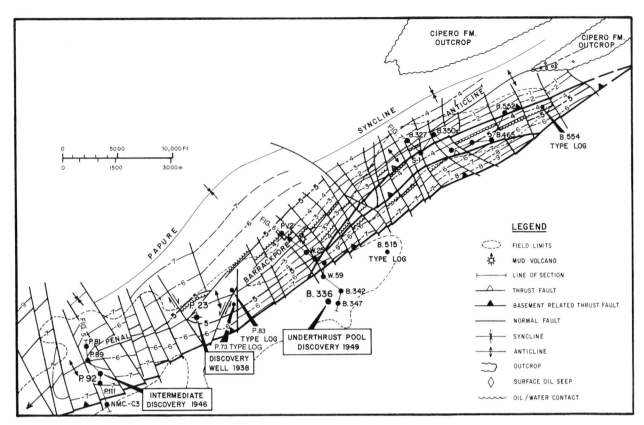

Figure 10. Structural contour map on marker above the overthrust Herrera sands. Contour interval, 1000 ft (305 m). Note oil seeps in Cipero Formation outcrop.

The cross section at the eastern end of the Barrackpore lease (Figure 7) illustrates structures associated with splayed reverse faults at the overthrust and intermediate structural levels. The entire anticlinal structure is intensely dissected by normal and tear faults that were formed mainly subperpendicular to the anticlinal axis. Dips at the broadly defined overthrust and intermediate structural levels vary considerably from gentle to subvertical to overturned. At the underthrust level, the structure is gently folded.

Oil-water contacts and oil-water transition zones vary by fault block, and this variance is primarily a function of selective hydrocarbon filling of fault blocks, the relative depths of the fault blocks, and the regional tilt of the structure (Figure 10).

The axis of the Pliocene fold is expressed at surface whereas fold axes at the Miocene structural levels are masked by Siparia synclinal feature and are primarily subsurface features.

STRATIGRAPHY

The Eocene to Miocene stratigraphy consists essentially of deep water marls and clays with pulses of turbiditic depositions of sands, silts, and mineral grains. Upper Miocene to Pliocene sediments of the study area are generally indicative of prodelta to freshwater delta plain environment and reflect a progressive shallowing of the Southern Trinidad basin.

Navet Formation (Eocene)

The Eocene Navet Formation consists mainly of light gray and greenish-gray, highly foraminiferal, sometimes weathered marls and marly clays. It embraces various isolated subunits including the Hospital Hill and Penitence Hill marls. It is sometimes siliceous and indurated with limestone bands and concretions.

Brasso Formation (Lower to Middle Miocene)

The lower to middle Miocene Brasso Formation is predominantly calcareous silts and clays, representing the neritic facies of the Cipero Formation (Figure 3). This formation is light bluish-gray in color and contains abundant micro- and megafossil faunas, glauconitic sandstone fragments of reef limestone, and conglomerates.

Cipero Formation (Oligocene to Miocene)

The Oligocene to Miocene Cipero Formation is characterized by deep water marls and clays, light blue to dark greenish-gray in color. This highly foraminiferal, glauconitic, sometimes limonitic and gypsiferous formation rests unconformably on the San Fernando Formation (Eocene) and unconformably beneath the clays and shales of the *Globorotalia mayeri* (Gg 7, Lengua/Karamat) zone. The Herrera and Karamat sandstone members were deposited within the Cipero Formation.

The Herrera member consists of deep water calcareous to noncalcareous, coarse- to very fine grained, quartz-rich sandstones and silty clays. These turbidites occur as conglomeratic sandstones and shale pebble beds, generally having a "salt and pepper" appearance because of the presence of quartz, feldspars, clays, and calcite (the light-colored constituents) and garnet, glauconite, barite, chlorite, zircon, tourmaline, and chert. The cherts comprise the majority of the dark-colored constituents of the sandstone.

The salient factors that contribute to the classification of Herrera and Karamat sandstones as turbidites include their presence within the deep water Cipero shales and the sedimentological characteristics of the recognized depositional cycles. The Herrera comprises the vast majority of the Miocene reservoir rock of the Penal/Barrackpore field. Coarse grains of the Herrera member are generally rounded, whereas the medium- to very fine grained constituents and the presence of texturally immature pre-Tertiary rock types suggest a nearby source for the Herrera sandstone.

Noncalcareous, greenish-gray clay, shales, sandstone, and silty sandstone are characteristics of the Karamat member. These sandstones were deposited in turbid water conditions in the chrono-interval typified by Gr 7a (*Globorotalia foshi robusta*) to Gg 7 (*Globorotalia mayeri*) faunal assemblages. This formation is transitional to the calcareous clays of the Lengua Formation, and also to the calcareous clays/shales of the Cipero Formation. It is characterized by benthonic shallow water fauna, and its sandstone lenses are lithologically similar to the Herrera sandstone. Where hydrocarbon-bearing, the Karamat sandstones are very prolific reservoirs.

Lengua Formation (Miocene)

Slightly to highly calcareous dark greenish marly clay, shale, and silts of the Lengua Formation are laterally transitional to the noncalcareous lower Cruse Formation (Figure 3). This formation is represented in places by rubble beds of older materials ranging from Oligocene to middle Miocene in age and is unconformable to the Cipero Formation in the crestal areas of the Penal/Barrackpore anticline.

Cruse Formation (Miocene to Pliocene)

Noncalcareous, micaceous sandstones and interbedded siltstones and clays are typical of the Cruse Formation. This formation is subdivided into the deep water clays and turbiditic sandstones of the noncalcareous lower Cruse Formation (Miocene), the delta-front middle Cruse sandstone (Pliocene), and the delta plain deposits of the upper Cruse (Pliocene).

Forest Formation (Pliocene)

Sandstone, siltstone, and clays of medium to darkgray color typify the Forest Formation (Figure 3). This formation is noncalcareous and sometimes limonitic, with interbedded brown siltstone and claystone. Along parts of the crestal areas of the Penal/Barrackpore anticline, the Forest Formation is found to lie unconformably on the lower Cruse Formation. The Forest sandstone is referred to as the Wilson sandstone in the former Wilson Lease.

Morne L'Enfer Formation (Pliocene)

The Morne L'Enfer Formation consists of an alternating series of equigranular quartz sandstones separated by bands of clay/shale and lignite. These sandstones were deposited in a freshwater environment.

TRAP

Hydrocarbon traps in the Miocene reservoirs at Penal/Barrackpore field are complex combination structural and stratigraphic traps. Structure is the dominant trapping mechanism of the field. Lengua and Karamat shales cover the middle Miocene unconformity surface where reservoir sandstones subcrop in some areas. These shales provide an effective seal.

Miocene and Pliocene Entrapment

Hydrocarbon trapping occurs in the crestal areas of the anticlines on a regional scale. Hydrocarbons have also migrated to the highest areas along strike of the thrusted structures, the highest areas being a function of the regional tilt.

Normal faults were created during and after the middle Miocene and late Pliocene compressional events. Northwest- and southeast-trending tear faults related to the late Pliocene transpression markedly influence the presence or absence of hydrocarbons in any given fault block (Figures 4 and 10). These normal and tear faults are mainly perpendicular or subperpendicular to the anticlinal axis and were formed prior to hydrocarbon migration via the associated thrust planes that connect to source rocks at greater depth.

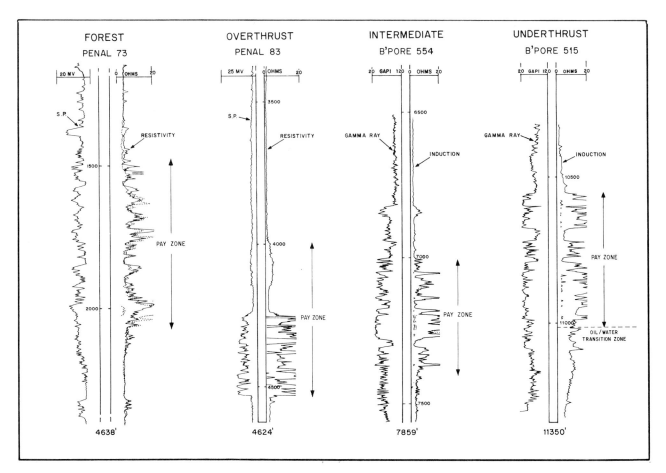

Figure 11. Type logs in Penal and Barrackpore fields. Locations of wells shown on Figures 9 and 10.

The normal and tear faults therefore acted as subtle controls on the hydrocarbons migrating up the thrust planes (Figures 4 and 10) and have resulted in wet fault blocks being structurally higher than juxtaposed oil-filled fault blocks in some places along the anticline; oil-water contacts and oil-water transition zones vary in the many fault blocks within the structural (thrusted) levels of the Miocene reservoirs and may be either a function of depth or of permeability (Figure 11). In the larger Pliocene structures associated with the Pliocene reservoirs, the major normal and tear faults may exert a similar effect.

Porosity pinch-outs resulting from small-scale variations in average grain size and permeability barriers created by calcite cementation (Griffiths, 1947) effect more subtle control on hydrocarbon distribution within the turbidite lenses in the anticlinal structure. The absence of Herrera and Karamat reservoirs at paleohighs, as indicated in Penal 92 (Figure 5), affects the presence of hydrocarbons on a localized scale (Figures 9 and 11).

Pliocene strata exhibit structural and stratigraphic trapping in highly lenticular deltaic reservoirs in the crest of folded structures of Pliocene age throughout the Southern basin. Reservoir sandstones lap onto the flanks of pre-existing middle Miocene highs, effectively trapping hydrocarbons migrating along the thrust faults and laterally along unconformities and the reservoir sandstones themselves. The lenticularity of the sandstones and their constituent grain sizes within the reservoirs subtly control the distribution of hydrocarbons, as observed in the selective fluid filling of sandstone lenses in some areas.

Faults

In the Penal/Barrackpore field area, thrust, tear, and normal faults have been identified. Middle Miocene compression resulted in basement-related thrust faulting that caused significant displacement of the beds within the Cipero Formation. Throw along these faults can exceed 1200 ft (366 m) (Figure 5).

Thrust faults have been identified by various methods, the most significant being the recognition of paleontological repeats of sections within wells. Dipmeter analyses were successful in further identification of these where there were significant changes in both direction and magnitude.

Besides major thrust planes previously identified (Figures 2 and 5) within the study area, associated thrust faults with lesser amounts of displacement have been recognized by means of fluid anomalies and by one or more of the aforementioned techniques. Thrusts associated with the middle Miocene structure sometimes splay and also are seen to merge laterally and horizontally in places (Figure 10).

In the late Pliocene compressional phase, normal faults bisected the entire Miocene–Pliocene section roughly perpendicular to the axis of the Penal/Barrackpore anticline. Displacement of the Miocene thrusts by these faults is noticeable in this field (Figure 10). Generally, normal faults that have significant throws form seals between adjacent reservoirs; these include the majority of normal faults found in the field. The larger of these faults can be identified by the extrapolation of surface fault trends (Figure 4). Fluid anomalies and contour miss-ties from lithostratigraphic correlations have also been used to map normal faults (Figure 10).

Tear faulting is not as ubiquitous as the normal and thrust faults formed during the late Pliocene transpressional phase. These faults are easily recognizable in both the surface and subsurface by the significant lateral offset of the general trend in the area. The faults formed effective barriers between juxtaposed reservoirs (Figure 10).

Recently reprocessed seismic lines have assisted in further defining both thrust and normal faults within the Penal/Barrackpore field.

Reservoirs

The Herrera and Karamat sandstones are classified as turbidites owing to the presence of the sandstones within deep water clays and shales (based on the presence of planktonics) formed south of the uplifted Cretaceous, Paleocene, and Eocene source (Figure 11).

X-ray diffraction analyses show that these sandstones are quartz-rich, contain varied amounts of calcite, clay, chert, and heavy minerals including garnet, glaucophane, epidote, chloritoid, staurolite, rutile, and chlorite. Many of these constituents indicate derivation from the low-grade metamorphics of the Northern Range. The total mineral content gives the sandstones a "salt and pepper" appearance. The recognition of repeated cycles of sedimentation, each with wide-ranging sedimentary structures and exhibiting some or all of the sequence of strata described by Bouma in 1961 (Poole, 1968), also indicate a turbidite depositional regime. Quartz occurs as detrital grains, secondary over-growths in pore throats, and as minute euhedral crystals coating detrital grains (SEM analysis by Callendar and Schroeder, 1983).

Great variations in pore shape and pore size are observed. This variation is caused by the relative presence or absence of calcite, clay, and quartz cement formed during diagenesis, and by the varied sorting of constituent grains that ranges from poor- to medium- to well-sorted (litharenites) (Griffiths, 1947). Porosities range from 16 to 34% and permeabilities from 0.5 to 120 md. They are directly related to the degree of diagenesis. Porosity occlusion between a depth range of 4500 to 12,000 ft (1373 to 3660 m) is generally low in the shallow (overthrust) Herrera and high in the more consolidated deep (underthrust) Herrera tested in the study area. Calcite is recognized as the most abundant cement, filling pore spaces and fractures. Clay cement and quartz overgrowths are also found (Callendar and Schroeder, 1983).

Oil-column heights are variable and exceed 300 ft (92 m) in places. Oil-water transition zones and oil-water contacts sometimes show marked variation from one fault block to another. Gravity segregation of hydrocarbons is a notable feature of the accumulations at the intermediate and underthrust structural levels. Moderately heavy crude (average 22° API) down flank in dipping reservoir beds is progressively succeeded updip by lighter crudes, condensate, and gas. At the intermediate level, segregation can be observed on a local scale at the Penal and Barrackpore leases. Segregation at the underthrust level is regional in scale; condensate at the southwest area of the Penal lease progressively gives way to heavier crudes eastward to a structural saddle in the underthrust sheet at the Barrackpore lease.

Biodegradation of oil at the Pliocene levels results from communication with surface waters in areas where the sandstones outcrop. Primary production is facilitated by a solution gas-drive mechanism for the majority of the reservoirs in the Penal/Barrackpore field. Water-drive assists production in the high-angled overturned (intermediate) limb at the Penal field.

Flow rates vary from 100 to 500 BOPD, and recovery efficiencies are 15% to greater than 20% of original oil in place (OOIP) for depletion/partial water drive reservoirs. API gravities range from 22 to 30°, and initial GORs average 1500 ft^3/bbl. Barriers and baffles to well flow include thrust planes (lateral flow), and normal and tear faults (developed perpendicular to oblique to the anticlinal axes), porosity loss and stratigraphic pinch-outs, low-permeability (tight) streaks, and wax formation.

Well spacing is controlled primarily by drainage radii that exhibit a variance of 4.5 to 30.0 ac (1.8 to 12.2 ha), as indicated by the mapped extent of the sandstone development at the objective horizon(s). Fault block size and porosity/permeability considerations also affect the effective drainage radius. In areas where thrust splays repeat the overthrust reservoirs and where the overthrusted level(s) directly overlie the intermediate structural level, multiple reservoirs can be encountered in one well. Additional wells, termed "pup" wells, have thus been drilled to test and exploit the shallower of such reservoirs.

Source and Migration

The source bed for Trinidad's oils are the mudstones (argillites) of the Upper Cretaceous

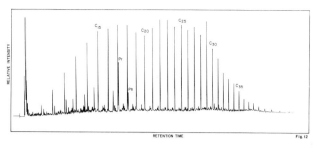

Figure 12. Gas chromatogram of the saturate fraction of 21° API crude oil from the intermediate Herrera reservoir (6700–7200 ft) in well Barrackpore 512.

Naparima Hill Formation (no units older than Oligocene are shown on stratigraphic charts). Total organic carbon ranges from 1.7 to 8.0% and averages 3.0%. Visual kerogen analyses show that the Naparima Hill Formation mudstones are dominated by amorphous kerogens, highly fluorescent in ultraviolet light, and presumably of algal or bacterial origin. Type II kerogens (hydrogen index = 234–641 mg HC/g TOC) with very good to excellent oil source potential (greater than 6 kg HC/ton rock) are indicated from pyrolysis. Using a vitrinite reflectance of 0.7–0.8%, an average hydrogen index of 490 mg HC/g TOC, and an average TOC of 3.0%, a potential yield of 128×10^6 bbl/km^3 can be calculated for the Naparima Hill source rocks (Rodrigues, 1988). A nonbiodegraded, medium-gravity crude oil, generated primarily from marine organic matter with some terrestrial influence, is indicated in the gas chromatogram of the saturate fraction of oil from the intermediate reservoir (Figure 12).

Peak oil generation and expulsion occurred relatively late in the evolution of the Southern basin of Trinidad, probably because of low heat flow associated with the Tertiary section. Thermal history modeling suggests that expulsion from source rocks and migration to structures occurred in the Pliocene, possibly as a result of placement of some areas of folded or subthrusted (footwall) source rocks into the oil window during the transpressional tectonic activity. Thus, deep-seated faults connecting Upper Cretaceous source rocks to Miocene turbidites and Pliocene deltaic sandstones would be the most likely migration pathways for oil accumulations in the Penal/Barrackpore field.

Hydrocarbons secondarily migrated up deep-seated thrust planes and also up the major normal and tear faults. Migration then continued along unconformities and throughout the sandstone reservoirs.

The presence of light hydrocarbons in Herrera sandstones subcropping on the middle Miocene synorogenic surface at the Penal lease (Figure 5) lends support to Pliocene migration and trapping.

The source rocks in and around the study area have not to date been penetrated or sampled. Source rocks, however, are expected to be mature based on extrapolation to approximately 17,000 ft (5185 m) and deeper where the Naparima Hill Formation may be found associated with basement-related faulting. The lithology of the Naparima Hill Formation varies from the typical light gray to olive gray calcareous siltstone/claystone ("argillite"), which is the most common lithology, to dark gray, noncalcareous shales and black, impure limestones and mudstones. Black bituminous material is common as coatings on planar surfaces and within fractures. Many of the Naparima Hill cores give off a faint to strong petroleum odor on freshly broken surfaces. (Rodrigues, 1988). The most oil-prone organic facies are the dark gray to olive gray, calcareous to noncalcareous siltstones and claystones.

EXPLORATION CONCEPTS

The Penal/Barrackpore field, in the southern Trinidad subbasin, is one of several east–northeast- to west–southwest-trending fold lines (seen as lineaments) formed during the peak period of compression during the Miocene, and considerably accentuated during the peak period of the late Pliocene transpressional phase (Figure 2).

Similarly deformed middle Miocene structures containing Nariva, Retrench, Karamat, and Herrera sandstone reservoirs are found south of the Central Range thrust (Figures 2 and 8). The Balata and Innis/Trinity fields and their associated middle Miocene structures are Southern basin analogues to the Miocene accumulations at the Penal/Barrackpore field (Figure 2).

The Balata subsurface anticline is one of the east–northeast- to west–southwest-trending structures formed during the maximum period of stress in what is referred to as the middle Miocene tectonic phase. In late Pliocene times, the sediments in the Balata area were refolded, thrusted, tear-faulted, and tilted, as exemplified in the surface anticline and its associated features. Mobile shale diapirs of Lengua to Cipero age are associated with high-angle reverse faults of Pliocene age (Figure 2). Regionally, the potentially giant oil fields recently discovered near Maturin in the Eastern Venezuela basin are structurally time-equivalent analogues to the middle Miocene features formed in the Trinidad basin at the height of the middle Miocene compression. The Furrial field, associated with the Anaco thrust, and other fields such as the Santa Ana, Santa Rosa, San Jacquin, and Toco fields are located along middle Miocene anticlines.

Analogues to the Penal/Barrackpore Pliocene play in the Eastern Venezuela basin include the Jusepin, Orocual, and Quiriquire fields in which sandstones of Pliocene age wedge out on the flanks of middle Miocene highs. The Forest Reserve field in the Southern basin, Trinidad, also provides an analogue to the Pliocene play at Penal/Barrackpore field. At Forest Reserve, one of the lineaments similar in orientation to the anticlinal axis is considered to be an older thrust fault rejuvenated during the late

Pliocene transpression. Diapiric shale was released along this lineament and along unconformities and bedding planes within the Cruse, Forest, and Morne L'Enfer formations (Figure 2). This suggests that tectonic activity may have occurred throughout the deposition of the upper Cruse, Forest, and Morne L'Enfer strata as they onlapped onto a middle Miocene high that was continually being accentuated.

Studies of the evolution of the Trinidad basin have indicated the general extent of the areas of potential sand deposition of Eocene to Pliocene age. Such highly integrated studies have enhanced plays associated with the Miocene compression on the basis of a plausible explanation for the existence of structures that do not contain hydrocarbons. The modification of seismic parameters and modifications to processing and reprocessing inputs have served to improve the resolution of thrusted Miocene structures containing essentially characterless shales, turbidite sandstones, shale diapirs, and complex cross-faulting. Some success has been achieved in those areas where the structures are not duplexed.

To date, the Penal/Barrackpore field is the only Trinidad oil field where hydrocarbons trapped in both late Pliocene and middle Miocene structures have been successfully explored and exploited. This illustrates the uniqueness of the play.

The northerly sourced Herrera and Karamat turbidites are the youngest and most southerly of those within the Cipero shales of the Central Range and Southern Trinidad subbasins. The younger Gr 7a sandstones of Herrera and Karamat age are seen to onlap onto and to be deposited around pre-existing highs at both the Penal/Barrackpore and Balata oil fields. The situation of these sandstones of Gr 7a-d age directly below sealing shales has resulted in significant accumulation within these reservoirs at both fields. These characteristics of middle Miocene reservoir development are an important guide for exploration and exploitation efforts in the Trinidad basin.

Accurate mapping of individual regional thrust planes and associated splays to properly delineate hydrocarbon trends and to detect lateral and vertical thrust fault mergers also proved to be useful in the construction of other maps and in the development of real models of structural conditions of the entrapments. This approach has since been used successfully to extend the productive intermediate Herrera structural level between eastern Barrackpore and Wilson leases (Dyer, 1987) and to make successful extensions in the Los Naranjos area.

ACKNOWLEDGMENTS

The author wishes to acknowledge the following persons for their contributions to this paper: Mr. Wayne G. Bertrand, Divisional Manager, Exploration and Production at the Trinidad and Tobago Oil Company Limited (TRINTOC), made possible the release of information used in this paper. Technical inputs were provided by Messrs. Llewllyn Tyson, Stephen Babb, Kirton Rodrigues, and Brian Baptiste of TRINTOC, and drafting services were provided by Alloy Fong and personnel of his staff at TRINTOC's cartographic department. The author also wishes to sincerely thank Brian Baptiste for editing and proofreading the text and Ms. Maria Michael for providing typing services.

REFERENCES CITED

Bitterli, P., 1958, Herrera subsurface structure of Penal Field, Trinidad, B.W.I.: AAPG Bulletin, v. 42, n. 1, p. 145-158.

Callendar, C. A., and P. A. Schroeder, 1983, Barrackpore Field (Trinidad), Geochemical Study: Internal Memo, Texaco Inc., Bellaire, Texas.

Dyer, B. L., 1985, Balata drilling prospects—a geological review: Internal Geological Report TG. 85/02.

Dyer, B. L., 1987, A geological analysis of the Intermediate Herrera at Mandingo, Barrackpore, Wilson and East Penal fields: Internal Geological Report 567.

Griffiths, J. C., 1947, Petrography of the Herrera Facies, Oligocene of South Trinidad: Internal Report GPS-304.

Poole, W. G., 1968, Some sedimentary features of the Herrera sands of the Clarke Road area, Barrackpore oilfield, Trinidad, West Indies: Trans. 4th Caribbean Geological Conference, Port of Spain, 1965, p. 101-110.

Rodrigues, K., 1988, Oil source bed recognition and crude oil correlation, Trinidad, West Indies: Advances in Organic Geochemistry 1987; Organic Geochemistry, v. 13, nos. 1-3, p. 365-371.

Tyson, L., B. L. Dyer, and S. Babb, 1989, An exploration study of onshore Trinidad: TRINTOC Internal Report.

Appendix 1. Field Description

Field name .... *Penal/Barrackpore field*
Ultimate recoverable reserves .... *127.9 MMBO; 628.8 BCFG*

Field location:
- **Country** .... *Trinidad*
- **State**
- **Basin/Province** .... *Southern basin, Trinidad*

Field discovery:
- Year first pay discovered .... *Oligocene Cipero Formation, Herrera sandstone member 1938*
- Year second pay discovered .... *Herrera sandstone (deeper, "intermediate lens") 1946*
- Year third pay discovered .... *Herrera sandstone, deep 1949*

Discovery well name and general location:
- First pay .... *Penal No. 23; Central Penal lease*
- Second pay .... *Penal No. 92; southwest Penal area*
- Third pay .... *Barrackpore No. 336; West Barrackpore*

Discovery well operator .... *United British Oilfields of Trinidad*
- Second pay .... *United British Oilfields of Trinidad*
- Third pay .... *Trinidad Leaseholds Limited*

IP:
- First pay .... *500 BOPD*
- Second pay .... *600 BOPD*
- Third pay .... *790 BOPD*

All other zones with shows of oil and gas in the field:

Age	Formation	Type of Show
Pliocene	Forest/Wilson	Oil

Geologic concept leading to discovery and method or methods used to delineate prospect
Wilson Forest Sandstone: surface geology, oil seeps; delineation: subsurface geology; discovery: nontechnical; delineation: subsurface geology, oil seeps, seismic surveys

Structure:

Province/basin type .... *Foreland basin (modified marginal sag basin); Bally 42, Klemme III Bb*

Tectonic history
The Penal/Barrackpore anticline formed as a result of middle Miocene compression that occurred in response to the oblique collision of the Caribbean and South American plates. A transpressional episode occurred during the late Pliocene causing rejuvenation of faults and additional folding.

Regional structure
One of a series of east-northeast–west-southwest-trending en echelon anticlinal structures found in the Southern basin of onshore Trinidad.

Local structure
Southwest-plunging Penal/Barrackpore anticline with a regional tilt of approximately 11° toward the southwest.

Trap:

Trap type(s) .... *Thrust-faulted anticlinal trap with multiple pays; numerous stratigraphic traps*

Chronostratigraphy	Formation	Depth to Top in ft (m)
Pliocene	Forest Formation	−1667 (508)
	Cruse Formation	−3583 (1093)
Miocene	Lengua Formation	−5583 (1703)
	Unconformity/Herrera (OT)	−6750 (2058)
	Fault/Herrera (Int)	−8417 (2567)
	Fault/Herrera (UT)	−9333 (2846)/ −9500 (2897)

Reservoir characteristics:

- **Number of reservoirs** 3
- **Formations** .... 1 (Cipero)
- **Ages** .... Mid-late Miocene
- **Depths to tops of reservoirs**
  Overthrust (OT), 2000–7000 ft (610–2135 m); Intermediate (Int), 6500–10,500 ft (1983–3203 m); Underthrust (UT), 8500–12,000 ft (2593–3660 m)
- **Gross thickness (top to bottom of producing interval)** .... OT, 500 ft (153 m); Int, 800–1000 ft (244–305 m); UT, 500–800 ft (153–244 m)
- **Net thickness—total thickness of producing zones**
  - Average .... OT, 200 ft (61 m); Int, 450 ft (137 m); UT, 350 ft (107 m)
  - Maximum .... OT 400 ft (222 m); Int, 550 ft (168 m); UT, 450 ft (137 m)
- **Lithology** .... Subrounded, poorly sorted, clear quartz and chert with varying grain sizes
- **Porosity type** .... Intergranular
- **Porosity** .... Ranges from 16 to 34%
- **Permeability** .... Ranges from 0.5 to 120 md

Seals:

- **Upper**
  - Formation, fault, or other feature .... Impermeable formation
  - Lithology .... Shale
- **Lateral**
  - Formation, fault, or other feature .... Impermeable formation, faults
  - Lithology .... Shale

Source:

- **Formation and age** .... Naparima Hill Formation (Turonian–Campanian)
- **Lithology** .... Mudstone (argillite)
- **Average total organic carbon (TOC)** 3.0%
- **Maximum TOC** 8.0%
- **Kerogen type (I, II, or III)** II
- **Vitrinite reflectance (maturation)** .... $R_o = 0.6$–0.8%
- **Time of hydrocarbon expulsion** .... Pliocene
- **Present depth to top of source** .... 15,000 ft (4575 m)
- **Thickness** .... 2000 ft (610 m)
- **Potential yield** .... 128×10^6 bbl/km^3

Appendix 2. Production Data

- **Field name** .... Penal/Barrackpore field
- **Field size:**
  - **Proved acres** .... 15,500 (6278 ha)
  - **Number of wells all years** 776

PENAL

Current number of wells	154 (producing)
Well spacing	4.5 to 30 ac (1.8 to 12.2 ha)
Ultimate recoverable	127.9 million bbl oil, 628.8 BCFG as of end of 1988
Cumulative production	114.4 million bbl oil, 606.2 BCFG as of end of 1988
Annual production	1.5 million bbl oil, 3.5 BCFG as of end of 1988
Present decline rate	NA
Initial decline rate	NA
Overall decline rate	NA
Annual water production	NA
In place, total reserves	NA
In place, per acre foot	NA
Primary recovery	127.9 million bbl oil, 628.8 BCFG as of end of 1988
Secondary recovery	NA
Enhanced recovery	NA
Cumulative water production	22.6 bbl as of end of 1988

Drilling and casing practices:

Amount of surface casing set	1000 ft (305 m)

Casing program
(1) Shallow Herrera: 9⅝-in. at 1000–1500 ft (305–458 m); 7-in. at TD 4500–7500 ft (1373–2288 m)
(2) Intermediate Herrera: 13⅜-in. at 1000–1200 ft (305–366 m); 9⅝-in. at 5000–5200 ft (1525–1586 m); 7-in. at 7000–11,000 ft (2135–3355 m)
(3) Deep Herrera: 18⅝-in. at 800 ft (244 m); 13⅜-in. at 3500–4000 ft (1068–1220 m); 9⅝-in. at 10,000–10,500 ft (3050–3202 m); 7-in. at 11,000–12,500 ft (3355–3812 m)

Drilling mud	Lignite-lignosulfonate water base
Bit program	24-in./17½-in./12¼-in./8½-in.
High pressure zones	Drilling through the Lengua Karamat (depth highly variable)

Completion practices:

Interval(s) perforated	Shallow, 2000–7500 ft (610–2288 m); intermediate, 6500–11,000 ft (1982–3355 m); deep, 8500–12,500 ft (2592–3812 m)
Well treatment	Nil

Formation evaluation:

Logging suites	GR-SP-IES/GR-FDC-CNL/GR-MSFL/shot, sonic
Testing practices	Production testing
Mud logging techniques	Conventional

Oil characteristics:

Type	Paraffinic
API gravity	20–30° API
Initial GOR	1500 SCF/bbl
Sulfur, wt%	0.5–0.8
Viscosity, SUS	40–140
Pour point	<0 to +80°F (−18 to 27°C)
Gas-oil distillate	20 to 40%

Field characteristics:

Average elevation	100 ft (30.5 m)
Initial pressure	3900–6900 psi (26,890–47,576 kPa)
Present pressure	500+ psi (3448 kPa)
Pressure gradient	0.65 psi/ft (14.69 kPa/m)
Temperature	150–205°F (66–96°C)
Geothermal gradient	0.012°F/ft (0.022°C/m)
Drive	Solution gas

Oil column thickness	*Avg. 350 ft (106.7 m)*
Oil-water contact	*Varies considerably*
Connate water	*20–40%*
Water salinity, TDS	*30,000 ppm*
Resistivity of water	*0.3 rm at 80°F*
Bulk volume water (%)	*NA*

Transportation method and market for oil and gas:

Transportation via pipeline for both oil and gas; crude oil sold to the National Refinery

Laojunmiao Field—People's Republic of China
Jiuquan Basin, Gansu Province

ZHAI GUANGMING
SONG JIANGUO
Scientific Research Institute of Petroleum Exploration and Development
Beijing, People's Republic of China

FIELD CLASSIFICATION

BASIN: Jiuquan
BASIN TYPE: Foredeep
RESERVOIR ROCK TYPE: Sandstone
RESERVOIR ENVIRONMENT OF DEPOSITION: Fluvial
RESERVOIR AGE: Miocene
PETROLEUM TYPE: Oil
TRAP TYPE: Faulted Anticline
TRAP DESCRIPTION: Thrusted asymmetric anticline broken up by imbricated minor thrust faults on forelimb

LOCATION

Laojunmiao oil field is located in the foothills of the Qilian Mountains with an elevation of 2000 to 3000 m, adjacent to Yuman city, Gansu Province, northwestern China (Figure 1). The climate is arid, the mean annual precipitation being 100 mm. The lowest temperature in winter is -30°C. The Shiyou River streams from the Qilian Mountains and flows into the Jiuquan basin from the south to the north (Figure 2). In addition to Laojunmiao, other fields in the area are Yaerxia, Shiyougou, Dan Bei, and Baiyanghe. The railway and highway linking Lanzhou city to the east (southeastern Gansu Province) and Wulumuqi city to the west makes the field's transportation convenient.

Ultimate recovery for the Laojunmiao field is expected to exceed 133 million bbl of oil.

HISTORY

In 1938, a field geological survey, headed by pioneering geologist Dr. Sun Jianchu, mapped the Jiuquan basin, Gansu Province, and discovered the Laojunmiao and Shiyougou anticlines. In March 1939, the No. 1 well (Figure 3) was drilled on the top of the Laojunmiao anticline and encountered the first oil reservoir, now termed "K," at a depth of 88 to 153 m in Miocene rocks. This pay initially produced 75 bbl/day of oil. The No. 4 well reached the second pay, termed "L," also in Miocene rocks, at a depth of 439 m and blew out strongly in April 1941. In 1944, an exploratory well, DH-1, was drilled on the top of the anticline, reaching the third Miocene pay termed "M," which produced 15 bbl/day of oil.

Exploration over 40 years has shown that the Laojunmiao field is a complex anticlinal trap associated with overthrusting and that its reservoir rocks are the sandstones of the Miocene Baiyanghe Formation. Development of the Laojunmiao oil field began in 1941, peripheral water-flooding began in 1955, and flooding from the top or flanks of the field in 1959. Currently, there are 807 producing wells and 220 injection wells, with average daily oil production of 7630 bbl.

Since the discovery of the Laojunmiao field, it has been recognized that the anticlinal zone in the southern Jiuquan basin is prospective for oil and gas accumulations. During the 1950s, exploration extended onto the Qingcaowan, Yaerxia, Shiyougou, and Baiyanghe anticlines, resulting in the discovery of the Yaerxia, Shiyougou, and Baiyanghe oil fields (Figure 4). Later exploration and deep drilling found the fractured reservoir of the Silurian beneath the Tertiary reservoir of the Yaerxia oil field and the Dan Bei oil field north of the Baiyanghe oil field.

DISCOVERY METHOD

In the light of the surface oil seepages and geological mapping, a wildcat well was sited on the top of the Laojunmiao anticline that discovered the Laojunmiao oil field in 1939. The structural complexity of the anticlinal zone has limited the effectiveness of

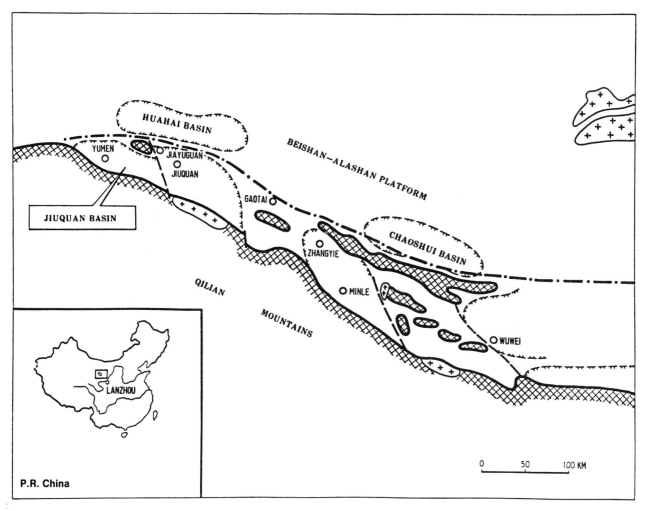

Figure 1. Location map showing the Jiuquan basin in the western part of the Gansu corridor along the front of the Qilian Mountains. The corridor contains several sedimentary basins.

seismic data. Geophysical surveys have not provided useful information for exploring deeper reservoirs and thrust sheets underlying the discovered oil fields.

STRUCTURE

Tectonic History

The Province/Basin type is classified as Bally 32, Klemme IIIA (St. John et al., 1984). The Jiuquan basin lies along the front of the Qilian Mountains, bounded by the Qilian Mountains to the south, the Kuantai and Hei Mountains to the north, and Hongliuxia Mountain to the west. It covers 2700 km². The sedimentary section comprises the Cenozoic-Mesozoic and the upper Paleozoic that overlies the folded and metamorphosed basement of the lower Paleozoic. The basin formation resulted from plate collision in western China that continued from the Mesozoic to the Cenozoic.

Hercynian Orogeny

The coal-bearing formations deposited in a transitional environment during the Upper Carboniferous and the continental sediments of the Permian comprise the first sedimentary cover of the Jiuquan basin. The marine regression might have been caused by the Hercynian orogeny.

Indo-Sinian Orogeny

During the Early Triassic, fluvial systems developed along the southern border of the Jiuquan basin. The Indo-Sinian orogeny at the end of the Late Triassic uplifted the basin and caused some folds to be formed and the Upper Triassic to be eroded.

Yanshanian Orogeny

During the Early Jurassic, the northern part of the basin was uplifted and the southern part subsided. The Longfengshan Formation of the Lower Jurassic, composed of swampy and fluvial sediments, unconformably overlies the Lower and Middle Triassic Chijinbao Formation. The Yanshanian orogeny of the Middle Jurassic caused the Longfeng-

Figure 2. LANDSAT image of the Jiuquan basin and surroundings, clearly showing the Quaternary alluvial fans derived from the Quilian Mountains that cover the entire basin. The Laojunmiao oil field is located in the southern part of the basin.

shan Formation to be folded, uplifted, and, hence, eroded. The regional extension resulting from the collision of south Tibet with north Tibet triggered the rifting stage of the Jiuquan basin. This stage lasted until the end of the Early Cretaceous. The rifted depressions were filled with the fluvial and lacustrine sediments of the Late Jurassic and the Early Cretaceous. The Yanshanian orogeny during the Late Cretaceous ended the rifting stage and uplifted the basin, resulting in the hiatus between the Lower Cretaceous and the Oligocene.

Himalayan Orogeny

During the early Oligocene, the basinward thrusting from the Qilian Mountains resulting from the collision of the Indian plate with the Eurasian plate caused the formation of the Cenozoic Jiuquan basin. Oligocene and Miocene sediments, termed the Houshaogou Formation and the Baiyanghe Formation, respectively, were deposited in fluvial and lacustrine environments. The thickness of this section decreased from the south to the north of the basin. In the Pliocene, the Qilian Mountains were rapidly elevated, and the depocenter migrated northward to the central basin where the fluvial sediments of the Shulehe Formation reached 1300 to 3000 m in thickness. The collision of the Indian plate with the Eurasian plate in the Neogene made the Qilian Mountains thrust northward, resulting in the formation of the folded and thrust zones in the southern basin, whereas in the central and northern basin, folding was weaker and synclinal and monoclinal structures formed.

Regional Structure

From the Neogene to the early Quaternary, the Himalayan orogeny strongly affected the Jiuquan basin and deformed its sedimentary cover. Three regional structural zones can be identified in this basin (Figure 4).

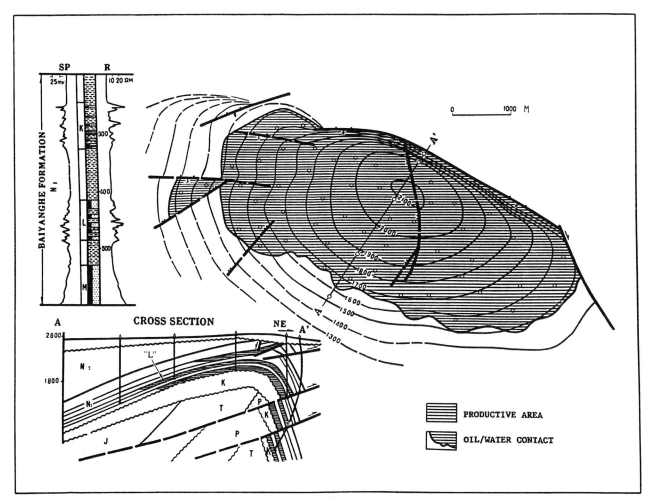

Figure 3. Structural map of the top of the Miocene Baiyanghe Formation "L" sandstone. Contour interval is 100 m (values above sea level). The type log shows the stratigraphic positions of the "K," "L," and "M" sandstones. The location of the cross section A-A' is shown on the structural map. The cross section demonstrates the asymmetry of the Laojunmiao anticline and demonstrates the underlying thrust sheets that have oil pools in the same reservoir units segregated from the main field in the anticline.

1. South Anticlinal Zone is elongated parallel to the Qilian Mountains with a length of 60 km and an area of 500 km². It comprises five anticlines trending east to west, which are the Qingcaowan, Yaerxia, Laojunmiao, Shiyougou, and Dahondquan, from west to east. The basement becomes deeper to the west.
2. Central Synclinal Zone, covering 1600 km², is located in the central basin. The basement of the synclinal zone is cut by normal faults trending north-northeast that are the result of the rifting stage during Late Jurassic and Early Cretaceous.
3. North Monoclinal Zone, covering an area of 600 km², is a gentle slope with several structural noses and flexures. Its Tertiary strata dip monoclinally south to southward.

Local Structures

There are two prominent types of local structures in the Jiuquan basin. First are the anticlines that occur in trend in the south anticlinal zone, created by basinward thrusting. These structures, formed during the Neogene, are all asymmetric, with a northern steep limb (60° to 80°) and a southern gentler limb (5° to 40°). The north limbs are usually broken by thrust faults. One of these structures is the Laojunmiao anticline, which lies in the middle part of the south anticlinal zone (Figure 3). The trap section of this paper provides detailed information about it.

Strong structural noses distributed in the north monoclinal zone comprise the second distinctive structural type of the area. They too were formed in the Neogene. Some of them are exposed, such as the Baiyanghe structural nose (Figure 4).

STRATIGRAPHY

The Jiuquan basin is mainly a Cenozoic-Mesozoic sedimentary basin overlying the folded basement of

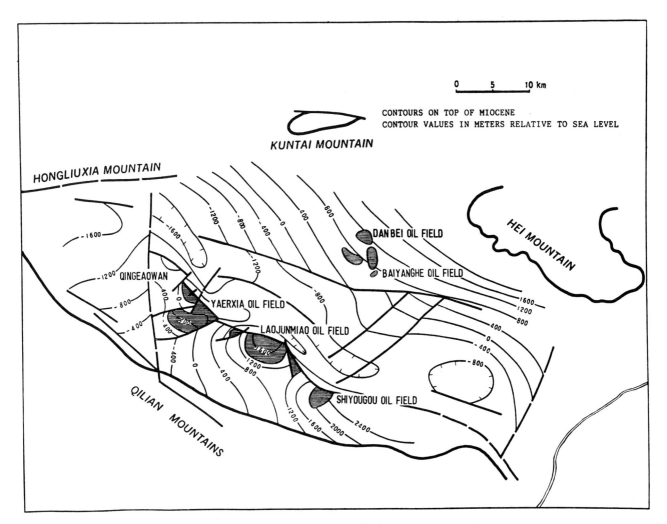

Figure 4. Structure contour map of the top of the Miocene of the Jiuquan basin showing the oil fields, the south anticlinal zone, the central synclinal zone, and the north monoclinal zone. Contour interval is 400 m.

the lower Paleozoic. Outcrops of the basement are extensive along the edge of the basin. Quaternary alluvial fans cover the entire basin floor.

The metamorphic lower Paleozoic is composed of marine clastic, carbonate, and volcanic rocks, with a thickness of 10,000 to 15,000 m. The upper Paleozoic, which is distributed in the southern basin, was deposited in transitional and continental environments. Its thickness ranges from 1000 to 2220 m.

The Cenozoic–Mesozoic is the main part of the sedimentary cover and comprises coal-bearing formations with a thickness of 4800 to 5600 m. It is this section that produces commercial oil in the Jiuquan basin. The generalized stratigraphic column above the Carboniferous system for the Jiuquan basin is shown in Figure 5. (The Permian is shown as included in the upper part of the Carboniferous system.) Structural-stratigraphic relationships of rock units of the Laojunmiao anticline are shown by Figure 3.

Silurian

The Silurian (lower Paleozoic, Figure 5) comprises colorful, fine-grained sandstones and shales interbedded with limestones. It contains graptolites and brachiopods. The Silurian rocks are slightly metamorphosed and form the basement of the basin. Fractured Silurian rocks are the reservoir of the Yaerxia oil field. The lower Paleozoic (Figure 5) is primarily Silurian.

Carboniferous

The main lithologies of the Carboniferous are dark gray mudstones, shales, and limestones interbedded with sandstones, conglomerates, and thin coal beds. It contains fusulinids, brachiopods, and plant fossils. Its thickness is about 220 m.

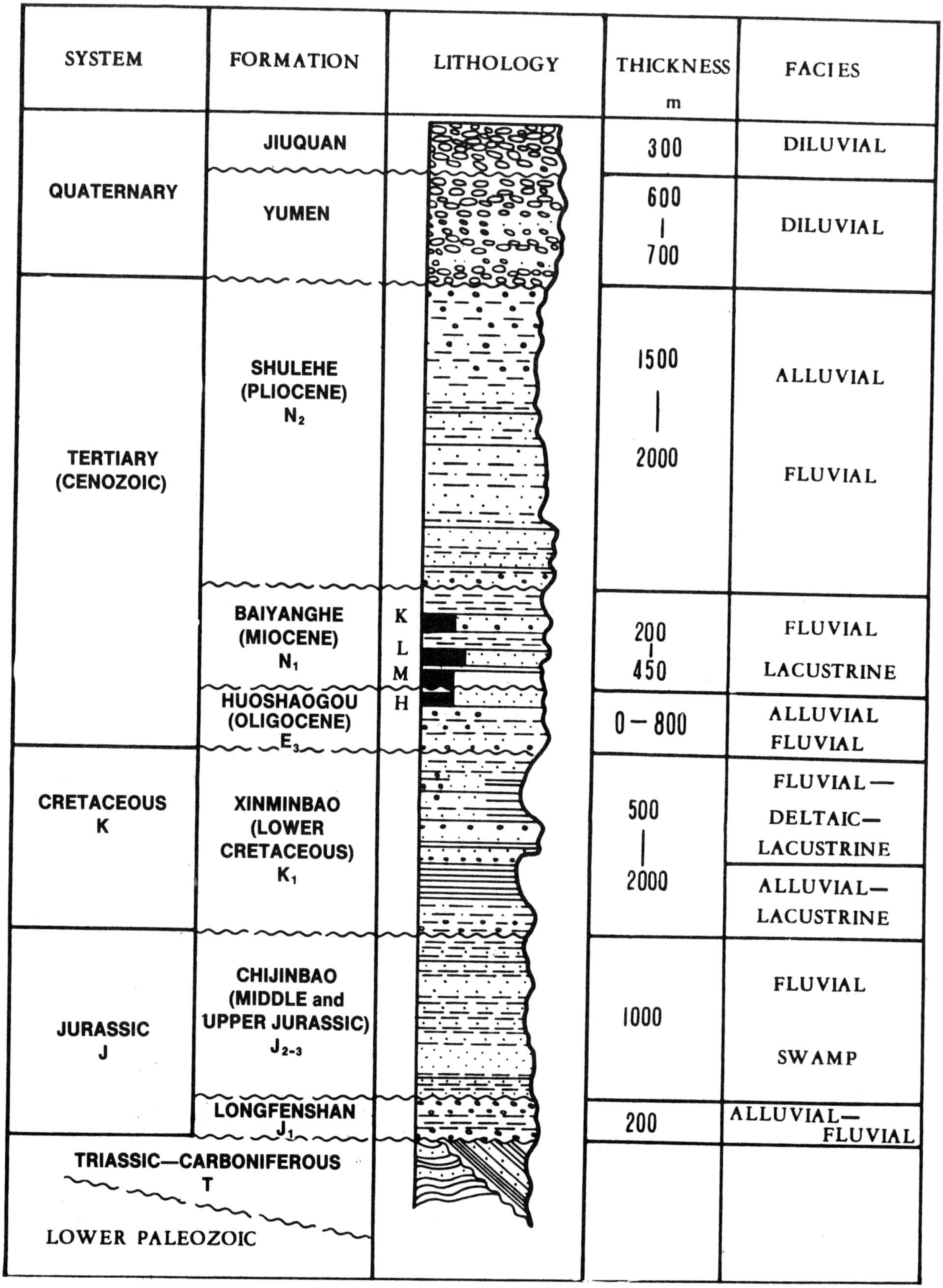

SYSTEM	FORMATION	LITHOLOGY	THICKNESS m	FACIES
QUATERNARY	JIUQUAN		300	DILUVIAL
	YUMEN		600 – 700	DILUVIAL
TERTIARY (CENOZOIC)	SHULEHE (PLIOCENE) N₂		1500 – 2000	ALLUVIAL / FLUVIAL
	BAIYANGHE (MIOCENE) N₁	K L M	200 – 450	FLUVIAL / LACUSTRINE
	HUOSHAOGOU (OLIGOCENE) E₃	H	0 – 800	ALLUVIAL FLUVIAL
CRETACEOUS K	XINMINBAO (LOWER CRETACEOUS) K₁		500 – 2000	FLUVIAL–DELTAIC–LACUSTRINE / ALLUVIAL–LACUSTRINE
JURASSIC J	CHIJINBAO (MIDDLE and UPPER JURASSIC) J₂₋₃		1000	FLUVIAL / SWAMP
	LONGFENSHAN J₁		200	ALLUVIAL–FLUVIAL
TRIASSIC–CARBONIFEROUS T				
LOWER PALEOZOIC				

Figure 5. Stratigraphic column of the Jiuquan basin. Baiyanghe Formation "K," "L," and "M" and Houshaogou Formation "H" sandstones are known Miocene oil reservoirs of the basin. The reservoir rocks were deposited mainly in fluvial environments. (In this columnar section, the Permian is included in the Carboniferous system.)

Permian

The Dahonggou Formation of the Lower Permian consists of sandstones interbedded with mudstones and thin coal beds and contains a large amount of plant fossils. It is 150 to 570 m thick.

Yaogou Group of the Upper Permian includes psephite and silicarenite interbedded with mudstone and shale as main lithologies. This group occurs in the southern and central basin, with a thickness ranging from 300 to 700 m.

Triassic (T)

Xidagou Formation of the Lower to Middle Triassic is composed of violet-red and light gray sandstones and conglomerates interbedded with mudstones, with prominent large-scale cross bedding. It is 400 to 1100 m thick and is distributed in the southern basin.

Jurassic

Longfengshan Formation of the Lower Jurassic (J11) contains violet-red and gray-green conglomerates, sandstones, and mudstones interbedded with coal. It is 200 m thick and was deposited in alluvial and fluvial environments. The formation is distributed in the southern basin and uncomformably overlies the Triassic.

The lower part of the Chijinbao Formation of the Middle to Upper Jurassic is sandstone and conglomerate; the middle part, dark-gray mudstones interbedded with siltstones; and the upper part, fluvial sandstones and conglomerates. The total thickness is 500 to 1000 m. The Chijinbao Formation is distributed in the northern and southern basin.

Cretaceous (K)

Xinminbao Formation of the Lower Cretaceous (K1) comprises dark-gray mudstones, shales, and marls interbedded with sandstones and conglomerates, which were deposited in lacustrine and fluvial environments. It is 500 to 2000 m thick and is the main source rock of the Jiuquan basin.

Tertiary

The Houshaogou Formation of the Oligocene (E3h) is composed of light-gray silicarenites interbedded with brown-red mudstones 0 to 800 m thick. It occurs in the northern basin and is the main reservoir rock in this area.

The Baiyanghe Formation of the Miocene (N1b) is distributed over the entire basin. It is composed of fluvial and lacustrine deposits and divided into three members from the bottom to the top:

1. Janquanzi Member comprises orange to red sandstones interbedded with red mudstones and gypsum 100 to 140 m thick and is the main reservoir rock of the Jiuquan basin. Included in the member are the productive K, L, and M sandstones.
2. Shiyougou Member comprises dark-red mudstones interbedded with sandstones and gypsum 44 to 105 m thick. It provides the regional caprock for the Jiuquan basin.
3. Ganyouquan Member comprises red mudstones and sandstones interbedded with rudaceous sandstones with a thickness of 260 to 280 m. It forms a secondary reservoir rock of the basin to which the "K" pay belongs.

The Shulehe Formation of the Pliocene (N2s) consists of light-gray sandstones, red mudstones, and conglomerates deposited in fluvial and diluvial environments. It occurs over the whole basin, with a thickness of 1500 to 2000 m.

TRAP

The Laojunmiao traps are in an asymmetric anticline resulting from basinward thrusting during the Neogene (Figure 3). The structure trends westnorthwest to east-southeast and has a length of 8 km and a width of 3 km. Anticlinal closure is about 800 m. The dip angle of the southern flank is 15–20° while that of the northern flank is 70–90°. The proved oil-bearing area is 15.34 km^2. The level of the oil-water contact (OWC) has been influenced by various factors, including facies changes and the faults, both normal and reverse, that cut the structure. On the less disturbed south flank of the anticline, the OWC at most places is at an elevation between 1500 and 1700 m. As shown in Figure 3, the Miocene reservoir rocks enclosed in thrust sheets beneath the anticlinal fold have trapped oil at levels far below the OWC of the main field.

RESERVOIR ROCKS

The Janquanzi Member of the Baiyanghe Formation, which had a source in the adjacent Qilian Mountains, is the main reservoir rock of the Jiuquan basin, including "K," "L," and "M" zones (Figure 5). The member was deposited in a fluvial environment, the paleoriver flowing through Shiyougou, Laojunmiao, and Yaerxia into the lake at Qingcaowan (Figure 6). The member is 100 to 400 m thick and is buried to a depth of 790 to 2400 m. The "L" pay comprises orange medium- to fine-grained sandstone (Figure 7) with a thickness of 60 m. The porosity averages 23%. The permeability ranges from 100 md to 500 md. Oil productive capacity per well during the initial stage was 225 to 750 bbl/day.

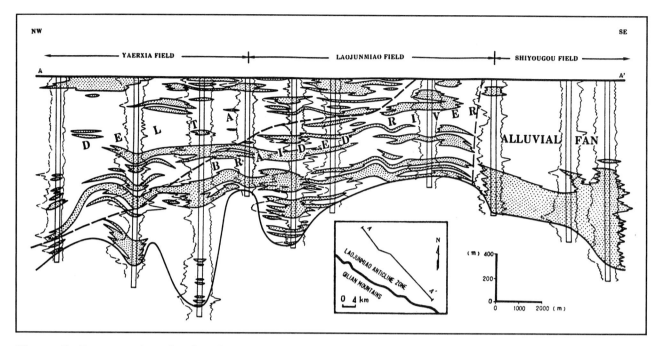

Figure 6. Cross section showing the lateral change of lithology and facies of the Baiyanghe Formation "L" sand along the south anticlinal zone. The alluvial fan system changes to braided river and lacustrine delta systems, in turn, from southeast to northwest.

FAULT

There are two groups of faults recognized in the Jiuquan basin (Figures 3 and 4).

1. One group comprises the overthrust faults that strike parallel to the Qilian Mountains. These include the overthrust faults along the north flank of the south anticlinal zone and the faults on the north monoclinal zone formed by the compressive stress from the Qilian Mountains during the Pliocene.
2. The outer group comprises normal faults that trend mainly north-northeast. They formed during the Early Cretaceous and controlled the lacustrine sedimentation during that time. The faults were reactivated during the Neogene and provided migration channels and traps for hydrocarbons from the oil-generating area to the Tertiary reservoir rocks.

SOURCE ROCKS

The dark lacustrine mudstones of the Lower Cretaceous Xinminbao Formation are the main hydrocarbon source rocks of the Jiuquan basin. The cumulative thickness of these mudstones reaches 500 to 1000 m. The total organic carbon is 0.98 to 1.26%, and kerogen is type II, with a vitrinite reflectance of 0.5 to 1.2%. The oil produced from the Laojunmiao oil field is paraffinic (Figure 8).

It has been recognized that the first petroleum migration occurred at the end of the Late Cretaceous and filled fault traps in the Cretaceous and the fractured reservoir rocks of the Silurian basement. The migration routes include interbedded fractures and faults (Figure 9). In the Neogene, the source rocks went into a high mature stage. Unconformities and faults were good migration channels. Petroleum migrated from the oil-generating area, through Yaerxia, Laojunmiao, to Shiyougou along the regional strike of the south anticlinal zone. The migration covered a distance of 20 km and more.

EXPLORATION CONCEPTS

The occurrence of hydrocarbon accumulations in the Jiuquan basin is closely related to the oil-generating area. On the left side (Figure 9) is the Qingxi depression, which is the oil-generating area. Hydrocarbons from the source rocks migrated laterally, first into the alluvial sandstones, forming accumulations in fault traps of the Cretaceous, and then into fractures in the Silurian basement, forming fractured lithology pools. Most hydrocarbons, however, migrated upward through faults into reservoir rocks of the Janquanzi Member of the Miocene Baiyanghe Formation. Lateral migration

Figure 7. SEM photos and photomicrograph of a thin section illustrating the pore types of the Baiyanghe Formation "L" sandstone. The main type of pore is intergranular with good connectivity. Kaolinite and dolomite are the main cement minerals. Size of the pore throats range from 2 to 3 μ. (A) SEM photo ×100 of "L" siltstone with intergranular porosity. (B) SEM photo ×100 of fine sandstone with intergranular porosity, good connectivity, Throat sizes 2 to 3 μ. (C) SEM photo ×340 of "L" sandstone with the interparticle porosity, good connectivity, filled by dolomite. (D) SEM photo ×100 of "L" sandstone showing kaolinite and dolomite cements. (E) Photomicrograph of "L" sandstone thin section showing intergranular porosity.

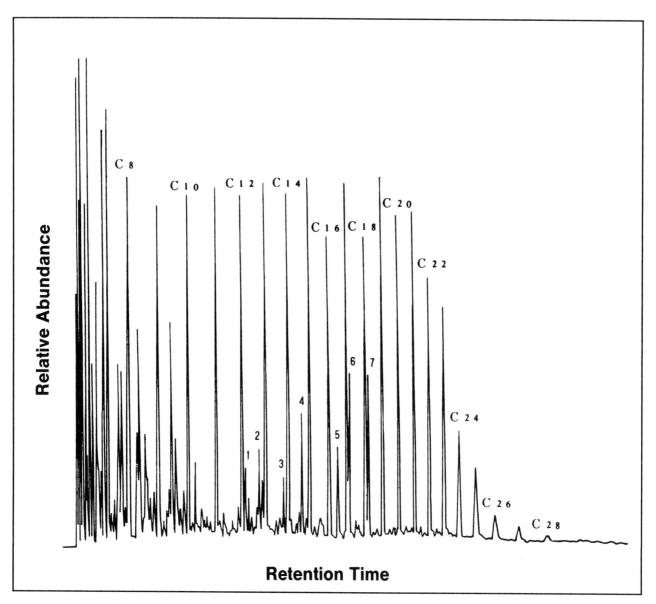

Figure 8. Representative gas chromatogram of "L" pay oil sample showing that the oil is paraffinic.

along the south anticlinal zone resulted in the formation of the Yaerxia, Laojunmiao, and Shiyougou oil fields.

As illustrated by Figure 9, the following conclusions have been reached:

1. The basement faults are vertical channels for hydrocarbon migration, but various kinds of oil and gas traps can form pools along these faults if there are favorable reservoir conditions or favorable entrapment conditions on either side of the migration channels.
2. The anticlinal zone adjacent to the oil-generating Qingxi depression is on the preferential migration route of the hydrocarbons so that the accumulation usually took place along this trend first.
3. The structural complexity and the sharp change in sedimentary facies limited the distance of hydrocarbon migration so that most oil fields occur near the oil-generating depression in the plain and near the source rocks in the sedimentary section.

These conclusions may provide an excellent model for exploration in similar basins.

EXPERIENCES

During the early stage of exploration in the Jiuquan basin, knowledge was limited, and while the Cretaceous was recognized as the petroleum source,

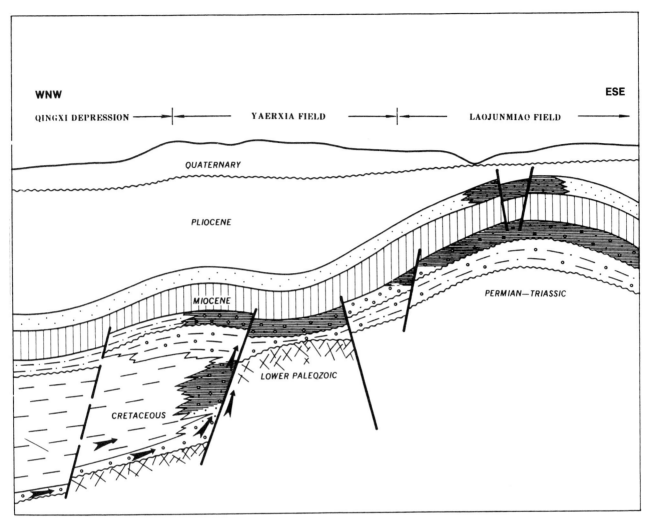

Figure 9. Cross section, west to east through the south anticlinal zone, illustrating hydrocarbon migration and accumulations. Unconformities and faults have provided main migration routes. The Miocene sandstones are favorable reservoirs for entrapment.

only the Tertiary was regarded as providing reservoirs and trapping conditions. Exploration for deeper Cretaceous and basement reservoirs was long delayed. A modern exploration program should begin with a basin analysis, including stratigraphic, geochemical, geophysical, and structural studies that will lead to earlier recognition of all of the potential reservoir beds and the criteria for recognition of conditions for petroleum entrapment, including fracturing and faults.

ACKNOWLEDGMENTS

We thank the Yuman Petroleum Administration and Scientific Research Institute of Petroleum Exploration and Development of China National Petroleum Corperation for supporting the presentation of this paper and for access to exploration data.

REFERENCE CITED

St. John, B., A. W. Bally, and H. D. Klemme, 1984, Sedimentary province of the world—hydrocarbon productive and nonproductive: American Association of Petroleum Geologists map series.

Appendix 1. Field Description

Field name .... *Laojunmiao field*
Ultimate recoverable reserves .... *133 MMbbl*
Field location:
    Country .... *People's Republic of China*
    State .... *Gansu Province*
    Basin/Province .... *Jiuquan basin*
Field discovery:
    Year first pay discovered .... *Miocene Baiyanghe Formation K ss   1939*
    Year second pay discovered .... *Baiyanghe Formation L ss (main reservoir)   1940*
    Third pay .... *Baiyanghe Formation M ss   1944*
Discovery well name and general location:
    First pay .... *No. 1 on top of Laojunmiao anticline*
    Second pay .... *No. 4 on top of anticline*
    Third pay .... *No. DH-1 on top of anticline*
IP in barrels per day and/or cubic feet or cubic meters per day:
    First pay .... *75 bbl/day*
    Second pay .... *100 m^3/day (3530 ft^3/day) of gas*
    Third pay .... *15 bbl/day*

All other zones with shows of oil and gas in the field:

Chronostratigraphic Unit	Formation	Type of Show
Tertiary	*Malianquan*	*Oil seepage*

Geologic concept leading to discovery and method or methods used to delineate prospect, e.g., surface geology, subsurface geology, seeps, magnetic data, gravity data, seismic data, seismic refraction, nontechnical:
Surface geology and drilling near oil seepages.

Structure:
    Province/basin type .... *Bally 32, Klemme IIIA*
    Tectonic history
    As a result of the collision of the Indian plate with the Eurasian plate during the late Tertiary, basinward overthrusting from the Qilian Mountains caused the formation of the anticlinal zone along the south border of the Jiuquan basin. One of the anticlines belonging to this zone forms the Laojunmiao oil field.
    Regional structure
    The Laojunmiao oil field lies in the south anticlinal zone of the basin.
    Local structure
    Asymmetric anticline trending west-northwest to east-southeast, with a steep north limb and a gentle south limb.

Trap:
    Trap type(s) .... *The trap is anticlinal with multiple pays*

Basin stratigraphy (major stratigraphic intervals from surface to deepest penetration in field):

Chronostratigraphy	Formation	Depth to Top in m
Tertiary	*Ganyouquan*	*329*
Tertiary	*Laojunmiao (K)*	*790*
Tertiary	*Malianquan (M)*	*810*

Reservoir characteristics:
  Number of reservoirs .... 3 pays of medium- and fine-grained sandstone
  Formations .... Baiyanghe
  Ages .... Miocene
  Depths to tops of reservoirs .... 790 m (L sandstone)
  Gross thickness (top to bottom of producing interval) .... 60 m
    Net thickness—total thickness of producing L sandstone
      Average .... 11.9 m
      Maximum .... 21 m
  Lithology .... Orange-red quartz sandstone, well sorted
  Porosity type .... Intergranular
  Average porosity 23%
  Average permeability .... 256 md

Seals:
  Upper
    Formation, fault, or other feature .... Shiyougou Formation
    Lithology .... Mudstone
  Lateral
    Formation, fault, or other feature .... Shiyougou Formation
    Lithology .... Mudstone

Source:
  Formation and age .... Xinminbao Formation of the Lower Cretaceous
  Lithology .... Mudstone
  Average total organic carbon (TOC) 1.19%
  Maximum TOC .... NA
  Kerogen type (I, II, or III) II
  Vitrinite reflectance (maturation) .... $R_o = 0.5-1.2\%$
  Time of hydrocarbon expulsion .... Neogene
  Present depth to top of source .... 2500-3000 m
  Thickness .... 535 m
  Potential yield .... 220-290 MMbbl/km^3

LAOJUNMIAO

Appendix 2. Production Data

Field name .... Laojunmiao field
Field size:
  Proved acres .... 15.34 km^3
  Number of wells all years 853
  Current number of wells 807
  Well spacing .... 200 m
  Ultimate recoverable .... 133 MMbbl
  Cumulative production .... 116 MMbbl
  Annual production .... 2.55 MMbbl
  Present decline rate 4%
    Initial decline rate .... NA
    Overall decline rate .... NA

LAOJUNMIAO

Annual water production	1.3 MMm³ (8.8 MMbbl)
In place, total reserves	380 MMbbl
In place, per acre-foot	NA
Primary recovery	2.2 MMbbl
Secondary recovery	115 MMbbl
Enhanced recovery	Cannot be estimated
Cumulative water production	24 MMm³ (150 MMbbl)

Drilling and casing practices:

Amount of surface casing set	
Casing program	12¾-in., 6⅝-in.
Drilling mud	Water-based mud
Bit program	Varies
High pressure zones	None

Completion practices:

Interval(s) perforated	Multiple intervals
Well treatment	Acidizing

Formation evaluation:

Logging suites	Older wells: gamma ray, resistivity; modern wells: gamma ray, neutron, sonic resistivity
Testing practices	Typically production tested
Mud logging techniques	Not typically used for infill wells

Oil characteristics:

Type	Paraffin
API gravity	32–34°
Base	NA
Initial GOR	70 m³/MT (337 ft³/bbl)
Sulfur, wt%	0.18
Viscosity, SUS	3.25 cp at 25°C
Pour point	15.5°C
Gas-oil distillate	NA

Field characteristics:

Average elevation	1900 m
Initial pressure	94.6 atm (1372 psi)
Present pressure	68.4 atm (992 psi)
Pressure gradient	1.2 atm/10 m (0.53 psi/ft)
Temperature	30°C
Geothermal gradient	3.3°C/100 m
Drive	Solution gas/water flood
Oil column thickness	600 m
Oil-water contact	Varies
Connate water	NA
Water salinity, TDS	70,000 mg/L
Resistivity of water	NA
Bulk volume water (%)	NA

Transportation method and market for oil and gas:
By train to Lanzhou refinery in Lanzhou city, Gansu Province.

Raman, Bati-Raman, and Garzan Fields—Turkey
Diyarbakir Basin, Southeast Turkey

RAFIK SALEM
Consulting Geologist
Fort Worth, Texas

HASAN OZBAHCECI
AYHAN UNGOR
MOSTAFA ISBILIR
Turkish Petroleum Corporation
Ankara, Turkey

FIELD CLASSIFICATION

BASIN: Diyarbakir
BASIN TYPE: Foredeep
RESERVOIR ROCK TYPE: Limestone
RESERVOIR ENVIRONMENT OF
 DEPOSITION: Carbonate Platform
RESERVOIR AGE: Cretaceous
PETROLEUM TYPE: Oil
TRAP TYPE: Two Anticlines, One Pinch-Out
TRAP DESCRIPTION: Elongate anticlines broken by numerous reverse faults and an updip pinch-out of reef and grainstone facies into nonporous facies

LOCATION

Three oil fields in southeast Turkey in Production District X (Figure 1) represent stratigraphic and structural hydrocarbon traps in late Maastrichtian Stage rocks (Figure 2). These are the Raman, Bati-Raman, and Garzan oil fields (Figure 3), located immediately south of the main front of the east–west-trending Miocene overthrust in southeast Turkey (Diyarbakir province), approximately 100 km (60 mi) north of the Syrian border. Other important oil and gas fields in District X are the Selmo oil field and the Dodan gas field (carbon dioxide) to the north of the Raman field and the Dincer and Camurlu oil fields to the south of Raman field, almost on the Syrian border.

The three oil fields described in this paper are in an area characterized by well-defined physiographic features including topographically high, narrow, elongated, and partly eroded anticlines. These alternate with broad, flat, synclinal areas. The trend of the structures is northwest-southeast in the north, changing strike to a more east-west direction southward away from the overthrust belt.

The estimated in-place oil reserves of the Raman oil field are 394,000,000 STB; Garzan oil field, 170,000,000 STB; and Bati-Raman oil field, 1,850,000,000 STB of oil (see Appendices 1 through 6). The productivities of Garzan and Bati-Raman fields are shown by Figures 4, 5, 6, and 7.

HISTORY

Pre-Discovery

The first surface geologic studies and mapping in the area were conducted by MTA geologists (Turkish Mineral Research Institute) in 1933. Their work delineated several important anticlinal structures. Among these, the Raman structure was selected to be tested for oil and gas first, followed by a test of the Garzan structure.

Discovery

Raman Oil Field

In 1939, MTA spudded the Raman #1 on the crestal part of the surface anticline. A cable tool rig was used to reach a total depth of 1048 m. The well, however, was completed as a freshwater well, with oil shows, in April 1940. In the ensuing years, MTA drilled the Raman #2 through #7 wells, all of which were completed at similar depths as freshwater wells with oil shows.

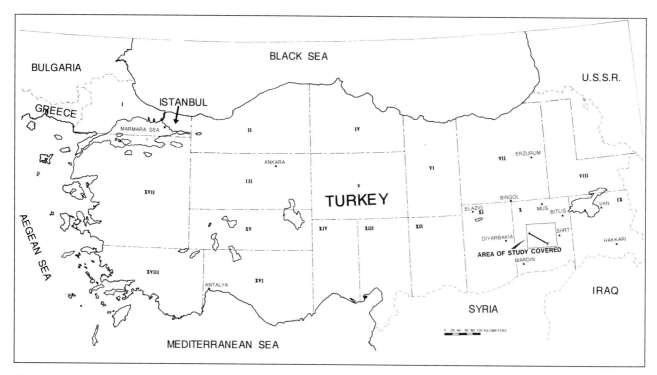

Figure 1. Index map showing area of study (box). Raman, Bati-Raman, and Garzan fields are in the southern half of the area of study. The line inside the box shows the location of the cross section in Figure 2.

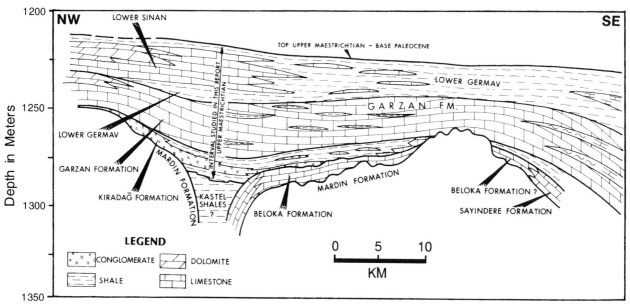

Figure 2. Diagramatic cross section to show the stratigraphic units above and below the producing horizon, the Garzan Formation of late Maastrichtian Stage. Scale on left side of the figure indicates the approximate drilling depth in meters of different units of the Maastrichtian. The Beloka is of the Campanian Stage and the Mardin ranges from Cenomanian to Turonian.

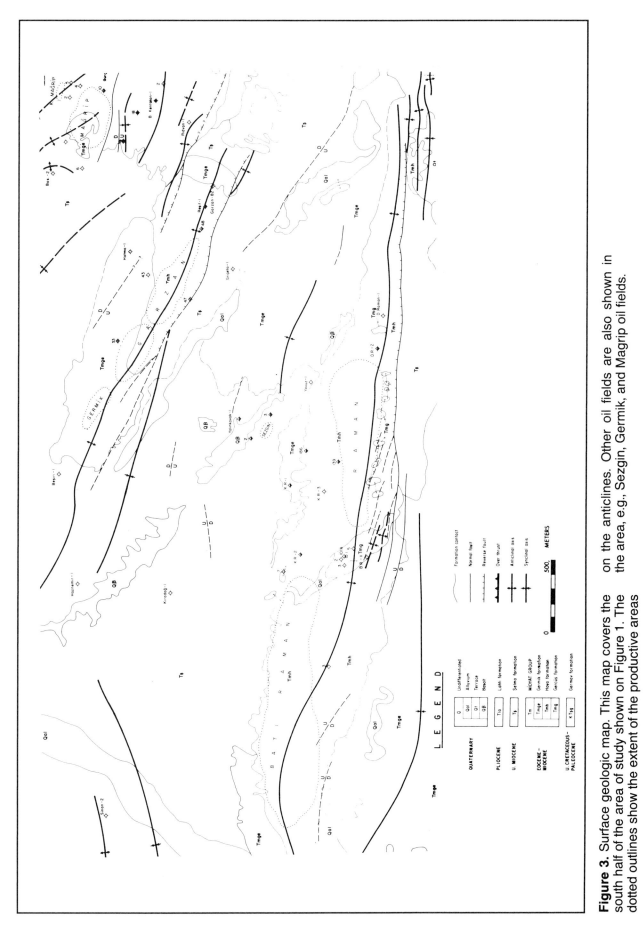

Figure 3. Surface geologic map. This map covers the south half of the area of study shown on Figure 1. The dotted outlines show the extent of the productive areas on the anticlines. Other oil fields are also shown in the area, e.g., Sezgin, Germik, and Magrip oil fields.

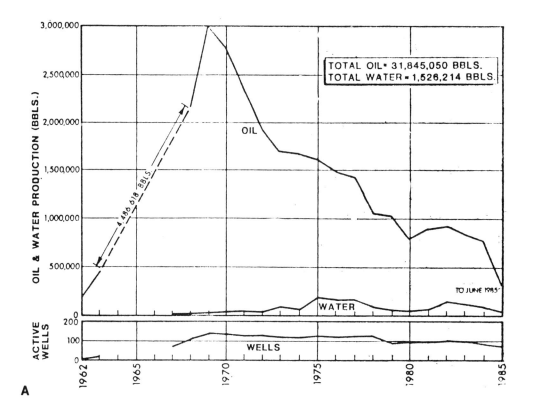

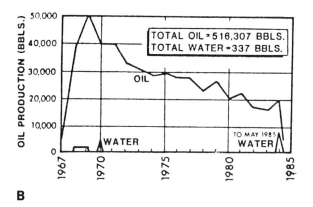

Figure 4. (A) Annual oil and water production and number of wells producing from the Garzan formation in the Bati-Raman field. Cumulative oil and water production to June 1985 is shown in the box.
(B) Bati-Raman #111, an example of an excellent well.
(C) Bati-Raman #121, an example of an average well.

It was not until 17 December 1946, when the Raman #1 well was drilled to approximately 1280 m into the Mardin Formation, that an oil discovery was made. The initial production was 40 bbl of oil per day (BOPD). Production was later improved to 300 BOPD after an acid treatment of the formation on 15 July 1948. The Mardin reservoir is composed of dolomitized and fractured carbonates of Cenomanian to Turonian Age.

Garzan Oil Field

The Garzan #1 well was spudded on the surface-mapped crestal part of the Garzan anticline in December 1944 by MTA, using a cable tool rig. After reaching a total depth in excess of 2000 m, the well was abandoned as a noncommercial well with oil shows on 9 March 1947. The Garzan #2 well was later drilled on the same structure to a total depth of approximately 2400 m. It was completed as the discovery well of the Garzan field in the Garzan Formation (shallower than the Mardin Formation that produces in the Raman structure; see Figure 2) in June 1951. MTA then drilled nine additional wells, all of which were completed as oil producers from the Garzan Formation. The Garzan reservoir rocks are composed of porous accumulations of small oyster

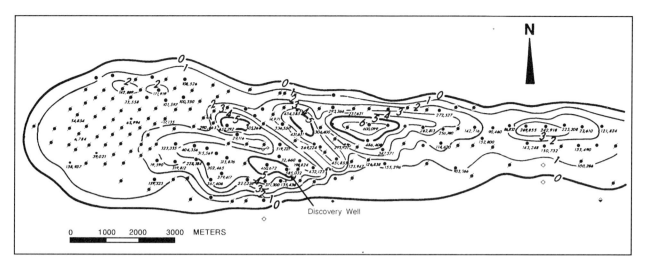

Figure 5. Equi-cumulative production map (by well) from the Garzan Formation in the Bati-Raman field. Cumulative oil volumes for each well were derived from estimates from monthly production tests made throughout the lives of the wells. Some wells are still producing and the estimated cumulative volumes from such wells have been increased accordingly. Contour interval, 100,000 bbl. (Conventional well symbols: solid circle, producing oil well; solid circle with diagonal bar, plugged oil well; open circle against cross, dry hole; open circle crossed by diagonal arrow, injection well.)

shell fragments interfingered with less porous sedimentary rock. Fortuitously, these reefoid buildups were found to occupy the crestal part of the Garzan structure.

In 1955, the new Turkish Oil Law required MTA to turn over to the Turkish Petroleum Corporation (T.P.A.O.) all its oil holdings and activities. T.P.A.O. proceeded to continue exploration, production, and development activities. Therefore, the commercial production of oil from the Garzan Formation was established in 1956, which was considered the discovery date.

Bati-Raman Oil Field

Based on the initial stratigraphic and structural studies conducted by T.P.A.O. geologists after 1955 and following the discovery of oil in the Garzan Formation in the Garzan structure, T.P.A.O. tested the Bati-Raman structure (West Raman). The Bati-Raman #1 well was drilled to approximately 1200 m and was completed as an oil producer from the Garzan Formation on 18 July 1961.

Concept and Summary of History

The original exploration concept prevailing at the time of drilling for oil in the first two fields was structural, mainly directed toward the drilling of crestal parts of obvious surface anticlinal features. Ultimately this resulted in the discovery of the Raman in 1948 and the Garzan field in 1956. Further structural and stratigraphic studies led to the realization of the importance of stratigraphic or, rather, combined structural-stratigraphic trap types. We now know that the Bati-Raman production (Salem et al., 1986) is from a reefoid "Garzan pay" on the northern flank and not from the crestal part of that structure. We also know that a complete and detailed understanding of stratigraphic and facies conditions did not exist at the time the structure was tested. These concepts were still developing at that time in Turkey and elsewhere. After examining available reports and interviewing several individuals, we still do not know the reasoning within T.P.A.O. prior to 1961 that resulted in testing the flank of the Bati-Raman structure. We can only guess that perhaps the reasons were based on the following points:

1. The "Garzan pay" had gained the reputation of being a "good pay" since the beginning of production from the Garzan oil field in 1956.
2. The prevailing philosophy was to drill structural closures.
3. The "Bati-Raman closure" had been well established.

Therefore, the Bati-Raman structure was chosen to test the "Garzan pay." Since the production is not from the crestal part, we also must guess that the choice of location of the discovery well was a fortuitous one which resulted in penetration of the oil column and discovery of the Bati-Raman oil field in 1961 on the northern flank of the structure.

More recent studies of facies distribution that are underway by T.P.A.O. geologists will undoubtedly bring to light the utility of facies distribution concepts. In addition, modern structural mapping (surface and subsurface) by T.P.A.O. geologists also may indicate some deeper possibilities for production associated with these structures.

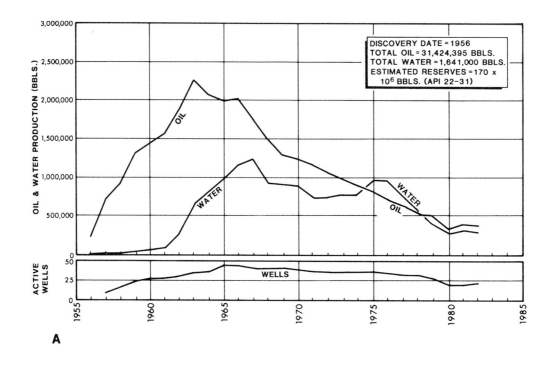
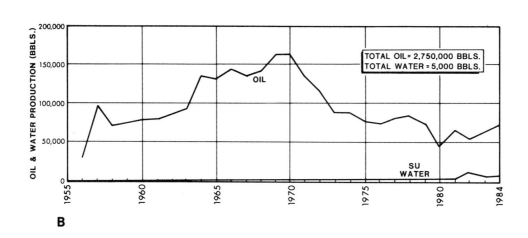
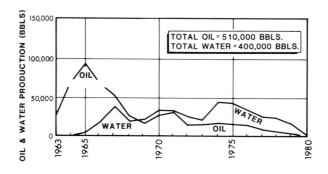

Figure 6. Annual production graphs for Garzan oil field. (A) Total field. (B) An excellent well, Garzan #15. (C) An average well, Garzan #6.

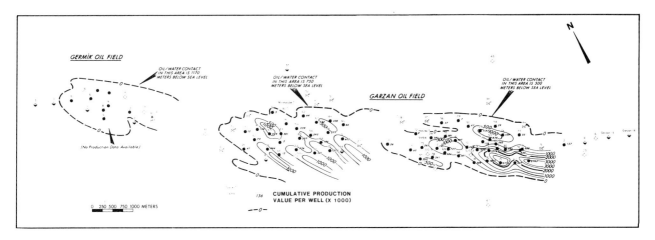

Figure 7. Garzan field equi-cumulative oil production map of recoveries by well from the Garzan Formation. Cumulative production in thousands of barrels from individual wells is based on monthly production tests of the wells throughout their productive life. Contour interval, 500,000 bbl.

Post-Discovery

T.P.A.O. has continued to develop the three fields by drilling additional wells and by using improved treatments and pressure maintenance as well as secondary recovery techniques, such as water injection, to enhance production. More recently, carbon dioxide transported by pipeline from the Dodan field has been injected into the reservoir.

T.P.A.O. has discovered several additional new fields in the area, such as Germik oil field.

DISCOVERY METHOD

The method of discovery of the three fields discussed in this study was by drilling anticlines previously mapped on the surface. Originally all reservoirs in the area were thought to be fractured carbonates. Recent studies (Salem et al., 1986) have shown that the reservoir porosities are primary intergranular and secondary diagenetic porosity in oyster-reefoid buildups, which are capped by shales of the Germav Formation. The understanding of the stratigraphic element in the trapping of hydrocarbons in these structures has added a great deal of insight as to additional new exploration approaches that should be used to discover more hydrocarbons in this area. Thus, a combination of sedimentologic, seismic-stratigraphic, and seismic-structural studies of the pay zones can be most rewarding. (See *Exploration Concepts* below.)

STRUCTURE

The area north of these oil fields has been subjected to several strong compressional or rather transpressional tectonic activities since the Late Cretaceous and after the deposition of the Garzan Formation. The latest of these tectonic activities took place in Miocene time and caused the main thrust belt in the area to the north of the oil fields. The Raman, Bati-Raman, and Garzan oil fields are located in the structural frontal zone (Diyarbakir province) immediately south of the main Miocene thrust belt that crosses southeast Turkey in a sublatitudinal direction. This frontal zone is characterized by the presence of numerous northwest-southeast-trending anticlines that are long and narrow and plunge to the northwest. They are asymmetric, with a gentler north flank and a steep to vertical south flank, bounded, in turn, by a reverse fault. On seismic lines, these anticlines are seen to be large and their crests are characterized by chaotic reflection zones; the structures also are shown to be cut by numerous normal faults.

The northwest orientation of the structural trend is more pronounced closer to the thrust belt, where the Garzan field is located (Figure 3). Farther south, the trend becomes more west-northwest (Bati-Raman and Raman fields). These anticlines formed during Miocene thrusting, verging to the southwest.

Raman and Bati-Raman structures are two separate closures on what appears to be one very long, east-southeast- to west-northwest-trending surface anticline. The closures are separated by a major cross fault or fault system. Raman field, which constitutes the eastern part of the structure (Figure 3), was discovered in 1946, and Bati-Raman, to the west, was discovered in 1961. Bati-Raman produces solely from Garzan oyster/reefoid carbonates, and Raman produces from Garzan carbonates as well as the underlying fractured and dolomitized Mardin carbonates.

The surface structures of Raman and Bati-Raman had been mapped by Turkish Petroleum Corporation geologists. Figure 8 presents a simplified subsurface

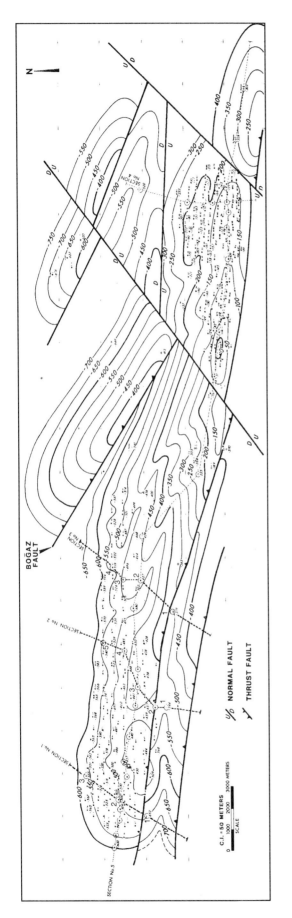
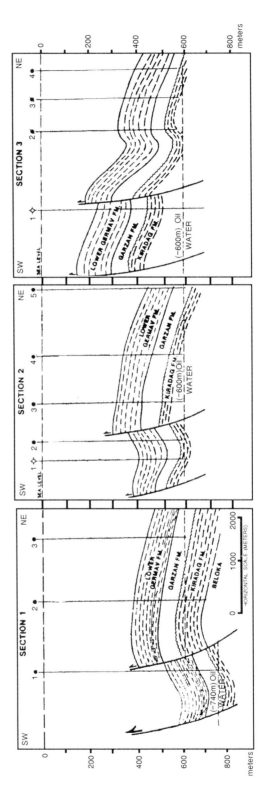

Figure 8. Structural interpretation the Raman/Bati-Raman area. The structural map shows the top of the Garzan Formation, which is the producing formation of the Bati-Raman field. Vertical depth scales of the cross sections are shown on the structural map; lines 4 and 5 are not shown in cross section in this study.

structural interpretation of the top of the producing Garzan Formation, constructed by the authors.

The structure of the Garzan field (Figure 9) resulted from the same Miocene compressional tectonics as Raman and Bati-Raman. It was also affected by a relaxation event after the compressional phase, resulting in numerous normal faults. Other normal faults that occurred during the deposition of the Garzan Formation are very important even though they do not cut the entire section. We feel that there are more syndepositional faults than are shown on the interpretation presented on Figure 9.

STRATIGRAPHY

The stratigraphic section encountered in the area ranges from Paleozoic to Quaternary, with several major unconformities. Table 1 summarizes the stratigraphy of the area.

The productive formations in the three fields are the lower Maastrichtian Garzan and/or the Cenomanian–Turonian Mardin. The approximate depths in meters relative to sea level of these formations in the fields are shown by Table 2.

We will discuss in detail the stratigraphy of the Maastrichtian only, which includes lower Germav, Garzan, and Kiradag formations.

The stratigraphy of the Garzan Formation in this area is represented on Figures 2, 10, and 11. The Garzan Formation is underlain by Kiradag shales, which occasionally contain sand and carbonate intercalations. It is overlain by the lower Germav shale caprock.

The lower Germav in the producing area is a greenish-gray, pyritic, glauconitic, and calcareous shale. To the west and northwest of this area, the lower Germav shales grade into and are replaced by a thick carbonate sequence called the lower Sinan Formation (Figure 2). The lower Germav shales show a drastic thinning over both the Raman and Bati-Raman oil field complex (Figure 12) and over the Garzan field (Figure 13). This thinning may have been caused by:

1. The presence of an older structure that predates deposition.
2. Depositional thinning over Garzan reefs that attained prominent sea-bottom relief during lower Germav deposition.
3. Differential compaction of the lower Germav shales over reefs and/or pre-existing structures.

The Garzan Formation contains the major reservoir rocks of these fields. In the District X area, it includes a variety of carbonates (micrites, biomicrites, pelmicrites, intramicrites, and argillaceous micrites). In some places these carbonates are interbedded with thin calcareous shales. Where productive, the Garzan Formation is composed of a variety of porous biocalcarenites (some with intraclasts). Fossil fragments include rudists, *Orbitoides*, *Loftusia*, and *Omphalocyclus*, and rare red algae and coral fragments. The porosity is intergranular, ranging from 12 to 16%. The porous biocalcarenites are interbedded with less porous biomicrites.

The biocalcarenite reservoir rocks in Bati-Raman and in the northwest flank of the Raman, as well as in the Garzan field, are complex reefoid buildups surrounded with fossil detritus (Figures 14, 15, and 16). The reef locations were probably controlled by pre-existing structural highs. The geometries of these buildups and associated detritus are not well defined.

The Kiradag Formation underlies the Garzan Formation and is composed predominantly of shales with occasional conglomeratic sandstone and limestone intercalations. The siliciclastics were sourced from the northwest. These conglomeratic sandstones are composed of weathered basaltic fragments, chlorite, chert, dolomite, and limestone fragments.

The Kiradag sediments were deposited in a littoral complex that included tidal flat, strandline, intertidal, lagoonal, tidal channel, and lacustrine facies (Salem et al., 1986). The formation varies in thickness from zero east of the Raman area to more than 125 m west of the Bati-Raman area. It also decreases in thickness from north to south over the structures. The Kiradag Formation is underlain by rocks of either the Beloka Formation or the Mardin Formation (Figure 2).

TRAP

The main reservoir in the Raman field is a structural accumulation in Mardin carbonates. The structure is a narrow, elongate anticline with a length of 12 km, an average width of 2.75 km, and an approximate vertical closure of 150 m. An oil-water contact (OWC) is encountered at 300 m below sea level (–300 m). The Garzan Formation on the Raman structure is productive on the north flank only and has the same OWC as that of the Mardin Formation.

The Bati-Raman field is a mainly stratigraphic trap, composed of agglomerations of small reefoid buildups surrounded by shell detritus and capped by shales. The southern limit of the productive interval is a gradation into impervious limestones and shales that were deposited at the same time as the porous reefoid reservoir facies. These impervious layers are structurally much higher (particularly toward the east end of the field) than the porous, productive, and time equivalent facies; that is to say, the impervious facies occupy the crestal nonproductive part of the Bati-Raman structure (Figures 8, 14, and 15). The OWC is the northern limit of the productive area. The Bati-Raman field is approximately 18 km long and 3.5 km wide, with an average structural closure of 160 m.

The average oil column is 80 m, ranging from 160 m in the south to none at the OWC in the north.

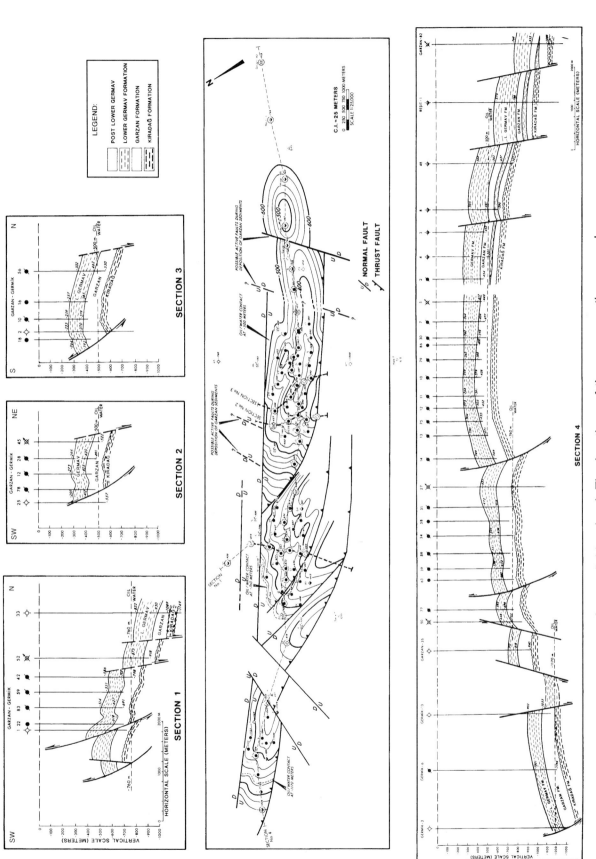

Figure 9. Structural interpretation of the Garzan field. Structural contours are of the top of the productive Garzan Formation. Vertical scales in meters below sea level. The locations of the cross sections are shown on the structural map.

Table 1. Stratigraphic column in the area of the Raman, Bati-Raman, and Garzan fields.

Chronostratigraphic Unit	Group or Formation	Lithology
Quaternary		Alluvium
Pliocene	Lahti	Conglomerates
Miocene	Selmo	Evaporites and carbonates
Miocene/Eocene	Firat/Midyat	Carbonates
Eocene	Gercus	Red shale and conglomerates
Upper Paleocene	Sinan	Carbonates
Lower Paleocene	Upper Germav	Shale
Upper Maastrichtian	Lower Germav	Shale
Lower Maastrichtian	Garzan	Carbonates
Lower Maastrichtian	Kiradag	Shale and sandstone
Campanian	Beloka	Carbonates
Cenomanian/Turonian	Mardin	Carbonates
Paleozoic	Bedinan	Shales and sandstone

Table 2. Approximate depth in meters relative to sea level of the Garzan and Mardin formations at the three fields.

	Raman Field	Bati-Raman Field	Garzan Field
Garzan Formation	−119	−485	−430
Mardin Formation	−279	−710	−730

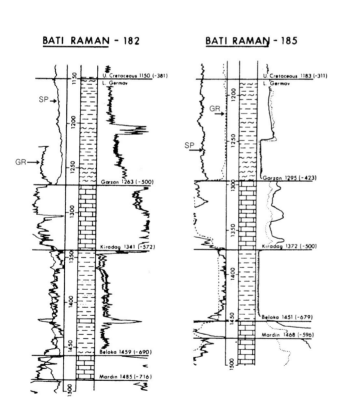

Figure 10. Examples of the stratigraphic units in an interval including the productive Garzan Formation in the Raman/Bati-Raman area. The curves to the right of the lithology logs are resistivity curves. Those to the left are gamma ray (GR) and self-potential (SP) curves. These unquantified curves are to show characteristic log signatures and correlations only.

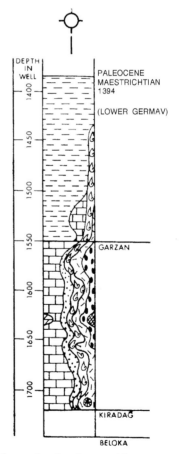

Figure 11. Example of petrographic composition of the Garzan pay horizon and adjacent beds in the Garzan field. Notice the petrographic composition includes: micrites (blocks); pellets (dots); fossils and fragments, including oysters and corals (cartoon representations); interclasts (black ovals); and sperry calcite cement (cross hatch).

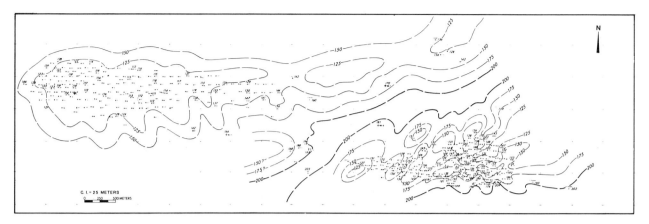

Figure 12. Isopach of the interval from the top of the Cretaceous section to the top of the Garzan Formation. In this area, this interval is represented by the lower Germav shale facies only, showing the thickness of the shale over the Raman field to the east and the Bati-Raman to the west.

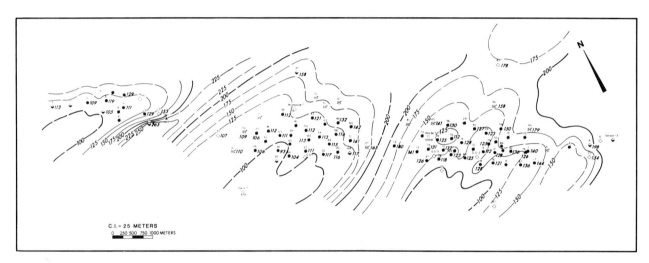

Figure 13. Isopach of the interval from the top of the Cretaceous to the top of the Garzan Formation, represented by lower Germav shale only in this Garzan oil field area. The shale is the upper seal of the Garzan reservoir. Notice the thinning of the lower Germav over the structural and reef accumulation; compare with Figure 16.

The main OWC is at –600 m. The OWC in a separate fault block is –740 m (Figure 8). In summary, the controlling factors in the entrapment of hydrocarbons in these two fields are mainly stratigraphic in the Bati-Raman and structural in the Raman field.

The Garzan oil field produces from a structural trap that is similar to that of the Raman field except that the producing horizon is the Garzan oyster-reefoid carbonates rather than the stratigraphically lower Mardin carbonates. The field consists of three separate closures. The two eastern ones constitute the Garzan oil field, and the small, westernmost one constitutes the Germik field (Figures 3 and 9). The Garzan field (the eastern two closures) is 11.5 km long and 1.5 km wide. Average vertical closure is 150 m. Average oil column is 75 m, ranging from 150 m to none at the outer edge of the OWC. The elevation of the main OWC in the easternmost closure is –500 m, while it is –740 m at the central closure. The third OWC, in the Germik field, is –1170 m.

The time of migration of oil into all of these traps is believed to have been during the Miocene or post-Miocene when the structures attained their present form.

Reservoir

This discussion will be restricted to the Garzan Formation reservoir rocks (Figures 2, 10, and 11 and Table 1). As mentioned earlier, the Garzan reservoir rocks consist of a highly complex mosaic of oyster-reefoid buildups and associated biocalcarenites. The intergranular porosity ranges from 12 to 16%, with

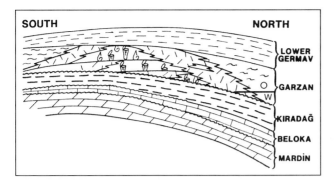

Figure 14. Generalized cross section through the Bati-Raman field to show possible facies relationships. Symbols as in Figures 11, 15, and 16. Notice the oil-water contact at the north end of the section. No scale intended.

subordinate amounts of intercrystalline porosity after diagenesis. The porous calcarenites and reefoid bodies are interbedded with less porous to nonporous micrites, biomicrites, and argillaceous micrites. The Garzan sediments were deposited on the flanks of pre-existing structural salients. Northeast-southwest-trending normal faults were active during the deposition of Garzan sediments. These syndepositional faults most probably continued to be active after the Garzan deposition and during the deposition of the lower Germav shales. The highs on the upthrow side became preferred sites for biogenic activities. The low areas on the downthrow side were most probably conduits for encroaching *fresh water* laden with clays and nutrients. It must be emphasized that the biogenic or reefoid buildups were small in size and do not represent major reefal accumulations. They display patchy accumulations similar to small oyster banks that proliferate in bays and estuaries or near shore lines (see depositional model of Garzan Formation in the Raman, Bati-Raman, and Garzan fields, Figures 14, 15, and 16).

Source Rocks

Recent geochemical work in District X has shown that the most probable source rock for the Garzan oil is the underlying Mardin Group carbonates. The Mardin Group is divided into several units. From base to top these units are the Sabunsuyo, the Derdere, and the Karababa on top. All these units were shown to be good to excellent source rocks. For example, the Derdere Formation has a TOC over 0.5% and shows an advanced state of maturity.

EXPLORATION CONCEPTS

Exploration in this area has paralleled the historical evolution of exploration concepts in other parts of the world. First, the anticlinal theory was applied. The result was the discovery of the Raman and Garzan fields. Stratigraphic concepts then came into play and the Bati-Raman was discovered. The advent of large-scale seismic exploration led to the discovery of other fields in the area, such as the Silivanka oil field. Presently, integrated sedimentologic analyses and seismic stratigraphy are in use by the Turkish Petroleum Corporation in its quest to discover more hydrocarbons. Other companies are also actively assisting the Turkish Petroleum Corporation in unraveling geological problems and in identifying the hidden hydrocarbon traps in District X in southeast Turkey. It is expected that much more oil will be discovered in this area.

The three oil fields discussed show clearly that no one idea, domain, or concept will lead to the discovery of all of the hydrocarbon accumulations in one area. Thus, in 1946 oil was discovered in the Mardin pay of the Raman anticline by drilling the crestal position of a structural closure. Yet after 27 years, in 1973, oil was discovered and is still being produced from the Garzan pay (above the Mardin pay) from the northern flank of the same Raman structure. This discovery of the Garzan pay resulted in more activity in the area and led to the emergence of another concept—that of stratigraphic, or rather, facies trapping of hydrocarbons. Undoubtedly as time passes, new concepts will be developed that will lead to the discovery of additional oil accumulations. We predict that the importance of diagenetic processes and their role in the trapping of hydrocarbons will become a part of the new concepts that will come into play.

Knowing what we know today, the following scheme is proposed for further exploration activities in the area to locate new accumulations of hydrocarbons.

1. Conduct regional as well as detailed facies analyses and develop facies models for productive areas and facies fairways extending beyond productive areas. These analyses must depend on extensive examinations of cores, sample cuttings, and thin sections, as well as mechanical well log responses.
2. Conduct seismic structural and stratigraphic analyses over productive areas and attempt to recognize, if possible, any seismic stratigraphic anomalies that are characteristic of those productive areas.
3. Combine items 1 and 2, developing criteria from both domains that characterize productive areas from both sedimentologic and seismic points of view.
4. Conduct a search for "look alike" anomalies in the established facies models and available seismic lines. This search should be tied to specific wells—whether dry or productive or with shows. The search might even include the shooting of new, selective seismic lines with parameters selected to enhance the appearance of such anomalies.

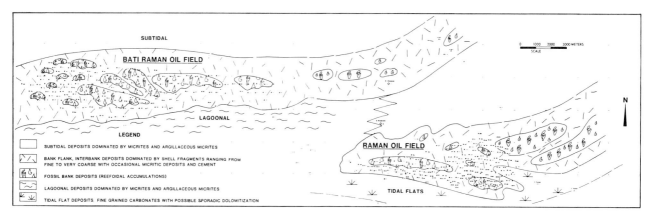

Figure 15. Hypothetical depositional model of the Garzan Formation in the Raman/Bati-Raman area.

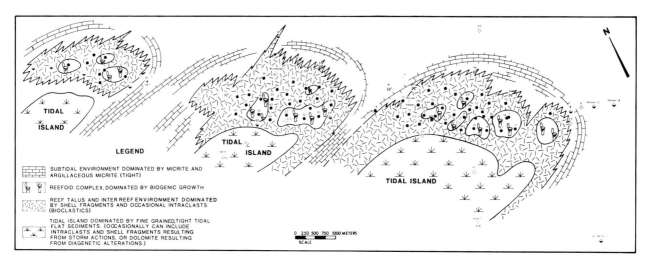

Figure 16. Depositional model of the Garzan Formation in the Garzan field area.

5. Analyze the "trap" conditions of such anomalies and make sure that all such factors as cap rock, water table, reservoir porosity, volume, and economics are sufficiently favorable as to warrant testing.

It is only by applying integrated systems that we may be able to decipher a small part of the secrets of any given area. New ideas should be tested in old producing areas.

ACKNOWLEDGMENTS

The authors are indebted to T.P.A.O. management, Mr. Ozer Altan, Mr. Ozan Sungurlu, and Mr. Dursun Acikbas for encouragement and permission to publish this paper. Special thanks to Mr. Hayrettin Okay and Mrs. Figen Yuksel of the Turkish Petroleum Research Lab for unlimited support and help during the work. All the work on which this paper is based was done by the authors as part of a study conducted for T.P.A.O. (Salem et al., 1986). During that study many other individuals helped in producing the report. Our deepest thanks to Mr. Abdurrahman Eren, Mr. Hasan Oncu, Mr. Mehmet Arac, and Mr. Zulfikar Becir and all other colleagues who helped in this endeavor.

REFERENCES

Acikbas, D., A. Akgul, and T. Erdogan, 1980, Baykan-Sirvan-Pervari Yoresinin Jeolojisi ve C. Dogu Anadolunun Hidrokarbon Olanaklari: Rapor No. 1543.

Dincer, A., and G. Kurt, 1983, Cudi Grubu icindeki Hazne Kaya Nitelikleri Seviyeleri Degerlendirmesi: Rapor No. 1807.

Salem, R., et al., 1986, Geologic and hydrocarbon evaluation of Maestrichtian sediments in Central District X, southeast Turkey: Special report, presented to T.P.A.O. in Ankara.

Sibal, J., 1977, Preliminary report on Early Mesozoic deposits in south-east Turkey: Rapor No. 1182.

St. John, B., A. W. Bally, and H. Klemme, 1984, Sedimentary provinces of the world—hydrocarbon productive and nonproductive: AAPG.

Unat, O. T., 1972, Guneydogu Turkiye, Kuzey Suriye ve Irakta Petrol Olusumu ve Gocme Sorunu: Rapor No. 749.

Appendix 1. Field Description

Field name .... *Bati-Raman field*
Ultimate recoverable reserves .... *500,000,000 STB*

Field location:
    Country .... *Turkey*
    State .... *District X*
    Basin/Province .... *Diyarbakir basin—south of Miocene overthrust belt*

Field discovery:
    Year first pay discovered .... *Lower Maastrichtian Garzan Fm. 1961*

Discovery well name and general location:
    First pay .... *Bati-Raman #1*

Discovery well operator .... *Turkish Petroleum Corporation*

IP:
    First pay .... *360 BOPD*

Geologic concept leading to discovery and method or methods used to delineate prospect
Surface geology and stratigraphy of a nearby older production.

Structure:
    Province/basin type .... *Inner shelf on continental margin*
    Tectonic history
Several compressional movements since Late Cretaceous and Cenozoic that ended with the Miocene thrust belt. To this belt are related anticlinal structures in the foothills, imbricate, and frontal zones.
    Regional structure
Several elongate, asymmetrical anticlines in the frontal zone.
    Local structure
West-northwest-east-southeast anticline with gentle dips on north flank (less than 18°); steep dips on south flanks bounded by high-angle reverse faults on the south side.

Trap:
    Trap type(s) .... *Porosity pinch-out updip (or onlap)*

Basin stratigraphy (major stratigraphic intervals from surface to deepest penetration in field):

Chronostratigraphy*	Formation	Depth/Sea Level (m)	Thickness (m)
Lower Eocene	Hoya	0/+765	270
Lower Eocene-upper Paleocene	Gercus	270/+495	230
Middle-upper Paleocene	Upper Germav	500/+265	650
Upper Maastrichtian	Lower Germav	1150/-380	100
Middle-upper Maastrichtian	Garzan	1250/-485	90
Lower Maastrichtian	Kiradag	1340/-575	110
Upper Campanian	Beloka	1450/-685	25
Cenomanian-Turonian	Mardin	1475/-710	350

*No wells in the field drilled below the Mardin Group.

Reservoir characteristics:
    Number of reservoirs *1*
    Formations .... *Garzan*
    Ages .... *Maastrichtian*
    Depths to tops of reservoirs .... *Approx. 1200 m*

Net thickness—total thickness of producing zones
   Average .... 70 m
   Maximum .... 160 m
Lithology
Porous oyster biocalcarenite with brackish water, large forams with subordinate amounts of sparry calcite cement or micrite matrix, occasionally contain intraclasts and/or pellets
Porosity type .... *Intergranular and intercrystalline, possible fractures*
Average porosity *18%*
Average permeability .... *58 md*

Seals:

Upper
   Formation, fault, or other feature .... *Lower Germav, Maastrichtian*
   Lithology .... *Shale and calcareous shale*
Lateral
   Formation, fault, or other feature .... *Garzan Formation facies change, becoming nonporous; OWC to the north*
   Lithology .... *From porous bioclastics to nonporous carbonates*

Source:

Formation and age .... *Different members of the Mardin Formation*
Lithology .... *Carbonates*
Average total organic carbon (TOC) *0.5%*
Maximum TOC *1.2%*
Kerogen type (I, II, or III) .... *NA*
Vitrinite reflectance (maturation) .... *NA*
Time of hydrocarbon expulsion .... *Miocene or post-Miocene*
Present depth to top of source .... *Approx. 1500 m*
Thickness .... *NA*
Potential yield .... *NA*

Appendix 2. Production Data

Field name .... *Bati-Raman field*
Field size:
   Proved acres *15,000*
   Number of wells all years *193*
   Current number of wells *75*
   Well spacing .... *NA*
   Ultimate recoverable .... *500,000,000 bbl*
   Cumulative production .... *Approx. 32,000,000 bbl*
   Annual production .... *Approx. 300,000 bbl*
   Present decline rate .... *NA*
      Initial decline rate .... *NA*
      Overall decline rate .... *NA*
   Annual water production .... *Approx. 200,000 bbl*
   In place, total reserves .... *1,850,000,000 bbl*
   In place, per acre-foot .... *145 bbl*
   Primary recovery and secondary recovery .... *500,000,000 bbl oil (carbon dioxide injection)*
   Enhanced recovery .... *NA (under consideration)*
   Cumulative water production .... *1,530,000 bbl*

Drilling and casing practices:
    Amount of surface casing set .... Approx. 1100 m
    Casing program .... NA
    Drilling mud .... Water base mud approx. 12.5–13 lb/gal weight
    Bit program .... NA
    High pressure zones .... None

Completion practices:
    Interval(s) perforated .... Intervals with sample and/or log shows in Garzan Fm.
    Well treatment .... Acidizing

Formation evaluation:
    Logging suites .... Dual induction, sonic, gamma ray
    Testing practices .... Occasional open-hole testing, then production testing through perfs.
    Mud logging techniques .... Continuous mud logging from surface to TD

Oil characteristics:
    Type .... NA
    API gravity .... 13.3°
    Base .... NA
    Initial GOR .... 18 scf/bbl
    Sulfur, wt% 5.65%
    Viscosity, SUS .... NA
    Pour point .... NA
    Gas-oil distillate .... NA

Field characteristics:
    Average elevation .... 965 m
    Initial pressure .... 1750 psi
    Present pressure .... NA
    Pressure gradient .... NA
    Temperature .... 129°F
    Geothermal gradient .... NA
    Drive .... Rock and fluid expansion drive
    Oil column thickness .... Avg. 25 m (80 ft)
    Oil-water contact .... 600 m below sea level
    Connate water .... NA
    Water salinity, TDS .... 40,000–100,000 ppm
    Resistivity of water .... NA
    Bulk volume water (%) .... NA

Transportation method and market for oil and gas:
Local pipeline system

Appendix 3. Field Description

Field name .... *Garzan field*
Ultimate recoverable reserves .... *70,000,000 STB*
Field location:
    Country .... *Turkey*
    State .... *District X*
    Basin/Province .... *Diyarbakir basin—south of Miocene overthrust belt*

Field discovery:
  Year first pay discovered .... *Lower Maastrichtian Garzan Fm. 1955*

Discovery well name and general location:
  First pay .... *Garzan #2*

Discovery well operator .... *Turkish Petroleum Corporation*

IP:
  First pay .... *400 BOPD*

Geologic concept leading to discovery and method or methods used to delineate prospect
Surface geology, surface mapping confirmed by drilling.

Structure:
  Province/basin type .... *Inner shelf on continental margin*

  Tectonic history
  Several compressional movements since Late Cretaceous and throughout the Cenozoic that ended with the Miocene thrust belt. This is associated with related anticlinal structures in the foothills, imbricate, and frontal zones.

  Regional structure
  Several elongate, asymmetrical anticlines in the frontal zone just south of the main thrust belt crossing southeast Turkey from east to west.

  Local structure
  Northwest-southeast elongate surface anticline, gentle dips on north flank (less than 18°); steep dips on south flanks bounded by high-angle reverse faults on the south side.

Trap:
  Trap type(s) .... *Structure, half dome*

Basin stratigraphy (major stratigraphic intervals from surface to deepest penetration in field):

Chronostratigraphy*	Formation	Depth/Sea Level (m)	Thickness (m)
Lower Eocene	Hoya	0/+970	290
Lower Eocene-upper Paleocene	Gercus	290/+680	310
Middle-upper Paleocene	Upper Germav	600/+370	685
Upper Maastrichtian	Lower Germav	1285/-315	115
Middle-upper Maastrichtian	Garzan	1400/-430	170
Lower Maastrichtian	Kiradag	1570/-600	105
Upper Campanian	Beloka	1675/-705	25
Cenomanian-Turonian	Mardin	1700/-730	400

*No wells in the field drilled below the Mardin Group.

Reservoir characteristics:
  Number of reservoirs *1*
  Formations .... *Garzan*
  Ages .... *Maastrichtian*
  Depths to tops of reservoirs .... *2400 m*
  Gross thickness (top to bottom of producing interval) .... *100 m*
  Net thickness—total thickness of producing zones
    Average .... *80 m*
    Maximum .... *NA*
  Lithology
  Fossil oyster fragments with subordinate amounts of sparry calcite cement or micrite matrix, occasional dolomite, occasional intraclasts and pellets
  Porosity type .... *Intergranular and intercrystalline*
  Average porosity .... *Approx. 15.5%*
  Average permeability .... *12 md*

Seals:
  Upper
     Formation, fault, or other feature .... *Maastrichtian lower Germav*
     Lithology .... *Shale and calcareous shale*
  Lateral
     Formation, fault, or other feature .... *Reverse fault to the south; normal faults and structural closure with water tables elsewhere*

Source:
  Formation and age .... *Different members of the Mardin Formation*
  Lithology .... *Carbonates*
  Average total organic carbon (TOC) *0.5%*
  Maximum TOC *1.2%*
  Kerogen type (I, II, or III) .... *NA*
  Vitrinite reflectance (maturation) .... *NA*
  Time of hydrocarbon expulsion .... *Miocene or post-Miocene*
  Present depth to top of source .... *2600 m*
  Thickness .... *NA*
  Potential yield .... *NA*

Appendix 4. Production Data

Field name .... *Garzan field*
Field size:
  Proved acres .... *2377 ac*
  Number of wells all years *102*
  Current number of wells *26*
  Well spacing .... *NA*
  Ultimate recoverable .... *70,000,000 bbl*
  Cumulative production .... *33,000,000 bbl*
  Annual production .... *140,000 bbl*
  Present decline rate .... *NA*
     Initial decline rate .... *NA*
     Overall decline rate .... *NA*
  Annual water production .... *120,000 bbl*
  In place, total reserves .... *170,000,000 bbl*
  In place, per acre-foot .... *NA*
  Primary and secondary recovery .... *70,000,000 bbl*
  Enhanced recovery .... *NA*
  Cumulative water production .... *NA*

Drilling and casing practices:
  Amount of surface casing set .... *1800 m*
  Drilling mud .... *Water base mud, 12.3–13 lb/gal weight*
  High pressure zones .... *None*

Completion practices:
  Interval(s) perforated .... *Intervals with sample and/or log show*
  Well treatment .... *Acidizing*

Formation evaluation:

    Logging suites .... *Dual induction, sonic, gamma ray*
    Testing practices .... *Occasional open-hole testing; production testing*
    Mud logging techniques .... *Continuous mud logging to TD*

Oil characteristics:

    Type .... *NA*
    API gravity .... *26.2°*
    Base .... *NA*
    Initial GOR .... *88.5 SCF/bbl*
    Sulfur, wt% *2.07%*
    Viscosity, SUS .... *NA*
    Pour point .... *NA*
    Gas-oil distillate .... *NA*

Field characteristics:

    Average elevation .... *970 m*
    Initial pressure .... *1350 psi and 1650 psi*
    Present pressure .... *NA*
    Pressure gradient .... *NA*
    Temperature .... *132°F*
    Geothermal gradient .... *NA*
    Drive .... *Dissolved gas and water drive*
    Oil column thickness .... *0–150 m (avg. 75 m)*
    Oil-water contact .... *510 and 760 m below sea level*
    Connate water .... *NA*
    Water salinity, TDS .... *20,000–60,000 ppm*
    Resistivity of water .... *NA*
    Bulk volume water (%) .... *NA*

Transportation method and market for oil and gas:
Local pipeline network

Appendix 5. Field Description

Field name .... *Raman field*
Ultimate recoverable reserves .... *100,000,000 STB*

Field location:

    Country .... *Turkey*
    State .... *District X*
    Basin/Province .... *Diyarbakir basin—south of Miocene overthrust belt*

Field discovery:

    Year first pay discovered .... *Cenomanian-Turonian Mardin Fm. 1946*
    Year second pay discovered .... *Middle and Upper Maastrichtian Garzan Fm. 1973*

Discovery well name and general location:

    First pay .... *Raman #8*
    Second pay .... *NA*

Discovery well operator .... *Turkish Mineral Institute (MTA)*
    Second pay: .... *Turkish Petroleum Corporation*

IP:
    First pay .... 300 BOPD Mardin
    Second pay .... 350 BOPD Garzan

Geologic concept leading to discovery and method or methods used to delineate prospect
Surface geology, surface mapping confirmed by drilling.

Structure:
   Province/basin type .... *Inner shelf on continental margin*
   Tectonic history
   Several compressional movements since Late Cretaceous and throughout the Cenozoic that ended with the Miocene thrust belt. This is associated with related anticlinal structures in the foothills, imbricate, and frontal zones.
   Regional structure
   Several elongate, asymmetrical anticlines in the frontal zone just south of the main thrust belt crossing southeast Turkey from east to west.
   Local structure
   West-northwest–east-southeast elongate surface anticline, gently dipping on north flank (18°); steep dipping to the south and bounded by high-angle reverse faults on the south side.

Trap:
   Trap type(s) .... *Structure, half dome*

Basin stratigraphy (major stratigraphic intervals from surface to deepest penetration in field):

Chronostratigraphy	Formation	Depth/Sea Level (m)	Thickness (m)
Lower Eocene	Hoya	0/+1231	318
Lower Eocene–upper Paleocene	Gercus	320/+911	280
Middle–upper Paleocene	Upper Germav	600/+631	600
Upper Maastrichtian	Lower Germav	1200/+31	150
Middle–upper Maastrichtian	Garzan	1350/−119	150
Lower Maastrichtian	Kiradag	1400/−169	5
Lower Campanian	Karababa	1405/−174	105
Cenomanian–Turonian	Mardin	1510/−279	320
Aptian–Albian	Areban	1830/−599	45
Lower–Middle Triassic	Bakuk	1875/−644	325

Reservoir characteristics:
   Number of reservoirs 2
   Formations .... Mardin and Garzan
   Ages .... Cenomanian/Turonian and Maastrichtian
   Depths to tops of reservoirs .... Mardin, 1280 m; Garzan, 1150 m
   Gross thickness (top to bottom of producing interval) .... 50 m and 70 m
   Net thickness—total thickness of producing zones
       Average .... 80 m
       Maximum .... 100 m
   Lithology
   Mardin Formation is dolomite; Garzan Formation is composed of oyster fragments with subordinate amounts of sparry cement and micrite matrix with occasional intraclasts and pellets
   Porosity type .... Intergranular and intercrystalline
   Average porosity .... Mardin, 12%; Garzan, 18%
   Average permeability .... Mardin 50 md; Garzan 58 md

Seals:
   Upper
       Formation, fault, or other feature .... Maastrichtian lower Germav

Lithology .... *Shale and calcareous shale*
Lateral
    Formation, fault, or other feature .... *Thrust fault on south side; high-angle normal faults east and west with OWC on north flank; lithofacies influence*

Source:

Formation and age .... *Different members of the Mardin Formation*
Lithology .... *Carbonates*
Average total organic carbon (TOC) *0.5%*
Maximum TOC *1.2%*
Kerogen type (I, II, or III) .... *NA*
Vitrinite reflectance (maturation) .... *NA*
Time of hydrocarbon expulsion .... *Miocene or post-Miocene*
Present depth to top of source .... *1300 m*
Thickness .... *NA*
Potential yield .... *NA*

Appendix 6. Production Data

Field name .... *Raman field*
Field size:

Proved acres .... *7600 ac*
Number of wells all years *222*
Current number of wells *65*
Ultimate recoverable .... *100,000,000 bbl*
Cumulative production (6/90) .... *50,500,000 bbl*
Annual production .... *NA*
Present decline rate .... *NA*
    Initial decline rate .... *NA*
    Overall decline rate .... *NA*
Annual water production .... *NA*
In place, total reserves .... *390,000,000 bbl*
In place, per acre-foot .... *NA*
Primary and secondary recovery .... *100,000,000 bbl*
Enhanced recovery .... *NA*
Cumulative water production .... *NA*

Drilling and casing practices:

Amount of surface casing set .... *1100 m*
Drilling mud .... *Water base mud, 12.3–13 lb/gal weight*
High pressure zones .... *None*

Completion practices:

Interval(s) perforated .... *Intervals with sample and/or log show*
Well treatment .... *Acidizing*

Formation evaluation:

Logging suites .... *Dual induction, sonic, gamma ray*
Testing practices .... *Occasional open-hole testing; then production testing through perfs*
Mud logging techniques .... *Continuous mud logging from surface to TD*

Oil characteristics:

Type	NA
API gravity	18°
Base	NA
Initial GOR	20 SCF/bbl
Sulfur, wt%	5.5%
Viscosity, SUS	NA
Pour point	NA
Gas-oil distillate	NA

Field characteristics:

Average elevation	1230 m
Initial pressure	1300 psi
Present pressure	NA
Pressure gradient	NA
Temperature	140°F
Geothermal gradient	NA
Drive	Rock and fluid expansion and water drive
Oil column thickness	100 m
Oil-water contact	300 m below sea level
Connate water	NA
Water salinity, TDS	40,000–100,000 ppm
Resistivity of water	NA
Bulk volume water (%)	NA

Transportation method and market for oil and gas:
Local pipeline network

Messoyakh Gas Field— Russia
West Siberian Basin

JAN KRASON
PATRICK D. FINLEY
Geoexplorers International, Inc.
Denver, Colorado

FIELD CLASSIFICATION

BASIN: West Siberia/Yenisei-Khatanga Trough
BASIN TYPE: Cratonic Sag
RESERVOIR ROCK TYPE: Sand and Partially Indurated Sand
RESERVOIR ENVIRONMENT OF DEPOSITION: Marine
TRAP DESCRIPTION: Gas sealed by overlying and lateral shales and gas-hydrate plugged sands; both free-gas and dissociated gas from hydrates produced

RESERVOIR AGE: Cretaceous Cenomanian
PETROLEUM TYPE: Gas
TRAP TYPE: Anticline

LOCATION

The Messoyakh gas field is in the northwestern part of the West Siberian basin of Russia. The field is about 250 km west-southwest of the city of Norilsk between the Ob Gulf and the Yenisei River (Figure 1). It is near the boundary between the West Siberian hydrocarbon province and the Yenisei Khatanga hydrocarbon province; literature sources have included the field in both provinces (e.g., Klemme, 1984; Semenovich et al., 1976). The exact location of the field is not clearly documented; locations cited in the literature vary in placement of the center of the field by as much as 27 km in latitudinal and 40 km in longitudinal directions (Table 1).

The Messoyakh gas field is a principal field of the Northwest gas district of the West Siberian basin (Meyerhoff, 1980). Other fields in the area include the Solenin, Pelyantin, Kazantsevo, and Zeminy (Figure 1). The supergiant Yamburg and Urengoy fields are about 300 km southwest of the Messoyakh gas field.

Reserve estimates for the Messoyakh gas field vary widely. Sumets (1974) estimated proven and probable reserves of 1.32 tcf (37.44 bcm). Carmalt and St. John (1986) ranked Messoyakh as the world's 99th largest oil or gas field with proven and probable reserves of 14 tcf (400 bcm). The range in reserve estimates reflects the uncertainties associated with producing from a hydrate reservoir.

Natural gas and water form a solid hydrate phase at high pressures and low temperatures. Gas hydrates clog pipelines and processing equipment in cold climates (Deaton and Frost, 1949). Hydrates occur naturally in low temperature sediments when sufficient natural gas is present. Hydrate-forming conditions exist in near-bottom marine sediments and beneath thick permafrost layers.

A portion of the reservoir at the Messoyakh gas field in western Siberia contains naturally occurring gas hydrates that have been produced commercially. Messoyakh serves as a case study of the problems encountered in exploration and production of shallow gas fields in arctic regions. The Messoyakh gas field illustrates that although hydrates complicate gas production, the in-place reserves of a field are increased by hydrate presence.

HISTORY

The Messoyakh gas field was discovered in 1967 (Makogon et al., 1971; Meyerhoff, 1980; Sapir et al., 1973; Sheshukov, 1973). The discovery well encountered the top of the gas-bearing Dolgan Formation (Figure 7) at 757 m below msl (mean sea level) and the gas water contact at 800 m (Sapir et at., 1973). The limits of the field were delineated by 11 exploration wells. At least 50 production wells were drilled in the 13 km by 20 km structure with spacing

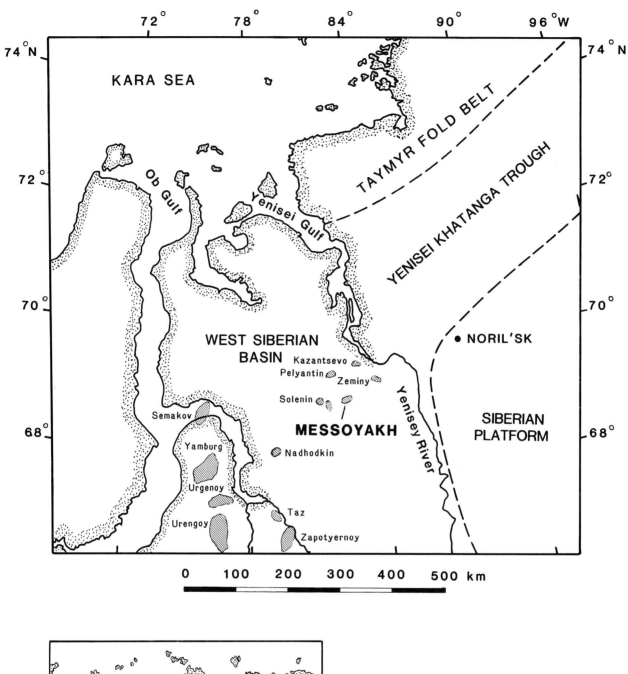
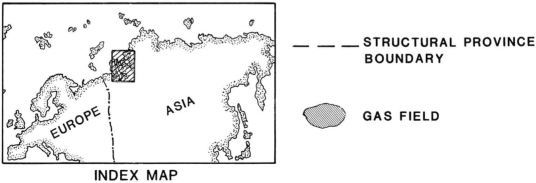

Figure 1. Map showing the Western Siberian basin, the Messoyakh field, other fields, and structural provinces.

Table 1. Location of Messoyakh gas field.

Source	Latitude N	Longitude E
Afanasenkov (1984)	69.17°	82.35°
Carmalt and St. John (1986)	69.20°	82.50°
CIA (1985)	68.98°	82.96°
Klemme (1984)	69.23°	82.64°
Salov et al. (1984)	69.22°	82.22°
Semenovich et al. (1976)	69.10°	81.97°

of 500 to 1000 m. Production was begun in 1970 and continued until 1977 at an average rate of 3 million m^3/day (110 mmcf/day) (Makogon, 1984). Following a four-year shut-in period, production was re-established at rates of 0.2 to 0.5 million m^3/day (7 to 18 mmcf/day) (Makogon, 1984, 1988).

Initial production of representative wells was reported by Meyerhoff (1980) as ranging from 3200 m^3/day (113 mcf/day) on a 3.25 mm choke to 180,000 m^3/day (6.4 mmcf/day) on a 12.7 mm choke. Sumets (1974) reported that the initial production rates of wells varied widely, with 27 of the wells averaging 150,000 m^3/day (5.3 mmcf/day) at a pressure drop of 0.2 MPa (29 psig) with another 16 wells producing only a small fraction of that amount. Initial production from wide-open wells was reported by Makogon et al. (1971) to range from 26,000 to 1,000,000 m^3/day (900 to 35,000 mcf/day), with the higher production rates coming from wells perforated in deeper sections of the reservoir formation.

DISCOVERY METHOD

Little information is available on the exploration concepts used by the Soviet petroleum geologists in the 1960s. Soviet geophysical teams conducted regional seismic surveys throughout the 1950s and early 1960s (Meyerhoff, 1980). Subsequent exploratory drilling of the mapped structures produced numerous discoveries of giant fields. Drilling was concentrated on the crests of closed anticlines, without consideration of possible stratigraphic traps.

While the discovery of Messoyakh in 1967 resulted from seismic lines, subsequent research has demonstrated a number of features of the deposit that may be of use in future exploration. A geochemical anomaly is present above the Messoyakh gas field. Gases enriched in methane and ethane were desorbed from snow samples collected from over the Messoyakh gas field (Bordukov et al., 1984b). The highest levels of hydrocarbons in snow were reported from the east end on the Messoyakh structure. Few details were presented on the Messoyakh geochemical study, but data from a nearby field suggest that the anomaly also exists in soil gases. A snow-gas anomaly similar to that at Messoyakh was documented at the Solenin field, 30 km to the west (Figure 1). Additionally, anomalous amounts of methane, ethane, and carbon dioxide were noted in gas from sediment collected from depths of 5 m at Solenin (Bordukov et al., 1984a).

The presence of methane and ethane in the snow cover over Messoyakh is surprising in view of the low permeability of the near-surface sediments. The reservoir rock at Messoyakh is overlain by 400 to 450 m of permafrost (Sheshukov et al., 1972) and a subjacent 200 to 250 m interval in which gas hydrates are stable. Both permafrost and interstitial gas hydrates dramatically reduce the permeability of sediments. Indeed, a 1 m thick section of methane hydrate-impregnated sediment can trap a gas column 50 m thick (de Boer et al., 1985). Geochemical anomalies at the Solenin field were well correlated with faults detected on magnetic surveys. The presence of analogous surface enrichment at Messoyakh suggests the presence of similar undocumented faults in the Quaternary cover of Messoyakh.

Messoyakh is associated with a negative magnetic anomaly. Kornev et al. (1982) noted the correspondence of magnetic anomalies and Western Siberian gas fields including the supergiant Urengoy and Yamburg structures. While the negative magnetic signature may be of use in predicting gas-bearing structures, Kornev et al. (1982) suggested using magnetic surveys as a way of low-grading prospects likely to produce gas in favor of more desirable oil-bearing structures.

Geomorphic methods have been useful in the mapping of the location and extent of Quaternary tectonic features in the Yenisei-Khatanga trough (Figure 1). Kulakov and Makhotina (1985) have detected 400 m of Quaternary uplift at Messoyakh by "analysis of elevations in the relief" of the area. The axis of the Neogene uplift is at the Malokhet arch some 40 km east of the Messoyakh field (Figure 2). The effect of Neogene uplifts on the configuration of the Messoyakh trap is not well documented; however, the nearby Zeminy gas field is sited along the crest of the Malokhet arch.

STRUCTURE

The Messoyakh gas field is part of the cratonic West Siberian basin (Figure 1). The anticlinal trap

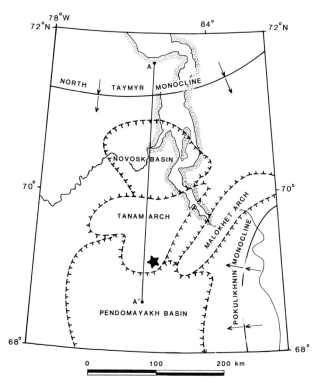

Figure 2. Major structural features of the Messoyakh region. Location of Messoyakh gas field shown by star. The Tanam arch, on which the Messoyakh is located, has also been referred to as the Nal'nuyo fold and the Messoyakh cupola. (Modified from Afanasenkov, 1984.)

is on the flank of the Tanam arch, updip from the Pendomayakh basin (Figure 2). Interpretations vary as to the extent of faulting in the trap and surrounding structures.

Tectonic History

The West Siberian basin is a cratonic basin containing up to 10 km of Jurassic to Quaternary sediments deposited on folded basement. The basement consists of terranes accreted onto the Archean Siberian platform in the Paleozoic and late Proterozoic. The basement underlying Messoyakh is probably composed of altered Baykalian (late Precambrian) miogeosynclinal rocks (Meyerhoff, 1980). Subsidence began in the Late Triassic. The sediment cover of the central and western parts of the basin averages 3 to 4 km thick. Sediments accumulated to much greater depths in the grabens in the northeastern portions of the basin and in the bordering Yenisei-Khatanga trough (Figure 1). Uplifts and differential subsidence concentrated in the Late Jurassic and Tertiary formed the productive structures of the region. Significant neotectonic activity has been noted in the Yenisei-Khatanga trough.

Regional Structure

The Messoyakh gas field is located on the southeast flank of a large arch trending about N60°E (Figure 2). The large positive feature has been called the Nal'nuyo fold (Sumets, 1974), the Messoyakh cupola (Sapir et al., 1973), and the Tanam arch (Semenovich et al., 1976; Afanasenkov, 1984).

Several interpretations of the structural configuration of the basement in the Messoyakh region have been advanced in the literature (Figure 3). Meyerhoff (1980), Rigassi (1986), and Surkov and Zhero (1981) show a horst bordered by well-defined basement faults underlying Messoyakh (Figures 3C and 3D). Semenovich et al. (1976) indicated a much broader uplift without obvious faults (Figure 3B). Makarenko et al. (1972) interpreted Messoyakh to be located on the northwest flank of a the Malokhet arch, without the horst or arch (Figure 3A) reported by Afanasenkov (1984), Meyerhoff (1980), Semenovich et al. (1976), Surkov and Zhero (1981), and others. Some of the inconsistency in interpretation of basement structure may have arisen from different operational definitions for the term *basement*.

The basement fault block underlying the Tanam arch is interpreted by Meyerhoff (1980) to measure 140 by 40 km. Rigassi (1986) and Surkov and Zhero (1981) indicated that movement on the basement faults during the Paleozoic was similar in style to the Jurassic and Tertiary deformation that produced the present arch. Differential subsidence of the region resulted in a thicker Jurassic sediment section to the southeast of the arch in the Pendomayakh (Afanasenkov, 1984) or Bol'shekhets (Semenovich et al., 1976) basin (Figure 4). The arch was uplifted 1.5 km in Late Jurassic to Early Cretaceous (Tithonian to Berriasian), producing an unconformity in the Middle to Late Jurassic sections over the arch (Figure 4). Following gradual differential subsidence in the Cretaceous, the arch was reactivated in the Tertiary, with about 1.3 km of uplift occurring between Oligocene and Pliocene time (Afanasenkov, 1984).

Local Structure

The Messoyakh gas field is located on a northeast-trending doubly plunging anticline measuring 20 by 13 km (Figure 5). Structural closure of 84 m is reported at the top of the Cenomanian reservoir formation (Sapir et al., 1973). The timing of the uplift is unclear, although the contours on a middle Quaternary marker suggest substantial recent uplift in the northwest portion of the structure (Figure 6).

No faults in the structure have been documented in published maps or cross sections, although Bordukov et al. (1984b) have reported surface leakage of gas from the east flank of the Messoyakh structure, presumably from a fault. The reported location of the field (Table 1) is not sufficiently precise to indicate whether the nearby basement faults mapped by Meyerhoff (1980) and Surkov and Zhero (1981) are related to the Messoyakh trap.

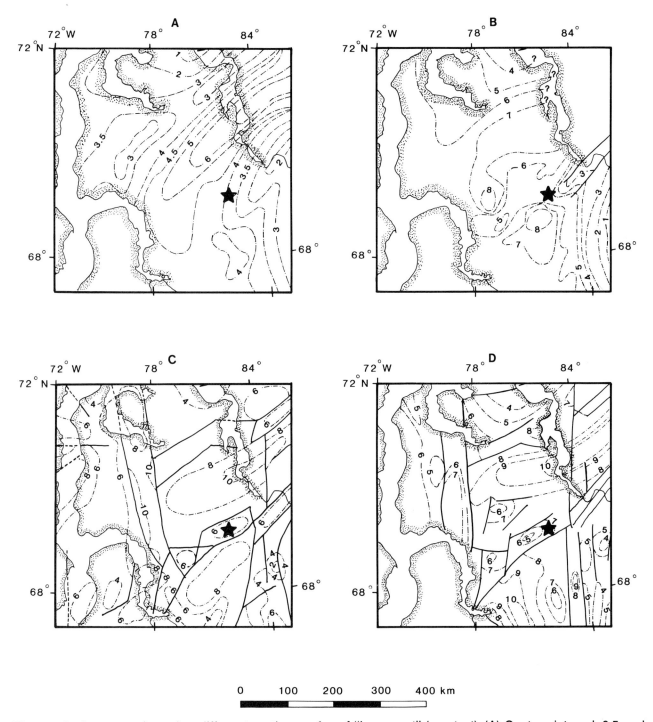

Figure 3. Interpretations by different authors of basement structure of the northeast part of the West Siberian basin. Basement depth contours in km. Faults shown by heavy lines. Approximate location of Messoyakh gas field shown by star. Some of the discrepancies may have arisen from differing definitions of "basement" (see text). (A) Contour interval, 0.5 and 1 km. Adapted from Makarenko et al. (1972). (B) Contour interval, 1 km. Adapted from Semenovich et al. (1976). (C) Contour interval, 2 km. Adapted from Meyerhoff (1980). (D) Contour interval, 1 km. Adapted from Surkov and Zhero (1981) and Rigassi (1986).

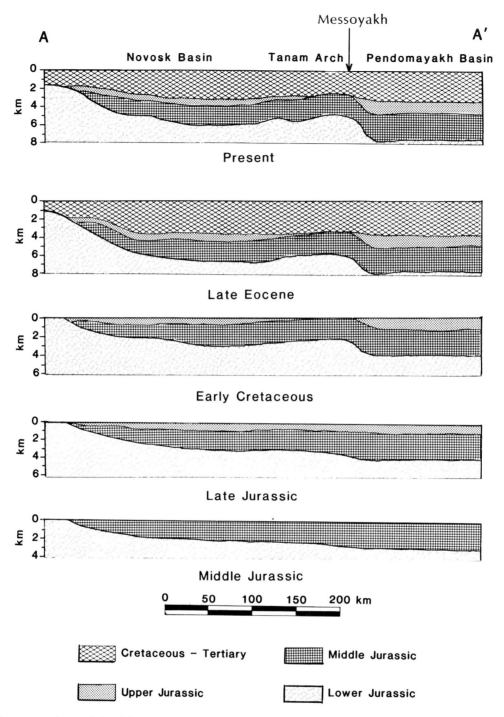

Figure 4. Cross sections of the Messoyakh gas field. Location of cross section shown on Figure 2. (Modified from Afanasenkov, 1984.)

STRATIGRAPHY

The Jurassic through Quaternary sedimentary section at Messoyakh totals 4000 to 5000 m (Afanasenkov, 1984). Approximately 1200 m of Lower Jurassic rocks and 800 m of Middle Jurassic rocks are overlain by 200 m of Upper Jurassic clastics (remaining after the Late Jurassic to Early Cretaceous regional uplift) (Afanasenkov, 1984). Some 2000 m of Cretaceous through Paleocene rocks are overlain by 70 m of Quaternary cover.

The nomenclature and gross lithology of the uppermost 1100 m of the sedimentary section at Messoyakh is shown in Figure 7. Detailed stratigraphic and lithological information on the strata is not available.

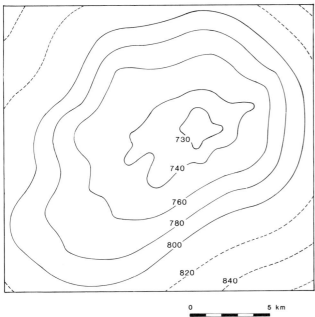
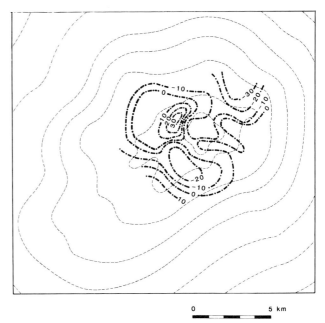

Figure 5. Structure contour map of the top of the Cenomanian Dolgan Formation, Messoyakh gas field. Contours in meters below mean sea level (dashed where inferred). Contour interval, 20 m. No location coordinates are shown because of wide discrepancies in descriptions of the location of the field (see Table 1 and Figure 10). (Adapted from Sapir et al., 1973.)

Figure 6. Structure of middle-late Quaternary marker bed, superimposed on structure map of Figure 5, Messoyakh gas field. Thin dashed lines are contours in meters relative to mean sea level (dashed with dots where inferred). Thick dashed dotted lines are contours on a middle Quaternary marker. Contour interval, 10 and 20 m, respectively. (Adapted from Sapir et al., 1973.)

TRAP

The trap at Messoyakh gas field is an anticline covering approximately 165 km² with 84 m of closure (Figure 5). The Dolgan reservoir is sealed by shales of the overlying Dorozhkov Formation (Figures 7 and 8). Soil-gas anomalies suggest the presence of faults along the eastern margin of the field (Bordukov et al., 1984b), but a possible role of faults in trapping has not been reported. Diminished permeability caused by the formation of natural gas hydrates in the upper part of the Dolgan Formation or the bottommost Dorozhkov Formation may have contributed to gas trapping.

Reservoir

The upper 74 m of the Dolgan Formation is the reservoir unit at Messoyakh. The reservoir formation is composed of sands and poorly indurated sandstone; interbeds of shale constitute up to 30% of the unit. The sandstone ranges in porosity from 16% to 38%, averaging 25% (Sapir et al., 1973). Permeability ranges from 10 to 1000 md with an average of 125 md (Sheshukov et al., 1972). Very low reservoir temperatures of 8 to 12°C have been recorded at Messoyakh (Makogon et al., 1971). Although the sand and shale sequences within the Dolgan Formation alternate frequently and cannot be correlated laterally, the sandstone units were reported to all be hydrodynamically connected (Sapir et al., 1973).

Gas Hydrates

The Messoyakh gas field differs from typical gas deposits in that a substantial portion of the natural gas in the reservoir is present in the form of hydrates. Hydrates are solid, ice-like compounds formed of gas and water at high pressure and low temperatures (Figure 9). The pressure and temperature conditions required for gas hydrate stability occur naturally in deep-sea marine sediments. Hydrates can also form in low-temperature sediments at the base of permafrost zones if the permafrost is adequately thick (at least 200 m) to exert sufficient hydrostatic pressure. The maximum formation temperature at which hydrates are stable (typically 9 to 12°C) defines the base of the hydrate stability zone. Assuming a 3°C/100 m gradient beneath permafrost, the gas hydrate stability zone would extend 300 to 400 m beneath the base of the permafrost.

Cherskii et al. (1982) compiled permafrost thickness and geothermal gradient data of the West Siberian basin to generate a map of the depth to the base of the gas hydrate stability zone (Figure 10). Gas reservoirs located at depths that are less than the depth of the base of the hydrate zone shown in Figure 10 should contain hydrates.

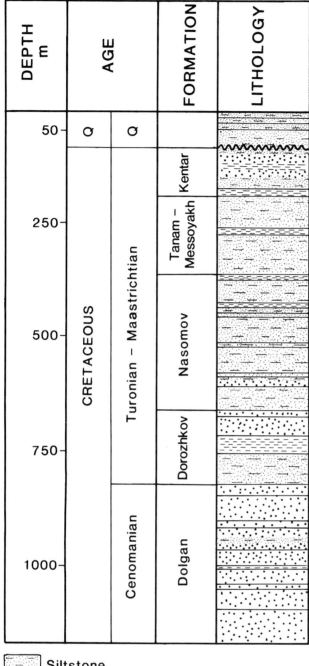

Figure 7. Schematic stratigraphic column, Messoyakh gas field. (After Makogon et al., 1971.)

The lower limit of gas hydrate stability occurs within the Dolgan Formation reservoir at Messoyakh (Figure 9). Hydrate is the stable form of pooled gas in the upper portion of the reservoir, while free gas occurs in the warmer conditions in the lower part of the reservoir. Rapid pressure declines during the initial production testing indicated that the in-place gas reserves were substantially less than the 40×10^9 m³ or 1.4×10^{12} ft³ originally estimated (Sheshukov, 1973). Since gas is released from the hydrate crystal lattice much more slowly than from open pore space, the anomalously rapid pressure decline was interpreted to indicate that a large portion of the reservoir was occupied by hydrates (Sapir et al., 1973). The presence of hydrates in the reservoir was subsequently indicated by well logs, drilling-rate anomalies, and an unusual depletion curve for the field.

The proportion of the reservoir that is occupied by hydrates has not been published. In calculating probable reserves, Makogon (1974) "conservatively estimated" that one-third of the reservoir is within the gas hydrate stability zone. Based on the gas hydrate stability curve (Figure 9), Makogon (1974, 1984, 1988) and Makogon et al. (1971) determined that the 10°C isotherm approximated the lower boundary of the gas hydrate stability zone at Messoyakh. Diagrammatic cross sections showing gross lithology and the 10° isotherm (Figure 11) suggest that considerably more than one-third of the sandstone reservoir is within the gas hydrate stability zone. However, the 10°C isotherm shows substantial relief; it fluctuates from -780 to -740 m in the 800 m between wells 130 and 131, and from -800 to -760 m in the 900 m separating wells 135 and 142. An absence of location information limits the volumetric interpretations that can be drawn from these diagrams.

Sapir et al. (1973) applied drilling rate data to infer the presence of hydrates in the Dolgan reservoir. A number of intervals with drilling rates similar to those measured in superjacent "dense shales" were shown to be clean sandstones when cored. They noted that the drilling rates through sandstone within the hydrate zone was analogous to those in permafrost, although the Dolgan reservoir at Messoyakh is 300 to 400 m beneath the base of permafrost.

Sapir et al. (1973), Sheshukov (1973), and Trofimuk et al. (1984) summarized the log responses attributable to hydrates at Messoyakh. These authors noted a lack of the expected negative SP response from intervals within the hydrate zone that were shown to be sandy by cuttings and cores. Owing to diminished permeability caused by hydrate-clogged pore throats, no mudcake was evident on caliper logs, and substantial caving was often noted. Similarly, microresistivity logs of hydrate-filled intervals indicate very low permeability in the hydrate zone. Gamma-ray logs from the hydrate intervals faithfully indicate the lithologies recorded in cuttings and cores. Low gamma values through the suspected hydrate zones indicate that the anomalously low permeability is not due to shale. Resistivity values in hydrate zones are greater than those of water-filled sand but less than those of a free-gas-bearing unit. Neutron values from hydrate-bearing sand zones are similar to water-bearing zones. The strong negative neutron kick associated with West Siberian Cenomanian gas

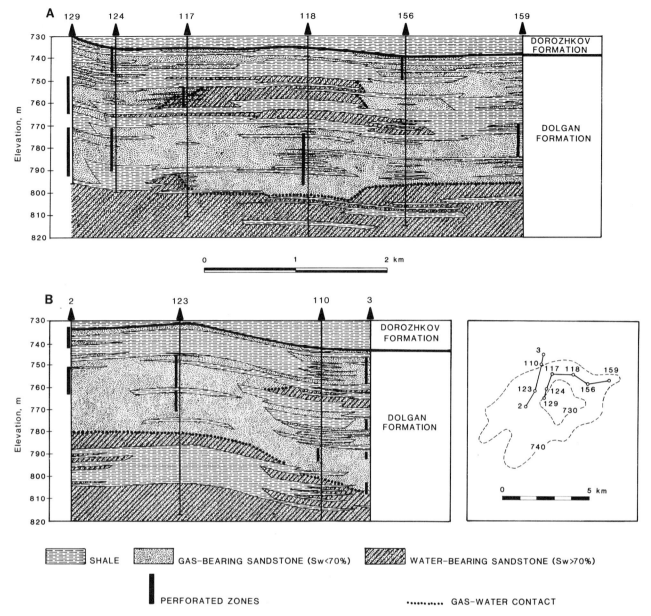

Figure 8. Cross sections of the Messoyakh gas field. Depths are subsea. (Adapted from Sapir et al., 1973.) (A) Dark vertical lines represent perforated zones. (B) Same as for A. Inset map from Figure 5.

deposits is absent within the hydrate zones, in spite of gas shows and production from the hydrate zones. All sources stressed that no single log parameter of hydrate presence exists; determination of hydrate presence requires corroboration of drilling rate data, temperature and pressure condition, and a full suite of logs.

Sapir et al. (1973) presented a sketch of log responses through the Dolgan reservoir at well 136 (Figure 12). The authors did not include core or drilling logs for the interval, or their interpretations of the well logs. The caliper and microresistivity curves show substantial caving and no mudcake above an elevation of −755 m, suggesting that the base of the hydrate stability zone is located at −755 m. However, the negative shift in SP and the positive resistivity and neutron kick at −782 m are characteristic of the transition from a hydrate-bearing zone to an underlying free-gas zone.

Gas-Water Contact

The interpretation of the depth and configuration of the gas-water contact at Messoyakh varies markedly among published sources. Meyerhoff (1980) reported the gas-water contact at −805 m. Sapir et al. (1973) reported that the "limit of the field" occurs at −803.5 m. Makogon (1984, 1988) and Trofimuk et al. (1984) report that the gas-water contact at −819 m has not changed during the 14-year production history of the deposit. However, their estimates of reservoir thickness and trap closure were not consistent with a gas-water contact at −819 m. Sapir

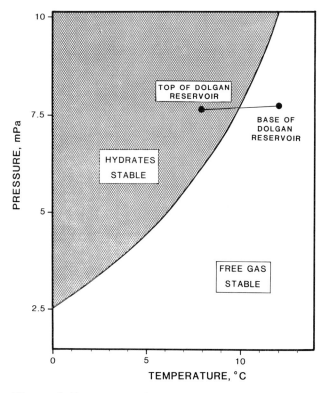

Figure 9. Pressure and temperature conditions of the Messoyakh reservoir. (Adapted from Makogon, 1988.)

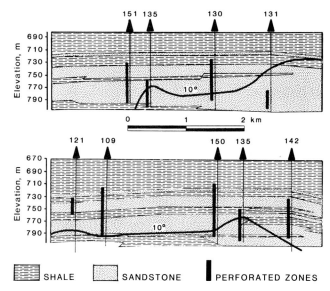

Figure 11. Cross sections of the Messoyakh gas field showing position of 10°C isotherm. Isotherm inferred to be the base of gas hydrate stability zone. Depths subsea. (After Makogon et al., 1971.) (Location not determined.)

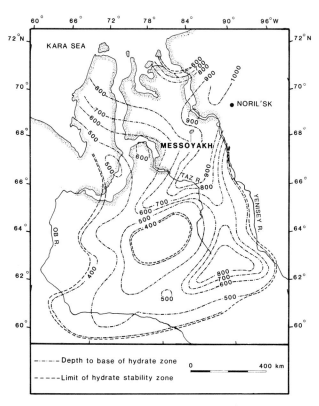

Figure 10. Distribution of the gas hydrate stability zone, West Siberian basin. Contour interval, 100 m. (After Cherskii et al., 1982.)

et al. (1973) reported that the gas-water contact varied in elevation from −779 to −811 m (Figure 8). A map of the gas-water contact presented by Sapir et al. (1973) shows that the contact generally mimics the structure of the trap (Figure 13).

Timing of Gas Accumulation

Comparison of the map of gas-water contact elevation (Figure 13) and the elevation of a Quaternary marker bed (Figure 6) shows a marked correspondence in the structural high 3 km west-northwest of the crest of the Messoyakh structure and the low located 4 km northeast of the crest. While Sapir et al. (1973) note the resemblance of the gas-water contact and the Quaternary bed, they minimized possible direct structural control or lithologic inhomogeneity of the reservoir as causes of the relief on the gas-water contact surface. Instead, they proposed that the irregular gas-water contact (Figures 8 and 13) resulted from hydrate formation in the deposit during gas migration and accumulation. Permeability changes within the reservoir caused by gas hydrate formation were proposed to be the principal control over the gas-water contact. Pores in some parts of the reservoir were thought to have been obstructed by hydrate formation during active gas migration. With some parts of the reservoir blocked by hydrates, subsequent migration of gas was focused elsewhere in the trap. Sapir et al. (1973) used their scenario of hydrate formation concurrent with migration to constrain timing of gas accumulation to the last 60,000 years when periglacial conditions existed in Siberia, and hydrate formation temperatures prevailed in the Dolgan reservoir rock at Messoyakh.

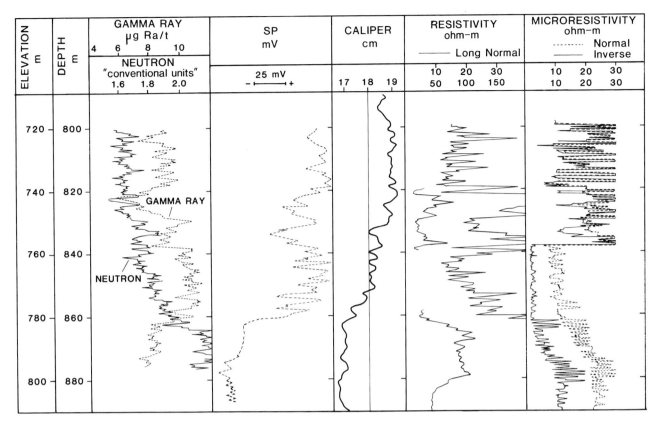

Figure 12. Logs of well 136, Messoyakh gas field. (Adapted from Sapir et al., 1973.) Caliper and microresistivity log curves suggest the base of hydrate stability zone at -755 m, but SP, resistivity, and neutron curves indicate transition from hydrate to free-gas at -782 m (see text, *Gas Hydrates*).

Greater degrees of hydrate pore occupancy in the flanks of the structures and an upward increase in hydrate pore occupancy near the axis of the fold may indicate that the deposit formed by a progressive cooling of an existing gas-saturated reservoir (Tsarev and Nenakhov, 1985). Alternatively, gas from a preexisting seep may have begun forming hydrates when permafrost developed in the Quaternary, developing a seal for further accumulation.

Source of Gas

The natural gas produced from Messoyakh averages 98.5% methane and about 0.1% heavier hydrocarbons (Table 2). The lack of heavier hydrocarbons suggests a biogenic source (Rice and Claypool, 1981; Schoell, 1983). A biogenic source has been reported for the natural gas from fields in the northwestern portion of the West Siberian basin including Messoyakh (Klemme, 1984). Most of the gas in Cretaceous reservoirs in the West Siberia basin is classified as biogenic or early diagenetic gas derived from coal beds in the Lower to middle Cretaceous Pokur series (Dolgan Formation). Methane $\delta^{13}C$ values of the coal-sourced gas averaged -60 ‰ (Meyerhoff, 1980; Yermakov et al., 1971).

The isotopic signature of the Messoyakh gas indicates a thermogenic source. Gas from Messoyakh was the heaviest of the suite of gases from the West

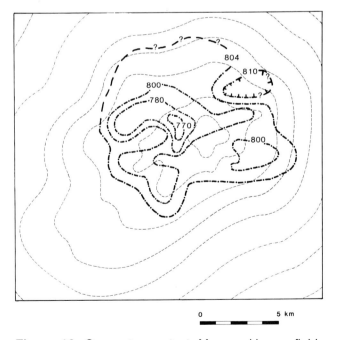

Figure 13. Gas-water contact, Messoyakh gas field. Superimposed on the structure previously shown by Figure 5, thin lines are contours in meters below mean sea level. Thick lines are contours of elevation of GWC. Contour intervals, 20 m and 10 m. (Adapted from Sapir et al., 1973.)

Table 2. Composition of Messoyakh gas

CH_4 (%)	C_2H_6 (%)	C_3H_8 (%)	C_4H_{10} (%)	C_5H_{10} (%)	CO_2 (%)	N_2 (%)	Reference
Cenomanian Reservoir:							
98.6	0.1	0.1	—	—	0.5	0.7	Makogon (1988)
98.04	0.05	—	—	—	0.95	0.95	Meyerhoff (1980)
98.22	0.0336	0.0015	0.00028	0.00027	0.50	0.22	Sapir et al. (1973)
98.8	trace	—	—	—	0.7	0.5	Sumets (1974)
Jurassic Reservoir:							
92.49	5.31	0.76	0.41	0.14	0.69	0.20	Meyerhoff (1980)

Siberian basin with methane $\delta^{13}C$ values of -40.0 to -42.5 ‰ (Prasolov et al., 1981). Yermakov et al. (1971) reported average methane $\delta^{13}C$ values of -47.9 ‰ compared with values ranging from -58.3 to -64.7 ‰ for methane from other Cenomanian gas reservoirs of the northeast West Siberian basin. Other anomalous geochemical features of the Messoyakh gas included elevated hydrogen content, and a $C_2:C_3$ ratio of 2 to 7 compared with 10 to 80 for the typical biogenic gas from the area.

Gavrilov et al. (1971) likewise found the isotopic signature of Messoyakh gas to be anomalously heavy, with a methane $\delta^{13}C$ value of -41.1 ‰ from well number P-9. Gas from the 833–839 m depth (approximately -750 to -760 m elevation) of well P-9 was also depleted in argon content with a value of 26 ppm. Gavrilov et al. (1971) stated that the Messoyakh gas was more similar to gases from Jurassic reservoirs than to gas from other Cenomanian reservoirs.

Oil stains were noted in cores and cuttings from some wells at Messoyakh. Disseminated heavy oil was reported from wells 109 and 148 by Yermakov et al. (1971). Sapir et al. (1973) reported that an oil and gas mixture was emitted from wells 123, 127, and 148 during testing. Oil was reported to have stained cores from 4 of 12 wells studied; oil content increased with depth reaching 25% of pore volume. A core from well 109 was "abundantly saturated" with oil at 10 m beneath the gas-water contact.

Thermogenic natural gas is present in Jurassic reservoirs at Messoyakh. Gas from Jurassic rocks at depths of 2614 to 2648 m contains more heavier hydrocarbons than gas from the Cenomanian Dolgan reservoir (Table 2).

Afanasenkov (1984) reported on the structural and geothermal history of the Yenisei-Khatanga trough from the Jurassic to the present. His study of the Messoyakh area shows that faults resulting from the uplift of the Tanam arch are or could be migration pathways for hydrocarbons generated in the basement. Migration of hydrocarbons updip along the flank of the Pendomayakh basin is another possible scenario for charging the Messoyakh trap with thermogenic hydrocarbons. (The cross section modified from Afanasenkov is shown by Figure 4 but fault zones are not indicated.)

Afanasenkov (1984) constructed a burial history plot for the Solenin gas field, which is adjacent to the Messoyakh field. We modified his plot to match more closely the sediment depths at Messoyakh (Figure 14). The isotherms used in Figure 14 are based on vitrinite reflectance results reported by Afanasenkov (1984). Afanasenkov (1984) used only formation temperature to estimate maturity rather than the integrated time and temperature approach that is more commonly applied in the West. The Lopatin method of thermal maturity estimation as described by Waples (1980) was used to prepare Figure 14.

The hydrocarbons in the Cenomanian Dolgan reservoir at Messoyakh migrated at least 2.5 km vertically if they were generated from source beds within the oil-generation window (Figure 14). Vertical migration distances of at least 4.7 km are required if the hydrocarbons were generated from the shallowest rocks within the wet gas window.

The Cenomanian Dolgan reservoir at Messoyakh contains a mixture of very dry thermogenic gas and heavy residual oil. Only minor amounts of liquid petroleum or condensate have been reported in the literature. This bipolar mixture of reservoired hydrocarbons complicates our assessment of the most probable potential source beds. The presence of residual oil in several of the wells at Messoyakh suggests that associated gas is present rather than, or in addition to, the deep-source gas expected on the basis of the low concentrations of ethane through pentane. Jurassic source rocks or source rocks in the undescribed pre-Jurassic basement are most probable if migration was primarily from sources immediately beneath the deposit. If the gas was generated in the adjoining Pendomayakh basin, the source is still most likely to be Jurassic rocks. However, the probable Jurassic sources in the Pendomayakh basin did not undergo the same degree of uplift at the Jurassic-Cretaceous boundary as in the Messoyakh area (Figures 4 and 14). Thus, the gas generation windows are significantly higher in the section in the Pendomayakh basin than over the Tanam arch near

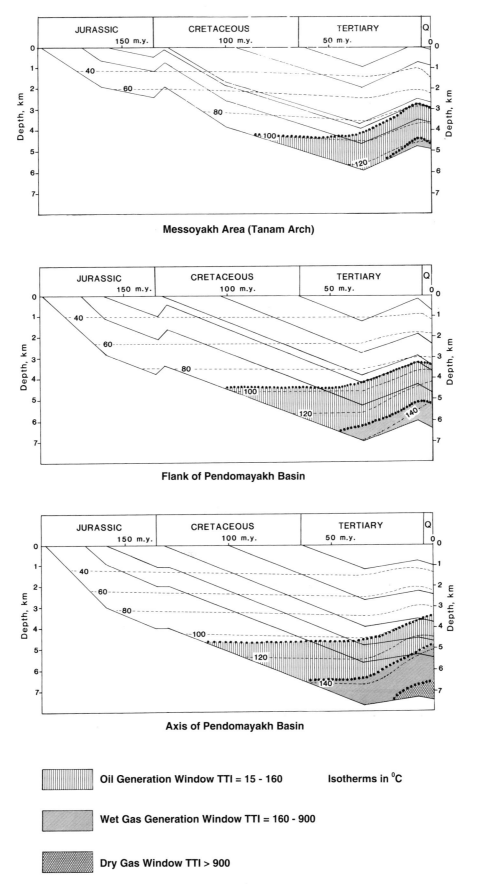

Figure 14. Burial history reconstructions of the Messoyakh region. Data from Afanasenkov, 1984, isotherms from R_o; modified by authors to Lopatin method (described by Waples, 1980).

Table 3. Gas production from selected wells (after Makogon et al., 1971).

Well	Proportion of Perforation in Hydrate Zone(%)	Wide Open Gas Flow $10^3 m^3/day$
121	100	26
109	100	133
142	100	285
151	100	300
150	81	413
135	20	626
130	82	629
131	0	1000

Messoyakh. If the gas produced at Messoyakh was generated in the basin, contributions from the pre-Jurassic basement units are not necessary.

The anomalous composition of the gas in the Dolgan Formation reservoir at Messoyakh conceivably could reflect the presence of hydrates in the reservoir. Trofimuk and Cherskii (1974) proposed that the unusually heavy isotopic signature of methane and argon at Messoyakh is due to isotopic fractionation during the process of hydrate formation. The water in gas hydrates is known to be enriched in ^{18}O (Davidson et al., 1983). Anomalously heavy methane is associated with hydrates at DSDP Site 570 (Claypool et al., 1985), but not at Site 533 (Claypool and Threlkeld, 1983). Trofimuk and Cherskii (1974) contended that the Dolgan reservoir was previously entirely within the gas hydrate stability zone and that all the gas in the Messoyakh reservoir was once in hydrate form. Warming in recent time has presumably shifted the base of the gas hydrate stability zone upward and caused the hydrates in the deeper portion of the reservoir to dissociate and release the isotopically enriched gas.

An alternate scenario of hydrate formation could address the lack of ethane through pentane in the Messoyakh gas. Mixtures of methane and ethane, propane, or iso-butane form hydrates more readily than pure methane (Deaton and Frost, 1949). Heavier hydrocarbons from natural gas may be preferentially incorporated into the hydrate phase, with the remaining free gas becoming correspondingly enriched in methane (Makogon, 1974; Tsarev and Nenakhov, 1985). Thus, one could reason that gas in the lower free-gas zone of the reservoir at Messoyakh has been depleted of hydrocarbon gases heavier than methane by formation of hydrates in the upper part of the reservoir. Sapir et al. (1973) indicated that gas collected from well 135 after methanol treatment was enriched in heavier hydrocarbon gases. However, the apparent enrichment was not great. This suggests that the composition of gas in the hydrate phase does not differ sufficiently from that of free gas to explain the very dry character of the Messoyakh gas.

Production

Wells completed in the hydrate zone at Messoyakh consistently produced less than nearby wells completed in the deeper, free-gas portion of the reservoir. The depth of perforated intervals relative to the base of the hydrate zone for nine wells in the Messoyakh field is shown in Figure 11. Approximate production rates are low in wells completed in the hydrate zone (Table 3).

Reported production levels from the wells have been normalized to account for the thickness of the perforated interval (Figure 15). A strong negative correlation of hydrate presence and well productivity is evident in Figure 15. Hydrates plug pore spaces in the reservoir, dramatically reducing the effective permeability. Thus, while the Dolgan reservoir rocks have adequate permeability (10–200 md), the hydrate zone constitutes a tight interval in the otherwise suitable reservoir rock.

Gas can be liberated from the hydrate in the upper portions of the reservoir by conventional production practices, but with difficulty. Opening the well-head valve decreases well-bore pressure below the gas hydrate stability pressure for the prevailing formation temperature. Hydrates present near the well bore dissociate to water and natural gas in the lowered pressure conditions (Figure 9).

Low production rates plague conventional production methods in the hydrate zone. Low permeability of hydrate-impregnated sediments limits production to rocks within a short radius from the well bore. Additionally, rapid hydrate dissociation from depressurizing the well decreases formation temperature owing to the latent heat of hydrate formation. Formation water released by dissociation of hydrate may freeze near the well bore at the lowered temperatures, further diminishing permeability.

In view of the difficulties in efficiently producing gas from hydrated reservoirs, two approaches were taken to maximize production at Messoyakh. Since the deeper portions of the Messoyakh reservoir in which free gas is present are similar to conventional gas reservoirs, wells were preferentially perforated beneath the gas hydrate stability zone. Stimulation of hydrate dissociation was attempted in wells completed in the hydrate zone.

Compounds that had been found effective in preventing hydrate formation in pipelines were injected into a number of wells at Messoyakh (Sheshukov et al., 1972; Sumets, 1974). Methanol and $CaCl_2$ were used alone and in mixtures. Methanol was found to be more effective at increasing gas production, but $CaCl_2$ cost less. The most economic treatments were mixtures of the two compounds.

Makogon et al. (1971) reported on methanol treatment of Wells 133 and 142. Similar production levels were obtained before and after methanol application, but higher well-head pressures could be maintained at a given production level subsequent to treatment. Production figures reported by Makogon et al. (1971) are normalized to account for

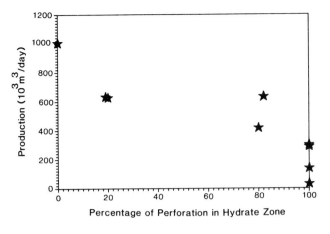

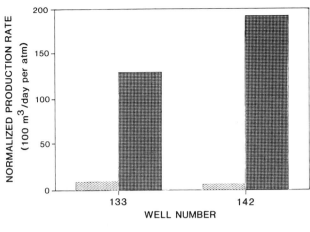

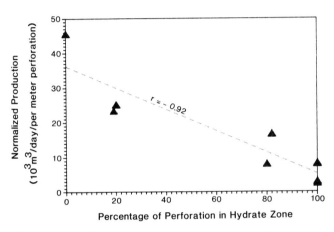

Figure 16. Effect of methanol injection on gas production rates. (Data from Makogon et al., 1971.)

Figure 15. Effect of hydrates on gas production rates. Data points correspond to wells shown in Figure 11. (Data from Makogon et al., 1971.)

the difference in well-head pressure and presented in Figure 16.

One limit on hydrate production rates is retrograde hydrate formation in some holes. The dissociation of hydrates by depressurizing lowers formation temperatures to the point that hydrates may reform at and clog formation pores. Sumets (1974) discussed methanol stimulation treatment methods that minimized retrograde hydrate formation. The most effective treatment involved injection of about 3 to 4 m^3 of methanol over a period of 24 to 26 hours under a pressure of 100 to 150 atm. Subsequently, air was forced into the perforated zone to drive the methanol deep into the formation. The methanol dissociated the hydrates and prevented hydrates from reforming in the reservoir during production. Production increases resulting from the methanol treatment are summarized in Table 4.

Makogon (1984, 1988) and Trofimuk et al. (1984) published production rate and formation pressure data from 13 years of production at Messoyakh (Figure 17). Production increased from 1970 when the field went on-stream to reach a peak rate of about 2.1 bcm/yr (74 bcf/yr) in 1972 (Table 5). Production diminished until 1978, when the field was shut in. Production was begun again in 1982, although the three principal sources for the production data (Makogon, 1984, 1988; Trofimuk et al., 1984) give conflicting estimates of rates for the more recent production cycle.

The contribution of hydrate gas to the total production from the Messoyakh field is shown in Figure 17. The dashed line on the formation pressure curve of Figure 17 indicates the pressure decline originally estimated for Messoyakh based on initial declines and calculated reservoir volume. The estimated decline shown by the dashed line assumes that the reservoir is occupied by free gas and water, with no hydrates present. The actual pressure curve diverges from the estimated pressure curve at a pressure of about 5.8 MPa when gas from dissociating hydrates began contributing substantially to formation pressure. The field was shut in from mid-1978 through mid-1981. During this time, hydrates continued to dissociate, releasing free gas into the reservoir and increasing the formation pressure to 6.2 MPa. Subsequent production has been accompanied by decreasing formation pressure as withdrawal rate exceeds the dissociation rate of the hydrates.

Makogon (1984, 1988) provided a series of comments on gas production at Messoyakh in addition to the diagrams reproduced in Figure 17. Makogon stated that up to 36% of the gas produced at Messoyakh is from wells perforated in the hydrate zone. He did not report which portion of the gas from the hydrate zone was produced by conventional depressurization and which by methanol and $CaCl_2$

Table 4. Effect of methanol treatment on gas production (after Sumets, 1974).

Well Number	Production Rate* Before Treatment 10³m³/day	Production Rate After Treatment 10³m³/day	Increase %
111	130	220	170
124	40	160	300
130	65	160	146
133	15	305	1900
134	100	390	290
135	100	250	150
142	10	390	3800
146	1	70	6900
150	50	155	210
151	1	120	11,900

*200 kPa pressure drop.

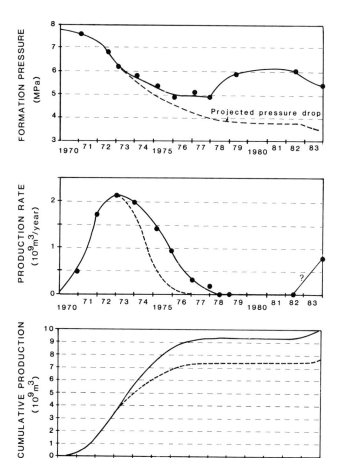

Figure 17. Production history of Messoyakh gas field. Dashed lines indicate projected formation pressure, production rate, and cumulative production from the deeper, free-gas zones of the reservoir. Solid lines are free-gas zones plus contributions from the hydrate gas zone above. Shut-in period, during which reservoir pressure increased, extended from mid-1978 to mid-1981. The pressure buildup and later increase in production rate resulted from the dissociation of hydrates to free gas (see text). (Adapted from Makogon, 1984, 1988.)

stimulation methods. Makogon (1984, 1988) stated that 2 bcm (70 bcf) of the gas withdrawn from the field during the first eight years of production was from hydrates. During the shut-in period, dissociation of hydrates added 3.17 bcm (112 bcf) of free gas to the reservoir. These statements on the magnitude of the hydrate contribution to the gas production at Messoyakh agree with the volumes calculated directly from the curves in Figure 17.

Relatively high production rates may be possible from the field in the future. Low production rates have been reported for the production period subsequent to the shut-in period. Makogon (1984) and Trofimuk et al. (1984) reported rates of 0.3 to 0.7 bcm/year (10 to 25 bcf/year) of gas, while Makogon (1988) indicated a gas production rate of about 0.2 bcm/year (7 bcf/year). During the shut-in period formation pressure increased to about 6.2 MPa. The annual gas production rate of about 2 bcm/year (7 bcf/year) was attained in 1973 when the formation pressure was in the range of 5.8 to 6.2 MPa. While other factors may dictate the low production rates of the most recent production phase, the reservoir pressure attained during the shut-in period is sufficient to permit much higher rates.

Reserves

The gas reserve estimates for Messoyakh range from 37 to 400 bcm (1.3 to 14 tcf). A paucity of available information on the field and disagreement on the contribution of hydrates to the total reserve base cause the wide range in published reserve estimates.

Hydrate Reserves

Gas hydrates are a compact mode of gas storage in a shallow reservoir since more natural gas can be stored as a hydrate form than in an equal volume of free gas and water. A unit volume of hydrate can contain 150 to 180 volumes of methane at standard conditions. At the typical conditions of the Mes-

Table 5. Gas production at Messoyakh (from Makogon, 1988).

Production (10^9 m^3)	Period	Comments
7.86	1969–1977	Free gas initially present in reservoir, initial production phase
1.96	1971–1977	Gas from hydrate dissociation, initial production phase
0.60	1981–1988	Gas from hydrate dissociation after shut-in period

soyakh reservoir (75 atm, 10°C), 1000 m^3 of methane occupies about 12 m^3 as free gas, but only about 6 m^3 as hydrate.

The actual difference in natural gas capacity of a unit volume of a hydrate reservoir compared to a free-gas reservoir may vary from the 2:1 ratio derived above. Owing to permeability changes caused by hydrate formation, naturally occurring gas hydrates rarely occupy the pore space completely (Finley and Krason, 1986b). When migration of natural gas into the gas hydrate stability zone is impeded by interstitial hydrates, the actual deposit consists of gas hydrates and excess pore water. The volumetric natural gas content of a reservoir consisting of hydrate and excess water is less than the gas content of a pure hydrate (about 160 m^3 gas per cubic meter of pore space).

When sufficient natural gas is present, the pore water of a low-temperature, high-pressure reservoir is entirely converted to hydrate, with the remaining pore space occupied by free gas. In these circumstances, a hydrate reservoir is a composite system consisting of gas stored in a free state and gas combined with the residual pore water as hydrate. Such a reservoir can contain substantially more natural gas than a conventional reservoir. The pore water (S_w) of a conventional reservoir simply occupies pore space; but when converted to hydrates, the pore water itself acts as a gas storage medium.

Sumets (1974) reported that the hydrate zone at Messoyakh is "oversaturated," or composed of free gas and hydrate with little if any liquid pore water. Sapir et al. (1973), Makogon et al. (1971), and Makogon (1974) have reported a S_w value of 40%. At that pore water content, 1 m^3 of pore space at 7.5 MPa and 10°C contains 50 m^3 of gas in the absence of hydrates (0.6 m^3 at 93 m^3 gas per m^3 pore space). However, if the water occupying 40% of pore space has been converted to hydrate, then the gas content of 1 m^3 of pore space is 118 m^3. The gas content of the composite hydrate and free-gas reservoir is the sum of 72 m^3 from hydrate (0.45 m^3 hydrate at 160 m^3 gas per m^3 hydrate) plus 46 m^3 of free gas from the remaining pore space (0.55 m^3 free gas at 93 m^3 gas per m^3 pore space). Water expands 10% upon conversion to hydrate, accounting for the increase from 40% S_w to 45% S_h. In the simplified case of the Messoyakh reservoir conditions, the presence of hydrates at the reported amounts results in a natural gas capacity that is 236% of that possible in the absence of hydrates.

Published Reserves

Halbouty et al. (1970) listed reserves from Messoyakh as 14.00 tcf (0.40 tcm) ultimate recoverable gas (Table 6). The same figure and its oil equivalent of 2333 million bbl was cited by Carmalt and St. John (1986) as recoverable reserves. Klemme (1984) reported recoverable reserves at Messoyakh of 2.3 BOE (bbl of oil equivalent) (370 bcm; 13 tcf) of which 0.7 BOE (110 bcm; 4 tcf) had been produced.

Meyerhoff (1980) claimed 3.1 tcf (88 bcm) of recoverable reserves from Messoyakh (Table 6). He did not mention hydrates or their possible contribution to reserves.

Sumets (1974) reported that Messoyakh contained 37.44 bcm (1.32 tcf) of proven and probable reserves. This figure apparently included both free gas and gas from hydrates (Table 6).

Sheshukov (1973) reported reserve estimates that took into account gas dissociated from hydrates in the reservoir (Table 6). The field was initially estimated to contain 40 bcm (1.4 tcf) based on estimates of the reservoir volume and water saturation. However, the rapid pressure decline recorded during initial production indicated reserves of 18 bcm (0.6 tcf). Assuming that the 55% of the originally estimated reservoir volume was instead filled with hydrates, Sheshukov (1973) calculated that 62 bcm (2.2 tcf) of gas was present in hydrate form in the upper portion of the reservoir. Including both free gas and hydrates, Sheshukov (1973) estimated total reserves at 80 bcm (2.8 tcf), but he cautioned that the hydrate reserves may not be producible.

Makogon (1974) did not directly report reserves for Messoyakh. However, he stated that when one-third of the reservoir is assumed to be occupied with hydrates, the calculated reserves are 54% greater than if only free gas were present. If the reserve estimate of Sheshukov (1973) for the Messoyakh structure filled with free gas is used (40 bcm; 1.4 tcf), Makogon's observation yields total reserves of 62 bcm (2.2 tcf) (Table 6).

Reserves from Production Data

We have calculated probable reserves from the production data that Makogon (1984, 1988) reported

Table 6. Reserves estimates for the Messoyakh field.

Source	Reserves bcm	Reserves tcf	Assumptions
Halbouty et al. (1970)	400	14	Ultimate producible gas
Carmalt and St. John (1986)	400	14	Recoverable reserves (1986)
Klemme (1984)	110	4	Produced reserves
	370	13	Proven and probable
Meyerhoff (1980)	88	3.1	Total recoverable reserves
Sumets (1974)	37.44	1.32	Proven and probable reserves
Sheshukov et al. (1972)	12	0.4	Total free gas reserves
	72–75	2.5–2.6	Total reserves including hydrates
Sheshukov (1973)	40	1.4	Total free gas reserves from volume of structure
	18	0.6	Total free gas reserves from initial pressure drop
	62	2.2	Total hydrate reserves in upper part of reservoir
Makogon (1974)	61 (?)	2.2 (?)	Reserves including hydrate are 54% greater than for free gas alone
CIA (1985)	<84	<3	Not classified as giant field

for the Messoyakh gas field. Total production of both gas and hydrate is about 10 bcm (0.35 tcf). That includes about 8 bcm (0.28 tcf) of gas originally present as free gas in the reservoir and about 2 bcm (70 bcf) of gas produced from dissociated hydrates (Table 5). Dissociation of gas hydrate in the reservoir serves to replenish the free-gas portion of the reservoir and maintain formation pressure. However, the four-year shut-in period showed that formation pressure is unlikely to again exceed 6.2 MPa.

We calculated the volume of the free-gas reservoir from the data from Makogon (1984, 1988). Based on his projected pressure curve (Figure 17), the free-gas reservoir would have declined from 7.8 to 4 MPa to produce 8 bcm of free gas in the absence of hydrates. Applying the gas state equation ($PV = ZnRT$) to that pressure drop indicates that in-place free gas originally totaled 15 bcm (530 bcf). At 78 atm and 10° to 12°C, the free gas originally present in the reservoir occupied 1.6×10^8 m^3 (5.6×10^9 ft^3). At a mean S_w of 40%, the free gas would occupy 2.5×10^8 m^3 of pore space or 1×10^9 m^3 of 25% porosity rock.

Based on the derived volume of the free-gas reservoir, we estimated the amount of gas in the hydrate portion of the reservoir. Assuming the reservoir at Messoyakh is two-thirds hydrate and one-third free gas, the hydrate portion of the reservoir occupies about 5×10^8 m^3 of pore space. Assuming that 40% of the reservoir pore space is occupied by hydrates, then each m^3 of pore space contains 72 m^3 of gas from the hydrate and 55 m^3 of gas in the remaining pore space for a total of 127 m^3. Based on these assumptions, the 2×10^9 m^3 hydrate portion of the reservoir originally contained 62 bcm (2.2 tcf) of gas in place in both hydrate and free gas. The total in-place gas at the Messoyakh gas field is thus 62 bcm (2.2 tcf) in the upper part of the reservoir and 15 bcm (530 bcf) in the lower part of the reservoir or 77 bcm (2.7 tcf) total. Meyerhoff reported a gas recovery factor of 65% for the Messoyakh gas field. Barkan and Voronov (1983) estimated gas recovery factors of 10% to 50% for hydrate deposits. Applying these factors to the derived in-place estimates yields recoverable reserves of 16 to 36 bcm (0.5 to 1.3 tcf). The accuracy of our derived figures is dependent on the accuracy of Makogon's data and numerous assumptions we applied during the derivation.

The recoverable reserves we calculated from the production data of Makogon (1984, 1988) are within the range of estimates published by Soviet sources in the 1970s (Table 6), particularly the 37.44 bcm (1.322 tcf) figure of Sumets (1974). Our estimated in-place figures are very close to the reserves estimates of Sheshukov et al. (1972) and Sheshukov (1973). Qualifications by Sheshukov (1973) that not all of the hydrate reserves may be producible suggest that his reserve figures may be in-place estimates. The higher estimates in Table 6 are not probable, at least for the Cenomanian reservoir. To contain 400 bcm (14 tcf) of in-place gas in the Messoyakh structure would require 176 m thickness of 25% porosity gas reservoir sandstone at $S_w = 40\%$, or 76 m thickness of gas hydrate reservoir sandstone. Recoverable reserves of 400 bcm (14 tcf) would require 270 m thickness of gas reservoir rock or 150 to 760 m thickness of gas hydrate rock, substantially greater than the 84 m closure of the structure.

Reserve estimates and production data can be combined to estimate the gas remaining at Messoyakh. Depending on the gas recovery factor for the hydrate intervals, probable recoverable reserves range from 16 to 50 bcm (0.56 to 1.8 tcf). Since about 10 bcm (350 bcf) of gas has been produced, remaining producible gas is 6 to 40 bcm (0.21 to 1.4 tcf), most in hydrates. The field is thus 20% to 60% depleted.

Economics may dictate that Messoyakh be abandoned before being fully depleted. The low production rates since 1982 (0.2 to 0.5 bcm; 7 to 18 bcf) suggest that 6 to 15 years may be needed to produce the 3 bcm (100 bcf) of gas released from hydrate during the four-year shut-in period. The maintenance costs to operate an arctic gas field may not be justifiable at such low production levels.

EXPLORATION CONCEPTS

Regional Play

Other gas fields similar in size to Messoyakh are located at the juncture of the West Siberian basin and the Yenisei-Khatanga trough. Afanasenkov (1984) mapped 15 "exploration areas" within a 100 km radius of Messoyakh. The Solenin, Zeminy, and Pelyantin fields are located in the general region of Messoyakh (Figure 1). However, Messoyakh differs from adjacent fields. The gas at Messoyakh is apparently thermogenic from fairly deep Jurassic or earlier sources, while the gas in most of the Cenomanian traps is derived from biogenic or diagenetic alteration of Lower Cretaceous coal.

Other shallow gas fields in the West Siberian basin contain deposits of hydrates. The hydrate stability zone is reported by Tsarev and Nenakhov (1985) to "encompass gas deposits" at the Russkoye and Malokhet fields, in addition to Messoyakh. Cherskii et al. (1982) reported that the base of the gas hydrate stability zone is 40 to 130 m above the top of the gas accumulations at Kharasavey, Bovanenkov, Neytin, Anktisn, and Zapolyar fields. Tsarev and Nenakhov (1985) proposed that gas deposits within such close proximity to the overlying gas hydrate stability zone may contain relict hydrate beds in overpressured zones. The upper 100 m of the gas reservoir at Russkoye is within the hydrate stability zone (Cherskii et al., 1982). Klemme (1984) reported that some or all of the gas reserves are present in hydrate form in ten giant gas fields in the West Siberian basin: Zapolyar, Bovanenkov, Kharasavey, Semakov, Messoyakh, Yuzhno Russkoye, Noviyport, Anktisn, Nahodkin, and Nietin. A biogenic gas accumulation in the hydrate zone above the supergiant Urengoy field is shown in a cross section by Klemme (1984).

Hydrate deposits exist in other hydrocarbon provinces of the Soviet Union. In-place gas volume in hydrate deposits in the Vilyui basin have been estimated by Cherskii et al. (1976) at 7300 bcm (260 tcf), with minimum recoverable reserves estimated at 780 bcm (28 tcf). Barkhan and Vorona (1983) have estimated that 75% of the Soviet gas hydrate reserves are located in the Vilyui basin or the eastern end of the Yenisei-Khatanga trough. They estimate that the West Siberian basin contains 15% of the hydrate reserves, while the Timano-Pechora basin contains 5%, and the Soviet Far East contains less than 2% of the total.

Gas hydrates have been detected beneath permafrost in Alaska and Canada. Sandy strata onshore and offshore of the Mackenzie delta have been shown to contain hydrates by well logs (Bily and Dick, 1974; Davidson et al., 1978; Weaver and Stewart, 1982). Additionally, sub-permafrost sediments of islands of the Canadian arctic are projected to have possible hydrate occurrences. Hydrates were drilled during exploratory drilling on the North Slope of Alaska. Drilling irregularities, well logs, and a hydrate-bearing core indicate hydrate presence in the shallow sediments above the Prudhoe Bay and Kuparuk River oil fields (Collett, 1987).

General Application of Geologic Parameters

In most aspects, the Messoyakh field is a textbook example of a conventional gas deposit. The anticlinal trap is located on the flank of a larger uplift, with nearby faults available to convey gas from a variety of potential source beds. The adjacent basin is also a possible source of hydrocarbons, with migration along inclined carrier beds supplementing vertical migration through fault systems.

The unusual feature of the Messoyakh field is the presence of hydrates within the reservoir. The cold arctic climate depresses formation temperatures within the shallow Dolgan Formation. Hydrates allow greater volumes of gas to be trapped within the closure volume of the trap than if only free gas were present.

Lessons

The discovery and production of the Messoyakh gas field generated valuable lessons in characterizing and producing shallow arctic gas deposits (Krason and Ciesnik, 1985). When faced with unusual production test results and conflicting core and well log data, the Soviet scientists and engineers combined previous theoretical work with knowledge gained from arctic pipelines to address the problem. Criteria for identifying hydrates in situ were developed. Methods of estimating hydrate reserves were derived at Messoyakh and later applied elsewhere in the Soviet Arctic. Stimulation methods were designed, tested, and apparently applied to the field.

Problems plagued production from hydrate intervals. However, much of the gas from the field was produced using conventional methods, since only the upper part of the reservoir at Messoyakh contains hydrates. The deeper sections of the reservoir were produced at rates comparable to conventional gas fields in the region. By scheduling shut-in periods to permit dissociating hydrates to regenerate formation pressure, the additional reserves present in the hydrate phase recharge the lower free-gas zone. Messoyakh showed that by such cyclical production, a composite hydrate and free-gas reservoir can be produced without resorting to the stimulation procedures needed to produce directly from the hydrate reservoir.

The lessons learned at Messoyakh may find direct application elsewhere in Russia and in Arctic regions of North America. The drilling and production methods may eventually be applied to the enormous potential gas resources of gas hydrates in near-bottom sediments of major offshore basins, e.g. Gulf of Mexico (Krason et al., 1985), Caribbean (Finley et al., 1987; Finley and Krason, 1986a), and Beaufort Sea (Finley and Krason, 1988).

ACKNOWLEDGMENTS

Initial research for this paper was supported by the U.S. Department of Energy Morgantown Energy Technology Center. Mark Ciesnik translated a number of key papers. Bertrand Gramont drafted the figures. Our Russian colleagues provided valuable clarification on the development and history of Messoyakh.

REFERENCES CITED

Afanasenkov, A. P., 1984, Catagenic alteration of disseminated organic matter of the Jurassic-Cretaceous sediments of the Yenisey-Khatanga oil-gas region: Petroleum Geology, v. 22, p. 206-216.

Barkan, Y. S., and A. N. Voronov, 1983, Ocienka resursov gaza v zonakh vozmozhnogo gidratoobrazovania (Evaluation of gas reserves in zones of possible hydrate formation): Sovietskaia Geologiia, v. 7, p. 37-41.

Bily, C., and J. W. L. Dick, 1974, Naturally occurring gas hydrates in the Mackenzie Delta, N.W.T.: Bulletin of Canadian Petroleum Geology, v. 32, p. 340-352.

Bordukov, Y. K., V. A. Lokshina, V. K. Palamarchuk, and S. B. Timkin, 1984a, Faults of the sedimentary cover and relation of hydrocarbon anomalies to them: Petroleum Geology, v. 22, n. 5, p. 200-202.

Bordukov, Y. K., V. I. Yefimov, and S. B. Timkin, 1984b, Results of a gas-biochemical survey of snow cover for direct exploration for hydrocarbon deposits in the Yenisey-Khatanga downwarp: Petroleum Geology, v. 22, n. 5, p. 203-205.

Carmalt, S. W., and B. St. John, 1986, Giant oil and gas fields, in Future petroleum provinces of the world: AAPG Memoir 40, p. 11-54.

Cherskii, N. V., V. P. Tsarev, and A. A. Solovev, 1976, Metodika otsenki prognoznykh zapasov gaza v zonakh gidratoobrazovania, na primiere Vilyuiskoi sineklizi (Methodology of estimation of gas reserves in gas hydrates zones, on the example of Vilyui Syneclise): Geologiia i Geofizika, v. 17, n. 9, p. 3-7.

Cherskii, N. V., V. P. Tsarev, and S. P. Nikitin, 1982, Investigation and prediction of conditions of accumulation of gas resources in gas-hydrate pools: Petroleum Geology, v. 21, n. 2, p. 65-89.

CIA, 1985, USSR Energy Atlas: Washington, U.S. Central Intelligence Agency, 79 p.

Claypool, G. E., and C. N. Threlkeld, 1983, Anoxic diagenesis and methane generation in sediments of the Black Outer Ridge, Deep Sea Drilling Project Site 533, Leg 76, in R. E. Sheridan, et al., Initial reports of the Deep Sea Drilling Project, v. 76: Washington, U.S. Government Printing Office, p. 391-402.

Claypool, G. E., C. N. Threlkeld, P. N. Mankiewicz, M. A. Arthur, and T. F. Anderson, 1985, Isotopic composition of interstitial fluids and origin of methane in slope sediment of the Middle America Trench, Deep Sea Drilling Project Leg 84, in R. von Huene, J. Aubouin, et al., Initial reports of the Deep Sea Drilling Project, v. 84: Washington, U.S. Government Printing Office, p. 683-692.

Collett, T., 1987, Geologic interrelations relative to gas hydrates within the North Slope of Alaska, in C. A. Komar, ed., Proceedings of the Unconventional Gas Recovery Contractors Review Meeting: U.S. Department of Energy, DOE/METC-87/6080 (DE87006490), p. 298-308.

Davidson, D. W., J. El-Defrawy, and A. S. Judge, 1978, Natural gas hydrates in northern Canada, in Proceedings of the Third International Permafrost Conference, v. 1, p. 938-943.

Davidson, D. W., D. G. Leaist, and R. Hesse, 1983, Oxygen-18 enrichment in water of a clathrate hydrate: Geochimica et Cosmochimica Acta, v. 47, p. 2293-2295.

Deaton, W. M., and E. M. Frost, 1949, Gas hydrates and their relation to the operation of natural gas pipelines: U.S. Bureau of Mines Monograph No. 8, 101 p.

de Boer, R. B., J. J. Houbold, and J. Lagrand, 1985, Formation of gas hydrates in a permeable medium: Geologie en Mijnbouw, v. 64, p. 245-249.

Finley, P., and J. Krason, 1986a, Basin analysis, formation and stability of gas hydrates in the Colombia basin; geological evolution and analysis of confirmed or suspected gas hydrate localities: U.S. Department of Energy, DOE/MC/21181-1950, Vol. 7 (DE86006637), 134 p.

Finley, P., and J. Krason, 1986b, Basin analysis, formation and stability of gas hydrates of the Middle America trench; geological evolution and analysis of confirmed or suspected gas hydrate localities: U.S. Department of Energy, DOE/MC/21181-1950, Vol. 9, 243 p.

Finley, P., and J. Krason, 1988, Basin analysis, formation and stability of gas hydrates of the Beaufort Sea; geological evolution and analysis of confirmed or suspected gas hydrate localities: U.S. Department of Energy, DOE/MC/21181-1950, Vol. 12, 212 p.

Finley, P., J. Krason, and K. Dominic, 1987, Evidence for natural gas hydrate occurrences in the Colombia basin: AAPG Bulletin, v. 71, n. 5, p. 555-556.

Gavrilov, Y. Y., Y. A. Zhurov, and Teplinskiy, 1971, Relation between isotopic composition of argon and carbon in natural gases: Doklady Akademii Nauk SSSR, v. 206, p. 208-210.

Halbouty, M. T., A. A. Meyerhoff, R. E. King, R. H. Dott, H. D. Klemme, and T. Shabad, 1970, World's giant oil and gas fields, geologic factors affecting their formation, and basin classification, in Geology of giant petroleum fields: AAPG Memoir 14, p. 502-555.

Klemme, H. D., 1984, Oil and gas maps and sections of the West Siberian Basin: USGS Open-File Report, 84-516.

Kornev, B. V., V. F. Nikonov, S. A. Sulima, and G. G. Yakovlev, 1982, Classification of hydrocarbon deposits of Siberia according to phase state: Petroleum Geology, v. 20, p. 273-280.

Krason, J., and M. Ciesnik, 1985, Gas hydrates in the Russian literature; geological evolution and analysis of confirmed or suspected gas hydrate localities: U.S. Department of Energy, DOE/MC/21181-1950, Vol. 5, (DE86006635), 164 p.

Krason, J., P. Finley, and B. Rudloff, 1985, Basin analysis, formation and stability of gas hydrates in the western Gulf of Mexico; geological evolution and analysis of confirmed or suspected gas hydrate localities: U.S. Department of Energy, DOE/MC/21181-1950, Vol. 3, (DE86001057), 168 p.

Kulakov, Y. N. and G. P. Makhotina, 1985, Recent tectonics of the Yenisey-Khatanga regional downwarp: Petroleum Geology, v. 23.

Makarenko, F. A., A. Y. Velyugo, G. B. Gavline, B. F. Mavritsky, V. A. Pokrovsky, B. G. Polak, Y. B. Smirnov, eds., 1972, Geotermicheskaya Karta SSSR (Geothermal map of the USSR): Scale 1:5,000,000: Geologicheskiy Institut Akademii Nauk SSSR, Moskva.

Makogon, Y. F., 1974, Gidrati prirodnykh gazov (Hydrates of natural gas) trans. by W. J. Cieslewicz; 1978, Geoexplorers Associates, Inc., Denver, 178 p.

Makogon, Y. F., 1984, Razrabotka gazogidratnoy zalezhi (Production from natural gas hydrate deposits): Gazovaya Promishlennost, v. 10, p. 24-26.

Makogon, Y. F., 1988, Natural gas hydrates—the state of study in the USSR and perspectives for its using: Paper presented at the Third Chemical Congress of North America, Toronto, Ontario, Canada, June 1988, 20 p.

Makogon, Y. F., F. A. Trebin, A. A. Trofimuk, and V. P. Cherskii, 1971, Obnaruzheniye zalezhi prirodnogo gaza v tverdom (gazogidratnom) sostoyanii (Detection of a pool of natural gas in a solid [hydrated state]): Doklady Akademii Nauk SSSR, v. 196, n. 1, p. 197-200.

Meyerhoff, A. A., 1980, Petroleum basins of the Soviet Arctic: Geological Magazine, v. 117, n. 2, p. 101-210.

Prasolov, E. M., I. L. Kamenskiy, A. P. Meshik, Y. S. Subbotin, L. N. Surovtseva, and O. N. Yakovkev, 1981, Formation of gas fields of the north of west Siberia from isotope data: Petroleum Geology, v. 19, p. 316-334.

Rice, D. D., and G. E. Claypool, 1981, Generation, accumulation, and resource potential of biogenic gas: AAPG Bulletin, v. 65, p. 5-24.

Rigassi, D. A., 1986, Wrench faults as a factor controlling petroleum occurrences in West Siberia, in Future petroleum provinces of the world: AAPG Memoir 40, p. 529-544.

Salov, V. M., V. N. Uklein, and V. A. Kanunnikov, 1984, Magnetotelluric sounding in the Yenisey-Khatanga and Noril'sk-Kharayelakh downwarps: Petroleum Geology, v. 22, n. 5, p. 193-195.

Sapir, M. H., E. N. Khramenkov, I. D. Yefremov, G. D. Ginzburg, A. E. Beniaminovich, S. M. Lenda, and V. L. Kislova, 1973, Geologicheskie i promislovo—geofizicheskie osobennosti gazogidratnoi zalezhi Messoiakhskogo gazovogo mestorozhdenia (Geologic and geophysical features of the gas hydrate deposits in the Messoiakh field): Geologiia Nefti i Gaza, v. 6, p. 26-34.

Schoell, M., 1983, Genetic characterization of natural gas: AAPG Bulletin, v. 67, p. 2225-2238.

Semenovich, B. B., G. K. Dikenshteyn, S. P. Makarov, P. P. Maksimov, et al., eds., 1976, Karta Neftegazonosnosty SSSR (Oil and gas map of the USSR), Scale 1:2,500,000: Ministerstvo Geologii SSSR, Moskva.

Sheshukov, N. L., 1973, Priznaki zalezhei gaza, soderzhazchikh gidrati (Features of gas bearing strata with the hydrates): Geologiia Nefti i Gaza, n. 6, p. 20-26.

Sheshukov, N. L., A. F. Beznosikov, Y. H. Kramenkov, and I. D. Yefremov, 1972, O zaleganii gaza v gidratnom sostoyanii na Mesoyakhskom mestorozhdenii (Occurrence of gas in the hydrate state in Messoyakh field): Gazovoe Delo, n. 6, p. 8-10.

Sumets, V. I., 1974, Predotvrashchenie gydratobrazovaniya v prizaboynoy zone (Prevention of hydrate formation in gas wells zone): Gazovaya Promishlennost, v. 2, p. 24-26.

Surkov, V. C., and O. G. Zhero, 1981, Fundament i razvitie platformennogo chekhla Zapadno Sibirskoy (Basement and evolution of the West Siberian platform): Moscow, Nedra, 143 p.

Trofimuk, A. A., and N. V. Cherskii, 1974, Mekhanizm razdeleniya izotopov vody i gazov v zonakh gidratoobrazovaniya zemnoy kory (Mechanism of fractionation of isotopes of water and gas in crustal zones of hydrate formation): Doklady Akademii Nauk SSR, v. 215, p. 210-212.

Trofimuk, A. A., Y. F. Makogon, M. V. Tolkachev, and N. V. Cherskii, 1984, Some distinctive features of the discovery, prospecting and exploitation of gas hydrate deposits: Geologiia i Geofizika, v. 25, n. 9, p. 1-7.

Tsarev, V. P., and V. A. Nenakhov, 1985, Formation conditions, genetic features, and methods of developing gas and gas hydrate deposits: Geologiia i Geofizika, v. 26, n. 10, p. 25-33.

Waples, D. W., 1980, Time and temperature in petroleum formation: application of Lopatin's method: AAPG Bulletin, v. 64, p. 916-926.

Weaver, J. S., and J. M. Stewart, 1982, In situ hydrates under the Beaufort Sea Shelf, in M. H. French, ed., Proceedings of the Fourth Canadian Permafrost Conference, 1981: National Research Council of Canada, p. 312-319.

Yermakov, V. I., N. K. Kulakhmetov, N. N. Nemchenko, and A. S. Rovenskaya, 1971, Genesis of Cenomanian gas and oil deposits in the northern part of western Siberia: Doklady Akademii Nauk SSR, v. 206, p. 223-224.

Appendix 1. Field Description

Field name .... *Messoyakh field*
Ultimate recoverable reserves .... *1–3 tcf (3×10^{10} to 1×10^{11} m^3)*
Field location:
    Country .... *USSR*
    State .... *Krasnoyarskiy Kray*
    Basin/Province .... *Yenisei-Khatanga trough, Western Siberian oil and gas province*
Field discovery:
    Year first pay discovered .... *Cenomanian Dolgan Formation 1967*
Discovery well name and general location:
    First pay .... *Well No. 1, 340 km west-southwest of Norilsk*
Discovery well operator .... *Ministry of Gas Industry, USSR*
IP:
    First pay .... *111 mcf, 9/32-in. choke, to 6275 mcf/day through ½-in. choke (3×10^3 m^3, 0.71 cm choke to 177×10^3 m^3 through 1.3 cm choke)*

All other zones with shows of oil and gas in the field:

Age	Formation	Type of Show
Cenomanian	*Dolgan*	*Heavy oil beneath GWC*
Jurassic	*NA*	*Condensate, gas*

Geologic concept leading to discovery and method or methods used to delineate prospect
Anticlinal theory, dome structure, surface geology, gravity, and seismic results

Structure:
    Province/basin type .... *Bally 1212; Klemme II B*
    Tectonic history
On the cratonic, folded basement were deposited up to 10,000 m of Jurassic through Quaternary sediments. Uplifts and differential subsidence of Jurassic and Tertiary age formed the oil and gas productive structures.
    Regional structure
The Messoyakh gas field is located on the flank of larger Tanam arch.
    Local structure
Locally there is an anticline measuring 20 by 13 km.

Trap:
    Trap type(s) .... *Structural (anticlinal) trap*

Basin stratigraphy (major stratigraphic intervals from surface to deepest penetration in field):

Chronostratigraphy	Formation	Depth to Top in m
Quaternary		*Surface*
Cretaceous:		
Maastrichtian	*Kentar*	*65*
	Tanam-Messoyakh	*200*
	Nasomov	*325*
Turonian	*Dorozhkov*	*680*
Cenomanian	*Dolgan*	*780*

Reservoir characteristics:
    Number of reservoirs *1*
    Formations .... *Dolgan*

Ages	*Upper Cretaceous–Cenomanian*
Depths to tops of reservoirs	*850 m (elevation: –730 m)*
Gross thickness (top to bottom of producing interval)	*78 m*
Net thickness—total thickness of producing zones	
Average	*20 m*
Maximum	*30 m*

Lithology
Marine sandstone interbedded with mudstone and siltstone

Porosity type	*Intergranular porosity*
Average porosity	*25%*
Average permeability	*125 md*

Seals:

  Upper

Formation, fault, or other feature	*Impermeable layers of Dolgan and Dorozhkov formations*
Lithology	*Shales and gas hydrate filled sandstones*

  Lateral

Formation, fault, or other feature	*Impermeable layers of Dolgan and Dorozhkov formations*
Lithology	*Shales*

Source:

Formation and age	*Jurassic and Lower Cretaceous*
Lithology	*Coal and organic matter-bearing shales*
Average total organic carbon (TOC)	*2%*
Maximum TOC	*NA*
Kerogen type (I, II, or III)	*II and III*
Vitrinite reflectance (maturation)	$R_o = 1.2$
Time of hydrocarbon expulsion	*NA (diverse possibilities; refer to text)*
Present depth to top of source	*Approx. 1500 m*
Thickness	*Approx. 300 m*
Potential yield	*NA*

Appendix 2. Production Data

Field name	*Messoyakh field*
Field size:	
Proved acres	*19,000 ha*
Number of wells all years	*Over 70*
Current number of wells	*NA*
Well spacing	*500 × 1000 m*
Ultimate recoverable	*1 to 3 tcf (3×10^{10} to 1×10^{11} m^3)*
Cumulative production	*330 bcf (9.3×10^9 m^3)*
Annual production	*0 to 70 bcf (0 to 2×10^9 m^3)*
Present decline rate	*NA*
Initial decline rate	*NA*
Overall decline rate	*NA*
Annual water production	*NA*
In place, total reserves	*NA*
In place, per acre foot	*NA*

Primary recovery	327 bcf (9.35×10^9 m^3)
Secondary recovery	21 bcf (6×10^8 m^3) through 1983
Enhanced recovery	NA
Cumulative water production	NA

Drilling and casing practices:

Amount of surface casing set	NA
Casing program	NA
Drilling mud	NA
Bit program	NA
High pressure zones	NA

Completion practices:

Interval(s) perforated .... *Dolgan sandstone (within and below gas hydrate zone)*

Well treatment

Open well, production with periodic hiatus for regeneration of formation pressure; experimental use of methanol and $CaCl_2$ to dissociate hydrates

Formation evaluation:

Logging suites	*Resistivity, micro-resistivity, SP, GR, sonic, neutron, caliper, temperature*
Testing practices	*DST*
Mud logging techniques	NA

Field characteristics:

Average elevation	*90 m*
Initial pressure	*7.9 MPa*
Present pressure	*5.0–6.1 MPa*
Pressure gradient	*10 kPa*
Temperature	*8–12°C*
Geothermal gradient	*0.013°C/m*
Drive	*Gas-water*
Oil column thickness	NA
Gas-water contact	*–770 to –803 m*
Connate water	NA
Water salinity, TDS	*1.5%; 15,000 ppm*
Resistivity of water	NA
Bulk volume water (%)	NA

Transportation method and market for oil and gas:

Pipeline to Norilsk (250 km)

South Belridge Field—U.S.A.
San Joaquin Basin, California

DONALD D. MILLER
Mobil Exploration and Producing U.S., Inc.
Denver, Colorado

JOHN G. McPHERSON
Mobil Research and Development Corporation
Dallas Research Laboratory
Dallas, Texas

FIELD CLASSIFICATION

BASIN: San Joaquin
BASIN TYPE: Wrench
RESERVOIR ROCK TYPE: Sand and Diatomite
RESERVOIR ENVIRONMENT OF DEPOSITION: Deltaic and Deep Water Diatomite
TRAP DESCRIPTION: Primarily anticlinal trap combined with pinch-outs, unconformity truncation, fractured chert, and tar seal

RESERVOIR AGE: Pleistocene and Miocene
PETROLEUM TYPE: Oil
TRAP TYPE: Anticline with Stratigraphic Components

LOCATION

The South Belridge field is located in an active tectonic setting between Los Angeles and San Francisco, in the San Joaquin Valley, the southern portion of the Central Valley of California, western United States. The field is 100 mi (161 km) north of Los Angeles and 40 mi (64 km) west of Bakersfield (Figure 1). South Belridge field is 12 mi (19 km) northeast of the basin-bounding San Andreas wrench fault zone, a prominent and important geologic feature.

South Belridge is one of numerous giant fields located in the western side of the San Joaquin Valley in oil-rich Kern County. Twenty neighboring fields in the San Joaquin Valley each contain reserves exceeding 100 million barrels (*Oil & Gas Journal*, 1990); 17 are located in Kern County. Two of these, Kern River and Midway-Sunset (Figure 1), have already produced in excess of 1 billion barrels. The deepest well in the state of California has been drilled nearby, 10 mi (16 km) southeast of South Belridge, to a depth of 24,426 ft (7445 m) in the Elk Hills field (Figure 1), where deep hydrocarbon potential was confirmed and metamorphic basement penetrated (Fishburn, 1990).

The projected ultimate oil recovery at South Belridge field exceeds 1 billion barrels. Cumulative field production through 1989 is approximately 700 million bbl of oil and 220 bcf of gas; remaining recoverable reserves are approximately 500 million bbl of oil (*Oil & Gas Journal*, 1990; California Division of Oil and Gas, 1987). The focus of field development and production has been the shallow Pleistocene Tulare sand reservoirs. With technological advances, interest in the deeper diatomite rocks has been increasing. Additional speculative diatomite reserves in the upper horizons of the Miocene Monterey are conservatively estimated as 550 million bbl of oil and 550 bcf of gas (Bowersox and Shore, 1990), which are recoverable using recently implemented water-flood techniques. Future recoverable reserve opportunities from this resource are contingent on higher oil prices and improved production technology. The resource base is conservatively estimated to exceed 10 billion barrels of oil (in place) in the diatomite and genetically related deeper rocks at South Belridge field.

The South Belridge daily production is approximately 165,000 bbl of oil and annual production is 60 million bbl of oil (*Oil and Gas Journal*, 1990), among the highest of any field in the United States. The field is a major producer of heavy oil using thermal recovery methods, similar to those used in many of the nearby fields (Figure 1). As a result, production rates have increased and declined historically,

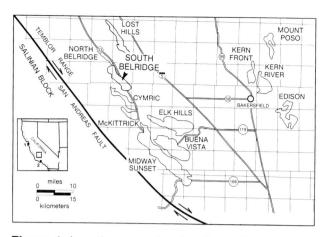

Figure 1. Location map showing South Belridge field, other giant oil fields that are heavy-oil producers, and the San Andreas fault in the southern San Joaquin Valley. (Modified from McPherson and Miller, 1990.)

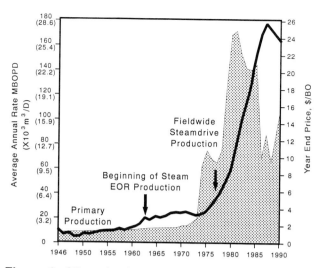

Figure 2. Oil production history and price (post-1945) of South Belridge field. The stippled curve represents the price per barrel of 13° API oil. Approximately 75% of this production is from the Tulare reservoirs. (After McPherson and Miller, 1990.)

following the trend in oil prices (Figure 2). Steamflood production of heavy oil from the shallow, unconsolidated sands accounts for approximately 75% of current production volumes. The diatomite and deeper horizons currently contribute only 25% of the production but contain vast reserves and resources for possible future production. The deepest well in the field was drilled by Shell to 14,565 ft (4439 m) (California Division of Oil and Gas, 1985).

HISTORY

Pre-Discovery

Small amounts of oil had been produced in California since 1864 (Knowles, 1959). Many years passed with no early California gushers comparable to the famous Texas gusher at Spindletop in 1901 to indicate that within two years (1903) California would be the nation's biggest oil-producing state. This changed when the Lake View Oil Company drilled the Lake View Gusher well in the Midway-Sunset field during March 1910 (Figure 1). The Gusher well was a blowout, which produced at peak rates reported as 125,000 bbl/day with an oil column 20 ft (6 m) in diameter spewing over the derrick (Knowles, 1959). It produced over 8 million bbl of oil during 18 months (Rintoul, 1978; Division of Oil and Gas, 1961). The spectacular success of the Gusher well led to the drilling of many new wells along the western side of the San Joaquin Valley.

Discovery

The frenzy of oil wildcatting after the Lake View Gusher well resulted in the unremarkable discovery of the South Belridge field. The shallow discovery well, Belridge Oil Company No. 101, Sec. 33, T28S, R21E, was located about 30 mi (51 km) northwest of the Lake View Gusher along the general trend of the local geology (Figure 3). It was sited next to an outcrop of oil-stained sand in a creek bed on a broad topographic expression of the subsurface anticline. The well was completed in April 1911 at a depth of 782 ft (238 m) (Ritzius, 1950). The initial production of 100 BOPD co-mingled heavy oil from the shallow Pleistocene Tulare sands and light oil from the underlying Miocene–Pliocene Belridge diatomite to produce a 25.3° API oil.

The South Belridge field has always attracted large investment capital. The land known as the Belridge property was purchased in January 1911 for approximately $1 million (California State Historic Marker, South Belridge discovery well) prior to drilling the discovery well in April 1911. The Belridge Oil Company grew as a one-field company with production and reserves large enough to rank it among the ten largest U.S. oil companies. In 1979, Shell Oil Company purchased Belridge Oil Company for the reported amount of $3.65 billion. The purchase included approximately two-thirds of the 8700 ac (35 km^2) of producing area, plus adjacent agricultural land. In 1985, a 480 ac (1.9 km^2) property (in Sec. 19, T28S, R21E) was purchased for $395 million by Celeron Oil & Gas Corporation (Russell, 1985). A 1987 Exxon acquisition valued this property at approximately $1 million per acre, making this the most expensive oil-field acreage in the world.

Property values throughout the history of the field were based on established oil reserves and upside potential of new technology and upon improved description of the reservoir. Field appraisal began with the 100 BOPD discovery well and continues today, with over 7000 wells in the field. The economic success of steamflood oil recovery processes in the shallow heavy-oil sands and the recognition of the production potential of the diatomite reservoir using secondary recovery projects further increased the property value. The vast oil and gas resource in the diatomite and deeper Monterey rocks is the focus of exploration and development research for future production.

Post-Discovery

The initial South Belridge development drilling campaign lasted about three years after the 1911 discovery well and resulted in approximately 100 wells (Ritzius, 1950). After three years of little activity, an active drilling campaign was waged during the years of World War I. One hundred wells were drilled during a two-year interval. Few wells were drilled and little additional acreage was added to the field from 1920 to 1942, except for 60 wells drilled from 1927 to 1929. Renewed field development along the structural crest began during 1942 and continued through 1949. The total number of wells increased by more than four times and the productive acreage increased by more than three times, extending the field nearly 5 mi (8 km) to the northwest and 1 mi (1.6 km) to the southeast. The history of field development is illustrated in Figure 3, where the field outline at the end of each of the first three decades is shown. Most of the wells were less than 1000 ft (305 m) deep, and primary production was from the heavy-oil Tulare sands.

Historically, development was slow and sporadic, as dictated by production technology and a fluctuating market demand for heavy, asphaltic crude oil. The shallow, heavy-oil reservoir sands were initially produced by primary methods and subsequently by in situ combustion, cyclic steaming, and steamflood methods. The first South Belridge field thermal recovery pilot was an in situ combustion thermal recovery experiment conducted from 1955 to 1958 (Gates and Ramey, 1958; Gates et al., 1978). This pilot demonstrated that viscous oil could be readily moved and oil recovery increased to between 40 and 60% compared with a primary recovery of approximately 10% of oil in place. Cyclic steaming was started in 1963, and a fieldwide continuous steam injection line-drive operation began in 1969. The steam-drive project was converted to steam patterns in the mid-1970s. Current production operations focus on steamflood development to produce the heavy (13–14° API) oil. The reader is referred to Small (1986), Dietrich (1988), and Miller et al. (1990) for details regarding completion and producing techniques tailored to the heavy-oil sands.

Fieldwide development and infill drilling accelerated around 1980 as a result of large oil price increases and a resulting combination of factors, including the engineering development of thermal combustion and

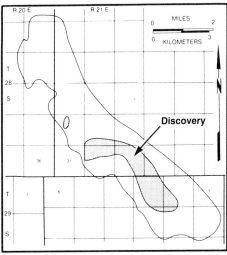

South Belridge Field 1920

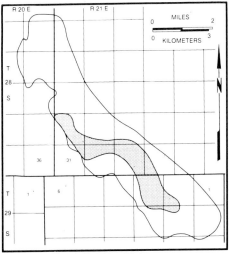

South Belridge Field 1930

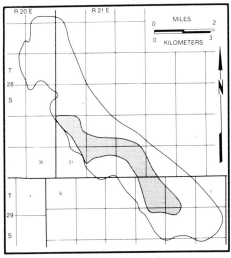

South Belridge Field 1940

Figure 3. Field development history at 10-year intervals illustrating areas of development and production relative to present-day field outline. Shading represents developed areas. Compare with Figure 4. (Modified from Ritzius, 1950.)

field-scale steamflood technology; successful hydraulic fracture stimulation of the tight diatomite rocks; improved geologic description; and the Shell Oil purchase of the major part of the field. Continued development resulted in 6100 active wells (Figure 4) with daily production of nearly 180,000 bbl of oil (California Division of Oil and Gas, 1987), and field total of 7541 wellbores (Conservation Committee of California Oil Producers, 1989).

Extensive development of the diatomite reservoir underlying the shallow, heavy-oil sand reservoir began in 1978 when Mobil Oil Corporation proved that hydraulic fracture stimulation was economically successful (Strubhar and Medlin, 1982). Production from the diatomite currently averages 30,000 BOPD and 60 MMCFGD from 1228 wells (Bowersox, 1990), with a peak rate in excess of 40,000 BOPD. Strickland (1982), Wendel et al. (1988), and Bowersox and Shore (1990) present diatomite reservoir production details.

DISCOVERY METHOD

The discovery well, Belridge Oil Company No. 101, was located next to an outcrop of oil-stained sand in a creek bed on a broad topographic expression of the subsurface anticline. Drilling near oil seeps, on topographic highs, and following trends were common methods used by the early wildcatters. These methods are valid in frontier basins today, making the large structure a target for today's explorer. However, the variability of the stratigraphic component, which controls the shallow reservoir distribution, may go overlooked using today's exploration technology.

STRUCTURE

Tectonic History

Regional tectonism strongly controls the character of central California sedimentary basins (Graham, 1987). The influence on basin infill and reservoirs is especially evident at South Belridge field. The basin's active tectonic setting is a fundamental large-scale control on the contrasting depositional settings and resulting character of the major reservoirs.

The San Joaquin basin has a polycyclic history as a transpressive successor basin of an earlier forearc setting. The elongate San Joaquin basin first developed as part of the Great Valley Mesozoic forearc basin on the western margin of North America (Ingersoll, 1979) and consisted mainly of a marine shelf and slope, open to the west (Clarke et al., 1975). Subsequent change from this convergent to a transform margin imposed a new structural character, taking the form of a basin increasingly closed off from the sea. This change, occurring during the late Cenozoic, resulted from strike-slip along the San Andreas fault system.

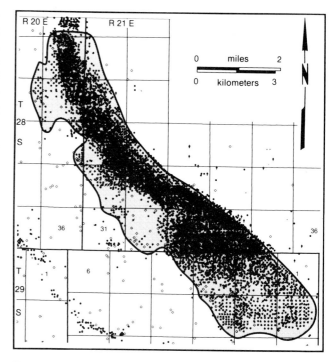

Figure 4. Well location map of South Belridge field showing locations of 6100 active wells (From Petroleum Information.)

A reorientation of the Pacific plate motion within the last 5 m.y. (Page and Engebretson, 1984) added a compressional component to uplift the basin and adjacent highlands. The combination of tectonic rearrangement and accompanying sedimentation controlled the final marine regression in the San Joaquin basin. These events, which began in the late Miocene and continued through the Pliocene–Pleistocene, produced a dramatic change of style in basin-fill sedimentation, described in more detail in the following paragraphs.

During the late Pliocene, the San Joaquin basin became landlocked (Bartow, 1987) and an extensive lacustrine system developed within the valley. For the duration of the Pleistocene, progradation and aggradation of sediments from both east and west sides of the basin outpaced subsidence (Lettis, 1982). Sedimentation was principally by means of coarse-grained delta systems (fan deltas and braid deltas, cf. McPherson et al., 1987). However, a fluvial-dominated, lobate delta system was the main depositional environment at South Belridge. This phase of sedimentation in the western San Joaquin basin was directly influenced by the mountains of the Salinian block and the rising Temblor and Diablo ranges. These uplands bordered the basin on the west (Lennon, 1976) and provided a provenance for the feldspar-rich sands and diatomite-clast gravels that characterize the shallow reservoir sands at South Belridge.

Strike-slip displacement along the San Andreas fault exerted a major influence on deposition at South

Belridge by reconfiguring the marine basin as an intermontane basin and providing variable source areas with moving sediment dispersal systems. The likely source area for the Pleistocene Tulare fluvial sediments deposited at South Belridge has been displaced by substantial right-lateral slip along the San Andreas fault system. An east-draining Pliocene basin is exposed on the the western side of the San Andreas fault (Galehouse, 1967). This Pliocene and subsequent Pleistocene basin is suggested as the drainage basin and source for the South Belridge sands.

Regional Structure

The South Belridge anticline and neighboring anticlines along the western side of the San Joaquin Valley have been considered classic examples of active wrench tectonics with en echelon folds associated with strike-slip along the San Andreas fault (Wilcox et al., 1973; Harding, 1976). The Miocene folds have an oblique orientation to the San Andreas, whereas the younger Pliocene–Pleistocene structures are nearly parallel (Mount and Suppe, 1987), compatible with a change in plate motions suggested by Page and Engebretson (1984) and associated fold-and-thrust deformation (Namson and Davis, 1988). Sedimentation was contemporaneous with basin development and the structural growth during the Miocene, Pliocene, and Pleistocene. Asymmetrical thickening of sediments on the east flank of the South Belridge anticline and the neighboring Lost Hills anticline indicates east-verging thrust faulting during deposition of the reservoir horizons. Interpretation of these anticlines as either fault-propagation folds or fault-bend folds is equivocal.

Local Structure

South Belridge field lies on the crest of a large southeast-plunging anticline, subparallel to the nearby San Andreas fault system (Figure 5). Surface expression of the anticline on the southeast nose is consistent with Harding's (1976) conclusion that these structures are presently growing basinward. Structural development has influenced paleotopography and the resultant distribution of the reservoir sediments in addition to controlling the present structural orientation of the field. The seismic profile shown on Figure 6 across the southern end of the anticline illustrates significant structural growth during deposition of the reservoir horizons. The onlap relationships on the flanks of the anticline and flattening upwards suggest growth fault folding. The axial surface of the anticlinal fold dips to the northeast.

The history of faulting within South Belridge field is complex. At least three episodes of Neogene normal faulting have been noted in the reservoir horizons. Early and middle Pliocene listric, normal faults with several hundred feet of offset and pervasive small

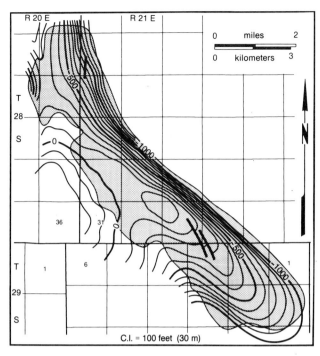

Figure 5. Simplified structural configuration of South Belridge field contoured on the top of diatomite unconformity, showing the southeasterly plunging anticline in the southern area of the field and a broad saddle in the northern area of the field. (Modified from California Department of Oil and Gas, 1985.)

offset vertical faults of late Pliocene age were identified by Bowersox (1990). Faults have also been observed in deeper wells, although fault interpretation is hampered by a lack of clear marker beds and an inability to attain satisfactory seismic profiles over the field. Recent faulting in this active tectonic setting may be overprinted by surface subsidence and reservoir compaction associated with production (Bowersox and Shore, 1990).

STRATIGRAPHY

The two principal reservoir units described are the Belridge diatomite of the Miocene–Pliocene Monterey Formation, and the overlying Pleistocene Tulare sands (Figure 7). They represent a deep-marine and a fluviodeltaic setting, respectively.

Two erosional unconformities, largely the product of the tectonic folding of the anticline, separate these reservoirs at South Belridge. Figure 8 illustrates these Neogene erosional unconformities and related tectonic events that affected the South Belridge reservoirs. A late Miocene unconformity is overlain by the Miocene–Pliocene Etchegoin and the Pliocene San Joaquin. The second unconformity occurs where the basal Tulare unconformity truncates the Etchegoin, the San Joaquin, and the Miocene Monterey formations. The Tulare was deposited unconformably on Pliocene and Miocene deposits on

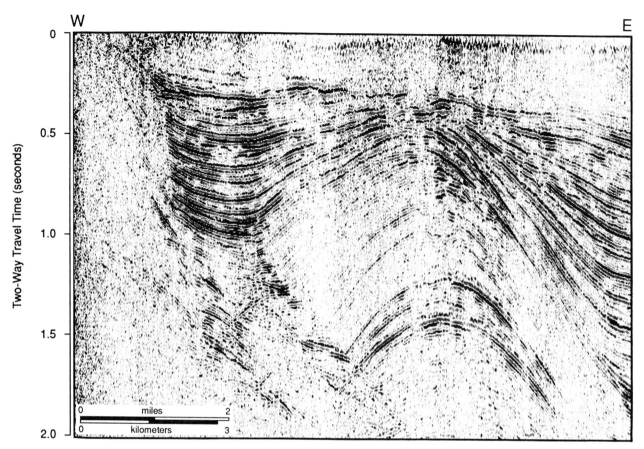

Figure 6. Seismic profile of the South Belridge anticline, illustrating stratigraphic onlap on both the northeastern and southwestern flanks. The reservoir horizons discussed in this paper are above 0.5 seconds. The line of the seismic profile is approximately east-west and is located near the southern end of the anticline, somewhat parallel to A–A' (Figure 9).

the structure but is believed to be conformable with Pliocene deposits toward the center of the basin to the east.

The Miocene Monterey Formation in the San Joaquin Valley overlies sandstones of the Oligocene–Miocene Temblor Formation in a time-transgressive relationship. It is, in turn, overlain by clastic-dominated, time-transgressive Etchegoin and equivalent formations (Graham and Williams, 1985) (Figures 7 and 8). The Monterey is a diverse pelagic-hemipelagic, biogenous sequence. Use of the term "Monterey Formation" in literature concerning the San Joaquin Valley follows that of Bramlette (1946). Stratigraphy of the Monterey Formation is complicated by time-transgressive provincial benthic foraminiferal stages (Graham and Williams, 1985), depositional facies transitions, and silica diagenesis that changes the mineralogy of biogenic opal-A to opal-CT and quartz during the heating attendant upon burial. In contrast to the Monterey of coastal and offshore reservoirs, the Monterey in the San Joaquin basin displays greater dilution of biogenic silica components by fine-grained terrigenous lithologies and less carbonate (Graham and Williams, 1985).

The Monterey rocks of the San Joaquin basin are of Relizian to late Mohnian age and occur in opal-A, opal-CT, and quartz phases (Graham and Williams, 1985). Locally, younger Delmontian diatomites (Miocene–Pliocene) are present at both the northern end (Schwartz, 1988) and the southern end of the structure, indicating continuous syndepositional growth of the anticline. At Belridge, the informal stratigraphic term "Belridge diatomite" is applied to the diatomaceous (opal-A) interval in the uppermost Monterey rocks.

The Etchegoin Group comprises the Jacalitos, Etchegoin, and San Joaquin formations (Loomis, 1990), only the latter two of which are recognized at South Belridge. The Etchegoin Formation was deposited regionally in a predominantly estuarine system and thins onto the flanks of the South Belridge anticline. The overlying San Joaquin Formation represents generally lacustrine and brackish-water environments (Barbat and Galloway, 1934). It also laps onto the anticline and overlaps the older deposits at the crest of the structure.

The Pleistocene Tulare Formation includes the youngest folded strata in the western San Joaquin basin. This nonmarine sequence (Woodring et al.,

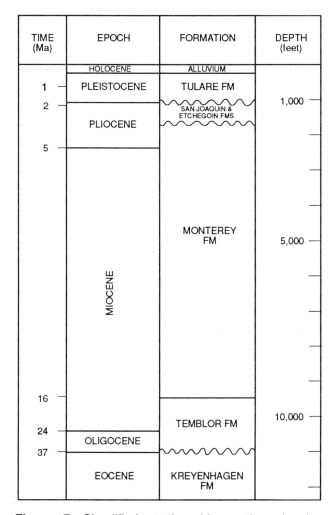

Figure 7. Simplified stratigraphic section showing geologic ages and unconformable contacts between the Pleistocene Tulare Formation, the Pliocene Etchegoin and San Joaquin formations, and the Miocene Monterey, containing the Belridge diatomite interval in South Belridge field. The Miocene, Pliocene, and Pleistocene are apparently conformable in the deeper parts of the basin to the east. In the stratigraphic column for South Belridge, the relationship of the Tulare Formation and Etchegoin Group to the underlying Monterey is simplified. For example, the Tulare directly overlies the Monterey on the crest of the South Belridge structure (Figure 9).

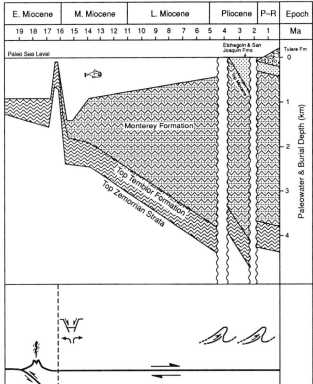

Figure 8. Paleobathymetry diagram and related tectonic events during deposition of the Miocene-Pliocene diatomite and the Pleistocene Tulare sand reservoirs. Data are from the nearby Chico-Martinez Creek outcrop section and South Belridge wells. Tectonic symbols at the bottom of the diagram depict the approximate timing of transpressive, forearc-successor basin tectonic events. (Modified after S. A. Graham, unpublished.)

1940; Stanton and Dodd, 1970) represents a final filling phase of sedimentation in the southern part of the basin during the Pleistocene. Structural deformation and erosion on the growing South Belridge anticline created a complex and irregular surface prior to Tulare deposition. The Tulare then was deposited unconformably on the erosionally truncated Pliocene and Miocene core of the structure, becoming conformable with upper Pliocene sediments toward the center of the basin to the east. Sediment was fed from nearby basin-margin highlands, via a fluvial-dominated lobate delta, into the extensive brackish to freshwater Tulare Lake (Miller et al., 1990). The Tulare at South Belridge comprises a wide variety of fluvial and deltaic facies.

TRAP

The trap is a combination of structural, stratigraphic, and tar seals in both the Tulare and diatomite reservoirs. The southeast-plunging Belridge anticline is the dominant control of the Tulare and the diatomite reservoirs traps (Figure 5). The hydrocarbon column in both is at nonhydrostatic equilibrium, suggested by surface tar seeps, tar seals, source rocks generating hydrocarbon in the oil window today, and a local increase in the gas-oil ratio with increasing depth. The petroleum accumulation at South Belridge is young and not at equilibrium when compared with many oil fields around the world.

The subcropping relationship of the Monterey Formation to the Tulare Formation in the axial

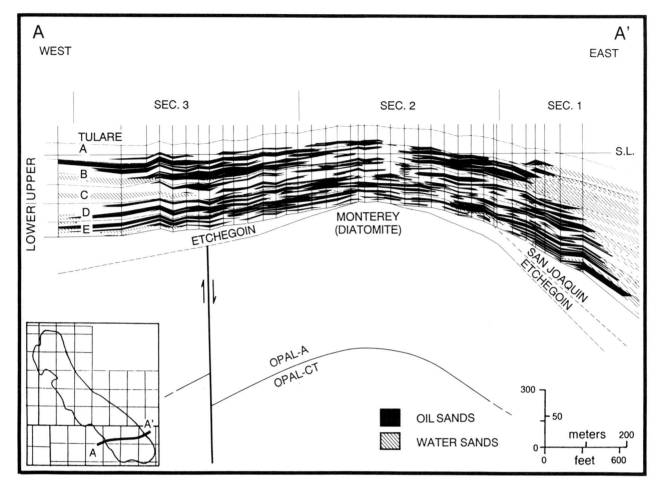

Figure 9. Cross section A-A' illustrating east-west fieldwide structural and stratigraphic relationships of Tulare reservoir sands overlying the diatomite (opal-A) reservoir. The diagenetic transition from opal-A to opal-CT occurs between 1500 and 1800 ft (457–549 m) in the southern area of the field. (Modified from McPherson and Miller, 1990.)

portion of the anticline provides the necessary pathway for hydrocarbon charge. The basal Tulare unconformity truncates Monterey rocks on the crest of the anticline, with the deeper Monterey Formation more sharply folded than the Tulare. Dips are commonly greater than 45°. Pre-Tulare and syndepositional hydrocarbon migration is suggested by tar mats in Pliocene Etchegoin and earliest Tulare sediments bounding the unconformity. The deeper Monterey is an organic-rich source rock, generating hydrocarbons at depth in the oil kitchen today.

In the Tulare, a basin-edge, updip stratigraphic pinch-out provides the principal trap boundary at the northwestern end of the field. At a smaller scale, individual "shoestring" channel-fill sand bodies encased in mud often have their own local oil-water contacts. The greatest area of hydrocarbon fill-up occurs at depth, resulting in a "Christmas tree" shape of numerous stacked reservoirs as seen in cross section (Figure 9).

The diatomite at South Belridge has various mechanisms that provide leaky seals for the reservoir through a combination of structural and stratigraphic relationships. The rocks of the Monterey Formation have the unusual capacity to serve as source, reservoir, and seal at the same time. The Tulare muds unconformably overlying the diatomite along the crest of the anticline provide the dominant seal. Laterally continuous clay-rich or silica-cemented lithologies occur in the diatomite. These serve as low-permeability leaky seals for vertically migrating hydrocarbons. This nonequilibrium reservoir benefits from ongoing petroleum migration through tectonic fractures. On the other hand, fracturing and local faulting may, in some cases, provide petroleum-trapping permeability barriers, resulting in local fluid-flow units within the diatomite reservoir horizon.

Tar-seals and fluid-level traps (Foss, 1972) are important in limited areas of the field. Biodegradation of Belridge diatomite oil near the overlying unconformity in the southern area of the field has dramatically increased the viscosity and effectively immobilized the oil in the low-permeability diatomite matrix at that horizon. Shallow Tulare fluid-level traps form where the oil cannot migrate upward beyond a surface of zero hydrostatic pressure, the groundwater table.

Reservoirs

The two contrasting reservoir horizons, separated by an unconformity along the crest of the structure, provide the major production and reserves at South Belridge field. Both are unconventional and exhibit unusual reservoir characteristics. The reservoirs are the product of two widely differing depositional settings that reflect major differences in basin tectonics, sediment supply, water depth, and climate. Each reservoir is described separately. The additional, nonproducing hydrocarbon resource known to exist in deeper Miocene Monterey rocks is beyond the scope of this paper.

The dynamic basin setting and paleotopography of evolving subsea structures during the Miocene influenced the distribution of depositional facies. Deep-marine deposition of diatomite and diatomaceous shale reservoirs produced offlap relationships along anticlinal axes and thickening of depositional units, with the more clastic-rich sediments being deposited in topographically low areas (Schwartz, 1988). The resulting variability of silica purity consequently influenced the distribution of the diatomite reservoir quality at South Belridge. The continuing structural growth of the anticline along the western basin margin and resulting paleotopography influenced the position of the Pleistocene shoreline. Thus, fieldwide distribution and deposition of the prograding fluviodeltaic sands that comprise the Tulare reservoir horizon were tectonically influenced.

The Belridge diatomite produces lighter oil (20-32° API) from diatomite and diatomaceous shales that had been deposited in a marine deep-water setting. The homogeneous appearing reservoir is characterized by extremely high porosity (60%), very low permeability (1 md, air, unstressed), and low primary oil recoveries estimated at 7%. Waterflood EOR projects are expected to recover at least the amount of oil produced with primary production.

In contrast, the shallow (1000 ft deep) overlying Tulare Formation produces heavy oil (13-14° API) from unconsolidated sands that had been deposited in a fluviodeltaic setting. Hundreds of vertically and laterally discontinuous reservoir sands are characterized by high porosity (35%) and permeability (darcys). Some operators predict oil average recoveries to be as high as 73% of the original oil in place using steamdrive recovery processes (Aalund, 1988), with small parts of the reservoir achieving 100% oil recovery (Miller et al., 1990).

Pleistocene Tulare Sands

The Tulare stratigraphy at South Belridge field is locally subdivided into upper and lower divisions, each containing several reservoir zones (Figures 9 and 10). The upper Tulare division includes zones A, B, and C, the lower division zones D and E and, in some places, F and G. These divisions are based on general engineering qualities and early mapping of oil-water contacts and are not formally recognized elsewhere in the basin. The reservoirs underlie a widespread lacustrine unit, Tulare Corcoran Clay, which is a stratigraphic marker.

The depth to the top of the Tulare pay zone in South Belridge is typically 500 ft (152 m) and the gross sand zone interval is 400 to 750 ft (122–229 m) thick. Oil-producing sands total as much as 150 ft (46 m) thick in the lower Tulare and 230 ft (70 m) in the upper Tulare. Gross reservoir thickness and pay zone thicknesses are variable because of oil-water contacts and the limits of sand distribution inherent to the depositional system.

The sands of the upper and lower Tulare are texturally and mineralogically immature. They are mostly lithic arkoses (classification after Folk et al., 1970) containing subequal amounts of feldspar and quartz (35–45%) and lesser amounts of rock fragments, mica, pyrite, carbonate, amorphous material, and clay minerals. Diatomite-clast gravels are particularly abundant in the fluvial-channel sequences of the upper Tulare zones. Diagenetic modification of the Tulare sands is insignificant owing to the young age and very shallow burial. Compaction has been minimal because maximum burial depths are generally less than 1000 ft (305 m). The sands, silts, and clays are unconsolidated and uncemented, except for local carbonate cemented paleosol horizons. Diagenesis induced by the steam-rock reactions during the steamflood process can be significant in the Tulare reservoir sands (Miller et al., 1990).

The clay mineral content in the reservoir facies is typically 5% and ranges from 1 to 8%. Clay minerals consist primarily of illite (60%) and smectite (40%), usually in mixed-layer morphology. These clays and other formation fines in the reservoir affect reservoir permeability and can create production problems. These problems include migration of fine clastics that plug pores and mineral dissolution and authigenesis under steam injection conditions.

The depositional model of the Tulare at South Belridge is that of a generally prograding fluviodeltaic system sourced in the nearby basin-margin highlands (Figure 11). Deposition was influenced by source-area tectonics, syndepositional structural growth, basinal subsidence rates, and adjustments to the level of Tulare lake. The lower Tulare reservoir sands (zones D and E) are mostly of delta-front and lower delta-plain origin, including distal-bar, distributary-mouth-bar, and distributary-channel sands (Figure 10).

The upper Tulare reservoir sands display a variety of fluvial styles. Somewhat more isolated meander-belt sands at the base (zone C) are overlain by more coalescing meanderbelt sands (lower zone B) which are in turn overlain by even less confined and more braided fluvial sands (upper zone B) (Figure 10). The zone A sands at the top of the zonal sequence,

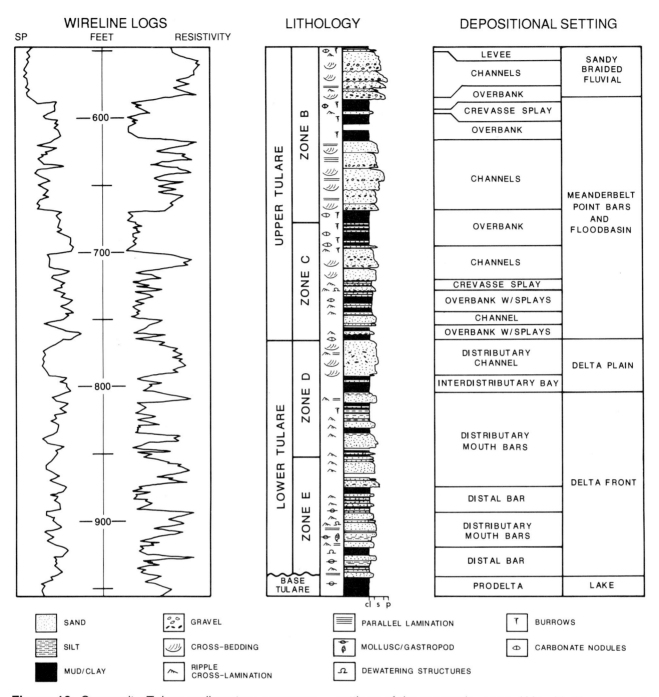

Figure 10. Composite Tulare sedimentary sequence and wireline log illustrating the prograding character and associated lithofacies with interpreted depositional settings of the reservoir zones. (After McPherson and Miller, 1990.)

although of less reservoir importance, establish a return to a meandering-channel system. The growing South Belridge anticline increasingly influenced the gross distribution of the deltaic sands and the overlying fluvial sand systems.

Specific depositional subenvironments of the fluviodeltaic setting in the Tulare include, (1) channels and bars, (2) crevasse-splays and levees, (3) overbank/interchannels, (4) distributary-channel-mouth bars, (5) distal bars, and (6) prodelta/lacustrine basin. These subenvironments are interpreted from their lithofacies associations (Table 1), which are described in detail by Miller et al. (1990), and McPherson and Miller (1990).

Channel and Bar Deposits

The upper Tulare reservoir in the southern portion of the field consists of channel and bar sands mostly of point-bar origin. They occur in units of stacked and amalgamated beds (Figure 10) averaging 10 ft

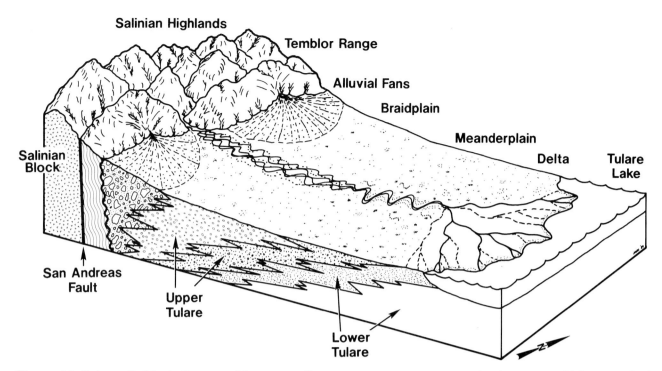

Figure 11. Schematic block diagram of the prograding fluviodeltaic depositional system for the Tulare at South Belridge field. Note the continuum between braidplain, meanderplain, and delta-plain systems. These three systems are present in the upper Tulare vertical sequence and define important sand body and reservoir geometries. (After McPherson et al., 1987.)

(3 m) thick. They comprise cross-bedded, pebbly, fine- to medium-grained, poorly sorted sands (lithofacies S_1 and S_2) at the base, with ripple cross-laminated, very fine grained sands (lithofacies S_3) at the top (Table 1).

Some of the channel and bar sands of the uppermost Tulare zones are coarser grained and pebbly, more thinly bedded, and contain multiple erosion surfaces. These sequences are interpreted as braided fluvial deposits.

Levee and Crevasse-Splay Deposits

These deposits consist mostly of interbedded sequences of ripple cross-laminated, very fine grained sand (lithofacies S_3) or silt (lithofacies Si) and mud (lithofacies M) (Table 1). Deposit thickness varies from 2 to 10 ft (1–3 m). Both sands and muds contain extensive bioturbation. Most splay sands are encased in overbank/interchannel mud, with occasional carbonate-cemented paleosol horizons.

Overbank Deposits

These interchannel deposits constitute a major component of the Tulare Formation in the southern portions of the field. The deposits comprise thin (less than 1 ft [30 cm] thick) muds and clays (lithofacies M) and minor silts (lithofacies Si) (Table 1) and were emplaced by interchannel sedimentation from flood-stage channel overtopping and by channel crevassing.

Distributary-Channel-Mouth-Bar Deposits

These deltaic sands are moderately well sorted, very fine to fine grained (lithofacies S_2 and S_3) (Table 1), and are mostly ripple cross-laminated, wavy bedded, or parallel-laminated. Mouth-bar sands that are sandwiched between distal-bar and prodelta sequences indicate lateral shifting of the distributary channel in the lower delta plain.

Distal-Bar Deposits

These deltaic deposits are composed of thin bedded, ripple-laminated sand (lithofacies S_3) interbedded with laminated silt (lithofacies Si) and mud (lithofacies M) interbeds (Table 1). Bioturbation is very common.

Prodelta/Lake Deposits

These deposits are composed primarily of thick muds, silty muds, and clays (M) (Table 1). Very thin lenticular beds of sandy silts in these deposits represent distal, delta-toe turbidite flows.

The reservoir sands exhibit excellent reservoir quality, with an average effective porosity of 35% and 3000 md permeability (Gates and Brewer, 1975) (Table 2). Porosity is mostly primary and intergranular in these unconsolidated sediments. Total porosity in the Tulare ranges from 32 to 42% and correlates poorly with permeability and lithofacies type (Figure 12). Ineffective porosity, up to 5% (absolute), is attributed to the clays and diatomaceous fragments that have been eroded from the underlying Miocene Monterey strata. Porosity is not a limiting factor in this unconsolidated reservoir. Core permeability values (measured to air at 300 psi confining pressure) range from less than 100 md to

Table 1. Lithofacies associations and their interpretations for the Tulare Formation. Lithofacies symbols, such as (G), are described in the text. (From McPherson and Miller, 1990.)

Lithofacies	Lithofacies Associations and Depositional Settings					
	Channel/Bar	Levee	Overbank	Mouth Bar	Distal Bar	Lake
Sandy pebble gravel (G)	○					
Cross-bedded gravelly sand (S_1)	●					
Fine-grained sand (S_2)	●	○		○		
Ripple-laminated vf sand (S_3)	○	●	○	●	○	
Silt (Si)		○	○		○	○
Mud and clay (M)		○	●	○	●	●

● dominant lithofacies.
○ associated lithofacies.

Table 2. Typical values for grain size, porosity, permeability, and oil saturation for the principal lithofacies of the Tulare Formation. (From Miller et al., 1990.)

Lithofacies	Grain Size (M_z in phi)	Porosity* (%) (ϕ)	Permeability** (md) (k_h)	(md) (k_v)	Oil Saturation, % (S_{oi})
Sandy pebble gravel (G)	-2.5	34	10,000	7,000	50
Cross-bedded pebbly sand (S_1)	1.1	34	10,000	8,000	80
Fine-grained sand (S_2)	2.3	35	6,000	5,000	75
Ripple-laminated vf sand (S_3)	3.3	36	700	300	60
Silt (Si)	4.5	35	50	10	15
Mud (M)	9.0	35	0.1	0.1	0

* effective porosities corrected to reservoir conditions.
** air permeabilities determined at 399 psi (2068 kPa).

10,000 md. These permeabilities display a strong correlation with lithofacies (Figure 13). This is primarily a function of grain size variability in the lithofacies; the coarser-grained lithofacies have the highest permeabilities (Figures 12 and 13) (cf. Beard and Weyl, 1973).

Vertical permeability on a reservoir-scale within the Tulare is greatly affected by lithologic heterogeneities. Discontinuous clay and silt interbeds act as baffles or local barriers to vertical flow. As a consequence, effective reservoir-scale permeability values can be as low as 1% of the measured core values (Dietrich, 1988) as calculated by the stochastic method of Begg et al. (1985). Lateral permeability estimates in the Tulare are similarly decreased by the tortuous fluid-flow path created as a product of the anastomosing of channel-fill sand bodies.

Original oil saturations are assumed to have averaged 76% in the Tulare reservoir (Gates and Brewer, 1975). Oil saturations range from a maximum of 85% in sands to less than 10% in silts. The saturations of the reservoir sands display no correlation with porosity (Figures 12 and 14), but there is a strong relationship of oil saturations to permeability, grain size, and lithofacies (Figures 12 and 15). The lower permeability sands have greater specific-surface areas and higher irreducible water saturations; the higher oil saturations correspond to the more permeable lithofacies. Vertical and horizontal distribution of these lithofacies thus controls overall oil storage capacity and producibility of the reservoir. This correspondence between permeability and oil saturation reflects the relative difficulty of oil migration into the lower permeability facies.

The total interval thickness of the lower Tulare reservoir ranges between 100 and 300 ft (30 and 91 m) and oil-producing reservoir sands total as much as 150 ft (46 m) thick (Figure 16). Productive reservoir sands are thicker and more continuous in the southern portion of the field. Nonproductive sands of greater thickness occur off structure toward the basin center and link the producing reservoir with an extensive, active aquifer (Figure 9). The lower Tulare sands are highly stratified and are distributed throughout the producing area of the field (Figure 16A) and regionally in the basin (Lennon, 1976).

The lower Tulare reservoir geometries are the product of distributary-channel, distributary-mouth-bar, and distal-bar deltaic facies (Figure 10). A typical lower reservoir unit comprises a 5 to 15 ft (2 to 5 m) thick sand body, with interbedded lacustrine and interdistributary-bay clays. Reservoir sands and overlying clay strata typically have good lateral continuity, as illustrated by each of the lower sand horizons having their own oil-water contact. The sand continuity is a product of the multichanneled, lobate nature of the river-dominated Tulare delta at South Belridge.

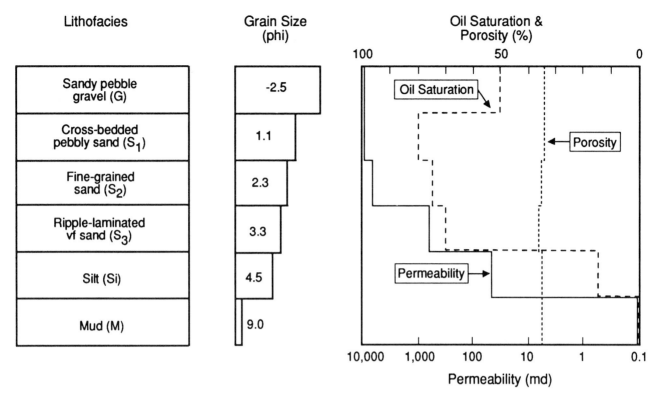

Figure 12. Typical reservoir properties of porosity, permeability, and oil saturation for the Tulare and their correlation with lithofacies. Oil saturation shows a strong correlation with lithofacies and permeability, but not porosity. (Modified from McPherson and Miller, 1990.)

The total interval thickness of the upper Tulare reservoir within the producing area ranges from 300 to 450 ft (91–137 m) and the reservoir sands total as much as 230 ft (70 m) thick (Figure 16B). The upper Tulare sand is somewhat restricted in distribution because of its fluvial origin. The thickest sands occur on both flanks of the anticline and thin over the crest in the southern portion of the field (Figures 9 and 16B). The sands thin and disappear in the northern area of the field. This reflects the general downslope trend of the fluvial sands and a paleotopographic influence of the developing structure on gross sediment distribution. Nonproductive, wet upper Tulare reservoir sands are thickest basinward to the east, similar to the lower Tulare reservoir sands.

Geologic description of Tulare reservoir geometries is critical to optimal steamflood field development. Many small-scale, geologically isolated reservoir sands would not be productive with more widely spaced wells. Flow-unit geometries are, thus, both defined and exploited by the high density of well penetrations necessary for optimal heavy-oil production. Geometries have been interpreted from wireline-log correlations, core facies analysis, log character, producing characteristics, steam pathways, and comparison to nearby Tulare outcrops. Preserved cores from 21 wells and information from unpreserved, disaggregated cores from 221 older wells support log correlations of 2700 wells in the field.

The geometries reflect the influences of structure and deposition. Within the large south-plunging anticlinal structure are complex, small-scale, reservoir geometries created largely by the depositional setting. Syndepositional growth of the South Belridge anticline further complicated the sand-body distributions and resulting reservoir geometries, illustrated in cross section (Figure 9) and map view (Figure 16). Both sand-body geometry and effective reservoir geometry of interconnected sand bodies show considerable vertical and lateral variation, reflecting temporal and spatial changes in the depositional setting.

The upper and lower Tulare reservoir sands consequently display quite different reservoir geometries. Reservoir continuity in the upper Tulare zones is potentially more restricted than for the lower Tulare zones because of the channelized, "shoestring" geometry of many individual sand bodies. A reservoir flow-unit typically consists of one or multiple stacked and amalgamated channel-fill sands, encased in overbank and interchannel muds (Figures 10 and 17).

On a smaller scale, three distinctive styles of fluvial reservoirs are recognized in the upper Tulare (McPherson and Miller, 1989). These reflect significant changes in the style of deposition with time and warrant different development considerations. They range from isolated channel fills to multiple stacked and amalgamated channels as part of a highly

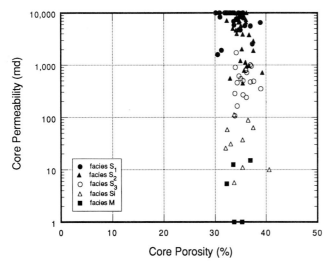

Figure 13. Core porosity and permeability values plotted by Tulare lithofacies. Note the strong association of permeability and lithofacies, but the independence of porosity and lithofacies. The porosity data have been corrected to reservoir conditions; permeability (to air) data were determined at 300 psi. (After McPherson and Miller, 1990.)

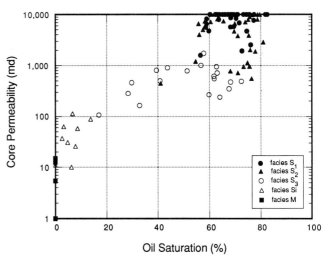

Figure 15. Core permeability and oil saturation plotted by lithofacies. Permeability (air) is determined at 300 psi confining pressure. (After Miller et al., 1990.)

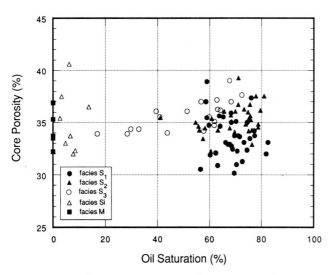

Figure 14. Core porosity and oil saturation plotted by lithofacies. The porosities have been corrected for reservoir conditions. (After Miller et al., 1990.)

channelized fluvial system (Figures 10 and 17). The zone C sands, which were deposited in a mud-rich meandering fluvial system, comprise relatively narrow and isolated channel-fill sand bodies encased in mud. Individual channel dimensions are consistent with the range calculated from paleohydraulic reconstructions (thickness = 10 ft [3 m]; width = 500 ft [152 m]). Vertical and lateral steam communication between individual sand bodies is poor owing to limited interconnectedness. The lower zone B sands were deposited in a sand-rich meandering system.

The dimensions of the individual channels are similar to those in the underlying zone C, but the channels are stacked vertically and laterally coalesced, with resultant increases to the effective reservoir flow-unit geometries (Figure 17). The upper zone B sands, which are the deposits of a more braided channel system, are coarser grained and the individual channels are wide and shallow. Amalgamation of these channel sands is well-developed, giving rise to a sheet-like reservoir flow-unit that has excellent continuity over many thousands of feet.

The produced oil gravity of the Tulare is 13 to 14° API (California Division of Oil and Gas, 1985). In general, the shallow Tulare reservoir oil is heavily biodegraded, with a transition to less biodegradation and larger GOR values in deeper diatomite horizons (Figure 18). The initial GOR reported for the Tulare was 35 SCF/STB, compared with the diatomite of 366 SCF/STB (California Division of Oil and Gas, 1985).

Rock and fluid properties in the Tulare change dramatically with the introduction of steam to the reservoir during steam-drive EOR. Oil saturations at South Belridge field attain residual saturations commonly less than 5% and as low as zero in steam-swept reservoir intervals. This is similar to other reported steam-drive projects (Ali, 1982; Traverse et al., 1983). Gravity override of steam to the top of reservoir flow units commonly results in the lowest residual oil in the upper portions of the reservoirs (Ali and Meldau, 1979). Oil saturations below these swept zones typically increase to more than 40% and retain nearly original saturations at the base.

The primary producing mechanism was solution-gas drive with some assistance from gravity drainage and aquifer support on the flanks of the field. Initial reservoir pressure was probably hydrostatic at approximately 400 psi in the Tulare reservoir.

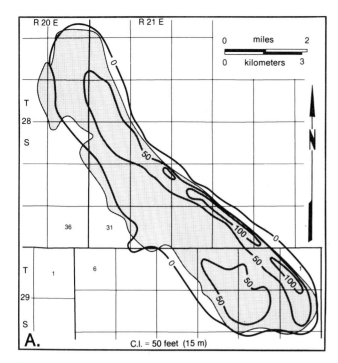
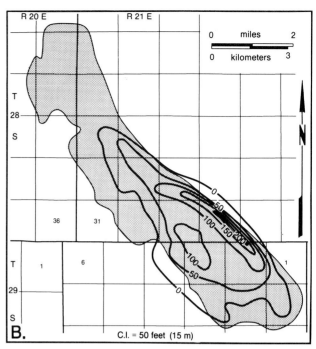

Figure 16. (A) Lower Tulare oil-sand thickness map indicating widespread distribution of sands on the flanks of the anticline in the southern part of South Belridge field. Sands thicken basinward to the northeast and disappear to the north and west, creating a stratigraphic trap at the northern end of the field. (B) Upper Tulare oil-sand thickness map indicating thickest sands on the flanks of the anticline in the southern part of South Belridge field. The contour interval is 50 ft (15 m), and the mapping utilized more than 1000 wells. The stipple represents the present field area. (After McPherson and Miller, 1990.)

Steamflood EOR processes currently provide the dominant drive mechanism, and pressures are variable. Two major Tulare reservoir development considerations are: (1) moving the high-viscosity oil, and (2) effectively contacting reservoir flow-units in the highly discontinuous reservoir sands. Fortunately, both factors are addressed by the large number of wells. More than 6000 closely spaced and shallow wells are the key to producing from hundreds of layered and laterally discontinuous reservoir sands that create laterally and vertically discontinuous reservoir flow-units.

Primary production, cyclic steaming, steam-drive injection, and in situ combustion (fireflood) methods have been variously used during the life of the field. Steamflood development dominates the heavy-oil recovery, with wells typically spaced 200 to 500 ft (61–152 m) apart in patterns and drilled to a total depth of 1000 ft (305 m).

Belridge Diatomite

The diatomite at South Belridge is an unusual reservoir with very high porosity and low permeability. The review presented here is designed to be a concise introduction at the expense of some important, yet equivocal, detail. Understanding the strong contrast of this Belridge diatomite reservoir with the overlying Tulare reservoir is considered paramount to appreciating the diversity and future potential of this field. The data presented here are largely from the southern area of the field. Excellent discussions of the diatomite in other areas of South Belridge field have been published by Schwartz (1988) and Bowersox (1990).

The productive diatomite reservoir is informally called the Belridge diatomite unit of the Reef Ridge Member of the Monterey Formation. This stratigraphic interval is characterized by amorphous biogenic silica (opal-A) above the diagenetic transition zone of opal-CT. The diatomite is overlain by the clastic Etchegoin and Tulare formations. Local operators use a more detailed stratigraphy subdivided on the basis of major productive intervals (Bowersox, 1990), depositional cycles (Schwartz, 1988), and combinations thereof.

The lithology is dominated by diatomite (>80 wt.% biogenic silica) (Figure 19) and diatomaceous shale (50–80 wt.% biogenic silica) (classification of Isaacs, 1981), with ubiquitous organic matter. Terrigenous-clastic-rich cycles within the Monterey diatomite reservoir at Belridge are attributed to distal turbidite flows (Schwartz, 1988) and correlated with Miocene submarine fan sequences deposited in the southern end of the basin (MacPherson, 1978; Webb, 1981).

Depth to the top of the Belridge diatomite pay zone typically ranges from 700 to 1200 ft (213–366 m). The

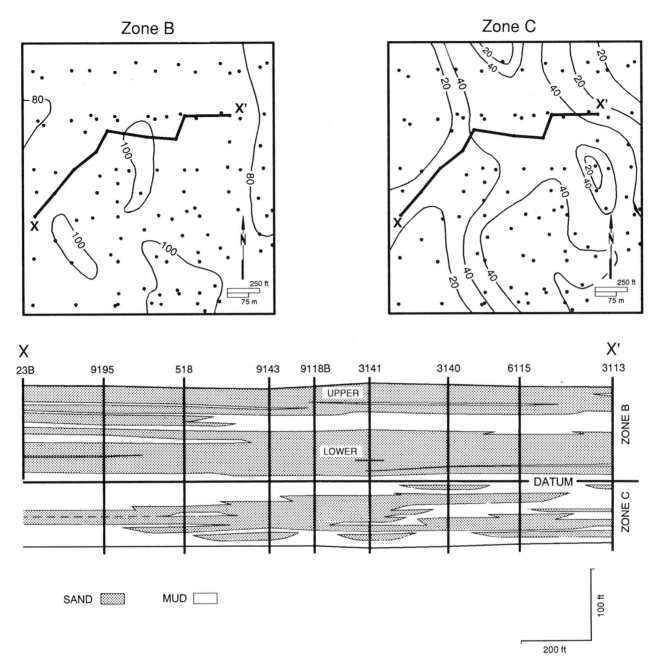

Figure 17. Detailed upper Tulare stratigraphic cross section and corresponding sand thickness maps of zones B and C, illustrating the scale and variability of vertical and lateral reservoir continuity. Note the discontinuous and highly channelized geometry of the fine-grained meanderbelt sands of zone C. By contrast, the overlying coarse-grained meanderbelt sands (lower zone B) and braided fluvial sands (upper zone B) have a sheet-like geometry with high lateral reservoir continuity. The corresponding net sand maps clearly show these differences in sand distribution. The contour interval is 20 ft (6 m), and the datum is the contact between the two zones. This cross section is located in the northeast quarter of Sec. 3, T29S, R21E. (After McPherson and Miller, 1990.)

gross pay-zone thickness is typically 800 to 1200 ft (244–366 m), all of which is potentially productive. The diatomaceous sediments at Belridge undergo diagenetic transformations to opal-CT at typical subsurface depths of 1500 to 1800 ft (457–549 m) in the southern portion of the field (Schwartz, 1988) and 2000 to 2300 ft (610–701 m) in the northern portions (Bowersox, 1990). Depth, reservoir thickness, and pay-zone thickness are highly variable as a result of syndepositional growth of the anticline, local differences in the depositional cycles, local oil characteristics, and a cross-cutting fracture system. The diagenetic transition from opal-A to opal-CT reduces the rocks' oil storage capacity, matrix permeability, and overall reservoir quality. The diagenesis of silica from opal-A (diatomite) to

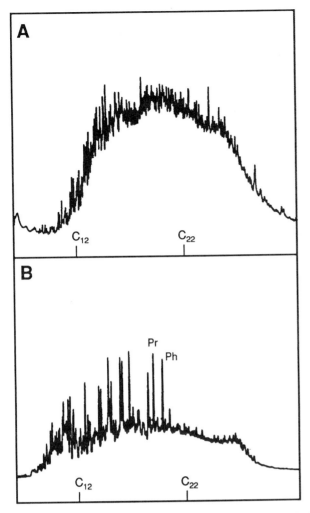

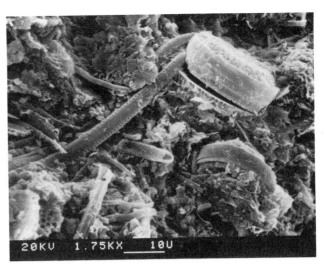

Figure 19. SEM photomicrograph showing the microporous nature of the diatomite, composed dominantly of opal-A silica in whole and fragmented diatoms. Clay in this sample ranges from 10 to 20 wt. % intermixed with delicate finer grained diatom fragments. Average pore size is generally less than 2 μ. Note the scale-bar of 10 μ. The sample is from 1250.3 ft deep in the Mobil 7005-3 well, Sec. 3, T29S, R21E. (Photo by Josh Cocker.)

Figure 18. Oil fingerprints of typical (A) biodegraded Tulare heavy oil, (B) less biodegraded diatomite oil. Carbon chain lengths and pristane and phytane are noted for reference. Both oil samples are thermal extraction chromatograms from core samples. High variability of oils exists in both the Tulare and diatomite reservoir horizons.

opal-CT and quartz is presented elsewhere in greater detail (Murata and Larsen, 1975; Schwartz et al., 1981; Isaacs, 1981; Williams, 1982, 1988).

The pelagic and hemipelagic Belridge diatomite was deposited in an anoxic deep-marine setting with water depths around 2000 to 3000 ft (610–914 m) (Graham and Williams, 1985) (Figure 20). Biogenic sedimentation was widespread during late-middle and late Miocene time because of the climatically induced explosion of diatom productivity (Ingle, 1981). The fine-grained terrigenous-clastic sediment cycles (turbidites) associated with submarine-fan packages to the south intermittently diluted diatomaceous biogenic muds (Graham and Williams, 1985; Schwartz, 1988). The cycles of clastic and biogenic sediments may also reflect biogenic productivity (seasonal diatom bloom and El Niño), dilution caused by tectonic uplift and earthquakes, and changing marine circulation patterns resulting from eustasy.

Massive diatomite, common in the upper interval in the southern area of the field, is attributed to bioturbation. The general depositional model for the Monterey Formation favors tectonically active basins with high organic (siliceous) richness, low oxygen and generally anoxic conditions, and relatively low terrigenous-clastic input.

Diatomite porosity is both intergranular and intragranular, with a very small fracture volume (<1%). Pore size is typically 5μ and less. The fine-grained nature of the pores (Figure 19) leads to high water saturations. However, fractures are very important to reservoir permeability and fluid-flow characteristics.

Two general diagenetic processes affect the vertical diatomite sequence. Compaction plays an important role, mechanically decreasing total porosity approximately 10% over the total interval. Nucleation of opal-CT crystallites increases with depth near the opal-A and opal-CT transition boundary. The onset of silica diagenesis increases density, decreases matrix permeability, and increases adsorbed water saturation.

A natural fracture system and related faulting forms the primary permeability system in the diatomite. A dual porosity system (fracture and matrix contributions) is indicated by reservoir performance characteristics of the diatomite. Mapping of induced fracture orientations, using surface tiltmeter arrays during well completions, indicates near vertical fractures with a preferred orientation of N45E in the northern part of the field (Bowersox, 1990) and N18E in the southern part of

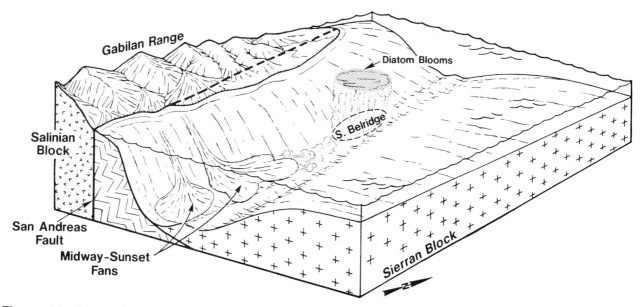

Figure 20. Schematic block diagram of the deep-marine setting of the diatomite depositional system in the southern San Joaquin basin. The deep-marine fan systems near the southern end of the basin episodically contributed clastics to the diatomite deposits at South Belridge. (Modified from Schwartz, 1988.)

the field. These orientations are similar to regional tectonic stresses (Mount and Suppe, 1987) and normal to the fold axis of the South Belridge anticline.

The diatomite exhibits extremely high porosity and low permeability, analogous to an oil shale or a chalk. Because these fine-grained rocks appear to be uniform macroscopically, they have often been regarded as homogeneous. However, the diatomite reservoir quality reflects a sensitivity to subtle differences during deposition or diagenesis. The highest reservoir quality is associated with the beds of greatest purity of diatomaceous silica. These beds have higher porosities, higher permeabilities, lower grain densities, and are more easily fractured than are the beds with terrigenous clastics. Reservoir quality is principally dependent on the interaction of matrix and fracture permeability. Brecciated, fractured diatomites make the best reservoirs.

Diatomite porosity of 50–70% generally decreases with increasing depth and varies with depositional cycles (Figures 21 and 22). Compaction appears to be the basic mechanism for the porosity decrease with depth. Typical permeabilities range from 0.1 to 5.0 md (unstressed air) and grain densities from 2.10 to 2.50 g/cc (Schwartz, 1988). Original oil saturations range from 10 to 60%, depending on lithology and position in the oil column. Oil saturation and porosity values from core analysis provide an oil resource of 1200 to 1700 bbl/ac-ft.

The total Belridge diatomite interval is approximately 1000 ft (305 m) thick, all of which is oil saturated on the crest of the structure. Wells are often completed over the entire Belridge diatomite interval, although a high percentage of oil production comes from a small percentage of the interval. Subtle differences in matrix permeability, fracture permeability, rock composition, oil properties, and scale of heterogeneities are important to define the better completion intervals. Anisotropic flow characteristics are caused by bedding and by fracture systems normal to bedding. This indicates that the pay zone is a function of rock subtleties and that it is responsive to the producing technology.

The produced oil gravity of the diatomite ranges from 20 to 30° API in a heterogeneous mixture. Stratified gravity segregation is common in some areas of the field where oil gravity decreases with depth (Bowersox, 1990). This model of oil distribution is common in Monterey reservoirs in California. The opposite relationship occurs in the southern end of the field, where biodegradation of the shallower oil results in higher viscosity and lower API gravity. In general, the GOR increases and effects of biodegradation decrease with increasing depth, on the flanks of the anticline, and down plunge. The GOR ranges from the 366 SCF/STB originally reported for shallow diatomite production (California Division of Oil and Gas, 1985) to 1700 to 5000 in the entire Belridge diatomite interval in the southern area of the field.

The primary producing mechanism is solution gas and compaction. Reservoir pressures are near hydrostatic. Important development considerations include: (1) understanding the fracture permeability system, (2) determining effective seals or nonequilibrium hydrocarbon traps, and (3) a predictable reservoir facies model.

SOURCE

The presumed source for the oil in both the Belridge diatomite and Tulare reservoirs is deeply buried portions of the Miocene Monterey Formation. The

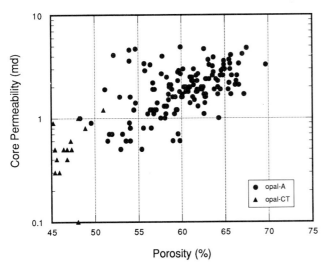

Figure 21. Porosity and matrix permeability (air) values of the Belridge diatomite reservoir from core. Samples were air dried. The data have been corrected to reservoir conditions. Fractured samples have been eliminated.

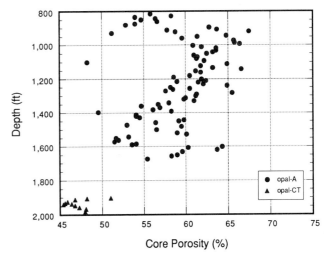

Figure 22. Porosity and depth plot, indicating porosity is a function of compaction and lithofacies. Core samples were air dried and corrected to reservoir conditions. The transition from opal-A to opal-CT indicated on density wireline logs occurs around 1700 ft (518 m) depth.

sequence is rich in source materials with a favorable kerogen type having total organic carbon (TOC) values commonly greater than 5% (weight) and ranging from 0.40 to 9.16% in the San Joaquin basin (Graham and Williams, 1985). The heavy oil in the shallow Tulare Formation is biodegraded and has characteristics of water washing. The diatomite oil is also biodegraded. Deeper horizons in the southern end of the field have less biodegraded oil with higher gas-oil ratios.

Published Monterey source rock studies in the San Joaquin basin typify the kerogens as strongly marine (type II) kerogen with terrestrial (type III) influence (Graham and Williams, 1985). The relative abundance of clay minerals in the San Joaquin basin Monterey rocks provides a buffering influence unlike the sulfur-rich oils of the offshore Monterey, which can generate oil at lower than usual temperatures (Orr, 1986). However, published vitrinite reflectance values (Graham and Williams, 1985) can be anomalously low relative to other thermal indices. Source-rock data from 11,100 to 11,500 ft (3383–3305 m) depths on the eastern flank of the Belridge anticline (Cal Canal field) have a thermal alteration index (TAI) range from 2 to 3 (avg 2.5), vitrinite reflectance from 0.25 to 0.44, and T_{max} of 423–439°C (Graham and Williams, 1985).

EXPLORATION CONCEPTS

Regional Play and General Applications of Geologic Parameters

The geologic setting in an active tectonic basin provides for great diversity of source rock, reservoir, and seal. In fact, rocks in the Monterey Formation can serve in all three capacities at the same time. Ongoing thermal maturation and petroleum generation, vertical and lateral migration pathways, trapping by structural, stratigraphic, and self-sealing tar mechanisms, and diverse reservoir rocks occur in this basinal setting.

Surface geologic mapping provides evidence for a giant fluviodeltaic depositional system that formerly was contiguous with the South Belridge Tulare field. Regionally offset on the other side of the San Andreas fault is a large, east-draining Pliocene basin (Galehouse, 1967). Restoring strike-slip displacement along the San Andreas fault suggests alignment of that basin with the major fluviodeltaic depositional system which comprises the Pleistocene Tulare reservoir at South Belridge.

Opportunities to discover other fields comparable to South Belridge are limited geographically to frontier areas. The prospects of discovering major fields by surface expressions of anticlines with oil seeps in creek beds, as was the case for South Belridge, are unlikely in this day. But shallow, heavy-oil reservoirs, such as the Tulare, and low-permeability high-porosity reservoirs, such as the diatomite, may await creative development. At South Belridge, wells that were unspectacular combined to produce more oil than any other field in the United States outside of Alaska (*Oil & Gas Journal*, 1990). Exploration and development persistence, geologic description, and new technology can develop a billion barrel field from an inauspicious discovery. Isolated reservoir sand bodies in the initial wells may provide only a hint of the development potential. A critical early step is to look carefully at the geologic scale of potential reservoir units. Isolated well performance resulting from highly variable scale of reservoir flow-unit geometries in the subtle, yet complex, diatomite and fluviodeltaic reservoirs provides little indication of EOR potential.

Opportunities for economic development of similar reservoirs with poor primary production benefit from comprehensive geologic characterization and applications of new technology. These qualities are vital to the economic development of the diatomite reservoir. Critical engineering studies necessary to exploit the full potential of the field will benefit from creative application of information, perhaps much different than that gathered for conventional reservoirs.

ACKNOWLEDGMENTS

This study is a product of the work of many people representing a multitude of petroleum-related disciplines. The authors especially thank T. E. Covington, J. D. Tucker, and E. Forrester, Jr., for sharing their expertise and data regarding the reservoirs. We acknowledge the worthy contributions of M. L. Barrett, J. D. Cocker, R. W. Mathias, and the South Belridge Reservoir Management Team. The manuscript benefited from reviews by J. M. Eagan, S.A. Graham, C. E. Tranter, and J. D. Tucker. Thanks are due to H. H. Mottern, R. J. Moiola, P. A. Sheetz, and T. C. Zeiner of Mobil Corporation for supporting this work. Thanks are also due Mobil Oil Corporation and Mobil Research and Development Corporation for permission to publish.

REFERENCES CITED

Aalund, L. R., 1988, EOR projects decline, but CO_2 pushes up production: Oil & Gas Journal, v. 86, n. 16 (April 18), p. 33-73.

Ali, F. A., 1982, Steam injection theories—a unified approach: Society of Petroleum Engineers, Paper No. 10746, p. 309-315.

Ali, F. A., and R. F. Meldau, 1979, Current steamflood technology: Journal of Petroleum Technology, October 1979, p. 1332-1341.

Barbat, W. F., and J. Galloway, 1934, San Joaquin Clay, California: AAPG Bulletin, v. 18, p. 476-499.

Bartow, J. A., 1987, The Cenozoic evolution of the San Joaquin valley, California: U.S. Geological Survey Open-File Report No. 87-581, 74 p.

Beard, D. C., and P. K. Weyl, 1973, Influence of texture on porosity and permeability of unconsolidated sand: AAPG Bulletin, v. 57, p. 349-369.

Begg, S. H., D. M. Chang, and H. H. Haldorsen, 1985, A simple statistical technique for calculating the effective vertical permeability of a reservoir region containing discontinuous shales, in Proceedings, 60th Annual Meeting of the Society of Petroleum Engineers, Paper No. 14271, 15 p.

Bowersox, J. R., 1990, Geology of the Belridge Diatomite, northern South Belridge field, Kern County, California, in J. G. Kuespert and S. A. Reid, eds., Structure, stratigraphy and hydrocarbon occurrences of the San Joaquin basin: Pacific Section SEPM, v. 64, p. 215-223.

Bowersox, J. R., and R. A. Shore, 1990, Reservoir compaction of the Belridge Diatomite and surface subsidence, South Belridge field, Kern County, California, in J. G. Kuespert and S. A. Reid, eds., Structure, stratigraphy and hydrocarbon occurrences of the San Joaquin basin: Pacific Section SEPM, v. 64, p. 225-230.

Bramlette, N. M., 1946, The Monterey Formation of California and the origin of siliceous rocks: U.S. Geological Survey Professional Paper 212, 57 p.

California Division of Oil and Gas, Department of Conservation, 1985, California Oil and Gas Fields, Central California Vol. I.

California Division of Oil and Gas, Department of Conservation, 1987, 72nd annual report of the state oil and gas supervisor 1986: v. 72, p. 57-58.

Clarke, S. H. Jr., D. G. Howell, and T. H. Nilsen, 1975, Paleogene geography of California, in D. W. Weaver, G. R. Hornaday, and A. Tipton, eds., Paleogene symposium and selected technical papers: Pacific Section AAPG Pacific Section of SEPM, Pacific Section of SEG, p. 121-154.

Conservation Committee of California Oil Producers, 1989, Annual Report.

Dietrich, J., 1988, Steamflooding in a waterdrive reservoir: Upper Tulare sands, South Belridge field, in Proceedings, 1988 California Regional Meeting, Society of Petroleum Engineers, Paper No. 17453, p. 479-494.

Division of Oil and Gas, 1961, California oil fields, summary of operations, Forty-Seventh Annual Report of the State Oil and Gas Supervisor: California Department of Natural Resources, Division of Oil and Gas, San Francisco, v. 47, n. 1, p. 29-38.

Fishburn, M. D., 1990, Results of deep drilling, Elk Hills field, Kern County, California, in J. G. Kuespert and S. A. Reid, eds., Structure, stratigraphy and hydrocarbon occurrences of the San Joaquin basin, California: Pacific Section SEPM, v. 64, p. 157-167.

Folk, R. L., P. B. Andrews, and D. W. Lewis, 1970, Detrital sedimentary rock classification and nomenclature for use in New Zealand: New Zealand Journal of Geology and Geophysics, v. 13, p. 937-968.

Foss, C. D., 1972, A note on the fluid level traps in the San Joaquin valley, in E.W. Rennie, Jr., ed., Geology and oil fields, west side central San Joaquin Valley: AAPG Pacific Section, Guidebook, p. 15.

Galehouse, J. S., 1967, Provenance and paleocurrents of the Paso Robles Formation, California: GSA Bulletin, v. 78, p. 951-978.

Gates, C. F., and S. W. Brewer, 1975, Steam injection into the D&E zones, Tulare Formation, South Belridge field, Kern County, California: Journal of Petroleum Technology, v. 27, p. 343-348.

Gates, C. F., and H. J. Ramey, Jr., 1958, Field results of South Belridge thermal recovery experiment: American Institute of Mining, Metallurgical, and Petroleum Engineering Petroleum Transactions, v. 213, p. 236-244.

Gates, C. F., K. D. Jung, and R. A. Surface, 1978, In situ combustion in the Tulare Formation, South Belridge Field, Kern County, California: Journal of Petroleum Technology, v. 30, p. 798-806.

Graham, S. A., 1987, Tectonic controls on petroleum occurrence in California, in R. V. Ingersoll and W. G. Ernst, eds., The geotectonic development of California, Rubey Volume VI: Englewood Cliffs, Prentice-Hall, p. 47-63.

Graham, S. A., and L. A. Williams, 1985, Tectonic, depositional, and diagenetic history of Monterey Formation (Miocene), central San Joaquin basin, California: AAPG Bulletin, v. 69, p. 385-411.

Harding, T. P., 1976, Tectonic significance and hydrocarbon trapping consequences of sequential folding synchronous with San Andreas faulting, San Joaquin Valley, California: AAPG Bulletin, v. 60, p. 356-378.

Ingersoll, R. V., 1979, Evolution of the Late Cretaceous forearc basin, northern and central California: GSA Bulletin, v. 90, p. 813-826.

Ingle, J. C., Jr., 1981, Origin of Neogene diatomites around the north Pacific rim, in the Monterey Formation and related siliceous rocks of California: Pacific Section SEPM, p. 159-180.

Isaacs, C. M., 1981, Field characterization of rocks in the Monterey Formation along the coast near Santa Barbara, California, in C. M. Isaacs, ed., Guide to the Monterey Formation in the California Coastal area, Ventura to San Luis Obispo: Pacific Section AAPG, v. 52, p. 39-54.

Knowles, R. S., 1959, The greatest gamblers: Norman, University of Oklahoma Press, 376 p.

Lennon, R. B., 1976, Geological factors in steam-soak projects on the west side of the San Joaquin basin: Journal of Petroleum Technology, v. 4, p. 741-748.

Lettis, W. R., 1982, Late Cenozoic stratigraphy and structure of the western margin of the central San Joaquin Valley, California: U.S. Geological Survey Open-File Report 82-526, 26 p.

Loomis, K. B., 1990, Depositional environments and sedimentary history of the Etchegoin group, west-central San Joaquin Valley, California, *in* J. G. Kuespert and S. A. Reid, eds., Structure, stratigraphy and hydrocarbon occurrences of the San Joaquin basin, California: Pacific Section SEPM, v. 64, p. 231-246.

MacPherson, B. A., 1978, Sedimentation and trapping mechanism in upper Miocene Stevens and older turbidite fans of southeastern San Joaquin Valley, California: AAPG Bulletin, v. 62, p. 2243-2274.

McPherson, J. G., and D. D. Miller, 1989, Three-dimensional geometry of fluvial reservoir sands: steam-drive case study (abs.): AAPG Bulletin, v. 73, p. 390.

McPherson, J. G., and D. D. Miller, 1990, Depositional settings and reservoir characteristics of the Plio-Pleistocene Tulare Formation, South Belridge field, San Joaquin Valley, California, *in* J.G. Kuespert and S. A. Reid, eds., Structure, stratigraphy and hydrocarbon occurrences of the San Joaquin basin, California: Pacific Section SEPM, v. 64, p. 205-214.

McPherson, J. G., G. Shanmugam, and R. J. Moiola., 1987, Fan-deltas and braid deltas: varieties of coarse-grained deltas: GSA Bulletin, v. 99, p. 331-340.

Miller, D. D., J. G. McPherson, and T. E. Covington, 1990, Fluviodeltaic reservoir, South Belridge field, San Joaquin Valley, California, *in* J. H. Barwis, J. G. McPherson, and J. R. J. Studlick, eds., Sandstone petroleum reservoirs: New York, Springer-Verlag, p. 109-130.

Mount, V. S., and J. Suppe, 1987, State of stress near the San Andreas fault: implications for wrench tectonics: Geology, v. 15, p. 1143-1146.

Murata, K. J., and R. R. Larsen, 1975, Diagenesis of Miocene siliceous shales, Temblor Range, California: U.S. Geological Survey Journal of Research, v. 3, p. 553-556.

Namson, J. S., and T. L. Davis, 1988, Seismically active fold and thrust belt in the San Joaquin Valley, central California: GSA Bulletin, v. 100, p. 257-273.

Oil & Gas Journal, 1990, U.S. fields with reserves exceeding 100 million bbl: v. 88, n. 5, p. 74-76.

Orr, W. L., 1986, Kerogen, asphaltenes/sulfur relationships in sulfur-rich Monterey oils: Organic Geochemistry, v. 10, p. 499-516.

Page, B. M., and D. C. Engebretson, 1984, Correlation between the geologic record and computed plate motions for central California: Tectonics, v. 3, p. 133-155.

Rintoul, W., 1978, "Spudding In": California Historical Society, Fresno Valley Publishers, 240 p.

Ritzius, D. E., 1950, South Belridge oil field: summary of operations: California Division of Oil and Gas, Department of Natural Resources, 36th annual report, state oil and gas supervisor: v. 36, p. 18-25.

Russel, M., 1985, Goodyear to buy oil properties for $395 million: Wall Street Journal, 4 December 1985.

Schwartz, D. E., 1988, Characterizing the lithology, petrophysical properties, and depositional setting of the Belridge Diatomite, South Belridge field, Kern County, California, *in* S. A. Graham, ed., Studies of the geology of the San Joaquin basin: Pacific Section SEPM, v. 60, p. 281-301.

Schwartz, D. E., W. E. Hottman, and S. O. Sears, 1981, Geology and diagenesis of Belridge Diatomite and Brown Shale, San Joaquin Valley, California (abs.): AAPG Bulletin, v. 65, p. 988-989.

Small, G. P., 1986, Steam-injection profile control using limited-entry perforations: Society of Petroleum Engineers Production Engineering (September), p. 388-394.

Stanton, R. J., and J. R. Dodd, 1970, Paleoecologic techniques—comparison of faunal and geochemical analyses of Pliocene paleo-environments, Kettleman Hills, California: Journal of Paleontology, v. 44, p. 1092-1121.

Strickland, F. G, 1982, Reasons for production decline in the diatomite, Belridge oil field: a rock mechanics view: SPE 10773, Proceedings, SPE California Regional Meeting, p. 581-589.

Strubhar, M. K., and W. L. Medlin, 1982, Fracturing results in diatomaceous earth formations, South Belridge field, California: SPE 10966, Proceedings, 57th Technical Conference, SPE, 12 p.

Traverse, E. F., A. D. Deibert, and A. J. Sustek, 1983, San Ardo—a case history of a successful steamflood: Society of Petroleum Engineers Paper No. 11737, 8 p.

Webb, G. W., 1981, Stevens and earlier Miocene turbidite sandstones, southern San Joaquin Valley, California: AAPG Bulletin, v. 65, p. 438-465.

Wendel, D. J., L. A. Kunkel, and G. S. Swanson, 1988, Waterflood potential of diatomite: new laboratory methods: SPE 17439, Proceedings, SPE California Regional Meeting, p. 373-381.

Wilcox, R. E., T. P. Harding, and D. R. Seely, 1973, Basic wrench tectonics: AAPG Bulletin, v. 57, p. 74-96.

Williams, L. A., 1982, Lithology of the Monterey Formation (Miocene) in the San Joaquin Valley of California, *in* L. A. Williams and S. A. Graham, eds., Monterey Formation and associated coarse clastic rocks, central San Joaquin basin, California: Pacific Section SEPM, p. 17-35.

Williams, L. A., 1988, Origins of reservoir complexity in the Miocene Monterey Formation of California, *in* S. A. Graham, ed., Studies of the geology of the San Joaquin basin: Pacific Section SEPM, v. 60, p. 261-279.

Woodring, W. P., R. Steward, and R. W. Richards, 1940, Geology of the Kettleman Hills oilfield, California: U.S. Geological Survey Professional Paper 195, 170 p.

Appendix 1. Field Description

Field name .... *South Belridge field*

Ultimate recoverable reserves .... *1.20 billion bbl (estimated);*
1.75 billion bbl (speculated)

Field location:
  Country .... *U.S.A.*
  State .... *California*
  Basin/Province .... *San Joaquin basin*

Field discovery:
  Year first pay discovered .... *Pleistocene Tulare Formation and Miocene Monterey Formation*
  (Belridge diatomite) 1911

Discovery well name and general location:
  First pay .... *Belridge Oil Company No. 101, Sec. 33, T28S, R21E*

Discovery well operator .... *Belridge Oil Company*

IP .... *100 BOPD (co-mingled Tulare and Belridge diatomite)*

All other zones with shows of oil and gas in the field:

Age	Formation	Type of Show
Pliocene	*Etchegoin*	*Oil*
Miocene	*Monterey (opal-CT)*	*Oil, gas*
	Monterey (quartz)	*Oil, gas*

Geologic concept leading to discovery and method or methods used to delineate prospect
Oil seep on surface expression, drilled during trend development.

Structure:
  Province/basin type .... *Transpressive, forearc-successor basin; Bally 332, Klemme III Bb*
  Tectonic history
  Forearc basin transformed to intermontane basin concurrent with San Andreas basin-bounding wrench faulting and associated with large en echelon anticlines.
  Regional structure
  Field lies along western margin of San Joaquin basin, subparallel to the San Andreas fault.
  Local structure
  Southeast-plunging anticline

Trap:
  Trap type(s)
  Combination traps for both reservoir horizons in order of significance are: (1) anticlinal trap, (2) stratigraphic pinch-out, (3) truncation unconformity trap, (4) fracture enhancement trap (Belridge diatomite only), (5) tar-seal trap, and (6) fluid-level trap (Tulare only).

Basin stratigraphy (major stratigraphic intervals from surface to deepest penetration in field):

Chronostratigraphy	Formation	Depth to Top in ft (m)
Pliocene–Pleistocene	*Tulare*	*Surface*
Pliocene	*San Joaquin*	*1000+ (305)*
	Etchegoin	*1020+ (311)*
Miocene	*Monterey*	*1050+ (320)*

Reservoir characteristics:
  Number of reservoirs .... *2 major zones*
  Formations .... *Tulare; Monterey (Belridge diatomite)*

Ages .... *Pleistocene; Miocene (Mohnian-Delmontian)*
Depths to tops of reservoirs .... *Tulare, 500 ft (153 m); Belridge diatomite,*
700 ft (214 m) in central portion of field
Gross thickness (top to bottom of producing interval) .... *Tulare, 400-700 ft (122-213 m);*
Belridge diatomite, 800-1200 ft (244-366 m)
Net thickness—total thickness of producing zones
  Average .... *Tulare, 200 ft (61 m); Belridge diatomite, 900 ft (274 m)*
  Maximum .... *Tulare, 400 ft (122 m); Belridge diatomite, 1200 ft (366 m)*
Lithology
Tulare sands are unconsolidated, texturally and mineralogically immature lithic feldsarenites; Belridge diatomites are diatom-rich fine-grained shales in primarily the amorphous (opal-A) silica phase
Porosity type .... *Tulare sands are essentially all primary intergranular; Belridge diatomite is principally intergranular and intragranular with a very small fracture volume*
Average porosity .... *Tulare, 35%; Belridge diatomite, 60%*
Average permeability .... *Tulare, 3000 md; Belridge diatomite, 1 md*

Seals:
  Upper
    Formation, fault, or other feature .... *Tulare Formation clays; late Miocene unconformity*
    Lithology .... *Clays and muds*
  Lateral
    Formation, fault, or other feature .... *Permeability loss in the Tulare due to basin-edge facies change; permeability loss in Belridge diatomite due to faulting and facies changes*

    Lithology .... *Tulare changes from sands to mud; Belridge diatomite changes from diatomite to clay-rich muds*

Source:
  Formation and age .... *Miocene Monterey Formation*
  Lithology .... *Biogenic siliceous rocks*
  Average total organic carbon (TOC) *1-5%*
  Maximum TOC .... *9% regionally*
  Kerogen type (I, II, or III) .... *Type II with some type III*
  Vitrinite reflectance (maturation) .... $R_o = 0.25 - 0.44$
  Time of hydrocarbon expulsion .... *Miocene to present day*
  Present depth to top of source .... *Diatomite at 700 ft (213 m) is immature source rock*
  Thickness .... *10,000 ft (3050 m)*
  Potential yield .... *NA*

Appendix 2. Production Data

Field name .... *South Belridge field*
Field size:
  Proved acres .... *8700 (3525 ha)*
  Number of wells all years .... *7541 total; current number of wells: 6100 active*
  Well spacing .... *Tulare, 10 ac (primary), 2.5 to 10 ac EOR patterns, 200-500 ft (61-152 m); Belridge diatomite, <1-10 ac spacing*
  Ultimate recoverable .... *1750 million bbl (estimated)*
  Cumulative production .... *Approximately 700 million bbl and 220 bcf*
  Annual production .... *64 million bbl (maximum to date)*

Present decline rate .... *Not appropriate for steamdrive*
    Initial decline rate .... *NA*
    Overall decline rate .... *NA*
Annual water production .... *NA*
In place, total reserves .... *NA*
In place, per acre foot .... *NA*
Primary, secondary, and enhanced recovery .... *Estimated ultimate recovery over 1 billion bbl of oil*
Cumulative water production .... *NA*

Drilling and casing practices:
    Amount of surface casing set .... *50 ft (15 m)*
    Casing program .... *NA*
    Drilling mud .... *NA*
    Bit program .... *NA*
    High pressure zones .... *Steam injected Tulare reservoirs*

Completion practices .... *NA*

Formation evaluation:
    Logging suites *Older wells have resistivity and SP; newer wells typically have induction-GR-SP, neutron-density, with occasional dielectric and high-resolution dipmeters*

    Testing practices
    Mud logging techniques *None on Tulare wells; hot-wire and chromatographs on Belridge diatomite wells*

Oil characteristics:
    Type .... *Naphthenic*
    API Gravity .... *Tulare 13–14°; Belridge diatomite 20–32°*
    Initial GOR .... *Tulare was 35 (SCFG/STBO); Belridge diatomite was 366 (SCFG/STBO)*
    Viscosity, SUS .... *Tulare, 1800 cp at 90°F (32°C); Belridge diatomite 0.5–100 cp at 120°F (49°C)*
    Pour Point .... *NA*
    Gas-oil distillate .... *NA*

Field characteristics:
    Average elevation .... *600 ft (183 m)*
    Initial pressure .... *Assume hydrostatic*
    Pressure gradient .... *NA*
    Temperature .... *Tulare, 95°F (34°C); Belridge diatomite, 120°F (49°C)*
    Geothermal gradient .... *NA*
    Drive .... *Tulare is solution-gas drive with limited gravity drainage and aquifer support (primary); Belridge diatomite is solution-gas drive and compaction*
    Oil column thickness .... *NA*
    Oil-water contact .... *Tulare has multiple contacts, oil-water contact not found in Belridge diatomite*
    Connate water .... *Tulare, 25%; Belridge diatomite, 40–90%*
    Water salinity, TDS *30,000*
    Resistivity of water .... *Tulare, 0.25–0.45 ohm; Belridge diatomite, 0.20–0.4*
    Bulk volume water (%) .... *NA*

Transportation method and market for oil and gas:
Pipeline for oil; gas is burned in steam generators and co-generation (electricity and steam) facilities.

Portachuelo Field—Peru
Talara Basin

HUGH HAY-ROE
Consultant
Kingwood, Texas

PAUL M. MILLER
Exxon Company International
Houston, Texas

FIELD CLASSIFICATION

BASIN: Talara
BASIN TYPE: Forearc
RESERVOIR ROCK TYPE: Sandstone and Quartzite
RESERVOIR AGE: Eocene and Pennsylvanian
PETROLEUM TYPE: Oil
TRAP TYPE: Faulted Anticline

RESERVOIR ENVIRONMENT OF DEPOSITION: Distal Facies of Submarine Fan Sequence
TRAP DESCRIPTION: Anticline with numerous low-angle normal faults; faulting occurred after oil and gas accumulated

LOCATION

Portachuelo field is one of only two fields found to date on the south flank of the Talara basin (Figure 1(A)). The basin lies within the northern part of the Sechura Desert, which is itself the northernmost sector of the great coastal desert that stretches over 2000 mi (3218 km) from central Chile.

Extending from near the mouth of the Chira River northward about 4 mi (6.5 km), Portachuelo is only 17 mi (27 km) southeast of Punta Balcones, the westernmost point in South America.

Geologically the field is located near the crest of the Portachuelo high (Figure 1B). This is a spur off the basement high that separates the Talara basin from the younger Sechura basin to the south (Figure 1C). The great Neogene graben known as the Lagunitos trough, with a structural relief exceeding 15,000 ft (4575 m), in effect separates the two southern oil fields from the rest of the prolific Talara basin.

HISTORY

Portachuelo field is a good example of the Talara geologist's complaint: "In this basin, every [expletive] well is a wildcat." The following record is not only a chronology of field development in the mechanical sense, but also a history of the evolution of geologic thought *about* the field and about subsurface geologic techniques developed in response to the surprises that came with the drilling of most of the wells (Travis, 1953; Youngquist, 1958).

For professionals in the detective work known as subsurface geology, Portachuelo field in northwest Peru has three aspects of special interest:

- *Entrapment history:* Fluid data (including pressures) from many of the rather closely spaced wells provided indispensable clues to the history of oil-trapping in the Eocene Salina pool. (Also described in Hay-Roe et al., 1983, on which the present study is partially based.)
- *Successful infill drilling:* The spectacularly complex faulting obliged reservoir geologists who worked in the Talara basin to develop special correlating and mapping techniques. Within Portachuelo field these techniques enabled us to locate undrained blocks that were economically attractive for infill wells on a spacing of only 13 ac (5 ha).
- *Evolving geologic picture:* Some fields are developed quickly and routinely, with few surprises as drilling proceeds. Portachuelo was not like that. It was developed over a period of more than 40 years, during which the subsurface picture (the geologic model) changed dramatically—not only because new information continued to be generated, but also because geologic interpretations became more sophisticated, reflecting the rapid advances in every area of earth science.

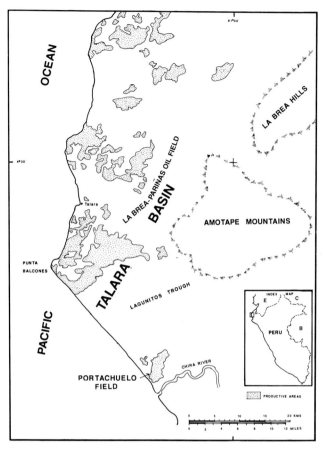

Figure 1A. Location map: southern Talara basin showing Portachuelo field, adjacent West Portachuelo pool, and pools of the LaBrea-Pariñas oil field. B, Brazil; C, Colombia; E, Ecuador.

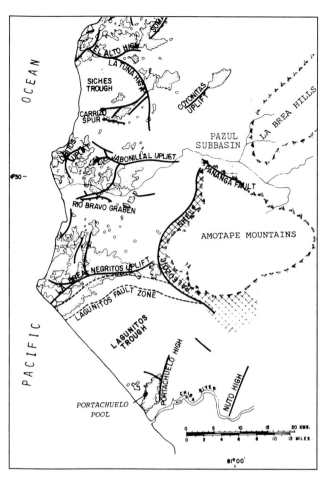

Figure 1B. Tectonic map of southern Talara basin. (After Hay-Roe et al., 1983, with permission of Journal of Petroleum Geology.)

This study is therefore presented as a case history in the use of (1) electric logs restored to the pre-fault sequence; dipmeters; detailed true-scale cross sections together with "straight-line" contour mapping to interpret the intensely faulted structure; and (2) pressure data and fluid-production data to reconstruct the general outlines of entrapment history in the Salina.

Early History

Portachuelo was discovered in 1931, but it is not an old field by Talara standards: The first oil wells in the basin were drilled in 1875. In fact, oil has been known and used there since prehistoric times. When the Spaniards under one of Pizarro's lieutenants arrived early in the sixteenth century, they found the natives mummifying their ancestors and waterproofing their containers with "brea" (pitch) from seeps that were given that name. The La Brea seeps are located some 10 mi (16 km) north of Portachuelo, near the foothills of the Amotape Mountains (Figure 1A).

The modern oil era in this basin began in 1914, when Toronto-based International Petroleum Co. (IPC) acquired the rights of a British firm, London and Pacific Petroleum. That firm had obtained a 99-year lease in 1888 from two Englishmen who had purchased outright both surface and subsurface rights in the "La Brea-Pariñas Estate"—essentially the southern half of the Talara basin—from a local citizen.

The correct taxes to be paid on this property, and even title to the property itself, were never fully agreed on by all parties. Whenever elements of the local or national government reached a settlement with the oil operator, there were always others who disputed the settlement. In 1922 the matter was submitted to an international tribunal headed by a Swiss jurist, and in 1924 that tribunal upheld IPC's right to La Brea-Pariñas.

Standard Oil of New Jersey (Exxon) became operator by acquiring IPC, although the Peruvian Congress never ratified the decision of the tribunal. Ongoing disagreement led eventually to the expropriation of all of IPC's operations in Peru (production,

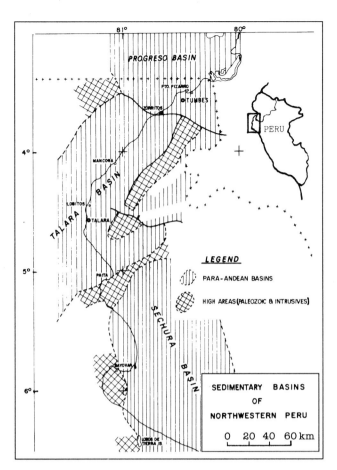

Figure 1C. Sedimentary basins of northwestern Peru. (After Hay-Roe et al., 1983, with permission of Journal of Petroleum Geology.)

refining, and marketing) by a socialist-oriented military government in 1968–1969.

The Discovery of Portachuelo

The discovery well, IPC No. 2570 (Figure 2), was located on a faulted anticlinal nose delineated by surface geologic mapping and subsurface data from a nearby dry hole (IPC No. 980). The discovery well, drilled in 1931, came in for 262 BOPD from sands in the Salina Formation (Figure 3).

But, except for a 7 ac (2.8 ha) south offset drilled in 1932, there was no follow-up for two decades. During the 1930s and 1940s, priorities for Talara geologists and development engineers were allocated through the concept of "rich areas" and "lean areas." In comparison with reservoirs in the "rich" central part of the basin (where wells in the Pariñas sandstone—stratigraphically between the Talara Shale and Pale Greda Formation but missing at Portachuelo—could produce up to 15,000 BOPD), Portachuelo was "lean." Further, there was no pipeline to the refinery, 25 mi (40 km) to the north, and the little wood-burning train ran only twice a week. In any case, more prolific fields closer to the refinery supplied all it could handle.

Ongoing Development

In the early 1950s, IPC returned to Portachuelo, prompted in part by the government's announcement that it was going to open the adjacent area on the south to applications. Seismic, gravity, and photogeologic work helped outline the Portachuelo high. A wildcat to the Pennsylvanian basement was drilled, but despite peaks on the gas detector, the well was not deemed worthy of testing. No reservoir rocks appeared to be present.

Pennsylvanian production was discovered in 1954 by wildcat No. 4610, on the northeast edge of the field (Figure 2). It had an initial producing rate of 26 BOPD from fractured Amotape quartzite (Figure 3).

Information from development wells of the early 1950s in the Salina sand yielded a very sketchy geologic picture. Geologists of that period quickly recognized that stratigraphically the Salina on the southern flank of the basin was different from that farther north: For one thing, it was only one-third as thick. Correlations from well to well proved difficult—with some wells, impossible. Geologists also recognized that fluid content of the sands was largely independent of present-day structural position.

An IPC in-house report written in 1952 stated:

> ... The [Salina] sands change thickness so rapidly laterally that it is difficult to correlate them from well to well ... The 1400-ft spacing of the wells is probably a greater length than the longest diameter of the sand bodies, with the result that a given sand may appear in only one well ... Sand lenses tend to achieve an average thickness, so that many of them look quite similar on the electric log, adding further to correlation difficulties.

On this basis, IPC's geologists developed a model (Youngquist, personal communication, April 1987) involving extremely lenticular sands, presumed to be deltaic, with very local hydrocarbon sources and very short range migration. Exploration largely consisted of guided mapping of sand/shale ratios. Seismic and gravity interpretations on the basin flank supported the concept of a small high area that helped localize the accumulation of hydrocarbons. Some geologists considered that the high had been present at the time of Salina deposition. Drilling established that the pool had a north-northeast trend, not east-west as was previously believed.

Breakthrough: The Composite Log and Detailed Mapping

In the early to mid-1950s, Portachuelo development gained momentum. John D. Tuohy (personal communication, April 1987) recalls, "When I got there a steam rig with a standard derrick was drilling

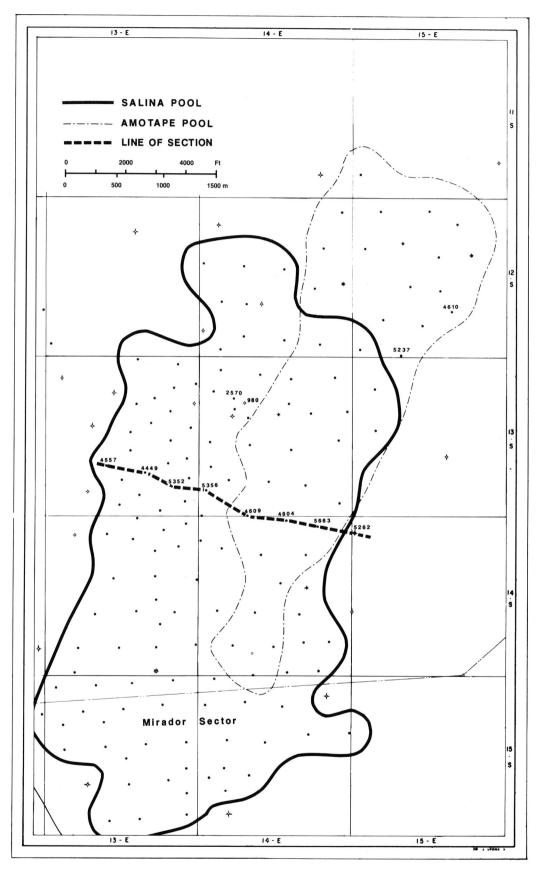

Figure 2. Pool outlines, Portachuelo field. Well numbers shown are those of wells mentioned in the text. Dashed line is cross section, Figure 6. Well 2570 is Salina discovery well, and 4610 is Amotape discovery well. (After Hay-Roe et al., 1983, with permission of Journal of Petroleum Geology.)

SYSTEM	SERIES	GROUP	FORMATION	THICKNESS feet
QUAT.	PL.		TABLAZO	
TERTIARY	EOCENE	LAGUNITOS	MIRADOR	1200+
			CHIRA-VERDUN	3800
		TALARA	TALARA SHALE	50-250
			PALE GREDA	200-1100
		SALINA	SALINA	1600
	PALEOCENE	MAL PASO	BALCONES	500-1500
CRET.	UPPER		REDONDO	100-1100
PENN.	MIDDLE	TARMA	AMOTAPE	6000

Figure 3. Generalized stratigraphic column of the Portachuelo area. (After Hay-Roe et al., 1983, with permission of Journal of Petroleum Geology.)

almost full-time. A couple of wells on the eastern edge of the field had production from the Amotape (Pennsylvanian) as well as from the Salina." In 1956 the field averaged 2678 BOPD from the Salina and 271 BOPD from the Amotape.

About this time a very powerful geologic tool was introduced: the *composite well log* (Figure 4). In the terminology of Talara geologists, this was not the kind of log showing a composite of several curves from the same well, but rather a composite of several wells using the same curves to restore a full stratigraphic section for the vicinity. This "restored" log represented a normal drilled sequence as it would have looked prior to fault shortening. It was used to identify individual lithologic units in other wells in the vicinity and thus to detect both the depths at which faults intersect any particular well in question and the resulting amount of section missing.

The area over which such a log can be used depends, of course, on the lateral persistence of distinctive curve characteristics. In Portachuelo, composite logs of the Salina Formation are most reliable within a radius of a mile or so, but individual stratal sequences with distinctive log features can be correlated over the entire field. For greater reliability, Talara geologists have worked with logs at 2 in. per 100 ft, laying a film transparency of one log over a paper copy of the other. In many areas of the basin the bed-by-bed agreement of curves is so precise (especially in shaly rocks) that the correlation is unchallengeable.

The first composite log developed in Portachuelo by Manuel Paredes (personal communication, March 1987), in the northwestern part of the field, led to new insights into the structural complexity of the area. It became apparent that at least some of the abrupt lateral variations in logged sequences were due *not* to lenticularity, as previously considered, but to fault-shortening.

In the late 1950s, R. B. Travis (personal communication, March 1987) set out to reassemble this three-dimensional jigsaw puzzle. He had to work without benefit of dipmeter surveys, for none were available in Portachuelo. It cost him two full years of complex and painstaking work with a close network of true-scale cross sections interlocked with structure contour maps at different levels within the Salina, but he succeeded in making it all fit together.

One factor that made this task even possible was the lack of any need for correlation above and below the Salina; Travis did not have to attempt to extend the mapping into the pre- and post-Salina intervals, because there was no effective control for mapping. (At the same time, that "advantage" made it impossible to confirm the suspected presence of growth faults or deeper faults.) This meticulous work not only demonstrated that lateral variations between wells were due mainly to faulting, but also guided the development drilling and served as an example for detailed studies elsewhere in the basin.

Subsurface mapping techniques used in the Talara basin involve simplifying assumptions, some of which would be hard to justify from strict conventional geological techniques. But the techniques have remained in use for one good reason: They work. Fault-blocks are treated as rigid and uncurved; contours on top of each block can thus be represented by straight lines (Figure 5), although both strike and dip can vary from one block to the next. Normal faults are likewise assumed to be uncurved planes unless evidence to the contrary is available. Drag on beds next to a fault is usually ignored (again, this assumption evidently works for mapping in this basin).

Both these assumptions permit the use of contour templates, referred to in Spanish as "la guitarra" because the series of evenly spaced, straight parallel lines look like the strings of a guitar. Using one transparent template for bedding dip and another for the plane of a fault, the geologist can quickly plot the trace of these intersecting surfaces. Another extremely handy tool is the apparent dip protractor, which permits rapid graphic solutions of the apparent dip. Cross sections are always constructed with no exaggeration (horizontal and vertical scales the same) so that dips and dip components are correct as shown.

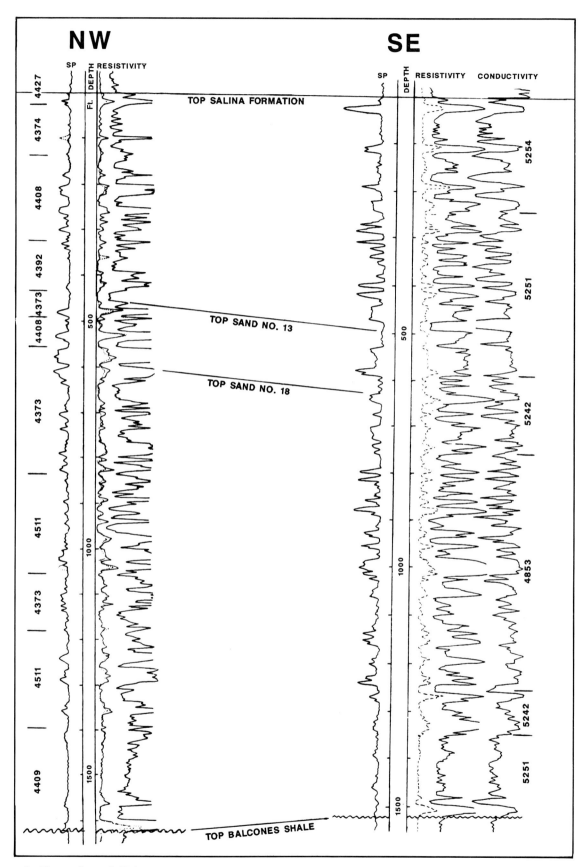

Figure 4. Examples of Talara-style composite logs (restored Salina sequence) from northwestern and southeastern sectors of Portachuelo pool. Numbers along left and right margins are well numbers of logs used to make up the composite. Depths in feet.

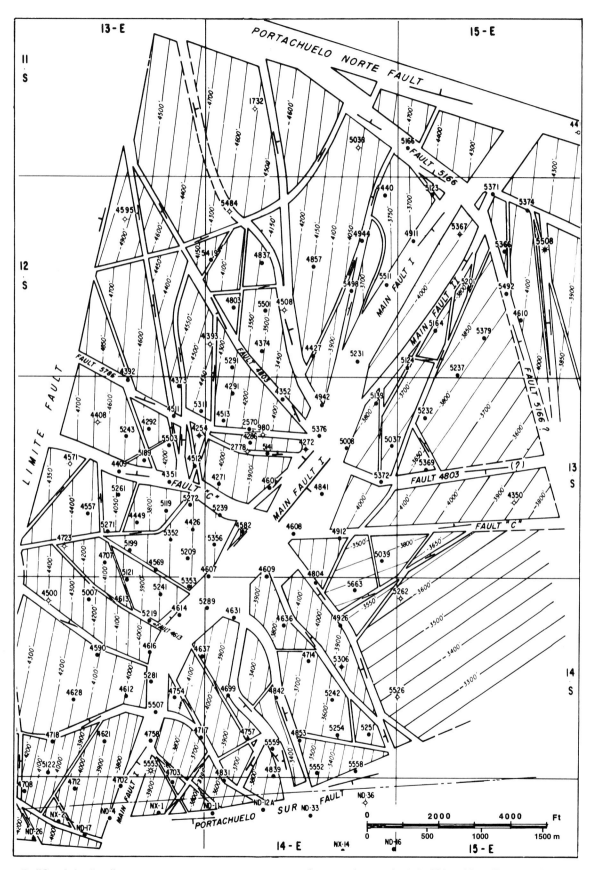

Figure 5. "Straight-line" structure contour map on top of Salina Sand No. 13, prepared by Petroleos del Peru. Contour interval 50 ft. (After Hay-Roe et al., 1983, with permission of Journal of Petroleum Geology.)

Still, because of the lack of wells on the east side of Portachuelo, problems of stratigraphic interpretation remained. A 1960 IPC internal report stated:

> ... Eastward [in] Portachuelo pool the Salina remains about the same thickness as over the higher parts of the structure. It is ... logical ... to explain the thinning [relative to the section along the western edge of the pool] by non-deposition over the eastern [i.e., now the central] part of the pool, due to contemporaneous activity of the Portachuelo High.

To illustrate this concept the report contained a "restored Salina isopach" that showed the original Salina thickness along the crest as only 800 ft, implying that the maximum thinning due to faulting still left 55% of the total original section.

Late Development

During the 1960s, data from the ongoing drilling, together with information from palynology and oil source studies, added new elements to the geologic picture of Portachuelo. The apparent eastward thinning of the Salina, once thought to be depositional, proved to be due to drastic fault-shortening (Figure 6). Beyond this former axial trend the faulting is again less intense, and the drilled Salina sequence is correspondingly thicker (Figure 7). This realization opened up a new area for profitable development on the east side of the field.

In the southernmost part of the field, Belco's (Belco Petroleum Corporation of Peru, which developed the southern sector of the field) drilling showed that the Salina sands were grading into shale southward, in what is now a structurally higher area. In the early 1960s the Salina sequence had been interpreted as thin, fluvial sheet sandstones alternating with shallow marine shales, but later work indicated that the Portachuelo Salina actually represents the distal part of a submarine fan sequence whose sediment source was to the north. Thus the increase in shaliness to the south would be expected.

Additional composite logs helped refine the Salina mapping, but dipmeters triggered a truly major overhaul by providing reliable dips. It turned out that in the previous comprehensive mapping described above, the assumed dip direction was actually 180° off. However, this did not mean that the two years' earlier work was wasted; the stratigraphic correlations and nearly all the fault "picks" in wells were still valid.

The increasingly detailed mapping made it possible to detect undrilled fault blocks with possible commercial size reservoirs, which could then be exploited by infill wells on 13 ac (5 ha) spacing. The tight control in areas of infill wells allowed increasing refinement of the mapping, leading to still more infill drilling and to an arrest of the decline in production. But the logging program of that period was never sophisticated enough to define fluid content with perfect reliability; some sands continued to provide a surprise when perforated.

As feasible sites for infill and edge wells were finally used up, the production from Portachuelo began to decline again. In the fault block fields of the Talara basin, all wells drain reservoir blocks of limited size. It is not unusual for a well to produce half its total reserves in the first 12 months (thereafter the decline is, of course, less drastic).

STRUCTURE

The extreme complexity of the Portachuelo field, like that of other fields in the basin, is due largely to normal faulting. Low-angle gravitational slide faults, while common in post-lower Eocene strata, are not evident in the reservoir units of Portachuelo. Lateral changes in lithology add to the complexity, but their effect is minor compared to the effect of the faulting.

Tectonic History

The Talara basin is predominantly an Eocene feature. In most parts of the basin Eocene strata account for more than 80% of the total thickness of post-Paleozoic sediments.

Since the early Mesozoic, the tectonic history of the coastal sector of northwestern Peru has been governed by its position near the leading edge of the South America plate. The entire basin, only 100 mi (160 km) long and about 30 mi (50 km) wide, represents a minor dimple in the upraised edge of the plate adjacent to the Peru-Chile trench, which marks the subduction zone of the Nazca plate.

Presumably this plate-edge position led to alternating periods of tension and compression in response to the nonhomogeneous westward movement of the plate. Evidence of at least one period of early Tertiary folding was largely erased subsequently by intense normal faulting. During the Oligocene or later time, instability of the upper Eocene sediments led to low-angle gravitational slide-faulting throughout the basin.

As might be expected, shorelines shifted frequently during Eocene deposition in this small basin. Continental, fluviodeltaic, and marine environments followed one another in quick succession. Neogene and Quaternary times have seen predominantly vertical movement, reflected in the multiple terrace deposits now uplifted along the coast.

Regional Structure

The Talara basin straddles the Pacific coastline, with nearly half of its 2800 mi² (7200 km²) offshore (Figure 1C). It is bounded by Paleozoic uplifts on the

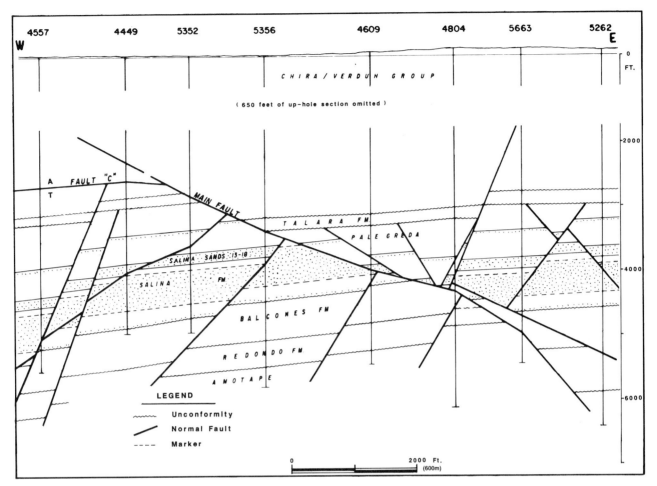

Figure 6. Structure cross section, Portachuelo Salina pool, prepared by Petroleos del Peru. No vertical exaggeration. See Figure 2 for location. Subsurface control inadequate to locate faults at or in the Amotape Formation. (After Hay-Roe et al., 1983, with permission of Journal of Petroleum Geology.)

north, east, and south, and presumably on the seaward side as well. This small, narrow basin is also quite deep: Pre-Tertiary rocks are interpreted to lie below 20,000 ft (6000 m) in the deepest parts.

Portachuelo field is located on a positive feature in the subsurface called the Portachuelo high (Figure 1B). This high is a spur off the major basement ridge to the south, the Paita uplift, which separates the Talara basin from the younger Sechura basin (Figure 1C). A major zone of normal faulting separates the raised southern flank of the Talara basin from the Lagunitos trough, a deep east-west graben to the north. The total relief between the upraised strata of the Portachuelo high and the deepest part of the trough is on the order of 15,000 ft (4600 m).

These two features—the Lagunitos trough and the Portachuelo high—are unusual in that they do not fit the pattern of the major transverse highs and lows of the Talara basin, all of which have a geographic expression: the highs are marked by coastal promontories, the lows by coastal embayments. In contrast, the coastline by the Lagunitos trough and Portachuelo high is remarkably straight (Figures 1A and 1B).

Local Structure

A discussion of the local structure associated with the Portachuelo field necessarily focuses on structure within the 1600 ft (490 m) Salina sequence, since that is where all detailed mapping was concentrated.

Portachuelo field formed as a simple north-plunging anticline, whose crestal area collapsed during intense normal faulting, while the flank areas were (relatively) raised. The resultant present-day structure, contoured conventionally and with faults omitted, is that of a west-dipping homocline with local closures of up to 400 ft (120 m) in parts of the field (Figure 8). The present-day axis at the top of the Pennsylvanian basement underlies the eastern part of the pool, but the crest at the Salina level, shifted by faulting and tilting, actually lies east of the study area.

Conventional contouring gives only a generalized configuration at the level of the mapping horizon—a smoothed average through the real, highly irregular surface, variously tilted and offset by faults.

The main characteristics of the normal faults cutting the Salina are these:

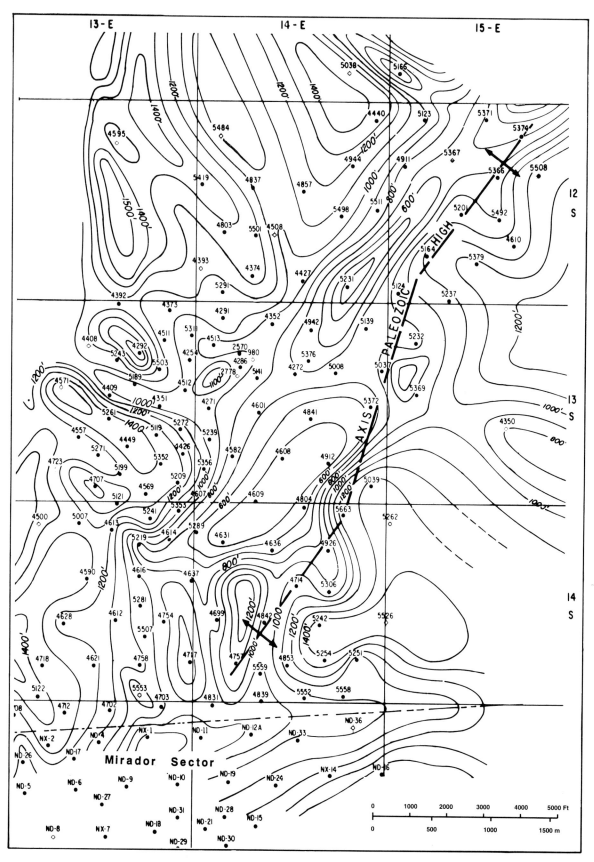

Figure 7. Isochore map, Salina Formation. Contour interval 100 ft. (After Hay-Roe et al., 1983, with permission of Journal of Petroleum Geology.)

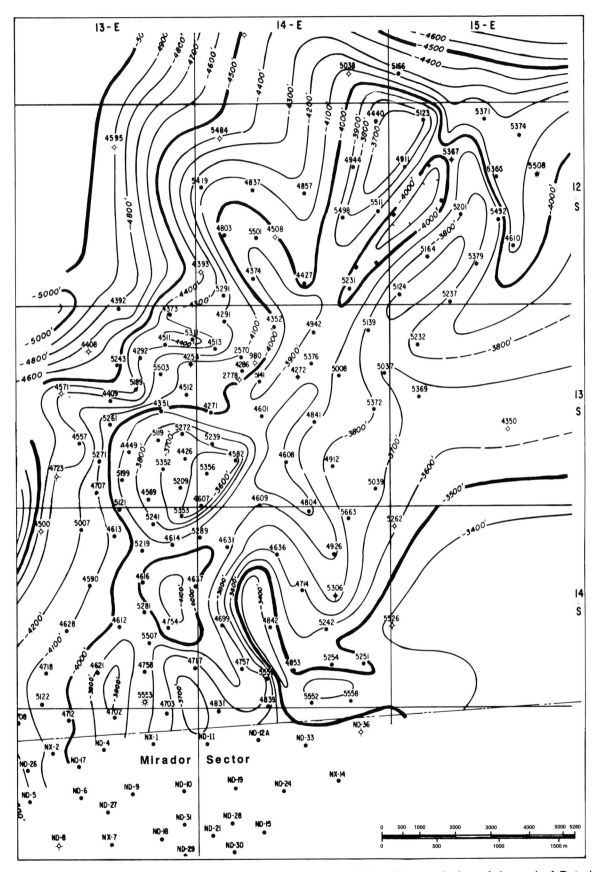

Figure 8. Conventional structure contour map on top of Salina Sand No. 13, prepared by Petroleos del Peru. Contour interval 100 ft. Faults omitted. (After Hay-Roe et al., 1983, with permission of Journal of Petroleum Geology.)

- Faults are normal, with dips mostly in the range of 50–60°.
- Stratigraphic throws range from less than 10 ft to about 1000 ft (300 m). In routine mapping, faults of less than 50 ft (15 m) have been ignored. This cutoff was supported by the next observation.
- Faults with throws of more than 50 ft appear to be effective barriers to fluid communication, whereas smaller faults may or may not be.
- A majority of the bigger faults (throws over 100 ft or 30 m) have a dip component opposite in direction to the prevailing dip of the Salina, thus compensating for the effect of bedding dip on the subsea elevation of any given stratum.
- The age of the main period of normal faulting has not been pinned down with any precision but is evidently latest Eocene or younger.

Compared with conventional mapping (Figure 8), structure mapping "Talara style" (Figure 5) gives a very different picture. The field now appears as an intricate mosaic of west- to northwest-dipping blocks separated by normal faults, including a major one (over 1000 ft or 300 m displacement) roughly parallel to the field axis, with another beyond the northern edge of the productive area.

Variations in the intensity of faulting are clearly reflected in the gross Salina thickness (Figure 7). In places along the east and west flanks of the basement high, where more than 1300 ft (400 m) of Salina is present, the faulting is minor; but near the axis of the basement high, the Salina has been reduced to only a third of its normal thickness, reflecting an increase in both the number and magnitude of normal faults.

From detailed east-west cross sections (such as Figure 6), it is possible to estimate the thinning and stretching of the Salina caused by normal faulting. The calculation is simplified by assuming no north-south component of the extension. If all the thinning and stretching takes place in the plane of the cross section, the cross-sectional area will be proportional to the volume of rock, which remains constant. On this basis the extension works out to an astonishing 52%; the *average* thinning is about 34%. As just mentioned, thinning is much less on the flanks and much more near the crest of the basement high.

STRATIGRAPHY

A stratigraphic column for the field (Figure 3) reflects Portachuelo's position close to the southern basin margin. Most Eocene units are thinned, and some that are present further north (Travis, 1953; Youngquist, 1958) are completely absent here.

The oldest rocks known in Portachuelo are the argillites, quartzites, and (less common) dark limestones of the Pennsylvanian Amotape Formation, which is considered basement throughout the Talara basin. The level of metamorphism has been characterized as "incipient to nonexistent" (Montoya, 1985, p. 153), although older units studied in outcrop do show higher levels.

The true thickness of the Amotape is unknown; it is complexly folded and may also be repeated by thrusting. As overturned sections are present in outcrop, they presumably prevent reliable thickness calculations in the subsurface as well. One Portachuelo deep test, No. 5237, penetrated 6025 ft (1835 m) of Amotape Formation without encountering different or older strata.

Above the major post-Amotape unconformity are two shale units: the Redondo shale of Campanian age, and—separated from it by another unconformity—the Paleocene Balcones shale. The Redondo, considered to be the source of both Salina and Amotape oil, varies in thickness from 100 to 1100 ft (30 to 335 m). Part of this variation may be due to faulting. Upper Cretaceous limestone units known elsewhere in the basin are almost entirely absent around Portachuelo. The Balcones in this area ranges from 500 to 1500 ft (150 to 455 m).

The lower Eocene Salina, which is discussed in the following section, is conformably overlain by a silty shale sequence of light gray color, known as the Pale Greda. In the field area it is 200 to 1100 ft (60 to 335 m) thick.

The post-Pale Greda hiatus in the field area is represented further north, in the central Talara basin fields, by close to 5000 ft (1525 m) of clastics—the prolific Pariñas sandstone and an overlying shale. These two units are entirely missing south of the Lagunitos trough. The succeeding middle Eocene Talara Shale, also nearly 5000 ft thick in the central part of the basin, is scarcely present in the report area, being only 50 to 250 ft (15 to 75 m) thick.

In the central part of the basin the Talara Shale is succeeded by about 2000 ft (610 m) of sand and shale; these units also are completely absent south of the Lagunitos trough. Directly on top of the post-Talara Shale unconformity the Portachuelo drilled sequence has around 3800 ft (1160 m) of late Eocene Chira and Verdun clastics, undifferentiated. They are conformably overlain by the youngest pre-Quaternary strata in the area, the fine-grained clastics of the Mirador Formation (locally over 1200 ft or 365 m thick).

HYDROCARBON TRAP

Like nearly all the oil pools in the Talara basin, the Portachuelo Salina pool has fault-trapping associated with a structural high. The deeper Amotape pool is considered a combination trap (a sub-unconformity accumulation on a structural high, modified by normal faulting).

Reservoirs

Reservoir Strata: Salina

If a reservoir is defined as a volume of permeable oil- or gas-bearing rock that is in pressure communication, then the Salina of Portachuelo is not a reservoir but, instead, hundreds of separate reservoirs, many of which will never be drained (untapped by existing wells, they are too small to develop). In this section we will, nevertheless, follow conventional usage in referring to "the Salina reservoir."

The 1600 ft (490 m) Salina sequence contains up to 45 sandstone units (Figure 4) that constitute the individual reservoirs of the Salina pool. They are predominantly fine-grained, pale gray to greenish gray quartzose sandstones. Some units are friable, others are calcite-cemented and fossiliferous, and still others are shaly and finely laminated or cross-laminated. Medium-grained to pebbly sandstones are less common; they also can be friable or calcite-cemented and fossiliferous. Glauconite is common.

Shales, which predominate in the section, are dark gray to black, silty to sandy, fissile in places, less commonly calcareous. Siltstone units ranging from 6 in. to 3 ft (0.15 to 1 m) in thickness (as seen in cores) are scattered throughout the section. The Salina of the study area is now interpreted as the distal portion of a submarine fan sequence. This interpretation accounts for the fact that floral zonules representing 5000 ft (1525 m) of the proximal section farther north are compressed into only 1600 ft (490 m) around Portachuelo.

Reflecting the lithologic variations mentioned above, sandstone reservoir characteristics also vary markedly, and the available averages do not give the whole picture. As an example, the average permeability of 20 md gives no hint that friable sands can reach values of 1000 md. The average porosity calculated from logs (20%) is one-third greater than the average measured in cores (15%).

During drilling, reservoir fluid content was monitored by cuttings fluorescence, supplemented by chromatography of the gas extracted from the drilling mud with a steam still. As a general rule, even the best producers never displayed spectacular cuttings fluorescence, but complete absence of fluorescence was a fairly reliable sign that the sand was wet.

The detailed geometry of the Salina reservoirs is extraordinarily complex. Since the 45 sands have been divided by faulting into blocks and smaller mini-blocks, the number of individual, separate reservoirs runs into the hundreds. Some have gas-oil contacts; some have oil-water contacts; some have neither, and some may have both. The actual picture becomes even more complicated when one realizes that faults acting as barriers under natural (pre-exploitation) conditions can leak when their entry pressure is exceeded under the high artificial pressure gradients created by flowing and pumping wells.

Reservoir Strata: Amotape

The Amotape Formation in the Portachuelo area consists of argillites and quartzites of Pennsylvanian age. Although the degree of metamorphism is very slight, intergranular porosity is commonly below the commercial range. Production from the Amotape is mostly associated with the highly fractured ortho-quartzite facies, although there are a few zones of argillite sufficiently competent to sustain the open fractures that can store and give up oil. Core studies also revealed some secondary vug porosity related to the fracturing, especially near the post-Amotape unconformity.

Total porosity in the quartzite is reported to average 10% (intergranular and fracture porosity combined); intergranular porosity in cores ranges from 0.7 to 5%. The higher figures were measured in essentially unmetamorphosed quartzitic sandstones. Silica is the most common cement, but calcite is also present in some medium to coarse sandstones. Calcite is likewise the most commonly observed fracture filling.

The average permeability is reported as 4 md, but initial producing rates (reflecting variations in effective transmissibility, as well as variations in bottom-hole pressure from 1250 to 3270 psi [8620 to 22,550 kPa]) range all the way from 20 to 1381 BOPD (Montoya, 1985).

Factors controlling the areal distribution of the quartzite facies (which is relatively local compared to the argillite) were never adequately determined. In the southern sector of the pool, Belco was not able to establish commercial production from the Amotape, in part because of the difficulty of finding the quartzite facies, but probably also because of the presence of limestone between the Amotape and the Redondo Shale (see following section).

Inferred Source Rocks

The inferred source of both Salina and Amotape oil is the Upper Cretaceous Redondo Shale, although Salina hydrocarbons could come in part from younger source rocks as well. Modern geochemical studies have never been stressed in the Talara basin. The few early studies, relatively unsophisticated by the latest standards, suggested that (1) hydrocarbons from the many different Eocene producing units all belonged to the same general family, and (2) the sedimentary section did not include any rocks consistently rich in organic matter (>2 to 3% TOC).

One empirical observation is important in working out the source and migration of oil now stored in the Amotape Formation: The Amotape produces commercially in the basin only where it is directly overlain by the Redondo Shale. For example, there is no commercial production from the Amotape in the southern sector of the pool, where it is separated from the Redondo by about 100 ft (30 m) of limestone.

The metamorphism of Amotape rocks presumably did not have any effect on the contained hydrocarbons. Thin-section study of argillite and quartzite from IPC No. 5237, which penetrated 6025 ft (1838 m) of Amotape, did not reveal any downward increase in the degree of metamorphism, which was only incipient despite the implication in the use of "argillite." In any case, indigenous hydrocarbons would presumably have leaked out during the great post-Pennsylvanian hiatus or would have been destroyed well before the metamorphic grade became high enough to visibly affect minerals seen in thin-section. It thus seems certain that the oil produced from the Amotape was not emplaced there until latest Cretaceous or early Tertiary time.

EXPLORATION AND DEVELOPMENT CONCEPTS

While Portachuelo field does not offer any great lessons for wildcatters (apart from the obvious one that a 21 million bbl field need not be left undeveloped for 20 years), for subsurface geologists its Salina pool provides important insights in two areas: determining optimum well spacing in discontinuous reservoirs and using reservoir-fluid data to work out the history of an accumulation.

In focusing almost entirely on the rocks, subsurface geologists may miss important clues provided by the contained fluids. A useful focus for reservoir studies is to consider the hydrocarbon reservoir as analogous to a plumbing system in which the rocks are the pipes and tanks, and the pressures recorded in the different fluids (including gases) are a measure of the system's stored energy. The payoff from applying this approach to the Portachuelo Salina pool is described in a following section.

Optimum Well-Spacing in a Highly Faulted Reservoir

A highly faulted sequence of thin sands separated by shales is inherently inefficient for recovery of stored oil. The inefficiency can be partly overcome by careful mapping to get an idea of optimum well spacing and to locate undrained fault blocks. Both these approaches were successfully applied in the Portachuelo Salina pool. The following discussion neglects possible reservoir limits caused by facies change, and considers only those limits resulting from faulting.

The average Salina fault-block has an area of much less than 40 ac (16 ha). For more economic drainage, closer well-spacing was required. This was accomplished in two ways. In the prolific western part of the field, already developed on 40 ac triangular spacing, infill wells were drilled on 13 ac (5 ha) centers to tap virgin blocks between the old wells. In the undrilled southeastern part of the field, the development grid was changed to 20 ac (8 ha) rectangular. A good indication of fault-block size is the average area enclosed by faults on a "straight-line" structure map of some horizon in the Salina (for example, Figure 5). Within the pool proper, the average block area is under 30 ac (12 ha) and drops further as smaller faults are included. The most detailed maps, not included in this paper, show an average block size of about 10 ac (4 ha).

The effective size of individual reservoirs does not depend on *all* of the faults, but only on those that act as seals. The kind of information that would be needed to determine precisely which faults are barriers to flow and which are permeable—such as detailed flowmeter and pressure surveys—has never been available. At most it seems feasible to make a few inferences indirectly. On the basis of structure mapping and production histories it appears that most of the faults that cut the Salina with more than 50 ft (15 m) of displacement act as seals, whereas smaller faults may or may not be effective barriers. The problem is not only that of the geometry of each individual fault (i.e., whether sand is faulted against shale to make a seal), but also whether the fault plane itself is cemented by calcite or sealed by tar.

The effective size of individual reservoirs does not depend on *all* of the faults, but only on those that act as seals. The kind of information that would be needed to determine precisely which faults are barriers to flow and which are permeable—such as detailed flowmeter and pressure surveys—have never been available. At most it seems feasible to make a few inferences indirectly. On the basis of structure mapping and production histories it appears that most of the faults that cut the Salina with more than 50 ft (15 m) of displacement act as seals, whereas smaller faults may or may not be effective barriers. The problem is not only that of the geometry of each individual fault (i.e., whether sand is faulted against shale to make a seal), but also whether the fault plane itself is cemented by calcite or sealed by tar.

Actual results confirmed the inefficiency of 40 ac spacing in two ways: First, original per-acre recoveries were too low; second, excellent production was obtained from infill wells on 13 ac (5 ha) spacing. Despite unpleasant surprises with certain wells, the overall program of infill drilling in the Salina pool was a commercial success. The program yielded one outstanding well for every three drilled—the same ratio found in the pool as a whole—and those outstanding infill wells added enough production to pay out the costs of the overall program (Figure 9). As an example, one infill well was located in the center of a triangle whose corner wells had collectively produced over 900,000 bbl and had declined to an average of less than 100 BOPD per well. The infill well came in for 1269 BOPD, paying out its own costs in three months. It produced 187,000 bbl in the first 12 months.

Fluid Distribution

A map of accumulated water production from the Salina (Figure 10) shows the distribution typical of an anticlinal trap: an axial trend of water-free producers, flanked by zones of intermediate water production, with the highest water-cut along the edges. High water production was seldom a problem, even in edge wells. It was often noted that a well starting out with high water-cut would clean up and

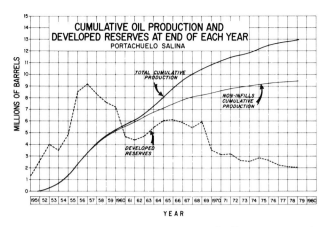

Figure 9. Production curves, Salina reservoir. "Developed reserves" = remaining reserves for sands that are open to production. (After Hay-Roe et al., 1983, with permission of Journal of Petroleum Geology.)

produce less water after continued flowing or pumping. The salinity of Salina formation-water is uniform within Portachuelo. It ranges from 15,000 to 18,000 ppm dissolved solids, with no apparent trends, either lateral or vertical.

Gas sands are more likely to be encountered near the top of the section (Salina sands 1 to 3 are commonly gassy), but there is no consistent level of gas-oil contacts anywhere in the pool. As would be expected, gassy Salina is more common within the trend of water-free oil production, and wells with a very low initial gas/oil ratio (IGOR) are mostly located near the western edge of the field. But maps of IGOR and initial gas production do not show a clear-cut picture like that of water production and are omitted from this report.

The authors consider that the more irregular distribution of gas is due to its higher mobility. At the time of faulting it was more widely redistributed than were oil and water. Also, since both known and presumed gas sands were always avoided in Salina well completions, initial gas production also depended on technical factors such as log-analysis results and inferences of fluid content from correlations with nearby producers.

Reservoir Pressures

Treated from a geologic rather than an engineering perspective, pressure data strongly support the conclusion that Salina hydrocarbons originally accumulated in an anticline.

The engineering treatment of initial pressures is shown in Figure 11—a standard plot of static BHP (bottom-hole pressure, adjusted to an arbitrary depth datum) vs. the time of measurement. Analysis of these values is greatly complicated by the fact that a well can penetrate up to 45 sands (each of which would, in an unfaulted field, be a separate reservoir) cut up into numerous fault-blocks. Some of the hundreds of resulting minireservoirs were drained by earlier wells, others not. Thus each data-point represents some kind of rough average for *all the sands, whether virgin or depleted, that were opened during the initial completion* of the well in question. The sloping dashed lines in Figure 11 enclose a trend that seems to suggest poolwide communication, with later wells coming in at lower initial pressures than early wells. However, four wells on regular spacing plot are above the "trend," and all but three of the infill wells are above it—many are far above. Clearly some other control is at work here.

A suite of maps of exploitation chronology (unpublished, Hay-Roe, 1965) provide a clue. They show how Salina development began near the northern end of the pool and proceeded southward and southeastward. But the mystery is still not cleared up until initial pressure distribution is displayed *geographically* (Figure 12). This map—a geological treatment of pressure data—has the distribution of values that would exist if an anticlinal accumulation were flattened by faulting, with *minimal readjustment of the pressures to their new depths*. The dozen wells marked by crosses had initial pressures 200 to 300 psi (1380 to 2070 kPa) lower than the normal pressure indicated for the vicinity and are judged to have been opened in one or more depleted or partly depleted sands.

Blocks located high on the former anticline (where the *absolute* pressure was less) were dropped to a lower position. Flank areas, which started off with a higher absolute pressure by virtue of their depth, were raised. Although some sands undoubtedly suffered pressure adjustment (i.e., fluid loss) during faulting, the majority still reflect the "fossil" pressure-distribution of a collapsed anticline.

In conjunction with this pressure distribution map, the exploitation chronology can now explain the apparent trend of initial bottom-hole pressures in Figure 11. Purely by chance, the discovery well for Portachuelo was in the north end of the pool, where initial pressures are high. Drilling proceeded south-southeastward toward the low pressure area whose center is well No. 4839. Thus, the *chronological* arrangement of wells completed through 1957 shown on Figure 11 indicates a consistent downward trend. Wells 4628 and 4804 are off the trend precisely because they are located on the higher-pressure east and west flanks, respectively.

Now it is clear why the wells drilled after 1960 do not fit the "trend" of Figure 11: They did not follow the geographic pattern of development. For example, well No. 5242 (right side of Figure 11) came in more than 1000 psi (7000 kPa) *above* that "trend." The fact that the initial pressure of No. 5242 is 350 psi (2450 kPa) *below* that predicted by Figure 12 suggests that some sands must have been partly drained by earlier wells.

Inferred History of Accumulation

The foregoing evidence supports the concept of an anticlinal accumulation in the Salina Formation of

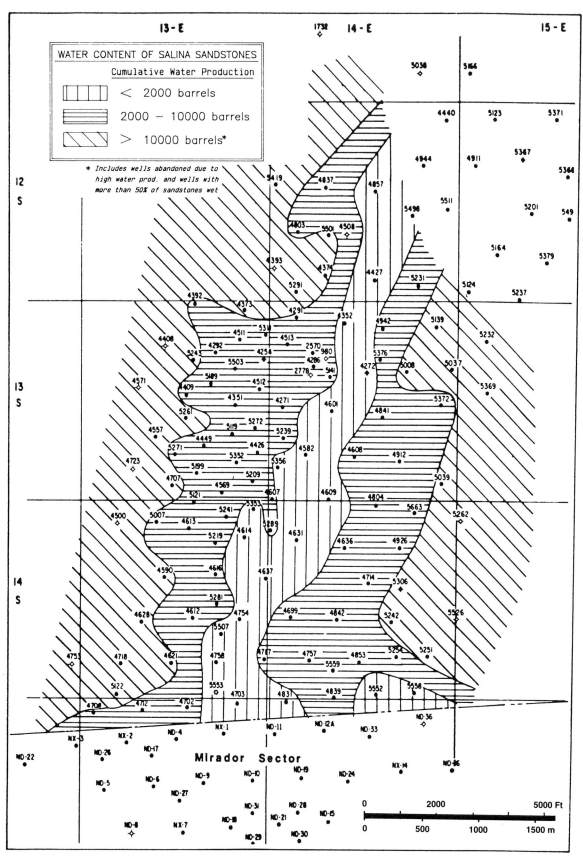

Figure 10. Cumulative water production, Portachuelo Salina pool. (After Hay-Roe et al., 1983, with permission of Journal of Petroleum Geology.)

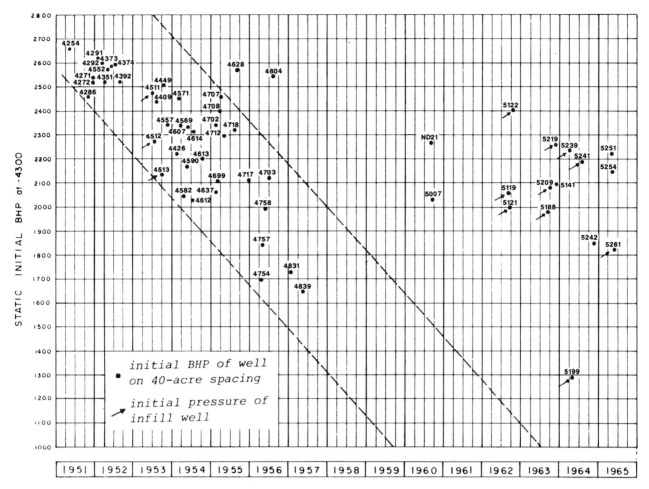

Figure 11. Initial bottom-hole pressures (psi), Salina Formation, vs. time of measurement. (After Hay-Roe et al., 1983, with permission of Journal of Petroleum Geology.)

Portachuelo (Figure 13A). The anticline, genetically related to the Paleozoic high, plunged north-northeast. By the time hydrocarbon migration into this structure took place, the shales separating the reservoir sands must have been sufficiently compacted to prevent appreciable vertical migration of the oil phase.

With cross-formational migration blocked, each of the 45 sands would act as a separate reservoir, having independent gas-oil and oil-water contacts whose positions would depend on the relative abundance of oil and gas that happened to migrate into that particular sand on the structure. The end result would be something like that represented in Figure 13A, which for simplicity portrays only a portion of the total Salina sequence.

This comparatively simple fluid distribution was enormously complicated by the post-early Eocene extensional faulting that flattened the Salina anticline until it was unrecognizable, although it did not totally eliminate the Portachuelo high. Presumably there was some escape of oil and gas (as well as water) along temporarily permeable fault planes, or directly across faults from one sand to another, during and immediately after each phase of the faulting. A very rough volumetric calculation suggests that approximately 40% of the original oil may have been lost from the reservoir. The redistribution of fluids into the present complex arrangement is suggested schematically in Figure 13B.

ACKNOWLEDGMENTS

Portachuelo's subsurface geology has been an ongoing challenge, a three-dimensional jigsaw puzzle with which many talented geoscientists have struggled. We particularly benefited from the work of former colleagues R. B. (Russ) Travis, Manuel Paredes, Lizardo Munoz, and Adolfo Perret; we also picked the brains of "old Talara hands" George H. Tappan, J. Fernando Zuñiga-Rivero, J. Roger Palomino, W. L. (Walt) Youngquist, John D. Tuohy,

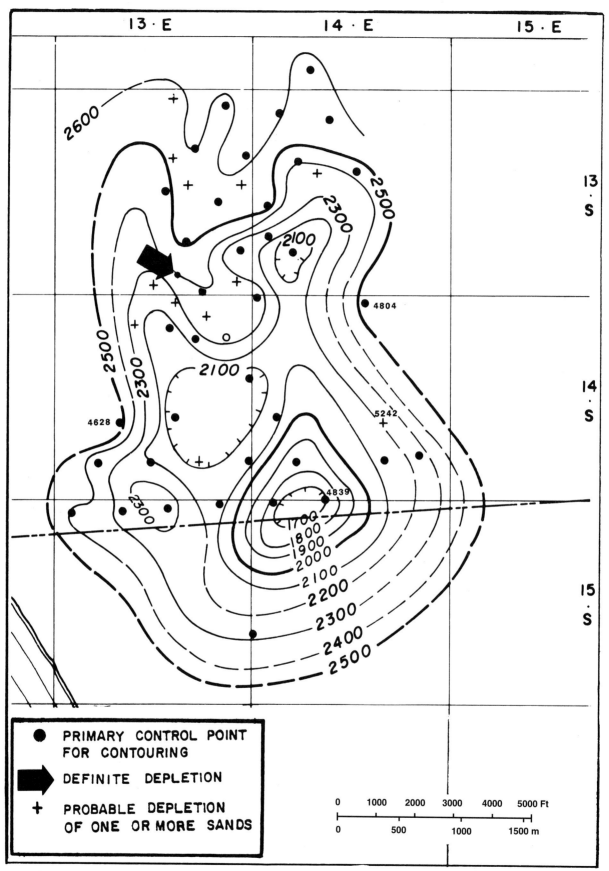

Figure 12. Initial bottom-hole pressure map, Salina Formation. Contour interval 100 psi (dashed where inferred). (After Hay-Roe et al., 1983, with permission of Journal of Petroleum Geology.)

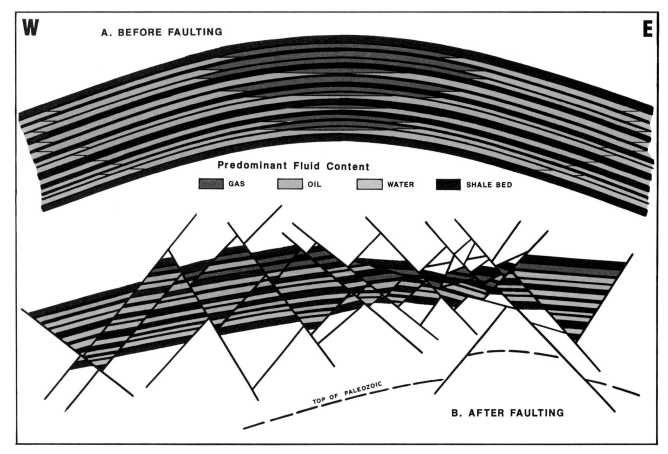

Figure 13. (A) Highly generalized schematic cross section of a portion of the Salina Formation of Portachuelo pool, prior to faulting. No scale. (B) Same strata as in A, after faulting and redistribution of fluids. No scale. Subsurface control inadequate to locate faults at or in the Amotape Formation.

Viktor Petters, Paul M. Tucker, and J. M. (Jim) Dorreen. Marco Antonio Raez provided valuable information on the southern sector of the field, developed in the late 1950s and 1960s by Belco Petroleum Corporation of Peru. To all these geoscientists, our thanks for their generous help.

Data for this study came from the files of Exxon Company International and Petroleos del Peru. We are grateful to Dave Amsbury, Bill St. John, and Melba (Jerry) Murray for their critical reading of the manuscript.

REFERENCES CITED

Hay-Roe, H., J. F. Zuñiga, and A. Montoya, 1983, Geology and entrapment history of the Portachuelo Salina pool, Talara basin, Peru: Journal of Petroleum Geology, v. 6, p. 139–164.

Montoya, A., 1985, Exploración y desarrollo del pre-cretáceo en la Cuenca Talara del noroeste del Perú: ARPEL Technical Bulletin, v. 14, n. 2, p. 145–159.

Travis, R. B., 1953, La Brea-Pariñas oilfield, NW Peru: AAPG Bulletin, v. 39, p. 2093–2118.

Youngquist, W., 1958, Controls of oil occurrence in La Brea-Pariñas oilfield, northern coastal Peru, in Habitat of oil: AAPG, p. 696–720.

Appendix 1. Field Description

Field name .... Portachuelo field
Ultimate recoverable reserves .... 21 million bbl

Field location:
    Country .... Peru
    State .... Department of Piura, northwestern Peru
    Basin/Province .... Talara basin

Field discovery:
    Year first pay discovered .... Salina Formation 1931
    Year second pay discovered .... Amotape Formation 1954

Discovery well name and general location:
    First pay .... IPC #2570, sq. mi 13S-14E, La Brea-Pariñas sector; 20 mi SSE of port of Talara

Discovery well operator .... International Petroleum Co. (IPC) (Exxon)

IP
    First pay .... 262 BOPD
    Second pay .... 26 BOPD

All other zones with shows of oil and gas in the field:

Age	Formation	Type of Show
Paleocene	Balcones Formation	Oil
Campanian	Redondo Formation	Oil

Geologic concept leading to discovery and methods used to delineate prospect

The 1931 discovery well, 2570, was based on surface geologic mapping plus limited subsurface data from IPC well #980, an early (pre-1920) abandonment. After the drilling of one 7 ac (2.8 ha) offset to the discovery well, development was suspended for 20 years. By the time it was resumed in 1951, gravimetry and photogeology had aided in outlining the structure. The second (deeper) pay was suspected from shows in bit cuttings and was confirmed by the drilling of well 4610 after the structural high was further delineated by seismic.

Structure:
    Province/basin type .... III Rift (Bally); III B b Transform rifted convergent margin (Klemme)
    Tectonic history

The Talara basin developed during the Paleogene near the leading edge of the South America plate where it overrides the Nazca plate. The main cycle of subsidence/deposition was in the Eocene. The Neogene and Pleistocene were marked by normal faulting and uplift.

    Regional Structure

The productive structure lies on the southern flank of the Talara basin, separated from the main part of the basin by the major transverse graben called the Lagunitos trough (Figure 1).

    Local Structure

The field lies high on the west flank of a highly faulted, asymmetric, northward-plunging nose.

Trap:
    Trap Types

The original trap was a simple anticline, subsequently broken up by intense normal faulting into numerous fault-block reservoirs that, for the most part, are not in pressure communication with one another.

Basin stratigraphy (major stratigraphic intervals from surface to deepest penetration in field):

Chronostratigraphy	Formation	Depth to Top in ft (m)
Upper Eocene	Lagunitos Group	0
Middle Eocene	Talara Shale	2700 (825)
Lower Eocene	Pale Greda (shale)	2900 (885)
	Salina (shale/sand)	3300 (1005)
Paleocene	Balcones Shale	4600 (1400)
Upper Cretaceous	Campanian Redondo Shale	5000 (1525)
Pennsylvanian	Amotape (quartzite/argillite)	5300 (1615)

Reservoir characteristics:

- **Number of reservoirs**
- **Formations** .... Salina, Amotape
- **Ages** .... Salina, early Eocene; Amotape, Pennsylvanian
- **Depths to tops of reservoirs** .... Salina, 2800-3800 ft (855-1160 m); Amotape, 4900-5800 ft (1495-1770 m)
- **Gross thickness (top to bottom of producing interval)** .... Salina, 1000 ft (305 m) (highly variable); Amotape, 500± ft (150± m)
- **Effective net thickness** .... Salina, variable; Amotape, variable
- **Lithology** .... Salina, friable, fine- to coarse-grained quartzose sandstones; Amotape, highly fractured orthoquartzite (some argillite)
- **Porosity type** .... Salina, intergranular; Amotape, fracture porosity (minor intergranular)
- **Average porosity** .... Salina, 15% from cores, 18-23% from logs; Amotape, less than 5% intergranular porosity
- **Average permeability** .... Salina, 15-25 md, locally up to 1000 md; Amotape, matrix perm. less than 0.5 md, no data on fracture perm.

Seals:

- **Upper**
    - Formation, fault, or other feature .... Salina, Pale Greda and intra-Salina shales; Amotape, Redondo shale
- **Lateral**
    - Formation, fault, or other feature .... Salina, normal faults; Amotape, normal faults

Source:

- **Formation and age** .... Salina, Redondo, Campanian; Amotape, Redondo, Campanian
- **Lithology** .... Shale
- **Average total organic carbon (TOC)** .... NA
- **Maximum TOC** .... NA
- **Kerogen type (I, II, or III)** .... NA
- **Vitrinite reflectance (maturation)** .... NA
- **Time of hydrocarbon expulsion** .... NA
- **Present depth to top of source** .... 5000 ft (1525 m)
- **Thickness** .... 300 ft (90 m)

Appendix 2. Production Data

- **Field name** .... Portachuelo field
- **Field size:**
    - **Proved acres** .... Salina, 3500 ac (1415 ha); Amotape, 1800 ac (730 ha)
    - **Number of wells all years** .... Salina, 103; Amotape, 18; Dual Zone, 24 (includes shut-in wells; total wells in field, including 19 dry holes: 164)

PORTACHUELO

Current number of wells	NA
Well spacing	40 ac triangular, with infill wells on 13 ac spacing
Ultimate recoverable	21,000,000 bbl (both reservoirs)
Cumulative production	Salina, 11,400,000; Amotape, 5,200,000; Dual Zone, 2,200,000 (exact figures not available)
Annual production	1,000,000 bbl (both reservoirs)
Decline rates of individual wells	Highly variable (a function of fault-block size, number of individual reservoirs perforated, and pressure in each reservoir)
Annual water production	NA
In place, total reserves	NA
Primary recovery	2,200,000 (both reservoirs)
Secondary recovery	No estimate
Enhanced recovery	No estimate
Cumulative water production	Salina, 450,000 bbl; Amotape, NA; Dual Zone, NA

Formation evaluation:

Logging suites	Induction electric log; sonic log with caliper; cased-hole neutron and casing/collar locator; logs supplemented by sidewall cores in sands of uncertain fluid content
Testing practices	Drillstem tests rare; zones to be completed normally determined from log analysis
Mud logging techniques	Drilling fluid samples in sealed containers taken in to lab for steam-still extraction and gas chromatography

Oil characteristics:

API gravity	Salina, 36° (range 29-39°); Amotape, 23° (range 19-30°)
Base	Salina, paraffin; Amotape, paraffin
Initial GOR	Salina, 300 to 3000; Amotape, NA
Sulfur, wt%	Salina, 0.06; Amotape, NA
Viscosity, SUS	NA
Pour point	Salina, 80°F; Amotape, NA
Gas composition	Methane, 96.0%; ethane, 2.2%; propane, 0.7%; butane, 0.2%; pentane, 0.7%; CO_2 0.2%

Field characteristics:

Average elevation	40 ft (12 m)
Initial pressure	Salina, 2660 psi at datum -4300 ft (18,300 kPa at -3110 m); Amotape, 3270 psi at datum -5000 ft (22,500 kPa at -1525 m)
Present pressure	Salina, highly variable; Amotape, NA
Temperature	Salina, 135°F (57°C); Amotape, 145°F (63°C) (variable)
Geothermal gradient	0.015°F/ft (0.027°C/m)
Drive	Solution gas (both reservoirs)
Oil column thickness	Highly variable (both reservoirs)
Oil-water contact	Salina, multiple; Amotape, -5200 to -6100 ft (-1585 to -1860 m)
Connate water	Salina, 50% (est.); Amotape, highly variable
Water salinity, TDS	Salina, 15,000-19,000 ppm; Amotape, 15,000-17,000 ppm

Transportation method and market for oil and gas:

Crude oil and produced gas move by company-owned pipeline 20 mi (32 km) north to company-owned refinery/petrochemical complex; products go to the domestic market..

Raguba Field—Libya
Sirte Basin

PHILIP BRENNAN
Consultant
Longboat Key, Florida

FIELD CLASSIFICATION

BASIN: Sirte
BASIN TYPE: Cratonic Sag on Earlier Rifted Basin
RESERVOIR ROCK TYPE: Sandstone, Limestone, and Orthoquartzite
RESERVOIR AGE: Cretaceous and Paleozoic
PETROLEUM TYPE: Oil
TRAP TYPE: Faulted Anticline
RESERVOIR ENVIRONMENT OF DEPOSITION: High Energy, Shallow Marine

LOCATION

The Raguba field lies within the North African coastal state of Libya (official name: Socialist People's Libyan Arab Republic). Libya has a long northern coastline on the Mediterranean Sea and is bounded on the east by Egypt and Sudan; on the west by Tunisia and Algeria; and on the south by Chad and Niger. Geologically, the Raguba field lies within the Sirte basin, a Cretaceous-Tertiary basin that occupies much of northeastern Libya and extends offshore into the Gulf of Sirte, a southerly embayment of the Mediterranean Sea (Figure 1). Other fields in the same part of the basin include the Nasser (originally Zelten), Dor Marada, and Haram oil fields and the Hateiba gas field (Figure 2).

Much of central and interior Libya is desert, and the Raguba field lies in typical desert terrain approximately 150 km (95 mi) southwest of the Gulf of Sirte oil terminal of Marsa el-Brega, to which the field is connected by pipeline. The Raguba field lies in the southeast corner of Petroleum Concession 20, Mining Zone II, originally granted in 1955 to Libyan American Oil Company (Figure 2). The field encompasses a surface area of approximately 15,000 ac (60 km^2; 23 mi^2), with surface elevations ranging from 62 to 100 m (200 to 330 ft). Estimates of recoverable reserves at Raguba range from 750 million bbl of oil (Clifford, 1984) to 1.083 billion bbl of oil equivalent (Carmalt and St. John, 1986), the latter figure comprising 1.0 billion bbl of oil plus reservoir gas converted to oil equivalent. The Raguba field was ranked as 216th among the largest known oil fields of the world by Carmalt and St. John (1986).

HISTORY

Pre-Discovery

In spite of the antiquity of human settlement along the littoral of present-day Libya, there are no known historical records of surface manifestations of hydrocarbons in the region. According to Waddams (1980), the first subsurface indication of hydrocarbons was a show of methane recorded in a water well drilled in western coastal Libya by the Italian colonial administration during 1914. In the years following World War I, the Italian authorities began a program of geologic mapping, and this culminated in the publication during 1934 of the first comprehensive geologic map of the then-colony. By 1937, the Italian state oil entity, Agenzia Generale Italiana Petroliche, or AGIP, had become interested in the oil and gas prospects, largely on the basis of continued shows of natural gas in wells drilled in search of water. AGIP divided Libya into 12 zones of varying promise, of which the "Sirtica basin" was considered to be the most prospective (Waddams, 1980).

The incipient Italian exploration effort was terminated by the outbreak of World War II. No petroleum exploration was carried out during the war years nor during the postwar period of military administration that preceded the emergence of Libya as an independent state in 1951. During 1953, the Libyan government became aware of interest on the part of the international oil industry and passed a basic Minerals Law under which qualified oil companies could undertake limited exploration programs, not including exploration drilling. Among

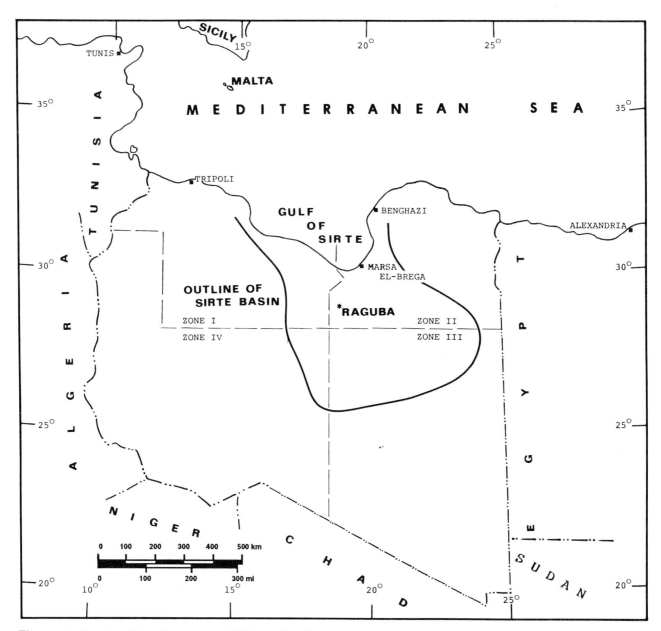

Figure 1. Regional location map of Libya, showing outline of Sirte basin, boundaries of Mining Zones, and location of Raguba field.

the oil companies taking advantage of this law were Standard Oil of New Jersey (later Esso and then Exxon) and Texas Gulf Producing Company, companies soon to be jointly concerned with the discovery and development of the Raguba field.

The subsequent Libyan Petroleum Law of 1955 divided the country into four Mining Zones (Figure 1), of which zones I and II covered the northern coastal region and extended inland to latitude 28°. These northerly zones contained the principal Libyan population centers and were substantially more accessible than the more remote interior desert areas of zones III and IV. The northern part of the country was also better known geologically from the earlier Italian mapping and the surface work carried out by the oil companies over the period 1953 to 1955.

These factors combined to ensure highly competitive bidding for acreage in zones I and II when concession applications were accepted by the Libyan authorities during 1955. At government instruction, the many conflicts and overlaps on these initial concession applications were resolved by intercompany negotiations (Eicher et al., 1975).

The first Libyan concessions were awarded at the end of 1955. Among them was Concession 20, Mining Zone II (Figure 2), awarded to Texas Gulf Producing Company, operating in Libya as Libyan American Oil Company or Liamco. Libyan American had previously applied for concessions in the hill country of northeastern Libya, where prominent surface structures were known to exist, and the company's application for concessions in the Sirte basin

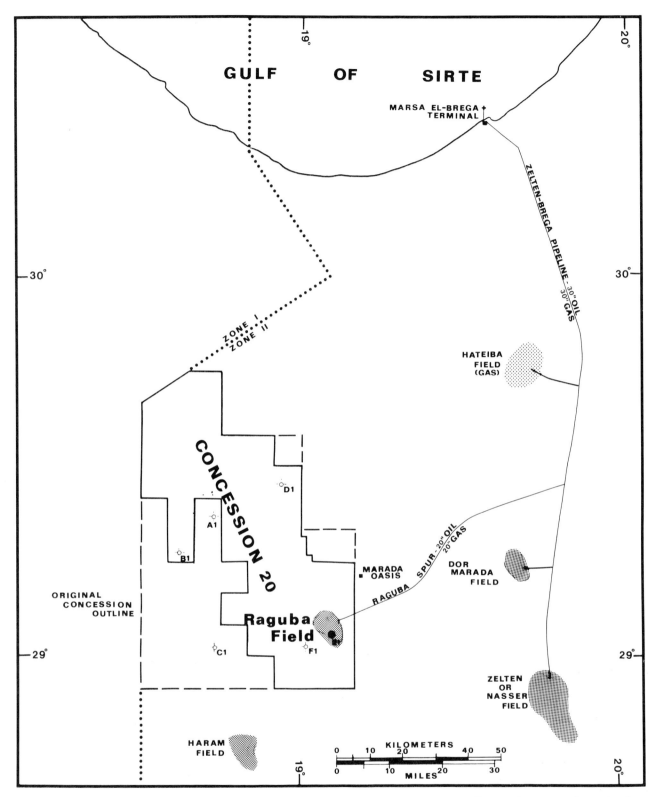

Figure 2. Position map of Concession 20, Mining Zone II, Sirte basin, showing location of Raguba and other fields, key wildcat wells, and connecting pipeline system.

(including Concession 20) was made on a speculative basis without benefit of any prior geologic investigation (Lester, 1990, personal communication).

Concession 20 covered a surface area of 1,162,876 ac (4708 km^2; 1817 mi^2). In common with other Libyan concessions, Concession 20 had a primary term of 50 years but was subject to a mandatory relinquishment of 25% of the original area after only five years. This latter provision was designed to encourage sustained exploration activity on the part of

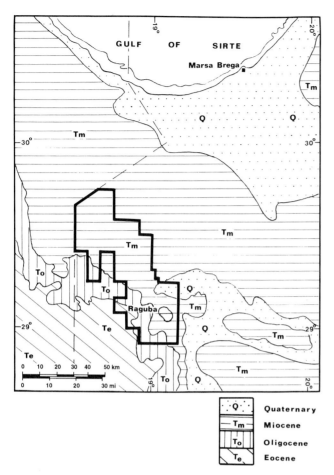

Figure 3. Surface geologic map of Concession 20 and adjacent areas of Sirte basin.

concessionaire companies. It was reinforced by a requirement that exploration on any given concession must commence within eight months of the date of award. In the case of Concession 20, awarded 12 December 1955, the deadline for the commencement of exploration operations became 12 August 1956.

Surface geologic mapping of Concession 20 began during 1956 and continued until the entire concession area had been covered on a reconnaissance basis. Bedrock throughout the concession area proved to be of Tertiary age, overlain by local developments of aeolian sand that included a substantial dune field in the southeastern part of the block. The Tertiary strata exposed ranged in age from middle Eocene through middle Miocene, striking northwest-southeast across the concession and dipping gently north and northeast at an average rate of less than 1° (Figure 3). Lithologies mapped included fossiliferous marls; coquinas; earthy, detrital and crystalline limestones; green shales and green gypseous clays; and occasional calcareous sandstones. No surface indications of hydrocarbons were noted in the course of the field work, nor was there any substantial evidence of surface structure, even over the site of what would become the Raguba field (although a surface nosing was mapped in the field area in the course of a later, more detailed, survey). It was clear from an early stage that seismic exploration would be required to guide any exploration drilling undertaken on the concession.

The initial seismic coverage of Concession 20 was of a reconnaissance nature, employing a single CDP crew with dynamite as the energy source. The survey was designed to produce a general structural interpretation of the entire concession block, including the outlining of subsurface prospects for seismic detailing and future wildcat drilling, and the identification of areas suitable for surrender at the end of the first five-year term in December 1960.

Seismic data from this reconnaissance survey ranged from fair to poor in quality and appeared to indicate that the greater part of Concession 20 comprised a platform or shelf on which subsurface structures were few and of low relief. Along the eastern edge of the concession, dips in the deeper part of the section steepened abruptly toward what was interpreted to be a hinge line between the platform or shelf area to the west and a depocenter lying off toward the east. This zone of steepened dip was termed the Marada flexure (Figure 4), after a nearby oasis of that name (Figure 2). The postulated depocenter to the east coincided with the site of a prominent northwest–southeast-trending Bouguer gravity minimum and was referred to as the Marada or Hagfa trough (Figure 4) (Brady et al., 1980).

The first three wildcat wells drilled on Concession 20 (A1-20, B1-20, C1-20) were all located on the shelf, in the central and southwestern parts of the block (Figure 2). All three wells penetrated a basically similar section of Tertiary carbonates, evaporites, and shales, underlain by minor thicknesses of Upper Cretaceous rocks, and all three wells bottomed in dense sedimentary quartzite of unknown age, at depths ranging from 1700 to 1820 m (5600 to 6000 ft). This bottom formation, informally known as the "Sirte quartzite," was accepted as economic basement in the area, and the three wildcat wells were completed dry and abandoned without encountering any significant shows of hydrocarbons.

A fourth wildcat test, D1-20, was drilled during 1960 in the northeastern part of the concession (Figure 2) on a minor seismic closure located adjacent to the zone of steepened dip considered to mark the eastern edge of the shelf. This well penetrated an expanded section of Tertiary rocks and also encountered a thin basal section of oil-stained bioclastic limestone immediately overlying the Sirte quartzite at total depth of 2245 m (7365 ft). This thin limestone interval contained abundant specimens of the benthonic foraminifer *Omphalocyclus macroporus*, indicating a Late Cretaceous age.

The Cretaceous oil show at D1-20 focused attention on an area 43 km (27 mi) to the southeast, where seismic data indicated the presence of a low-relief subsurface anticlinal structure, extending out to the very edge of the shelf. It was reasoned that a structure in this position should be favorably located to trap hydrocarbons migrating up the steep gradient from the depocenter lying immediately to the east; this

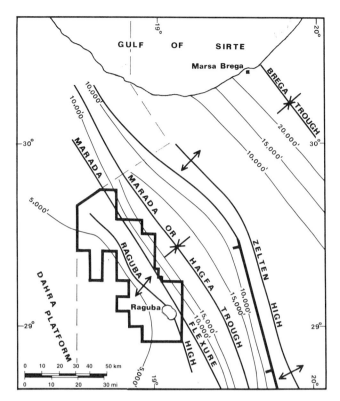

Figure 4. Structural setting of Concession 20. The Raguba high lies along the eastern edge of the Dahra platform, adjacent to the depocenter of the Marada/Hagfa trough. (Partially after Brady et al., 1980.)

was the Marada/Hagfa trough, already known from wildcat drilling by other operators to contain substantial thicknesses of Tertiary and Cretaceous shales. The presence of an effective seal or cap rock, in the form of basal Tertiary shales, had already been demonstrated in wells drilled elsewhere on Concession 20 and could be predicted to occur across the seismic anomaly in the southeast corner of the block. This seismic feature was programmed for drilling as the fifth wildcat on Concession 20; it was named the Raguba prospect and the test well was designated E1-20, in accordance with Libyan requirements. E1-20 was to be the discovery well of the Raguba field (Figure 2).

Discovery

The E1-20 well was located at latitude 29°04′33″ N, longitude 19°05′10″ E, in an uninhabited desert area 130 km (80 mi) south of the Mediterranean shore and 70 km (43 mi) northwest of the Zelten (Nasser) Paleocene oil field, discovered in 1959. The nearest human settlement to the E1-20 location was the oasis of Marada, 20 km (12 mi) to the northeast (Figure 2). Ground elevation at the well site was 77 m (251 ft) and K.B. elevation 81 m (265 ft). The well was operated by Esso Sirte Inc., which had acquired a 50% interest in the concession during 1959. Libyan American Oil Company, the original concessionaire,

now held a 25.5% interest; the other minority interest participant was W.R. Grace & Company with 24.5%.

E1-20 was spudded during November 1960. Beneath a thin cover of aeolian sand, the well encountered fossiliferous limestones of early Miocene age and thereafter penetrated the predicted sequence of Miocene to Paleocene carbonates, evaporites, and shales. An oil show was recorded in middle Eocene carbonates and a gas show in a carbonate interval developed within a Paleocene shale section; neither show appeared to offer the prospect of commercial production.

At a drilling depth of 1610 (-1528) m (5279, -5014 ft), the E1-20 well entered a section of bioclastic limestone with minor amounts of quartz sandstone; the section was porous and oil-stained and flowed gas and light crude oil on drill-stem test. Continuous coring established that the clastic carbonate section extended to a depth of 1680 (-1600) m (5512, -5247 ft), at which depth the well encountered quartzitic sandstones/sedimentary quartzites, dense in appearance but carrying live oil in well-developed fractures. Drilling was terminated at 1895 m (5563 ft), 15 m (51 ft) into the quartzitic sandstone section; however, it was clear from the results of continuous testing that the E1-20 well was still within the oil column at total depth (subsea elevation 1615 m, 5298 ft).

The 71 m (233 ft) of oil-bearing bioclastic limestone present was clearly of shallow-water origin and contained a benthonic microfauna indicating a Late Cretaceous age. The section was assigned to the Zmam formation of intracompany terminology and was informally termed the "Raguba clastics" during the early phase of field development (Figure 5). The underlying section of dense quartzitic sandstone was informally termed the "Sirte quartzite" (Figure 5); it was accepted as being of probable Paleozoic age, with its upper surface representing an important regional unconformity (refer to section on *Stratigraphy*).

Following completion of drilling and coring, the Raguba E1-20 discovery well was logged, cased to bottom, and perforated for production over the 12 m (40 ft) interval 1658–1670 m (5440–5480 ft). The well was then tested at various flow rates on various chokes; the official initial potential established was 2250 BOPD on ½-in. choke, gravity 40.5° API. Testing of the discovery well was completed on 18 January 1961.

Post-Discovery

Following the successful completion of the Raguba E1-20 discovery well, a long stepout was drilled 10 km (6 mi) to the northwest. This well, E2-20 (Figure 6), was clearly located on the downthrown side of a fault lying somewhere to the northwest of E1-20 and not readily apparent on the original seismic. E2-20 drilled an expanded section of lower Paleocene shale, failed to establish the presence of the Late Cretaceous reservoir section, and encountered the surface of the

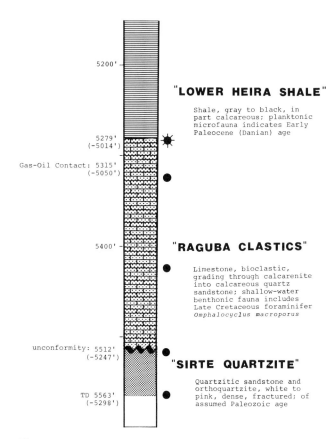

Figure 5. Reservoir section encountered in E1-20 (Raguba) discovery well (terminology shown represents original field usage—refer to Figure 8).

underlying quartzitic sandstone section at a subsea elevation of 1888 m (6195 ft), 290 m (948 ft) low to its occurrence at E1-20.

A second stepout, E3-20 (Figure 6), was then drilled 3.2 km (2 mi) to the northeast of the discovery; this well encountered no Upper Cretaceous clastics but passed straight from lower Paleocene shale into quartzitic sandstone at a subsea elevation of 1522 m (4995 ft), 77 m (252 ft) high to the E1-20 subsea elevation for this horizon of 1600 m (5247 ft). The upper part of the quartzitic sandstone section at E3-20 proved to be oil-bearing, apparently in fracture porosity, and the section was drilled and tested down to a subsea elevation of 1690 m (5546 ft), where saline water was encountered, suggesting the presence of an oil-water contact.

A third stepout well, E4-20 (Figure 6), was drilled 3.2 km (2 mi) southwest of, and presumably downdip from, E1-20. The E4-20 well encountered the top of the Upper Cretaceous "Raguba clastics" at a subsea elevation of 1637 m (5371 ft), well within the vertical limits of the oil column already proved by the E1-20 well. At E4-20, however, the Upper Cretaceous section was tight down to a subsea elevation of approximately 1707 m (5600 ft), below which depth a limited porosity development tested salt water (and was in turn underlain by additional tight section, still of Late Cretaceous age, above the top of the quartzitic sandstone section at -1786 m (-5859 ft).

The presence of tight section within the anticipated reservoir interval at E4-20 was discouraging, but the excellent producing characteristics of the "Raguba clastics" at the E1-20 discovery well clearly dictated a need for further appraisal drilling.

A fifth well, located 3.2 km (2 mi) south-southeast of the E1-20 discovery, was more successful. E5-20 (Figure 6) encountered the top of the target "Raguba clastics" at a subsea elevation of 1588 m (5210 ft), with 50 m (164 ft) of oil-bearing section present above the surface of the quartzitic sandstone section at -1638 m (-5374 ft). A sixth well was then drilled 2.8 km (1.75 mi) northeast of the E3-20 "quartzite completion" (Figure 6), to confirm rollover on the Raguba anticline and to determine whether the Upper Cretaceous clastic section might be present on the northeast flank of the structure. This well, E6-20 (Figure 6), confirmed the anticipated northeast dip, encountering the surface of the quartzitic sandstones at a subsea elevation of 1616 m (5302 ft), 94 m (307 ft) low to the same point in E3-20. But E6-20 encountered no Upper Cretaceous section and was abandoned after finding the upper part of the quartzitic sandstone section tight. It was clear that, on an immediate basis, development drilling would have to be confined to the southwest flank of the Raguba oil field structure.

By early 1962, the Raguba field contained a total of 20 wells, comprising 14 commercial oil completions and 6 dry holes. Of the 14 oil wells, 12 had established production in the Upper Cretaceous "Raguba clastics," while two produced from the quartzitic sandstone section, now generally accepted as being of Paleozoic age. Testing had confirmed the presence of a primary gas cap, with a gas-oil contact at a subsea elevation of 1539 m (5050 ft) (Figure 6); an oil-water contact had been established at a subsea elevation of 1677 m (5502 ft) (Figure 6). This indicated an oil column 138 m (452 ft) in thickness; the total hydrocarbon column, gas plus oil, established at this time was 207 m (679 ft).

The position of the major normal fault cutting across the trend of the structure and bounding the producing area to the northwest had been indicated by a later generation of seismic control and confirmed by drilling (Figure 6); the throw on this fault appeared to be in the order of 244 m (800 ft). At least one additional normal fault, of lesser throw, was suggested by seismic and drilling to be present in the southeastern part of the structure (Figure 6).

An unsuccessful wildcat test, F1-20 (Figure 6), drilled approximately 5 km (3 mi) southwest of the E4-20 dry hole, had proved the Upper Cretaceous section to exceed 365 m (1200 ft) in thickness. Of this section, some 127 m (418 ft) appeared to correlate directly with the "Raguba clastics," the remainder apparently representing "older" Upper Cretaceous. That the wedge of "Raguba clastics" tapered gradually up the southwest flank against the rising surface of the underlying quartzitic sandstones had been demonstrated by the development drilling; E11-20, located northeast of the E1-20 discovery well,

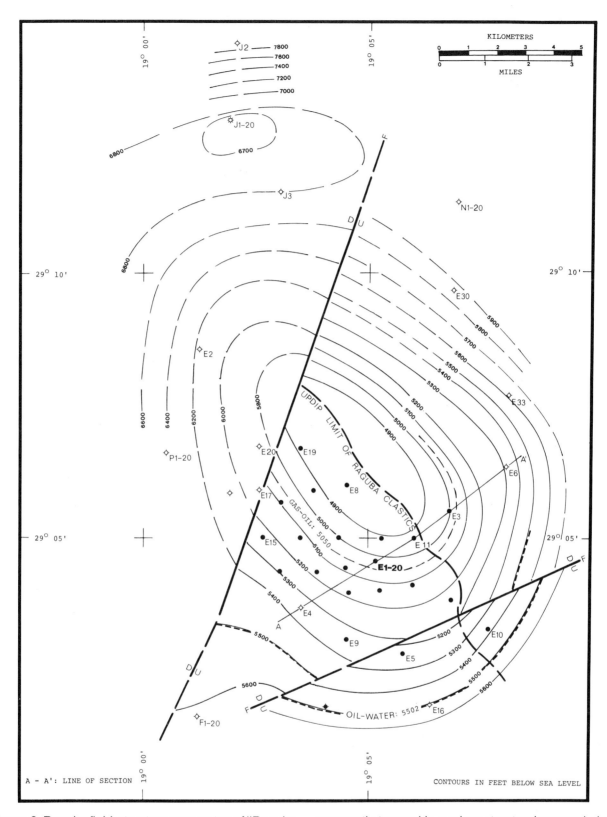

Figure 6. Raguba field; structure map on top of "Raguba clastics" (Kalash Formation)/Hofra quartzitic sandstones, based on results of early drilling. An unrelated gas accumulation is present at the J1-20 location, apparently trapped in a minor structural reversal along the downfaulted northwesterly plunge of the Raguba structure. Contour interval, 100 and 200 ft.

appeared to lie virtually at the feather-edge of the Upper Cretaceous section (Figure 6).

Volumetric calculations based on the results of this drilling suggested an oil-in-place reserve of 1.25 to 1.5 billion stock tank barrels (STB), with an indicated recoverable reserve of 460 to 550 million STB. The Raguba field was declared commercial, and a decision was reached to move the production to tidewater via a spur pipeline running northeast from the field to connect with the existing 76 cm (30-in.) pipeline system linking the Zelten field (later renamed the Nasser field) with the Marsa el-Brega export terminal on the Gulf of Sirte (Figure 2). The 51 cm (20-in.) Raguba spur was laid during the second half of 1962, and the Raguba field was ready for initial production at the end of January 1963, two years from the date of completion of the E1-20 discovery well.

The Raguba field went on production during February 1963; during the balance of that year, the field produced at an average rate of 48,000 BOPD. Within four years, average daily oil production had reached 100,000 bbl, and cumulative production had exceeded 100 million bbl. The field reached its peak production level of 128,000 BOPD during 1968 and 1969 and was considered fully developed by 1972, when a total of 79 wells had been drilled within the field limits, of which 67 were capable of commercial oil production. Through 1972, cumulative production of oil exceeded 350 million bbl and cumulative production of associated natural gas reached 325 bcf.

Initially, the increasing volume of associated gas produced with the crude oil was flared at the field site, since plans for re-injection of gas into the reservoir remained incomplete and no alternate use for the gas existed. During this period, rates of production of associated gas rose from 40 MMcfpd to more than 100 MMcfpd. At a later date, this associated gas production was piped to a gas-liquefaction plant at the Marsa el-Brega terminal via a new 51 cm (20-in.) gas pipeline laid alongside the existing oil line connecting the Raguba field with the Zelten (Nasser)-Brega system (Figure 2). With the commissioning of the Brega gas plant during 1971, flaring of produced gas at Raguba field effectively ceased.

By the end of 1978, 16 years after the start of production, the Raguba field had produced 500 million bbl of oil and approximately 470 bcf of natural gas; this amounted to two-thirds of the estimated ultimate recoverable reserve of the field. By that same year, the daily rate of oil production had fallen to 65,000 bbl and continued to show a steepening decline to about 45,000 BOPD during 1980 and to less than 20,000 BOPD during 1982. However, the excessively low production rates for 1981 and 1982 (Table 1) also reflected operational difficulties, as a Libyan state oil company assumed control of the now-nationalized field from the foreign concessionaire operator.

The new operating company for the Raguba field, Sirte Oil Company, has elected not to report production figures for individual Libyan fields, so that the performance of the Raguba field over the past several years is difficult to assess. However, the field's cumulative production through 1982 was 556 million bbl of oil and 550 bcf of associated gas, so that large future increases in field production rates should probably not be anticipated.

DISCOVERY METHOD

The Raguba hydrocarbon discovery of 1961 was the result of wildcat drilling on a subsurface anomaly outlined by reconnaissance seismic surveys and subsequently detailed by additional seismic coverage. The concession had originally been acquired on a speculative basis; subsequently (and prior to the drilling of the Raguba discovery well), the concession had been covered by reconnaissance surface geologic mapping, supplemented by limited programs of aerial photographic, airborne magnetometer, and gravity meter coverage. No surface structure was readily apparent in the immediate area, where the Tertiary bedrock was at least partially obscured by deposits of aeolian sand. The limited gravity data available indicated that the Raguba seismic anomaly was associated with a fairly prominent positive gravity trend.

The Raguba subsurface prospect was identified in the course of rapid seismic reconnaissance coverage of a large concession lying within a "new" basin that was undergoing active exploration but about which relatively little was known. By the time a decision was reached to drill at Raguba, the pre-Tertiary form of the Sirte basin had begun to be understood, and oil had been discovered in the basin in both Paleocene and Cretaceous horizons. By this time also, well data from Concession 20 itself were suggesting a possible play at Cretaceous level along the flexure or hinge line between what appeared to be a stable shelf underlying the western part of the concession and a prominent trough or depocenter lying toward the east and northeast (Figure 4). It was recognized that the Raguba prospect was favorably located along this indicated hinge line play. A component of regional geologic thinking therefore reinforced the geophysical interpretation in preparing the way for the successful exploration drilling of the Raguba subsurface prospect.

STRUCTURE

Geologic Setting

The Sirte basin of northeastern Libya (Figure 1) is a late Mesozoic–Cenozoic fault or rift basin developed on a surface of Precambrian and eroded Paleozoic rocks. The basin is situated on the northern foreland of the central African craton (Barr, 1972) and contains in excess of 7600 m (25,000 ft) of predominantly marine upper Mesozoic and Cenozoic

Table 1. Raguba field production statistics.

Year	Annual Prod./bbl	Avg. BOPD	Cumulative bbl.
1963	15,977,596	43,774	15,977,596
1964	26,660,618	73,043	42,638,214
1965	34,839,142	95,450	77,477,356
1966	34,898,809	95,613	112,376,156
1967	39,110,499	107,152	151,846,664
1968	46,804,056	128,230	198,290,720
1969	46,668,350	127,859	244,959,070
1970	44,299,032	121,367	289,258,102
1971	35,839,327	98,190	325,097,429
1972	34,076,591	93,361	359,174,020
1973	30,192,320	82,719	389,366,340
1974	20,586,000	56,400	409,952,340
1975	18,182,325	49,705	428,094,665
1976	24,519,072	67,175	452,613,737
1977	24,856,500	68,100	477,470,237
1978	23,724,833	65,000	501,195,070
1979	22,098,272	60,543	523,293,342
1980	16,270,970	44,578	539,564,312
1981	9,744,658	26,698	549,308,970
1982	6,454,000	17,682	555,762,970
1983	7,300,000*	20,000*	563,062,970
1984			
1985			
1986			
1987			
1988			
1989			
1990			

*Production for 1983 is estimated figure; no figures for the field have been released by the operator for years 1982 through 1990.

sediments in its deeper segments (Clifford et al., 1980).

The area of the present Sirte basin was evidently emergent from late Paleozoic through much of Early Cretaceous time (Barr, 1972), with continuous uplift having created a major positive element known as the Sirte arch (Figure 7). From the beginning of Late Cretaceous time, extensional tectonics caused a partial collapse of the Sirte arch (Hea, 1971); this collapse created a system of horsts and grabens, mainly following a northwest-southeast grain. One of the linear structural lows resulting from these movements was the Marada or Hagfa trough located immediately east and northeast of the Raguba field (Figure 4).

Marine incursion from the north began during early Late Cretaceous time, at first following the troughs or grabens and later transgressing on to the intervening and structurally higher areas. Synsedimentary movements along at least some of the fault zones contributed to differences in thickness and facies of Upper Cretaceous strata (Roberts, 1970); some of the original or reactivated fault blocks must have remained emergent as islands in the transgressing Late Cretaceous seas until the end of Cretaceous time.

Widespread submergence followed the Cretaceous transgression, resulting in the deposition of thick sections of lower Paleocene shales across much of the western Sirte basin and especially in structural lows such as the Marada/Hagfa trough (Figure 4). Movement continued along major pre-Cretaceous fault lines, but structure at Cretaceous level was progressively modified by the effects of post-Cretaceous sedimentation; in the area of Concession 20 and the adjacent Marada/Hagfa trough, the "deep" structure was effectively obscured by the end of Eocene time.

Raguba Structure

The structure containing the Raguba hydrocarbon accumulation is an ovoid anticline measuring approximately 23 × 18 km (14 × 11 mi) at Cretaceous level. The axis of the structure is oriented northwest-southeast, paralleling the trend of the adjacent Marada flexure and reflecting the dominant structural grain within this part of the Sirte basin (Figure 4). Structural relief may exceed 300 m (1000 ft) on all flanks; a minimum of 240 m (790 ft) of relief has been demonstrated along the critical southwest flank, where dips on top of the Raguba clastics range between 1° and 2°. The northwestern third of the structure has been downfaulted toward the northwest along a prominent normal fault that trends northeast-southwest across the axial trend, with a throw of approximately 244 m (800 ft); the main

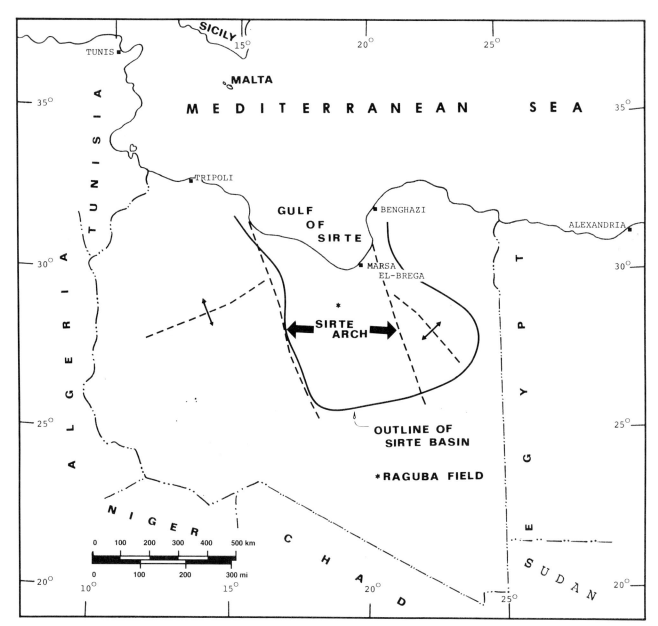

Figure 7. Outline of Sirte basin, Libya, showing position of pre-Mesozoic Sirte arch. The partial foundering of the Sirte arch during Late Cretaceous time created a number of linear troughs, including the Marada/Hagfa trough (Figure 4).

producing area of the field lies south and east of this fault. A smaller, but still important, normal fault cuts across the southeast portion of the producing area on an east-northeast to south-southwest trend; this is an up-to-the-southeast fault, with throws in the range of 30–45 m (100–150 ft) (Figure 6).

The Raguba oil field structure is located on a generalized high trend, running from northwest to southeast through Concession 20, west of and more or less parallel to the Marada flexure (Figure 4). The Raguba structure has a core of quartzitic sandstones and sedimentary quartzites of Paleozoic age, the weathered surface of which represents an important regional unconformity (Figure 8); the quartzitic sandstone core of the Raguba structure was probably a topographic high on this erosional surface.

Throughout most of the Late Cretaceous transgression, the Raguba area remained emergent, persisting as an island with deep water deposits being laid down in the trough to the east and shallow water transgressive deposits accumulating preferentially along the island's western and southwestern flanks. These transgressive Late Cretaceous units included the high-energy deposits of bioclastic limestones with subordinate quartz sandstones informally termed the "Raguba clastics" (Figures 5, 8, and *Stratigraphy* section). These onlapping Upper Cretaceous sediments do not appear to have reached the apex of the pre-existing Paleozoic high.

Major structural growth at Raguba occurred during early Cenozoic time, probably in response to increased downwarping of the adjacent Marada/

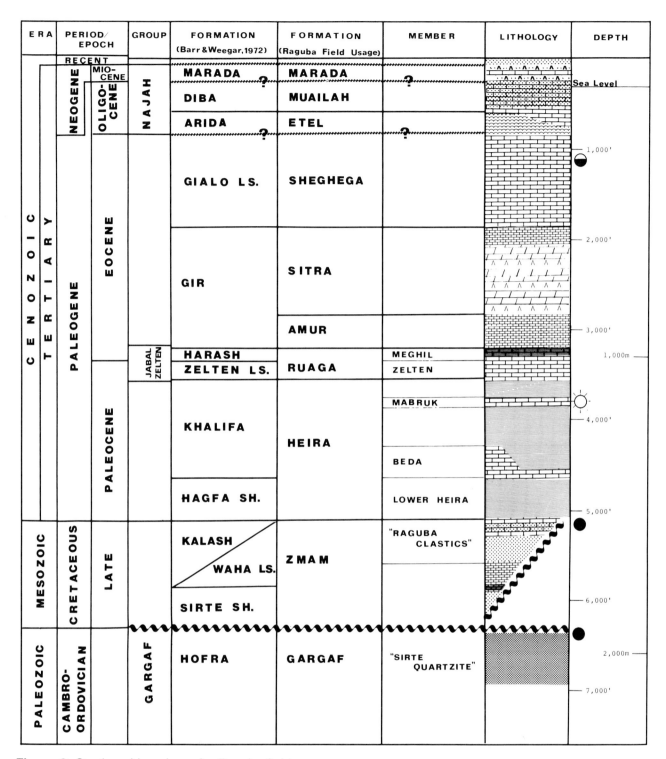

Figure 8. Stratigraphic column for Raguba field area, relating early "field" terminology to terminology published by Barr and Weegar (1972).

Hagfa trough; the principal faults affecting the structure also became active at this time. The sediments being laid down across the growing structure were shales, so that the elements of the hydrocarbon trap at Raguba were in place before the end of the Paleocene.

As deposition continued, structural relief over the buried high became less well defined with time, a situation compounded by regional tilting that had the effect of displacing the Raguba crest progressively toward the southwest (Figure 9). This tilting, and the increasingly restricted environment of deposition,

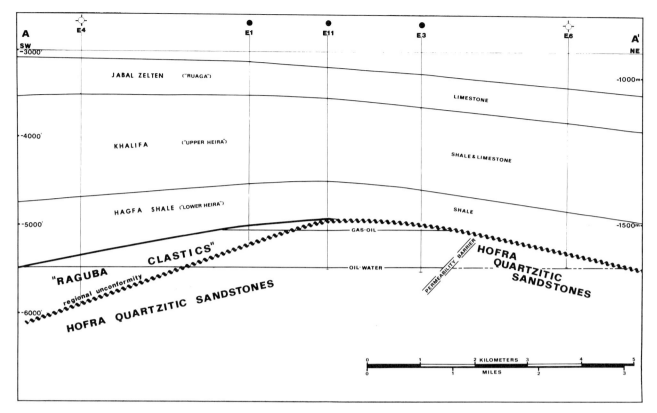

Figure 9. Southwest-northeast structural cross section through Raguba field, showing relationship of "Raguba clastics" (Kalash Formation) reservoir to underlying Hofra Formation quartzitic sandstone reservoir.

heralded the general retreat of the seas from the greater part of the Sirte basin during later Cenozoic time. The progressive loss of structural relief persisted throughout Paleogene time, so that the Raguba structure had generally lost its identity by Neogene time. The lack of structural expression above the buried high at the present-day land surface has already been noted.

STRATIGRAPHY

In the crestal area of the Raguba subsurface structure, 1615 m (5300 ft) of Cenozoic strata unconformably overlie pre-Mesozoic section that has been penetrated to a depth of only 180 m (600 ft). The Cenozoic section thickens gently off the southwest flank, where it is conformably underlain by a wedge of Upper Cretaceous strata; this Upper Cretaceous wedge thickens from a feather edge high on the flank to 380 m (1255 ft) at the F1-20 well control point, 5 km (3 mi) southwest of the field area. At this point, the combined Cenozoic-Mesozoic section is approximately 2210 m (7250 ft) thick. An expanded Cenozoic section, some 1980 m (6500 ft) thick, is developed northwest of the main producing area, where the extension of the Raguba structure has been downfaulted toward the northwest (Figure 6). The Cenozoic section thickens abruptly across the Marada flexure into the Marada/Hagfa trough (Figure 4).

The pre-Mesozoic rocks subcropping basal Paleocene shales across the crest of the Raguba structure comprise white, gray, and pink quartzitic sandstones and orthoquartzites; these rocks are fine grained and generally dense, except where subaerial weathering has created minor secondary porosity. Local fracturing is also present in these rocks across the crestal area, presumably induced during uplift and folding of the structure.

Originally given the informal name "Sirte quartzite," these quartzitic sandstones are now correlated with the Hofra Formation, which has its type section in the Mobil Hofra A1-11 well 100 km (63 mi) northwest of Raguba; the Hofra Formation has been assigned a probable Cambrian-Ordovician age and is included in the Gargaf Group (Barr and Weegar, 1972) (Figure 8). The eroded surface of the Hofra Formation at Raguba and elsewhere in the Sirte basin is a major regional unconformity, the result of a cycle of erosion initiated toward the close of Paleozoic time. At the crest of the Raguba high, this Paleozoic erosion surface appears to have remained exposed until the end of Cretaceous time. The hydrocarbon column of the Raguba field extends into the upper part of the Hofra Formation quartzitic sandstones, and this interval is discussed in greater detail in the section on *Reservoirs*.

Late Cretaceous deposits at Raguba comprise a wedge of bioclastic limestones and quartz sandstones directly overlying the eroded surface of the Hofra Formation along the flanks of the structure. This section, informally termed the "Raguba clastics," forms the principal reservoir section in the Raguba field and is treated in greater detail in the section on *Reservoir Stratigraphy*. Early practice was to include these "Raguba clastics," together with additional underlying Upper Cretaceous section, in the Zmam Formation; the Zmam appears to be the equivalent of the Kalash Limestone Formation and the underlying Waha Limestone Formation of Barr and Weegar (1972) (Figure 8).

The marked deepening of the sedimentary environment at the beginning of Paleogene time resulted in the widespread deposition of gray marine shales carrying a planktonic microfauna of early Paleocene age. The contact between these shales and the underlying Kalash Formation ("Raguba clastics") appears essentially conformable; the contact is clearly unconformable where the Upper Cretaceous clastic sediments are absent and the basal Paleocene shales lie directly upon the eroded surface of the Hofra Formation quartzitic sandstones (Figure 9). The lower Paleocene shales pass conformably upward into a thick section of similar shales interbedded with crystalline to bioclastic limestones, also indicated to be of Paleocene age.

This entire Paleocene section, up to 600 m (2000 ft) thick in the Raguba field area, was originally named the Heira Formation, the lower shale section being termed the "lower Heira" or "Danian shale." The name Heira Formation was never formalized, but has appeared in the literature (Terry and Williams, 1969; Bebout and Poindexter, 1975; Gumati and Kanes, 1985). The "lower Heira" shale section, which ranges in thickness from 90 to 245 m (300 to 800 ft), correlates with the Hagfa Formation of Barr and Weegar (1972), and the "upper Heira" shale-limestone section with the Khalifa Formation of the same authors (Figure 8).

One of the more prominent limestone units within the "upper Heira" or Khalifa Formation carries dry gas in the Raguba field area. Regionally, this gas-bearing limestone clearly correlates with the "Mabruk member" or "first Mabruk pay," named for its occurrence in the Mabruk field of Concession 17, 200 km (125 mi) northwest of Raguba. This gas-bearing section will be treated in greater detail in the section on *Trap*.

The "upper Heira"/Khalifa shale-limestone section is conformably overlain by shallow water limestones with subordinate shales, ranging in thickness from 120 to 150 m (400 to 500 ft) and carrying a benthonic microfauna indicative of a very late Paleocene to very early Eocene age. This section was named the Ruaga Formation at Raguba and is clearly the equivalent of the Ruaga Limestone Formation of Bebout and Poindexter (1975), who divide it into the Zelten (lower) and Meghil members. These in turn appear to equate with the Zelten Limestone and Harash formations of the Jabal Zelten Group of Barr and Weegar (1972) (Figure 8).

The Jabal Zelten/Ruaga carbonates are conformably overlain by a thick (up to 600 m; 2000 ft) section of nummulitic limestones with a central section of evaporites. Of early Eocene age, this section was originally referred to as two formations, the Amur and Sitra, but appears to correlate with the Gir Formation of Barr and Weegar (1972). Shallow marine conditions persisted through middle Eocene time, and the Gir is succeeded by up to 300 m (1000 ft) of nummulitic limestones, originally named the Sheghega Formation but apparently equivalent to the Gialo Formation of Barr and Weegar (1972). The Gialo is overlain, with possible disconformity, by shallow-marine limestones of Oligocene age, interbedded with clay shales at the base and with glauconitic sandstones at the top. This section was assigned to the Etel and Muailah formations of field terminology and appears to equate with the Arida and Diba formations of Barr and Weegar (1972). The latter authors include these two formations, together with the overlying Marada Formation of Miocene age, in their Najah Group (Figure 8).

The Marada Formation, which lies unconformably on the Oligocene Diba Formation, comprises a 60–90 m (200–300 ft) section of sandstones, sandy limestones, shales, and gypseous clays, indicating a restricted shallow marine to continental environment. The formation forms the bedrock in the Raguba field area, beneath a variable but generally thin cover of Recent aeolian sand.

The combined Cretaceous–Tertiary section in the Concession 20 area of the Sirte basin represents a single sedimentary cycle, which commenced with transgression in early Late Cretaceous time and ended in regression during early Miocene time. The Late Cretaceous marine transgression came from the north, from the ancestral Mediterranean (Tethyan) basin. The early Miocene regression was back toward the north, and the youngest part of the Sirte basin extends offshore into the present-day Mediterranean Sea along the bight of the Gulf of Sirte (Figure 1).

TRAP

The hydrocarbon trap at Raguba is structural, an ovoid anticline with its longer axis oriented northwest-southeast. The structure was subjected to moderate regional tilting toward the northeast during Cenozoic time, so that critical closure would be expected to lie along the southwest flank; however, the demonstrated structural relief along the flank, between the crest and the F1-20 control point, exceeds the height of the hydrocarbon column. Adequate closure also exists along the northeast and southeast flanks, where dips average close to 3°. Toward the northwest, critical closure is provided by a down-to-the-northwest normal fault with a throw exceeding the height of the hydrocarbon column. Vertical and lateral seals are provided by Paleocene shales, which directly overlie the reservoir across most of the Raguba structure and which are downfaulted against the producing section along the original northwestern plunge. A permeability barrier occurs within the reservoir along the northeast flank, owing to decreased intensity of fracturing within the Hofra quartzitic sandstone section (Figures 6 and 9).

The area within the total structural closure at Raguba is large, probably exceeding 50,000 ac (200 km^2; 78 mi^2). At the projected limits of the oil-water contact, the structure measures 14 by up to 12 km (8.5 by up to 7.5 mi), and encompasses approximately 30,000 ac (120 km^2; 47 mi^2). However, not all of this area is productive; the principal producing area lies along the southwest flank, east of the major down-to-the-northwest fault and extending southeast along the southeast plunge of the anticline (Figure 6).

The Raguba high apparently had some topographic expression on the Paleozoic erosion surface prior to the deposition of the transgressive Upper Cretaceous section. By the close of Cretaceous time, Cretaceous sedimentation along the southwest flank had effectively buried any pre-existing relief. Substantial structural growth by uplift, rather than compression, took place during early Paleogene time, when much of the faulting affecting the structure was initiated (or rejuvenated). Because of this early growth, the Raguba structure was in a position to trap migrating hydrocarbons following deposition of the early Paleocene Hagfa Formation ("lower Heira") cap rock shale section.

The intensity of folding decreased upward through the section; by the end of the Paleocene into early Eocene time, the structure had assumed the form of a gentle dome, centered northward of the crest of the deeper subsurface high. The domal folding in the upper part of the Paleocene section created a secondary structural trap in a limestone unit ("Mabruk member") of the Khalifa Formation or "upper Heira" (Figure 8), at a subsea elevation of approximately 1060 m (3500 ft). This trap contains dry gas only and, in its present form, covers some 13 km^2 (5 mi^2), with a gas column of about 30 m (100 ft).

Reservoirs

The producing intervals at Raguba field are Late Cretaceous to Paleozoic in age and consist of a series of bioclastic and siliciclastic rocks ("Raguba clastics" of the Zmam/Kalash Formation) overlying a section of indurated quartzitic sandstones ("Sirte quartzite," now Hofra Formation of the Gargaf Group) (Figures 5 and 8). The contact between the two principal rock types within the common hydrocarbon reservoir is an unconformity of regional significance in the Sirte basin.

The lower and older portion of the reservoir (Hofra Formation) consists of white, gray, and pink fine-grained quartzitic sandstone, highly indurated and grading locally into orthoquartzite. The section is predominantly dense but has been extensively weathered in its upper part, creating a fissured and possibly rubbly surface zone containing limited developments of secondary porosity. Some tectonic fracturing is also present, particularly in the crestal area of the structure. This fracture system effectively controls oil production from the quartzitic sandstone portion of the reservoir and may extend deep into the section; at E3-20 (Figure 6), fluids were still being tested 168 m (551 ft) into the quartzitic sandstones, at a bottom-hole elevation only 2 m (6 ft) above the projected field oil-water contact.

It is difficult to assign accurate reservoir parameters to this Hofra Formation quartzitic sandstone lithology. During the initial reserve calculations, an average porosity value of 2% was used, together with an average irreducible water saturation of 50%. At the crest of the Raguba structure, the gross thickness of Hofra section above the projected oil-water contact is approximately 180 m (600 ft), although not all the large body of rock involved is uniformly fractured and thereby capable of oil production.

The younger and more prolific component of the Raguba field hydrocarbon reservoir consists of up to 137 m (450 ft) of clastic rocks of Late Cretaceous age. These "Raguba clastics" thin to a feather edge toward the crestal area of the Raguba high; down flank, they successively overlie the eroded surface of the Hofra Formation quartzitic sandstones and a development of older Upper Cretaceous rocks of nonreservoir character.

Lithologies within this portion of the reservoir range from bioclastic limestones with only minor amounts of quartz sandstone to almost pure, fine- to medium-grained quartz sandstones with only minor admixtures of bioclastic carbonate material. A subsidiary rock type within the reservoir section consists of calcilutite with varying amounts of carbonate skeletal debris and fine-grained quartz sandstone. The quartz sandstones (and floating sand grains) within the reservoir section are clearly derived from the underlying Hofra quartzitic sandstone section; they are preferentially developed in the northern and northwestern areas of the field, the bioclastic limestones being more prominent in the southern and southeastern sectors (Figure 10).

These siliciclastic/bioclastic units make up the entire thickness of "Raguba clastics" along the higher flank areas of the structure (Figure 10) and commonly exhibit superior reservoir qualities. Porosities within the section range from 10% to 25%, with an average of about 17%; permeabilities, both horizontal and vertical, are high, ranging up to 600 md with an average in the area of 200 md. Productivity indices (PIs)* were from the beginning known to be much higher than in the quartzitic sandstone portion of the reservoir and were capable of further improvement by stimulation procedures. Thus, in the case

---

*Productivity index (PI): measure of oil production in bbl per psi of drawdown between SIBHP and FBHP.

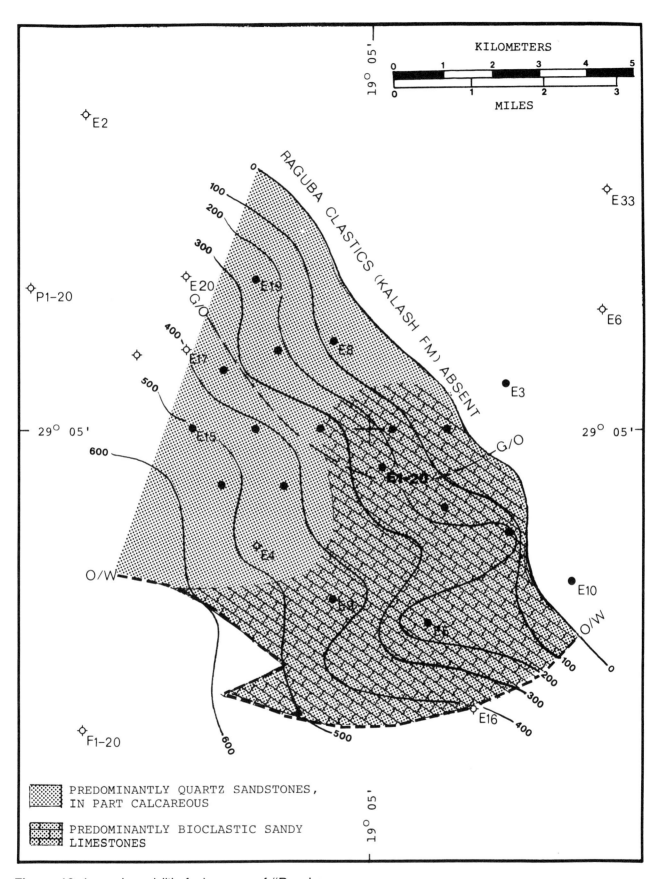

Figure 10. Isopach and lithofacies map of "Raguba clastics" reservoir along southwestern flank of Raguba oil field structure. Contour interval, 100 ft.

of E1-20 (Figure 6), an initial PI of 14 rose to 103 following acid treatment; in the case of E12-20, the improvement was from a PI of 19 to one of 93.

The high percentages of bioclastic carbonate material in the reservoir sections of both these wells (and in other wells in the southeastern part of the field) obviously facilitated the effectiveness of acid stimulation; the "Raguba clastics" section in E1-20 is approximately 65% carbonate, in E12-20 between 50 and 60%. In the northwestern part of the field, the reservoir section of "Raguba clastics" may contain from 60 to 90% quartz sandstone, significantly reducing the response of the section to acid treatment. But these quartz sandstones are frequently clean and well sorted, and so may have very good reservoir qualities in their natural (untreated) state.

The calcilutitic (occasionally shaly or marly) component of the "Raguba clastics" is developed in the upper part of the reservoir section at some downflank locations. It appears to become a more widespread development further down dip toward the southwest and may be a precursor of the deepening of the environment that marked the beginning of Tertiary time. This type of section is only spottily porous and has limited permeability; when developed within the limits of the oil column, it can result in noncommercial oil completions, as at E4-20 (Figure 6).

At the time the Raguba field went on production, it was estimated that the "Raguba clastics" section contained approximately 80% of the recoverable oil. Wells failing to encounter the clastic section and establishing production in the quartzitic sandstones were completed in that zone; however, subsequent field development programs were always designed to exploit the reserves along the "Raguba clastics fairway" between the gas-oil and oil-water contacts along the southwestern flank of the structure (Figures 6, 9, and 10). The bulk of the cumulative production of oil has been drawn from this area of the field.

While differing widely in lithologies and producing characteristics, the "Raguba clastics" and underlying Hofra quartzitic sandstones are parts of a common reservoir containing a single hydrocarbon accumulation. The uppermost part of the reservoir is occupied by a primary gas cap, and, prior to the start of fieldwide production, a gas-oil contact had been confirmed at a subsea elevation of 1539 m (5050 ft). The oil reservoir occupied the body of rock lying between this gas-oil contact and an apparently horizontal oil-water contact established at a subsea elevation of 1677 m (5502 ft). The Raguba oil column was therefore 138 m (452 ft) in thickness and was underlain by highly saline bottom water containing up to 150,000 TDS.

Crude oil present in the Raguba field reservoir proved to range from 41.5° to 43° API, with an average of about 42°; the average sulfur content of the oil was 0.27 wt. %. The crude oil in the reservoir was saturated and had an initial producing gas-oil ratio (GOR) of 935 cf/bbl and a formation volume factor (FVF) of 1.58, the latter indicating a fairly high reservoir shrinkage situation. The original reservoir pressure was 2374 psig at 5082 ft subsea (167 kg/cm^2 at 1549 m subsea).

Prior to the start of production, it was already understood that porosities in the "Raguba clastics" might decrease in a downflank direction. This suggested the possibility of only a limited water drive as a primary producing mechanism. On the other hand, a primary gas cap was known to be present, and vertical permeabilities were good through much of the clastic reservoir section. Initial calculations suggested that a natural depletion mechanism involving solution gas drive plus gravity segregation would support a withdrawal rate of up to 6% per year and would result in the recovery of approximately 37% of the original oil in place.

Partial pressure maintenance by gas injection appeared likely to increase this recovery to more than 40%. This was an attractive option since, at least initially, it would involve the reinjection of produced gas only. Partial pressure maintenance by water injection appeared likely to be at least as effective as gas injection in increasing the level of primary recovery. Although associated costs were likely to be higher, a potential source of injection water had already been located in carbonate section of early Eocene age in the central area of the oil field structure.

Pressure maintenance by gas injection was never attempted. As previously noted, the greater part of the Raguba field production of associated gas was eventually piped to a gas-liquefaction plant located at the Marsa el-Brega export terminal (Figure 2), with a smaller amount being utilized within the field in artificial-lift operations as reservoir pressure declined.

Commencing in 1967, pressure maintenance by water injection into the aquifer was initiated in the southeastern sector of the field, utilizing an injection well drilled in the immediate vicinity of the E-16 dry hole (Figure 6), in which the top of the clastic reservoir section lay at a subsea elevation of 1696 m (5564 ft), 19 m (62 ft) below the field oil-water contact at 1677 m (5502 ft) subsea. High rates of water injection were quickly attained, but, at least over the short term, no significant improvement in bottom-hole pressure was noted in nearby producing wells. Subsequent reservoir studies, however, indicated the presence of an extensive aquifer and suggested that a favorable pressure response with time could be anticipated. As a consequence of the reservoir analysis, injection of water into the aquifer was resumed at a later date in both the southeastern and southwestern sectors of the field. The effectiveness of this pressure-maintenance program is currently being evaluated.

When the Raguba field was first readied for production at the end of 1962, estimates of recoverable oil ranged from 460 million to 550 million STB. At a withdrawal rate of 6% per year, these reserves should have supported a daily production rate of between 75,000 and 90,000 bbl, and such rates were actually maintained during the first few years of production (Table 1). As field development proceeded, estimates of the oil in place and oil ultimately recoverable increased, to 1.875 billion and 750 million bbl, respectively. At its maximum producing rate of 128,000 BOPD, attained during 1968-1969 (Table 1), the Raguba field was performing at an annual withdrawal rate of 6.1% of the ultimately recoverable oil in the reservoir, very close to the original estimate.

Source

It has been remarked (Brady et al., 1980) that source rocks are plentiful in the Sirte basin. Certainly, the sedimentary section present contains substantial thicknesses of marine shales, developed over wide areas of the basin in the case of Tertiary deposits, more restricted to basin deeps in the case of Cretaceous deposits. Brady et al. (1980) listed the Kheir Formation marls and the Sheterat Formation shales of the Tertiary and the Rakb Formation shales of the Cretaceous as potential source rocks; of these, the Sheterat is essentially equivalent to the Hagfa and Khalifa formations ("Heira Formation") and the Rakb to the Sirte Shale Formation of the Concession 20 and adjacent Marada/Hagfa trough areas (Figure 8).

The Sirte Shale Formation is of Late Cretaceous age and, where present, conformably underlies the Kalash Formation or the Waha Limestone Formation (Barr and Weegar, 1972) (Figure 8). The Sirte Shale Formation is not present in the immediate Raguba field area, where the Kalash/Waha equivalent, the "Raguba clastics," directly overlies the eroded surface of Hofra Formation quartzitic sandstones. But the Sirte Shale Formation may be represented in the thicker Upper Cretaceous section developed southwest of the field area and is known from well control to be present in the Marada/Hagfa trough to the east and northeast. Clifford (1984) has noted that "the Upper Cretaceous shales thicken markedly into the troughs and are thought to provide the best source rocks, with TOCs averaging 2%."

The shale sections of the Paleocene Hagfa and Khalifa formations (Heira of original field terminology, Figure 8, Sheterat of Brady et al., 1980) also thicken into the troughs, the lower Paleocene Hagfa shale section to more than 300 m (1000 ft) (Barr and Weegar, 1972). These Paleocene shales also have TOC levels of 2% or greater.

Gumati and Schamel (1988) have calculated that the greater part of the Tertiary section in the trough areas of the Sirte basin is immature to a depth of approximately 3000 m (10,000 ft). Below that depth, thermal maturity rises rapidly, so that at 3600 m (12,000 ft) basal Paleocene and Upper Cretaceous shales exhibit mature vitrinite reflectance (R_o) values in the range 0.65-1.21%. Gumati and Schamel (1988) quote data from a deep well in the Marada/Hagfa trough which indicate that the Kalash Formation is submature ($R_o = 0.5\%$), while the shales of the underlying Sirte Shale Formation (Figure 8) should fall within the "oil window" with R_o values of 0.7% or greater (Figure 11). On the platform area to the west, the Paleocene shale section immediately overlying the Raguba field structure contains adequate levels of TOC but is too immature to have generated the hydrocarbons in the underlying trap (Figure 11). The source of the Raguba oil appears to lie in the Marada/Hagfa trough, with the Sirte Shale Formation and (possibly) the Hagfa Formation being the most likely source rocks.

The sedimentary section overlying the post-Paleozoic unconformity is up to three times as thick in the Marada/Hagfa trough as on the adjacent shelf toward the west (Dahra platform, Figure 4). The Upper Cretaceous shales in the trough appear to have reached thermal maturity by early Tertiary time and would be in a position to feed hydrocarbons to shelf-edge structures via the zone of steep dip (Marada flexure, Figure 4) linking the trough to the platform or shelf; the post-Paleozoic unconformity (and any associated mantling clastics) may have provided a conduit for the migrating gas and oil. The Raguba structure was in a position to receive and retain hydrocarbons by the end of Paleocene time, and it seems likely that the generation and entrapment of the Raguba gas and oil accumulation was a late Paleogene to early Neogene event (Figure 11).

EXPLORATION AND DEVELOPMENT CONCEPTS

At the time of the original grant of Concession 20 at the end of 1955, no exploration wells had been drilled in the Sirte basin, the onshore area of which covers approximately 360,000 km^2 (140,000 mi^2) (Figure 1). Over much of the central part of the basin, including the area of Concession 20, surface geologic mapping provided little evidence of structure involving the Tertiary bedrock, while the pre-Tertiary form of the basin remained conjectural. It was clear from an early stage that geophysics would be the primary exploration tool in the basin.

"The granting of concessions signalled the beginning of exploration methods other than geological mapping" (Eicher et al., 1975). These took the form of both regional programs of aerial photographic and airborne magnetometric coverage and more "local" programs of surface gravity meter and seismic work on individual concessions, often conducted in concert with continuing programs of surface geologic mapping. This rapid tempo of exploration activity reflected the specific work requirements on the large concession blocks, including the need to outline areas suitable for the mandatory 25% relinquishment of acreage due at the end of the first five-year period of concession life.

Regional gravity and magnetometer data from the central part of the Sirte basin revealed a linear

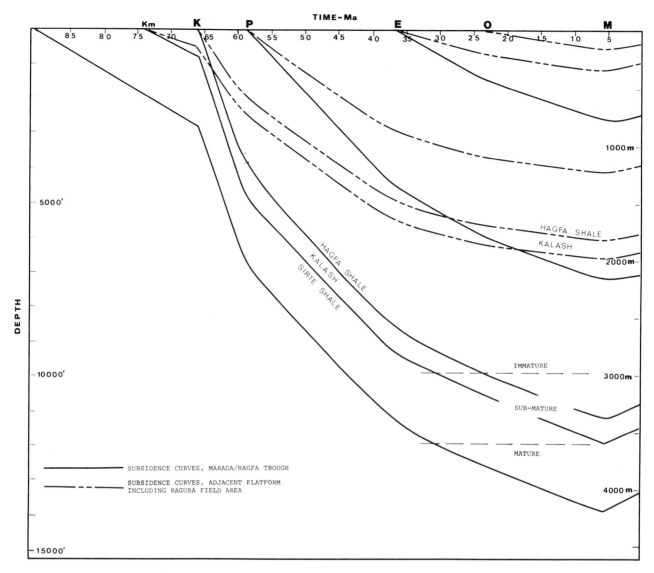

Figure 11. Subsidence curves for Raguba field area of Dahra platform and for adjacent Marada/Hagfa trough. The source for the Raguba field crude oil is believed to lie within the Marada/Hagfa trough depocenter, probably in the Upper Cretaceous Sirte Shale Formation (Figure 8).

pattern of positive and negative anomalies; these bore no visible relationship to the near-surface sedimentary section and were taken as indicative of basin structure at older Tertiary or pre-Tertiary level. The basin had originally been recognized as a Tertiary feature, presumably underlain by Mesozoic and Paleozoic formations (Eicher et al., 1975). As an understanding emerged of the rifted nature of the deeper section, the possibility of pre-Tertiary plays, associated with fault-structures at depth, became readily apparent.

In fact, oil was discovered in both Tertiary and pre-Tertiary horizons in the Sirte basin during 1958 and 1959, at such fields as Zelten (Figure 2), Dahra-Hofra (northwest of Raguba), Waha (south of Zelten), and Amal (east of Raguba). By the end of 1959, some 40 exploratory wells had been drilled in the basin; in the general area of Concession 20, well data had confirmed the Marada/Hagfa trough as a Late Cretaceous-Tertiary depocenter, while seismic was beginning to delineate the hinge line linking this with the shelf area lying to the west (Figure 4). During 1960, the drilling of the D1-20 well (Figure 2) would confirm the possibility of a Cretaceous oil play along this hinge line, setting the stage for the drilling of the Raguba E1-20 discovery well.

The Raguba prospect was mapped as an anticline on what were interpreted to be early Tertiary or pre-Tertiary seismic horizons. When drilled, the subsurface structure proved to have a core of Cambrian-Ordovician quartzitic sandstones, developed on a post-Paleozoic erosional surface and overlain with pronounced unconformity along its flanks by transgressive clastic rocks of Late Cretaceous age. The complex feature had been uplifted and faulted during Paleogene time and was completely capped by Paleocene shales. The hydrocarbon column in the structure was continuous across the unconformity from the Upper Cretaceous clastic into the Paleozoic quartzitic sandstone section.

Following discovery, the Raguba field did not prove easy to develop. The quality of the original seismic data had permitted the identification of the Raguba subsurface structural anomaly but had yielded little in the way of stratigraphic resolution at reservoir level. Delineation of the field was therefore complicated by a lack of information on the distribution, and to some degree the reservoir quality, of the Upper Cretaceous clastics, and on the nature and extent of the fracture system affecting the Paleozoic quartzitic sandstones forming the core of the structure. More detailed seismic control, mainly reflection but with a small amount of refraction shooting, was obtained while delineation of the field area was actually in progress; this provided essential information on the position and throw of faults bounding or affecting the productive area. But, from the beginning, the successful development of the Raguba field was a matter of detailed subsurface geology.

ACKNOWLEDGMENTS

The author wishes to thank AAPG reviewers J. Glenn Cole and Howard H. Lester for constructive criticism of the text, and Messrs. Shailendra K. Goel, Roger S. Plummer Jr., James A. Simons, and Ward Wheatall for additional helpful comments and suggestions.

REFERENCES CITED

Barr, F. T., 1972, Cretaceous biostratigraphy and planktonic foraminifera of Libya: Micropaleontology, v. 18, n. 1, p. 1-46.

Barr, F. T., and A. A. Weegar, 1972, Stratigraphic nomenclature of the Sirte basin, Libya: Petroleum Exploration Society of Libya, Tripoli, Libya, 179 p.

Bebout, D. G., and C. Poindexter, 1975, Secondary carbonate porosity as related to Early Tertiary depositional facies, Zelten field, Libya: AAPG Bulletin, v. 59, n. 4, p. 665-693.

Brady, T. J., N. D. J. Campbell, and C. E. Maher, 1980, Intisar 'D' oil field, Libya, in M. T. Halbouty, ed., Giant oil and gas fields of the decade 1968-1978: AAPG Memoir 30, p. 543-564.

Carmalt, S. W., and B. St. John, 1986, Giant oil and gas fields, in M. T. Halbouty, ed., Future petroleum provinces of the world: AAPG Memoir 40, p. 11-54.

Clifford, A. C., 1984, African oil—past, present and future, in M. T. Halbouty, ed., Future petroleum provinces of the world: AAPG Memoir 40, p. 339-372.

Clifford, H. J., R. Grund, and H. Musrati, 1980, Geology of a stratigraphic giant: Messla oil field, Libya, in M. T. Halbouty, ed., Giant oil and gas fields of the decade 1968-1978: AAPG Memoir 30, p. 507-524.

Eicher, D. B., J. S. Royds, and J. F. Mason, 1975, Africa, Part II, Libya, in E. W. Owen, ed., The trek of the oil finders: AAPG Memoir 6, p. 1436-1445.

Gumati, Y. D., and W. H. Kanes, 1985, Early Tertiary subsidence and sedimentary facies—northern Sirte basin, Libya: AAPG Bulletin, v. 69, n. 1., p. 39-52.

Gumati, Y. D., and S. Schamel, 1988, Thermal maturation history of the Sirte basin, Libya: Journal of Petroleum Geology, v. 11, n. 2, p. 205-218.

Hea, J. P., 1971, Petrography of the Paleozoic-Mesozoic sandstones of the southern Sirte basin, Libya, in Carlyle and Gray, eds., Geology of Libya Symposium, Tripoli, Libya, p. 107.

Roberts, J. M., 1970, Amal field, Libya, in M. T. Halbouty, ed., Geology of giant petroleum fields: AAPG Memoir 14, p. 438-448.

Terry, C. E., and J. J. Williams, 1969, The Idris "A" bioherm and oilfield, Sirte basin, Libya—its commercial development, regional Paleocene geologic setting and stratigraphy, in Exploration for Petroleum in Europe and North Africa: London, Institute of Petroleum, p. 31-68.

Waddams, F. C., 1980, The Libyan oil industry: London, Croom Helm, 338 p.

Appendix 1. Field Description

Field name .... Raguba field
Ultimate recoverable reserves .... 750,000,000 bbl

Field location:
    Country .... Libya
    State .... Cyrenaica (Petroleum Mining Zone II)
    Basin/Province .... Sirte

Field discovery:
    Year first pay discovered ... Upper Cretaceous Kalash Fm. and Cambrian-Ordovician Hofra Fm. 1961

Discovery well name and general location:
    First pay .... Esso Sirte-Libyan American E1-20 (Raguba); 29°04'33"N, 19°05'10"E
    Discovery well operator .... Esso Sirte Inc.

IP:
    First pay .... 2250 BOPD on ½-in. choke, 40.5° API

All other zones with shows of oil and gas in the field:

Age	Formation	Type of Show
Eocene	Gialo ("Sheghega")	Oil
Paleocene	Khalifa ("upper Heira")	Gas

Geologic concept leading to discovery and method or methods used to delineate prospect
Regional gravity meter/surface magnetometer surveys gave clues to pre-Tertiary structure; prospect outlined by reconnaissance reflection seismic in area which results of earlier wildcat drilling indicated was favorable for Cretaceous play.

Structure:
    Province/basin type .... Bally 1211; Klemme III A
    Tectonic history
Regional uplift, erosion and partial foundering of Paleozoic surface, followed by Late Cretaceous transgression along fault troughs, with pre-Cretaceous sea floor topography drowned by end of Cretaceous time. Deepening of entire basin through Tertiary time until start of regression during Miocene time.
    Regional structure
On eastern edge of Dahra platform adjacent to flexure or hinge line (Marada flexure) between platform and adjacent basin deep (Marada/Hagfa trough).
    Local structure
Faulted anticline trending northwest-southeast; simple dip on northeast and southwest flanks; minor faulting on southeast; major faulting on northwest plunge.

Trap:
    Trap type(s)
Anticlinal trap involving fault closure on northwest flank, with lesser production from fractured, weathered quartzitic sandstone below major unconformity, and with onlapped bioclastic sandstone reservoir wedging out on structure yielding majority of production.

Basin stratigraphy (major stratigraphic intervals from surface to deepest penetration in field):

Chronostratigraphy	Formation	Depth to Top in ft (m)
Recent	Aeolian sand	0
Miocene	Marada	1-100 (0-30)
Oligocene	Diba	300 (90)
	Arida	700 (215)
Eocene	Gialo	1000 (300)
	Gir	2000 (610)
Eocene-Paleocene	Jabal Zelten	3500 (1065)
Paleocene	Khalifa	3900 (1190)
	Hagfa	4900 (1495)
Upper Cretaceous	Kalash	5100 (1555)
	Sirte*	6000 (1830)*
Cambrian-Ordovician	Hofra	5200 (1585)

* Sirte shale not present across productive area of Raguba structure.

Reservoir characteristics:

- **Number of reservoirs** 2
- **Formations** .... Kalash Formation ("Raguba clastics"); Hofra Formation ("Gargaf quartzite")
- **Ages** .... Kalash, Late Cretaceous; Hofra, Cambrian-Ordovician
- **Depths to tops of reservoirs** .... 5100-5200 ft (1555-1585 m) at crest of structure
- **Gross thickness (top to bottom of producing interval)** .... 700 ft (215 m)
- **Net thickness—total thickness of producing zones**
    - Average .... 250 ft (75 m)
    - Maximum .... 300+ ft (90+ m)
- **Lithology**
Bioclastic limestone grading through sandy limestone/calcareous sandstone to quartz sand and sandstone; underlain by weathered and fractured quartzitic sandstones/orthoquartzites
- **Porosity type** .... Kalash, intergranular to intercrystalline; Hofra, fracture and minor secondary
- **Average porosity** .... Kalash (clastics), 17%; Hofra, 2%
- **Average permeability** .... Kalash, 200 md; Hofra largely impermeable except where fractured

Seals:

Upper
- **Formation, fault, or other feature** .... Hagfa Formation
- **Lithology** .... Marine shale

Lateral
- **Formation, fault, or other feature** .... Hagfa Formation (shale) as above; fault closure on northwest plunge of structure; permeability barrier on northeast flank

Source:

- **Formation and age** .... Sirte shale, Late Cretaceous-Campanian
- **Lithology** .... Marine shale, dark gray/brown, carbonaceous
- **Average total organic carbon (TOC)** 2%
- **Maximum TOC** .... NA
- **Kerogen type (I, II, or III)** II
- **Vitrinite reflectance (maturation)** .... $R_o = 0.65$ to 1.21%
- **Time of hydrocarbon expulsion** .... Probably late Paleogene to early Neogene
- **Present depth to top of source** .... 10,000+ ft (3050+ m)
- **Thickness** .... 600+ ft (180+ m)
- **Potential yield** .... NA

Appendix 2. Production Data

Field name .... Raguba field
Field size:
    Proved acres .... 15,000 ac (23 m²; 60 km²)
    Number of wells all years 79
    Current number of wells .... 40 (last report)
    Well spacing .... Variable
    Ultimate recoverable .... 750 million bbl (119 million m³)
    Cumulative production .... 556 million bbl (88 million m³) (through 1982; plus 550 bcf gas)
    Annual production .... 7 million bbl (1.1 million m³) (last available figure)
    Present decline rate .... NA
        Initial decline rate .... NA
        Overall decline rate .... NA
    Annual water production .... NA
    In place, total reserves .... 1875 million bbl (298 million m³)
    In place, per acre foot .... 500 bbl (80 m³)
    Primary recovery .... 750 million bbl (119 million m³)
    Secondary recovery .... NA
    Enhanced recovery .... NA
    Cumulative water production .... NA

Drilling and casing practices:
    Amount of surface casing set .... 800-1000 ft (250-300 m)
    Casing program
    13⅜-in. into Sheghega/Gialo; 9⅝-in. into Heira/Khalifa; 7-in. or 5½-in. through clastic pay section in Hofra
    Drilling mud .... NA
    Bit program .... NA
    High pressure zones .... None present

Completion practices:
    Interval(s) perforated .... 30-50 ft (10-15 m) casing perforations
    Well treatment .... Acid stimulation of bioclastic section

Formation evaluation:
    Logging suites .... ES-induction, laterolog, sonic; MLL in some early wells
    Testing practices .... Conventional drill-stem testing in conjunction with coring in early wells;
                                                                                                        through-casing in development wells
    Mud logging techniques .... Conventional well-site procedures

Oil characteristics:
    API gravity .... 41.5-43°, average 42° API
    Base .... Mixed
    Initial GOR .... 935 SCF/bbl
    Sulfur, wt% 0.27
    Viscosity, SUS .... Kin: cSt at 70°F 5.58, at 100°F 3.56
    Pour point .... +30°F
    Gas-oil distillate .... NA

Field characteristics:
    Average elevation .... 270 ft (82 m)
    Initial pressure .... 2374 psig at -5082 ft (167 kg/cm² at -1549 m)
    Present pressure .... NA
    Pressure gradient .... NA

Temperature .... NA
Geothermal gradient .... 1.2°F/100 ft (approx. 2.2°C/100 m)
Drive .... Gas-cap expansion, gravity segregation
Oil column thickness .... 452 ft (138 m)
Oil-water contact .... −5502 ft (−1677 m)
Connate water .... 10–25% in clastics, est. 50% in underlying quartzitic sandstone
Water salinity, TDS .... 110,000–150,000 ppm/TDS avg. 135,000 ppm
Resistivity of water .... 0.08 ohm-m at 75°F
Bulk volume water (%) .... NA

Transportation method and market for oil and gas:
By pipeline to export terminal at Mediterranean shore, thence by tanker to world market

Tintaburra Field—Australia
Eromanga Basin, Queensland

C. B. NEWTON
Petrocorp Exploration Indonesia, Ltd.
Jakarta, Indonesia

FIELD CLASSIFICATION

BASIN: Eromanga
BASIN TYPE: Cratonic Sag
RESERVOIR ROCK TYPE: Sandstone
RESERVOIR ENVIRONMENT OF
  DEPOSITION: Fluvio-Deltaic

RESERVOIR AGE: Jurassic Callovian and
  Cretaceous Neocomian
PETROLEUM TYPE: Oil
TRAP TYPE: Anticline

TRAP DESCRIPTION: Multiple pays within closure on upthrown side of high-angle reverse faulted anticline with stratigraphic component in lesser reservoirs

LOCATION

The Tintaburra oil field is situated within Petroleum Lease 29 (previously within Authority to Prospect 299P Part 2), toward the northern end of the Thargomindah shelf on the southeast margin of the Jurassic-Cretaceous Eromanga basin, Queensland, Australia (Figures 1A and 1B). It is located 22 mi (35 km) southwest of the township of Eromanga (see Eromanga 1 well) and 620 mi (1000 km) west of the Queensland state capital, Brisbane. The field is located 62 mi (100 km) northeast of the Jackson oil field, the largest field in the Queensland portion of the Eromanga basin, and 19.9 mi (32 km) west of the Talgeberry oil field (Figure 1B).

The Toobunyah oil field (Figure 2) is located 1.2 mi (2 km) south of Tintaburra and is in fact the southern culmination of the Tintaburra field, the two being connected by a shallow saddle at the main reservoir horizon, the Hutton Sandstone (Figures 3 and 4).

Total areal extent of the field is approximately 500 ac (203 ha) at the level of the Hutton Sandstone reservoir. The Toobunyah culmination to the south covers a similar area.

Estimated ultimate oil recovery from the Tintaburra culmination alone is 1.1 MMBO with approximately 98% of the reserves being in the top Hutton Sandstone reservoir. Oil reserves in the Toobunyah culmination total 1.3 MMBO, with 89% being in the top Hutton and 11% in the Birkhead Formation.

HISTORY

Pre-Discovery

Initial petroleum exploration activity on the Thargomindah shelf comprised broad reconnaissance gravity surveys in 1952 by Shell (Queensland) Development Pty Ltd. In 1956–1957, gravity traverses were conducted in the Eromanga area for Tallyabra Oil Pty Ltd, and these were followed, in 1958, by reconnaissance aeromagnetic surveys conducted by the Bureau of Mineral Resources (BMR).

In 1959, the first seismic exploration in the Queensland portion of the Eromanga basin was carried out by Smart Oil Exploration (SOE). The Grey Range Seismic Survey recorded single fold data over the Orient anticline (Figure 20). The BMR also recorded seismic data along the Quilpie-Eromanga Road (runs east-west along north boundary of ATP 299-A) in 1959 and provided the first subsurface coverage of the Harkaway and Tallyabra trends (Figure 20).

Exploration drilling activity commenced on 14 April 1959 with SOE Scout Bore 3 (Gumbla), located 5.6 mi (9 km) northwest of Tintaburra (Figure 1B). The well was drilled to test the Harkaway anticline inferred from gravity and surface geological evidence. The percussion rig reached a total depth of 1935 ft (590 m), close to the Toolebuc Formation (Figure 4). A viscous brown grease recovered from a depth of 1155 ft (352 m) (Winton Formation?) was analyzed

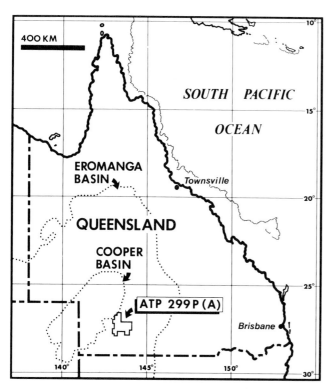

Figure 1A. Location of the Eromanga basin and ATP 299P (Part 2) (Authority to Prospect area).

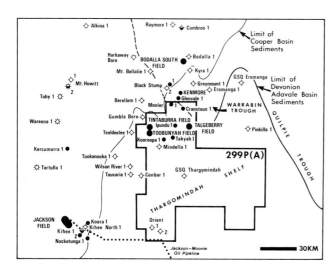

Figure 1B. Tintaburra field location map and basin setting.

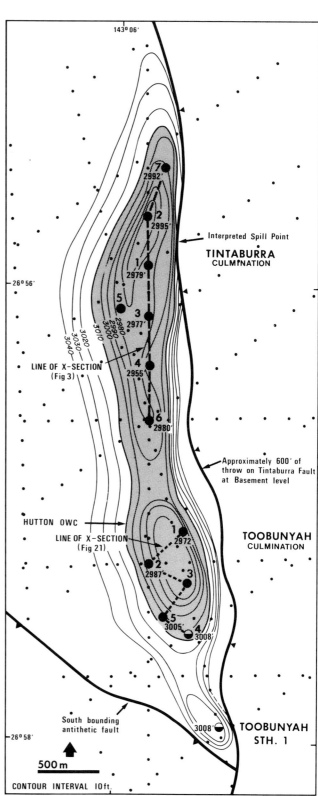

Figure 2. Structure on top of porosity, Hutton Sandstone reservoir. The locations of seismic shotpoints are shown. Depths shown are subsea in feet.

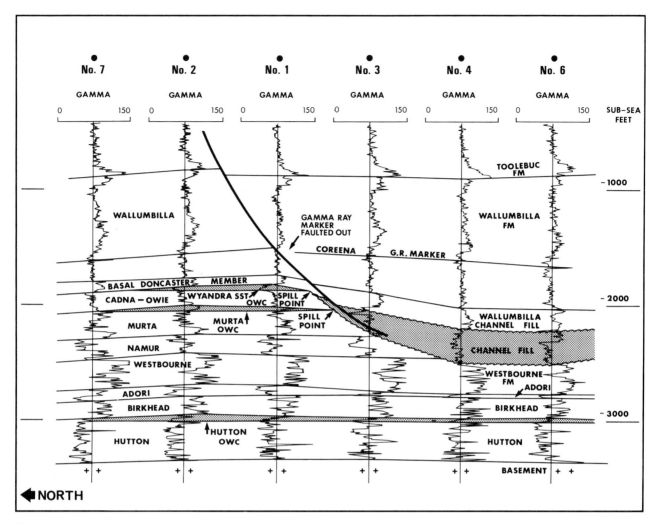

Figure 3. Tintaburra field north-south structural cross section. See Figure 2. for location.

at the Government Chemical Laboratory; on chloroform extract it gave 74% brown mineral oil of the consistency of vaseline, which fluoresced lilac. Gas shows containing up to 73% methane were also recorded (Denmead, 1960). The bore was gamma ray logged to 1438 ft (438 m) by the BMR in 1960.

SOE spudded Orient 1 (Figure 1B) scout hole (50 mi or 80 km south of Tintaburra) close to the crest of the seismically defined Orient Anticline (Figure 20) on 23 May 1960 and with the use of a second rig reached a total depth of 3200 ft (976 m) in the Hutton Sandstone in May 1961.

Slight gas shows were recorded while drilling from the Murta Member (Figure 4) to total depth, and a weak oil show was reported from the base of the Westbourne Formation. The bore was gamma ray logged to 2800 ft (854 m) by the BMR in 1968 (in the Westbourne Formation).

SOE Orient 2 was spudded on 7 August 1961 on the flank of the Orient anticline to test Mesozoic and Paleozoic sediments thought to be present. Orient 2 reached a total depth of 3535 ft (1078 m) in basement metasediments after penetrating a full Eromanga basin sequence. Fluorescence was reported from a core in the Wyandra sandstone member (Figure 4), and minor gas shows were reported throughout the sequence.

The Orient bores are important in that they confirm the southern extent of the Hutton Sandstone reservoir and the Birkhead seal.

Modern exploration began in 1982, spurred on by the discovery of oil in Eromanga basin sediments at Jackson 1, 62 mi (100 km) to the southwest (Figure 1B). Lennard Oil NL, as operator, shot the 209 mi (337 km) Harkaway Seismic Survey, confined mainly to an area of the surface-defined Harkaway anticline (Figure 20). Hartogen Energy Limited became operator for a much expanded joint venture at the end of this stage.

The surface Harkaway anticline was seen on seismic data to comprise two parallel, reverse-faulted (down-to-the-east) anticlines. A number of culminations were defined along both the western (Tintaburra) and eastern (Harkaway) trends (Figure 20).

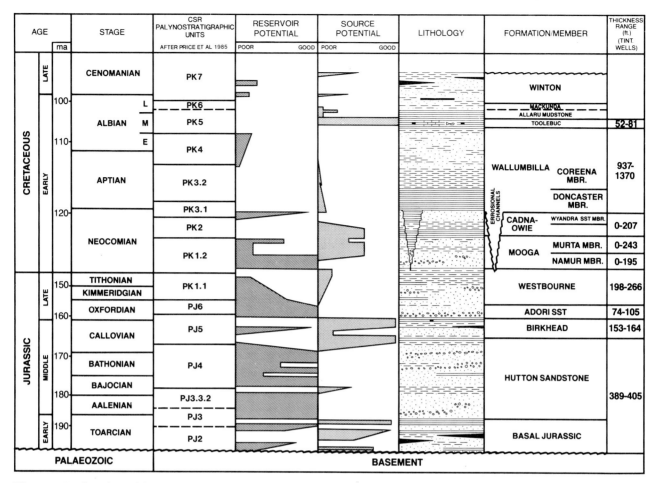

Figure 4. Stratigraphic summary chart, Eromanga basin sequence. "C" seismic horizon, base Wallumbilla Formation.

Discovery

Initial prospect mapping of the Tintaburra area at 1:100,000 scale was completed in 1982 using the 12-fold 1982 Vibroseis data that comprised a 2.2 × 3.1 mi (3.5 × 5 km) grid. Tintaburra appeared as a fault closure (Figure 5) on the western upthrown side of a high-angle reverse fault. Evidence for Tertiary age of faulting was its surface expression, displacement of shallow events, and relatively uniform isopach intervals. A minor down-to-the-west reverse fault was mapped on the west flank.

Fault closure was mapped at basement level and at the "C" horizon (base of the Wallumbilla Formation, Figure 4), with separate north and south (time) culminations separated by a large Neocomian channel, represented by a 150 millisecond (ms) "cut" into Upper Jurassic sediments. The nature of channel fill was unknown; however, its seismic character indicated it to be similar to Wallumbilla Formation lithologies. Depression of Jurassic reflectors beneath the channel indicated the low velocity of the fill.

The 1983 Minedilla Seismic Survey recorded 15 mi (24 km) of infill seismic over Tintaburra, which reduced the dip grid to 0.9 mi (1.5 km). The Varisource (Geosource copyright) recording technique provided 2400% Vibroseis coverage for a 1200% field effort.

An integrated interpretation of Minedilla and Harkaway data delineated a drilling target at Tintaburra. Mapping at 1:50,000 scale defined a fault closure at basement level of 35 ms over an area of 10 mi² (26 km²) (Figure 6). Fifteen milliseconds of fault-independent closure was mapped over 3.3 mi² (8.5 km²), again with separate north and south (time) culminations.

Two separate closures were identified at the base Wallumbilla level ("C" seismic horizon), separated by an east–west-trending channel. Fifty milliseconds of fault and channel closure was interpreted on the northern closure over 5 mi² (13 km²), of which 30 ms was "independent" closure over 2.2 mi² (5.6 km²) (Figure 7).

The stratigraphic prognosis for Tintaburra 1 was obtained from a seismic tie with BP Bodalla 1, 40 mi (65 km) to the north-northwest (Figure 1B), which was the closest available exploration well at the time.

Tintaburra 1, located on the time crest at Jurassic and Cretaceous levels of the Tintaburra North culmination, was designed to test the Eromanga basin sequence within fault-independent closure on

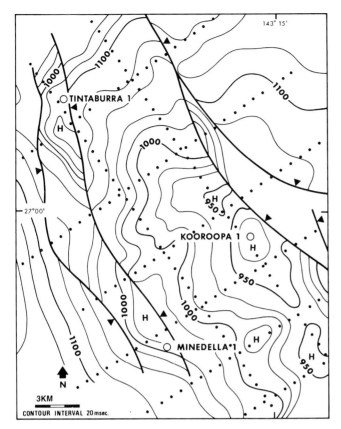

Figure 5. Time structure map at basement level, northwest ATP 299P (Part 2). 1982 Harkaway Seismic Survey.

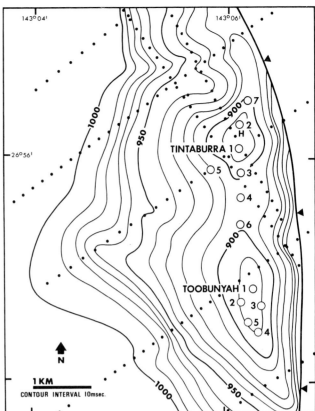

Figure 6. Tintaburra field: time structure map at basement level, 1983 Minedilla Seismic Survey.

the upthrown side of a Tertiary high-angle reverse fault. Permian Cooper basin sediments, if present, were a secondary objective. The well spudded on 17 December 1983 and was cased and completed as a dual zone oil producer on 11 January 1984.

Drill-stem test (DST) 2 tested excellent shows from the Wyandra sandstone member of the Cretaceous Cadna-owie Formation and resulted in a flow of 85 BOPD and 340 BWPD through a 1-in. choke (Figure 8). This represented the first flow of oil from the Wyandra sandstone member in the Eromanga basin. DST 3 tested shows from sands in the top Murta Member, and resulted in a recovery of 10 ft (3.1 m) of oil and 460 ft (140 m) of muddy water.

Oil shows were encountered throughout the Jurassic Westbourne and Birkhead formations. DST 4 tested the entire top Hutton oil column (Figure 9) and resulted in a flow of 1750 BOPD through a 1-in. choke, with a surface flowing pressure of 40 psi (275.8 kPa). The well reached its total depth of 3975 ft (1212 m) in metamorphic basement. The lower Murta Member of the Early Cretaceous Mooga Formation and Westbourne Formation tested water at 120 BPD and 160 BPD, respectively. DST results for Tintaburra 1 and subsequent appraisal wells are documented in Table 1.

Log analysis of the shaly Wyandra sandstone member was difficult owing to finely interlaminated, shaly sandstones and siltstones, and DST 7 was located above a possible oil-water contact (OWC), resulting in a recovery of 26 bbl of clean oil. A 40 ft (12.2 m) oil column was delineated by the test results. DST 8 recovered 60 ft (18.3 m) of oil from a shaly top of the Hutton Sandstone and indicated a 64 ft (19.5 m) gross oil column. The well was cased and completed for dual zone oil production.

Post-Discovery

Following Tintaburra 1, consideration was given to depth mapping of the "C" horizon and basement seismic events. Mapping of the main Hutton reservoir was not practical because of the poor seismic reflection associated with the Hutton-Birkhead interface.

The velocity distortion (pushdown) beneath the "C" horizon, the result of thick, low-velocity channel fill above the "C" horizon, was immediately apparent when seismic line EH83-02 (see Newton, 1986, figure 5, p. 338) was referenced to the Toolebuc seismic marker as a datum. Recognition of velocity-induced

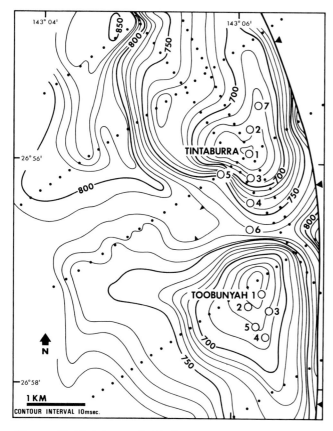

Figure 7. Tintaburra field: time structure map at "C" seismic horizon, 1983 Minedilla Seismic Survey.

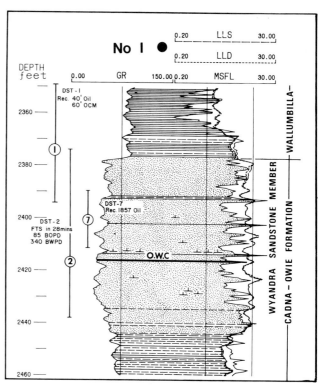

Figure 8. Tintaburra 1, Wyandra sandstone member log signature, lithology, and test results.

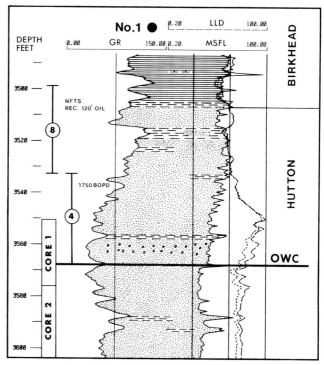

Figure 9. Tintaburra 1, top of the Hutton Sandstone log signature, lithology, and test results.

distortion beneath the channel made it important to determine whether Tintaburra was in fact a single closure at the top of the Hutton level. A single closure was considered structurally more credible and consistent with fault conjunction.

Tintaburra 1 did not intersect the channel-fill sequence but provided information on the vertical velocity gradients within both Cretaceous and Jurassic sections. In order to accommodate the rapid lateral velocity and thickness variations arising from the channeling, seismic interval velocities were represented by linear velocity functions (LVF) (Blackburn, 1985). Using these velocity functions ($V = Vo + kz$) for depth conversion greatly reduced the (time) depression observed beneath the channeling on basement and Hutton horizons. Smoothing of the depth values resulted in one culmination being mapped at the top of the Hutton level with 65 ft (19.8 m) of fault-independent closure over an area of 1.6 mi² (4.1 km²) (Figure 10).

A 64 ft (19.5 m) Hutton oil column in Tintaburra 1 supported the hypothesis that the structure, as mapped in depth, was oil-filled to fault-independent closure.

Following depth mapping, a third round of seismic recording was made over Tintaburra and Toobunyah (known at the time as Tintaburra South) during 1984 (Kalboora Seismic Survey). The 22 mi (35 km) of 20- fold (single sweep recorded) Vibroseis data complemented 1982 and 1983 data to provide a grid 0.5 × 0.6 mi (0.8 × 1 km). Upholes were recorded at 0.9– 1.2 mi (1.5–2 km) spacing to constrain seismic static corrections.

Table 1. Tintaburra field drill-stem test results.

Well	DST	Formation	Interval	Choke (in.)	Result	Recovery (ft)
1	1	Cadna-owie	2346–2396	¾	NFTS	40 oil 60 OCM
1	2	Cadna-owie	2328–2438	1	85 BOPD 340 BWPD	
1	3	Murta	2587–2622	1	NFTS	10 oil 460 MW
1	4	Hutton	3526–3568	1	1750 BOPD	
1	5	Murta	2614–2654	1	120 BWPD	
1	6	Westbourne	3164–3204	1	160 BWPD	
1	7	Cadna-owie	2390–2412	1	NFTS	1857 oil
1	8	Upper Hutton	3498–3528	1	NFTS	60 oil 60 OCM
1	9	Birkhead	3362–3392	1	NFTS	6 mud
2	1	Murta	3585–3634	1	NFTS	187 oil
2	3	Namur	2818–2829	1	960 BWPD	—
2	4	Westbourne	3122–3172	1	1100 BWPD	—
2	5	Westbourne	3178–3212	1	NFTS	40 oil 430 W 1758 MW
2	6	Hutton	3510–3565	1	2520 BOPD 72 BWPD	—
2	7	Hutton	3509–3541	1	250 BOPD	—
2	8	Murta	2582–2626	1	NFTS	35 oil 967 OCM
3	1	Hutton	2509–2559	1	1320 BOPD	
3	2	Basal Jurassic	3882–3967	1	475 BWPD	
4	1	Hutton	3516–3576	1	165 BOPD 935 BWPD	
4	2	Hutton	3519–3566	1	405 BOPD 135 BWPD	
5	1	Hutton	3542–3564	1	NFTS	671 oil
6	1	Hutton	3512–3575	1	1425 BOPD 675 BWPD	
6	4	Birkhead	3422–3465	1	NFTS	30 mud
7	1	Murta	2771–2852	1	275 BWPD	
7	2	Hutton	3539–3580	1	1812 BOPD 18 BWPD	

NFTS, no fluid to surface
BOPD, barrels oil per day
BWPD, barrels water per day
OCM, oil cut mud
MW, muddy water
W, water

The 1982 and 1983 data were reprocessed and phase rotations applied to resolve the consistent miss ties between the three vintages of data. Line 84-05, a north-south axial line, was the most significant, dramatically illustrating the channel with associated listric faulting and a velocity-induced saddle (Figure 11). Time mapping at basement level confirmed Tintaburra to be an anticline with fault-independent north and south culminations (Figure 12). At the south end of the field a small antithetic fault enhanced closure in the critical southern direction. Separate north and south closures were again mapped at the base of the Wallumbilla level (Figure 13).

Depth mapping incorporating 1984 seismic data was carried out using the LVF method, and attempts were made to account for variations in subweathering and weathering velocities. The problem of channel-induced ray path distortion was recognized but not quantified.

Resolution of the LVF technique was considered to be within 30 ft (9.2 m). With this constraint it was concluded that the crestal depression mapped in time at basement level was negated in depth (Figure 14). Separate north and south closures were mapped at the base of the Wallumbilla level (Figure 15).

Tintaburra 2

Results of depth mapping influenced the location of Tintaburra 2. The well, located 1312 ft (400 m) north of Tintaburra 1, was drilled in August 1984 to a total depth of 4000 ft (1220 m). A slightly lower than predicted velocity to the base of the Wallumbilla resulted in the tight Wyandra sandstone member being intersected 35 ft (10.7 m) lower than predicted. The error decreased with depth and basement was intersected 4 ft (1.2 m) high to prognosis, an error of 0.1%.

The Hutton Sandstone was found to be 16 ft (4.9 m) thicker than at Tintaburra 1. The top of the formation was intersected 20 ft (6.1 m) high to prognosis and only 2 ft (0.6 m) downdip of the discovery well.

No shows were recorded from a core in the Wyandra sandstone member although it was intersected above the OWC at Tintaburra 1. Oil was recovered from Murta sandstones and the Westbourne Formation; however, the top of Namur and the top of Westbourne sandstones were water saturated (see Table 1).

A 57 ft (17.4 m) oil column was intersected at the top of the Hutton Sandstone, and a DST of the complete column flowed at 2520 BOPD accompanied by 72 BWPD. The OWC, which was clearly defined both on logs and in core, was intersected 7 ft (2 m) high to that at Tintaburra 1. There was a significant improvement in the upper Hutton sequence at Tintaburra 2 and it was completed for 250 BOPD.

Tintaburra 3 to 7

Tintaburra 3, drilled immediately following Tintaburra 2, was located 1312 ft (400 m) south of Tintaburra 1. The well was located on the edge of the "C" horizon channel to provide the first information on the nature of channel fill material, critical to assessment of the reserves potential of the southern part of the field.

The base of the Wallumbilla Formation was intersected within 0.5% (12 ft or 3.7 m high) of prognosis. The Cadna-owie Formation and most of the Murta were absent due to channel erosion (Figure 3), and a channel fill sequence was intersected. The stratigraphic sequence below the Murta Member was as predicted and the top of the Hutton and basement horizons were intersected 26 ft (7.9 m) and 54 ft (16.5 m) high to prognosis, respectively. Facies changes reduced the reservoir quality and pay thickness in the top Hutton, and 1320 BOPD flow was recorded on test.

The 1.4% error in the predicted depth to basement at Tintaburra 3 was confirmed by the velocity survey to be due to ray path distortion associated with steep channel sides and listric faulting. Assumed velocities were vindicated, placing the source of the error on the seismic *times*; this was not unexpected, because migrated seismic sections disregard ray path refractions.

Velocity information gathered from Tintaburra 3 enabled seismic time distortion to be quantified, and new basement time maps were constructed. An adjusted basement time map (Figure 16), serving as a quasi-depth map, formed the basis on which Tintaburra 4, 5, and 6 were located.

Wells 4 and 6 penetrated successively thicker channel sequences and the top of the Hutton was intersected within 0.5% (9 ft or 2.7 m) of prognosis.

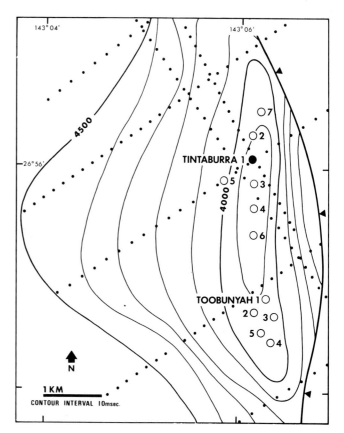

Figure 10. Tintaburra 1 smoothed depth structure map at basement level, post-Tintaburra 1.

Oil columns in excess of 45 ft (13.7 m) were penetrated.

Tintaburra 5 was located on the flank of the main channel on the western edge of the field (Figure 17). All horizons were intersected low to prognosis with an error of 1.4% (55 ft or 16.8 m) at basement. Unmigrated seismic data were used, and this was found to be a contributing factor to the error in depth prognosis.

A further round of depth mapping was carried out prior to drilling Tintaburra 7. Data from the first six wells were used to constrain the velocity functions used in the LVF method but *not* to constrain depths. Figure 18 shows that the method resulted in a good agreement between actual and predicted depths with the largest discrepancies in channel flank wells where ray path distortion is greatest.

North of Tintaburra 2, the top of the Hutton was interpreted to stay high relative to basement, and a northerly improvement in reservoir quality was also predicted. The success of Tintaburra 7 confirmed both concepts.

Additional seismic data were recorded over the Tintaburra South culmination (Toobunyah) during 1985, and depth mapping again indicated Tintaburra and Toobunyah to be one closure at the basement and the top of the Hutton levels.

Toobunyah 1 was located on the interpreted crest of the Toobunyah culmination, 0.6 mi (1 km) south-

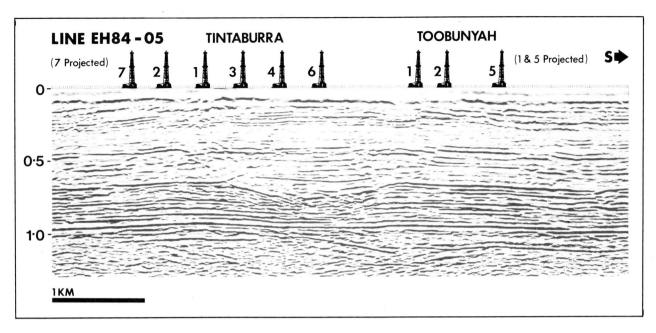

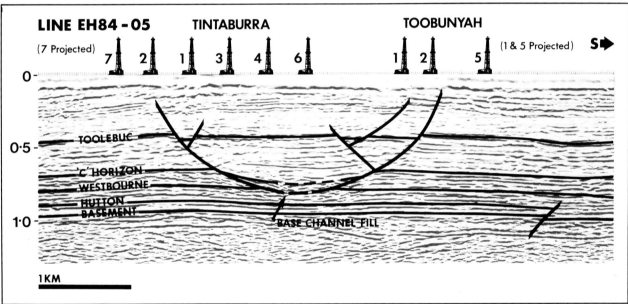

Figure 11. Single sweep processed seismic line illustrating the channel with associated listric faulting and velocity induced saddle. "C" seismic horizon, base Wallumbilla Formation.

southwest of Tintaburra 6. The primary objective was the Hutton Sandstone. The well spudded on 23 October 1985 and was cased and completed as a dual zone oil producer on 14 November 1985. Upper and lower Birkhead Formation sandstones successfully tested oil, and new pool discoveries were recorded (Figure 19). The top of the Hutton Sandstone was encountered 5 ft structurally low to Tintaburra 6. A 39 ft (11.9 m) oil column was penetrated in the top Hutton Sandstone, and a DST of the best quality sands flowed oil at a rate of 1030 BOPD. The Hutton Sandstone OWC was intersected at 3007 ft (917 m) subsea (ss) in the middle of the range of OWCs at Tintaburra, compelling evidence that the two pools were connected.

Velocity information gained from Toobunyah 1 was used to refine the depth map at basement and the top of the Hutton and to locate Toobunyah 2, 3, 4, and 5 appraisal wells.

The Tintaburra depth mapping exercise differs (philosophically and practically) from the commonly used technique of forcing depth maps to fit well intersections. If well data are used as velocity constraints, discrepancies between depth estimates and actual values provide a useful measure of uncertainty inherent in depth estimation. Forcing depth estimates to actual data provides no additional data *of predictive value* away from wells.

Following the discovery of oil in the Eromanga basin sequence at Tintaburra, some 12 exploration

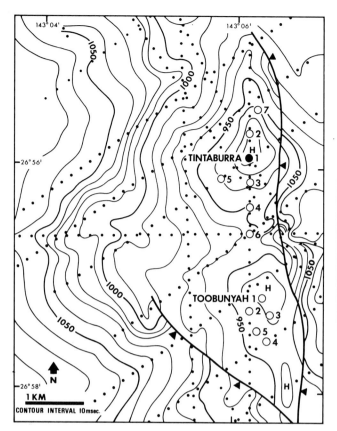

Figure 12. Tintaburra field time structure map at basement level, 1984 Kalboora Seismic Survey.

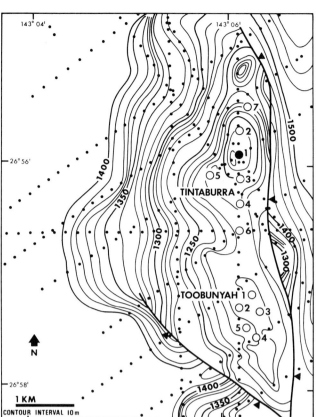

Figure 14. Tintaburra field depth structure map at basement level, pre-Tintaburra 2.

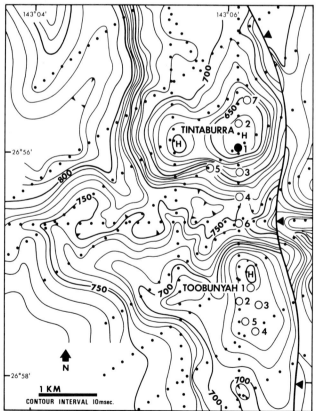

Figure 13. Tintaburra field time structure map at "C" seismic horizon (base Wallumbilla Formation), 1984 Kalboora Seismic Survey.

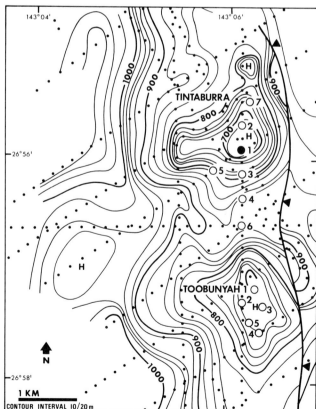

Figure 15. Tintaburra field depth structure map at "C" seismic horizon (base Wallumbilla Formation), pre-Tintaburra 2.

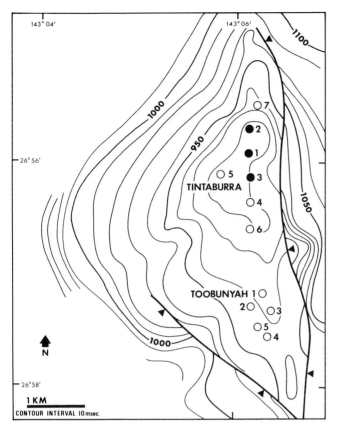

Figure 16. Tintaburra field basement time structure map adjusted for the measured (velocity induced) time distortion, post-Tintaburra 3.

wells have been drilled within the confines of the original ATP 299P(2). Of these, nine exploration wells have resulted in new field discoveries, many with multiple pay horizons. The 75% success rate is remarkable by any standard and confirms the area as a significant oil province, although the average field size is small (less than 5 MMSTB OIP).

The first exploration well to be drilled post-Tintaburra 1 was Minedella 1, 13.7 mi (22 km) southeast of Tintaburra 1 along the Tintaburra fault trend. The well tested the Eromanga basin sequence on the upthrown side of the Tintaburra fault within fault-dependent closure. All prospective horizons were water wet, and this gave strong support to the theory that fault-independent closure was a prerequisite for hydrocarbon trapping in the area.

Figure 1B shows the discoveries in the area to date, with the most significant being Talgeberry (Wyandra, Murta, and Birkhead), Cranstoun (Birkhead), Ipundu (Wyandra), Bodalla South (Hutton and basal Jurassic), and Kenmore (Hutton). All have structural styles similar to that described at Tintaburra; however, most are not affected by the Neocomian submarine channeling. Original oil in place in the above and adjacent fields is estimated at 100 MMSTB (3P category).

DISCOVERY METHOD

The road to discovery, as described above, commenced with the recognition of a number of parallel north-northwest/south-southeast anticlinal trends on surface geological maps and from Landsat imagery.

Early seismic surveys confirmed the potential for structural traps along the anticlinal trends, and successive rounds of seismic were recorded in order to define a simple structural closure within the Jurassic and Lower Cretaceous. Well control was sparse and seismic ties to exploration wells long and tedious; however, the main reservoir and seal horizons were successfully identified predrill. The source of hydrocarbons in the basin remains conjectural and the Tintaburra anticline, located on the edge of the Thargomindah shelf, was chosen for drilling because it could access both Mesozoic and Permian sourced hydrocarbons.

The geological concept behind Tintaburra 1 was straightforward. The well tested the hydrocarbon potential of the Hutton Sandstone, sealed by the overlying Birkhead Formation shale within fault-independent closure. The sandy nature of the Jurassic sequence made delineation of a fault-independent closure a priority. The play was considered untested in the area.

Today's explorationists, in finding fields of this type, have the following problems to contend with.

1. Although the top Hutton accumulation, the main oil reservoir in the field, is dominantly structurally controlled, rapid facies variations occur at the top of the reservoir/base of the overlying Birkhead seal that have important implications to trap integrity. These variations are generally not resolvable seismically, but seismic inversion is proving a powerful tool in the hands of the modern explorationist in other areas.

2. The Hutton Sandstone reservoir is considered full to fault-independent closure. The low fault-independent structural relief demands a depth conversion method where errors can be kept to less than 1%, notwithstanding the geophysical complexities associated with the "C" horizon erosion. Modern seismic data, higher fold with higher resolution, improved statics, and greater knowledge of velocities (from seismic and wells), enhance the chance of the modern explorer in solving this problem.

3. Tintaburra and other fields in the area have some form of surface expression caused by the Tertiary faulting and this enabled early seismic to be more effectively located. As exploration proceeds to the more subtle structural traps and to structural/stratigraphic traps, seismic and an increased understanding of the distribution/association of reservoir and seal units will be the main tool of the explorationist.

Technological advances are not the only advantages the modern explorer has over his predecessor.

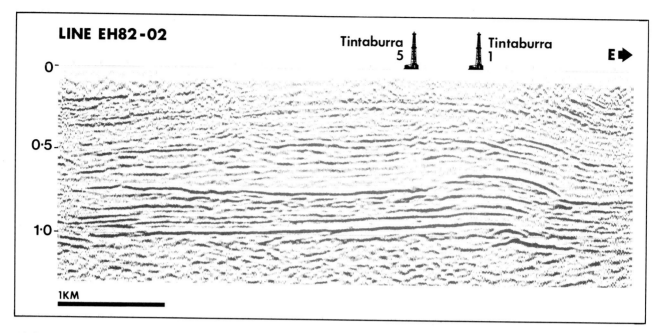

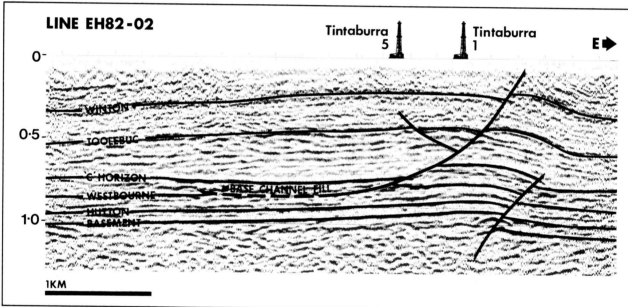

Figure 17. Seismic section showing Tintaburra anticline, channeling, and listric faulting. "C" seismic horizon, base Wallumbilla Formation.

Although acceptance of a Jurassic source for hydrocarbons on the Thargomindah shelf following the Tintaburra discovery is by no means universal, most modern explorers will not limit themselves to plays within the Permian edge. If they did, a lot of oil would have remained undiscovered.

Surface Manifestations

The Thargomindah shelf, in common with adjacent areas of southwest Queensland, contains good physiographic criteria on which structural analysis of the underlying sediments can be based (Senior, 1978).

The Cainozoic geological history of the area has given rise to distinctive erosional and depositional landforms. Senior (1978) showed that structural axes in the area were clearly mirrored in relief patterns at all scales. Large synclines are occupied by broad alluvial lowlands; e.g., Wilson River syncline, Cooper Creek syncline (a north-south syncline just east of Jackson field, Figure 1B), and Bulloo River syncline (Figure 20).

Erosional landforms at all evolutionary stages are present, from simple doming of the duricrust to

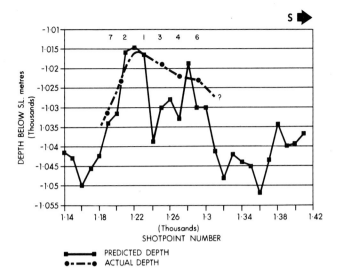

Figure 18. Tintaburra field, EH84-05, chart showing actual and predicted depths to basement. Largest discrepancies occur in channel flank well where ray path distortion is greatest.

breaching of the anticlines; capture and gathering of axial drainage is the final evolutionary stage.

Landsat Four false color imagery clearly shows the trend of both the Tintaburra anticline and Harkaway fault. Physiographically, the Tintaburra anticline is at an advanced stage of evolution, consisting of an elongate, slightly sinuous core zone comprising dark-toned soils developed on Winton Formation sedimentary rocks. The soil plain is surrounded by dissected duricrust. Strongly defined linear features to the northwest indicate the trend of the Harkaway fault on the east flank of the Harkaway anticline. Drainage exits are concentrated on the southwest flank of the field due to the steeper dips associated with a northwest-southeast-trending antithetic fault.

The Wilson River is confined where it traverses at right angles the southern nose of the Harkaway and Tintaburra anticlines.

STRUCTURE

Tectonic History

The Jurassic-Cretaceous Eromanga basin is a pericratonic basin and is classified as a 121 under the scheme of Bally and a IIIA under the scheme of Klemme. The irregular lobate shape of the Eromanga basin is typical of pericratonic basins where underlying older Paleozoic basins have influenced the deposition of younger sequences.

The Tintaburra field is located south and east of the subcrop edges of the Permian-Triassic Cooper and Devonian-Carboniferous Adavale basins where Jurassic Eromanga basin sediments directly overlie

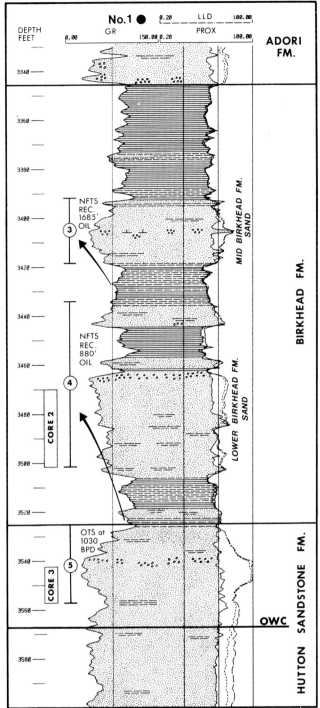

Figure 19. Toobunyah 1 Birkhead Formation and top of the Hutton Sandstone log signature, lithology, and test results.

basement lithologies on the Thargomindah shelf (Figure 1B).

Major angular unconformities exist between Eromanga and Cooper basin sediments and between Cooper and Adavale basin sediments.

Two major phases of structural deformation can be recognized on the northern Thargomindah shelf.

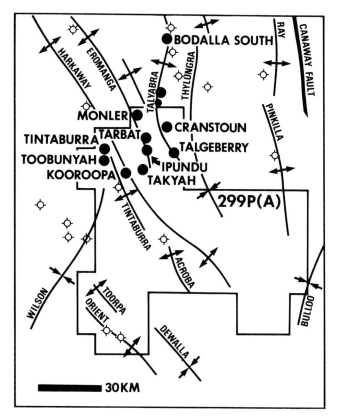

Figure 20. Tectonic elements as mapped at the "C" horizon seismic reflector (base Wallumbilla Formation) and seen at the surface.

1. Late Triassic: Structuring was intermittent throughout the Permian and culminated at the end of the Triassic with compressional deformation, uplift, and erosion resulting in removal of much of the Permian and Triassic Cooper basin sequence that underlies the Eromanga basin. The compressional deformation is manifested by folding wrench and thrust faulting.

Jurassic and Cretaceous time was a period of gentle cratonic downwarp and relative tectonic quiescence. Oblique continent/island arc collision between Australia and Papua New Guinea was responsible for the second major phase of structural deformation.

2. Late Tertiary/Miocene: Continent/island arc collision resulted in compressional deformation on the Australian craton and is manifested in the Eromanga basin by subtle basement folds and high-angle reverse faulted anticlines. Anticlinal trends in the Eromanga basin sequence are strongly influenced by underlying Paleozoic structural grain.

Regional Structure

Basement involved compressional deformation of late Tertiary age forms the dominant structural style on the northern Thargomindah shelf and is manifested by a number of north-northwest/south-southeast-trending anticlines (Figure 20). The Tintaburra anticline parallels the prominent surface-defined Harkaway anticline 4.3 mi (7 km) to the east. Both anticlines are bounded at depth by high-angle reverse faults that generally displace the Jurassic strata and die out in the Cretaceous sequence.

Local Structure

Tintaburra is a culmination on the western upthrown side of a mid-Tertiary age, high-angle, reverse-faulted anticline that plunges gently to the northwest (Figure 2). A small northwest–southeast-trending antithetic fault, also of Tertiary age, enhances closure on the Tintaburra/Toobunyah structure in the critical southerly direction (Figure 14). Figure 2 shows that the anticline comprises two culminations, separated by a shallow saddle at basement and Jurassic levels. The northern culmination was drilled first and comprises the Tintaburra field, while the southern culmination represents the Toobunyah field. The saddle between the two culminations has a depth of approximately 40 ft (12.2 m) at the top Hutton Sandstone level, less than the gross oil column. The two fields are connected at the top Hutton level.

Figure 2 shows structural spill for the two culminations at the top Hutton level is north of Tintaburra 2 where closure becomes fault-dependent.

Associated with the Tertiary high-angle reverse faulting is detached extensional deformation in the form of listric faulting within the Cretaceous. The listric faulting formed in response to tensional stress on the crest of the anticline during folding. The faults sole out near the base of the marine Cretaceous sediments (Figure 3).

STRATIGRAPHY

The Jurassic-Cretaceous Eromanga basin covers an area of 401,440 mi^2 (1.04 million km^2) of central and eastern Australia and spans four states. The Eromanga basin forms approximately 60% of the Great Artesian basin, one of the largest artesian basins in the world.

The Jurassic-Cretaceous stratigraphy of the northern Thargomindah shelf is summarized in Figure 4. Sedimentation began in Early Jurassic with deposition of nonmarine and locally lagoonal-lacustrine units. They consist of fluvial quartzose sandstone, thin carbonaceous shales, siltstones, and minor coals deposited on the eroded surface of the Cooper basin, Adavale basin, and older, nonsedimentary basement. The shales within this sequence, known as the basal Jurassic, are thicker and more organic rich in the basin depocenter to the north where they are considered marginally mature to mature oil-prone source rocks. The basal Jurassic sequence onlaps southwards onto the Thargomindah shelf and is represented at Tintaburra/Toobunyah by some 30 ft (9.2 m) of sandstone and silty shales no older than PJ2 (Figure 4). Continued cratonic downwarp in the basin led to widespread deposition of the clean quartzose fluvial units of the Hutton Sandstone.

Jurassic units gradually onlap the Thargomindah shelf to the southeast. The Hutton Sandstone thins by approximately 200 ft (61 m) between Tintaburra and Orient, some 47.2 mi (76 km) to the south.

A transgression resulted in a change in deposition from fluvial quartz sands of the Hutton Sandstone to lagoonal-lacustrine and lower delta plain siltstones and shales of the Birkhead Formation. These organic-rich shales and siltstones provide the seal to hydrocarbons reservoired at the top of the Hutton Sandstone. Hydrocarbons have also been found in channel and progradational sandstones within the Birkhead Formation.

Reservoir sandstone distribution within the Birkhead is sporadic. Sand content and reservoir quality increase, and ability to act as a seal decreases, with increasing fluvial character to the south and southeast onto the Thargomindah shelf. Juxtaposition of the Hutton Sandstone reservoir and the Birkhead seal is responsible for entrapment of approximately 80% (author's estimate) of oil reserves in the Eromanga basin.

The Birkhead Formation is considered a good potential source rock. Thickness, source quality, and thermal maturity increase to the northwest into the basin depocenter.

The Hutton–base of the Birkhead boundary is diachronous and strongly facies controlled. These facies variations account for most of the rapid variation in thickness of the two units and have important implications for trap integrity.

A relative regression resulted in re-establishment of high energy fluvial-nearshore deposition in the area. Clean quartzose sandstones of the Adori Formation were deposited prior to being transgressed by low-energy lacustrine facies of the Westbourne Formation. The Adori Formation exhibits good to excellent reservoir properties.

Interbedded sandstone and subordinate siltstone characterize the Westbourne Formation. Claystones are rare and are confined to the top and to a lesser extent the middle of the unit. The mineralogical content of the fine-grained Westbourne Formation sandstones makes the gamma ray log response of the unit anomalous. Concentrations of mica, potassium, feldspar, and other radioactive minerals cause the formation to appear to be much shalier on the gamma ray than it actually is.

On the Thargomindah shelf, the Westbourne Formation is essentially sand-dominated and only the top 50-100 ft (15.3-30.5 m) of the formation has potential to act as a seal. Economically significant hydrocarbons are therefore only found beneath this top Westbourne Formation seal and not within the lower Westbourne or Adori formations where sealing units are generally absent. Reservoir quality in the Westbourne Formation varies from fair to excellent.

Cretaceous sedimentation (Figure 4) began with deposition of high-energy quartzose fluvial sandstone (Namur sandstone member) of the Mooga Formation. These braided fluvial sands pass up into progradational sand bodies of a nearshore environment. The Namur sandstone member is an excellent reservoir rock and a major artesian aquifer. Hydrocarbon accumulations, however, are rare as the Namur and base of the overlying Murta Member generally lack laterally continuous effective seals. Upper Namur sandstone member deposition is gradational into lacustrine to marginal marine Murta Member of the Mooga Formation, consisting of organic-rich shales, silts, and fine-grained sandstones of variable reservoir quality.

The Murta Member is considered a fair to moderate potential oil source rock throughout most of the Eromanga basin and is considered by the majority of workers to be the source for most of the oil in the Murta and overlying Wyandra sandstone member of the Cadna-owie Formation.

Oil accumulations found in the Murta to date have been confined to top porosity, immediately below the laterally continuous shale seal at the top of the Murta.

A basin-wide regression resulted in deposition of a coarsening-upward sequence of paralic/deltaic shales, siltstones, and fine-grained sandstones of the Cadna-owie Formation. Shales of the lower Cadna-owie Formation have moderate source potential and along with the Murta probably provide the source for oil reservoirs of the Wyandra sandstone member. On the Thargomindah shelf, deltaic progradation prior to the major Aptian transgression resulted in deposition of the Wyandra sandstone member at the top of the Cadna-owie Formation (Figure 8). The Cadna-owie Formation can be widely correlated across the southwest Eromanga basin and exhibits a strong southeasterly provenance. Fluvial equivalents of nearshore deltaic deposition on the Thargomindah shelf are recognized on the Cunnamulla shelf (southeast of area), and in northern New South Wales (Bungil Formation).

The Wyandra sandstone member is stratigraphically the highest major aquifer in the Eromanga basin, and major supplies of artesian water come from the unit on the Cunnamulla shelf, where fluvial nearshore deposition resulted in excellent reservoir quality sands.

In the basin depocenter northwest of Tintaburra, delta front sands were deposited at the top of the Cadna-owie Formation and as such the Wyandra sandstone has limited reservoir potential.

The impedance contrast between the high-velocity, sometimes cemented Wyandra sandstone member and low-velocity, Wallumbilla shales produces a characteristic "C" horizon seismic reflector that persists over much of the Eromanga basin.

The Wallumbilla Formation provides a regional seal to the Jurassic and Early Cretaceous artesian reservoirs. The formation is organically lean and is generally immature by conventional standards.

The nature of the "C" horizon reflector over the northwest portion of ATP 299P(2) and adjoining permits is distinctive because it is the base of major submarine channel erosion. Where channeling occurs, the Wallumbilla Formation is strongly unconformable on the underlying Cadna-owie Formation. In places, submarine channeling has eroded up to 950 ft (289.7 m) of Cretaceous and Late Jurassic units, leaving the Wallumbilla Formation unconformably overlying units stratigraphically as low as the Westbourne Formation.

Initiation of submarine channel erosion is thought to have resulted from an Aptian sea level rise, with consequent submergence of the Thargomindah shelf. Sediment starvation on the outer shelf terminated shelf and slope construction, and erosion of the shelf and slope began.

World-wide, submarine canyon erosion is a phenomenon that occurs locally during marine onlap (deposition less than subsidence); it coincides with reduced sediment supply to platform areas and deposition of condensed sequences or with complete sediment starvation and local erosion (Brown, 1985).

The channels in the Tintaburra area exhibit a sinuous bifurcating pattern, trending northwest toward the Cretaceous depocenter in the Windorah trough. Maximum channel erosion is 950 ft (289.7 m), which implies a similar minimum water depth in the depocenter to the northwest.

Deposition of the Wallumbilla Formation shales and mudstones marked the beginning of distal marine sedimentation in the basin in response to the early Aptian sea level rise. The Wallumbilla Formation thickens regionally along with other Cretaceous and Jurassic units into a depocenter to the northwest. The Wallumbilla Formation also exhibits dramatic local thickness variations as it infills the deep submarine channels with low-velocity sediments. The effect of this low-velocity channel fill is seen in Figure 11, where a distinct velocity sag is evident under the axis of a channel.

Two types of channel fill are known from seismic and well data. First, chaotic fill consisting of gravity- and-density-transported fine-grained sandstone and siltstone, and second, marine claystones of the Wallumbilla Formation. Following infill of the channels, deposition of the remaining Cretaceous units is characterized by relative tectonic quiescence and slow regional subsidence. The Early Cretaceous Toolebuc, Allaru, Mackunda, and Winton formations (Figure 4) are all recognized on the northern Thargomindah shelf and show only minor lateral variation (in contrast to underlying Jurassic fluvio-lacustrine units).

The Toolebuc Formation is an excellent potential source rock, extremely rich in oil-prone organic matter; however, the unit is considered conventionally immature throughout most of the basin.

Cretaceous deposition ceased in late Cenomanian and the top of the Winton Formation is an erosional unconformity strongly weathered during the Late Cretaceous and Tertiary.

A thin veneer of early Tertiary alluvials and duricrusts (Senior and Mabbutt, 1979) disconformably overlies the Early/Late Cretaceous Winton Formation over much of the basin.

TRAP

Trap Type

The Tintaburra oil field is an anticlinal trap with multiple pays. The main pay, the Hutton sandstone, is dominantly structurally controlled. The gross oil column in the top of the Hutton is 64 ft (19.5 m), which conforms with the mapped fault-independent closure at that level, the reservoir being full to structural spill. The OWC varies within a few feet over the field, the average being 3010 ft (918 m) ss. The reservoir is sealed by shales and mudstones of the overlying Birkhead Formation. Integrity of the base of the Birkhead seal is maintained across the Tintaburra/Toobunyah field; however, this is not the case in other fields in the area where oil, if present, is trapped at the top of porosity within the Birkhead Formation.

Sandstones within the Birkhead Formation are poorly developed and tight over Tintaburra, but good reservoir quality sands are present within the mid- and lower Birkhead Formation on the Toobunyah culmination (Figure 21).

There is a strong stratigraphic component to the Birkhead Formation trap, and the oil column in the Birkhead Formation exceeds the 64 ft (19.5 m) of mapped fault-independent closure. This is explained by the stratigraphic pinch-out of the Birkhead reservoir northward onto the Tintaburra culmination (Figures 21 and 22) where the closure mapped in depth at the top of the Hutton (Figures 2 and 4) and the top of the Birkhead Formation levels opens out into the fault.

The Murta Member and Wyandra sandstone member oil accumulations are partly stratigraphically controlled in that the seals to these reservoirs are breached by Neocomian submarine channeling. A 40 ft (12.2 m) oil column is present at the top of porosity in the Murta Member sealed by the regionally extensive Murta Member shale.

Spill point of the Murta oil accumulation is coincident with the level where submarine channeling seen at the "C" horizon breaches the top Murta seal (Figures 3 and 11). The sandy base channel fill does not seal, and the zone is full to channel-independent closure.

Only Tintaburra 1 and 2 intersect top of porosity in the Murta Member above the field OWC.

Hydrocarbons are reservoired within the Wyandra sandstone member at the top of porosity below the Wallumbilla Formation regional seal. Figure 3 indicates that the spill point for the Wyandra

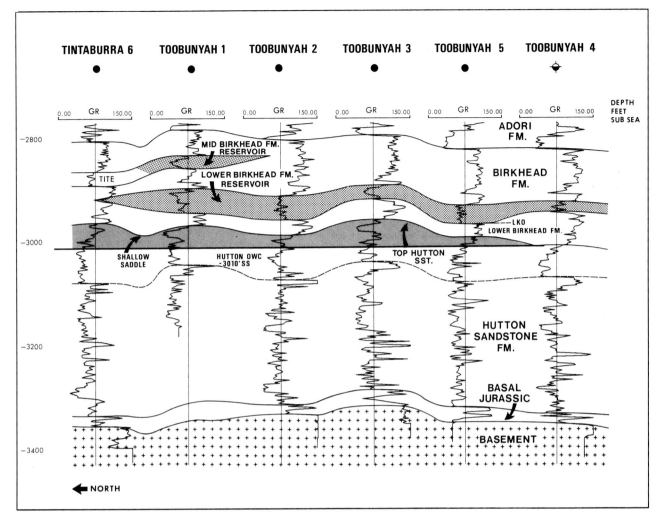

Figure 21. Toobunyah culmination north-south structural cross section, Adori Formation to basement showing distribution of Birkhead Formation reservoirs. See Figure 2. for location.

sandstone member is coincident with the breach of seal by channel erosion limiting the oil column to 40 ft (12.2 m) (as with the Murta Member oil column). Only Tintaburra 1 intersected the top of porosity in the Wyandra sandstone above the OWC.

Oil was recovered on test from the Westbourne Formation at Tintaburra 2, crestally located at the mid-Westbourne Formation level. Figure 23 shows the oil to be recovered from a 14 ft (4.3 m) thick shaly sand underlying a 3 ft (0.9 m) shale break. Despite approximately 60 ft (18.3 m) of mapped closure at this level, the observed oil column is only a few feet in extent, probably reflecting lateral discontinuity of the seal.

Structural closure at Tintaburra/Toobunyah is a result of the mid-Tertiary compressional episode that gave rise to the reverse-faulted Tintaburra anticline. No closure would have existed prior to the mid-Tertiary structuring.

Reservoirs

Hutton Reservoir

In Tintaburra this unit ranges in age from PJ3 to upper PJ5 (Figure 4) and averages 380 ft (116 m) in thickness, thickening regionally to the north. The Hutton Sandstone comprises well-sorted, quartzose sandstones (Figure 24) with minor siltstone interbeds occurring toward the middle and top of the unit. The net oil reservoir sands consist of medium-grained, well-sorted sandstones with mono-crystalline quartz (Figure 24) forming between 60% and 80% of the framework grains; metamorphic rock fragments and feldspars constitute the remainder. Primary intergranular porosity is dominant, interconnected with minor secondary dissolution porosity.

Authigenic kaolinite and silica (Figure 24) are the principal precipitates in the pore system. Only trace quantities of expandable smectite are observed, and

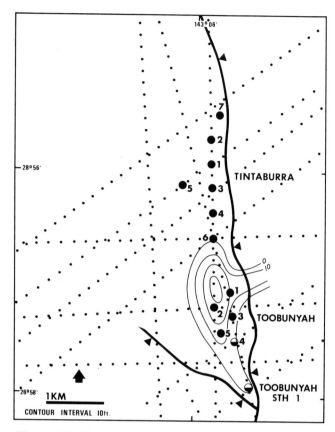

Figure 22. Tintaburra/Toobunyah field: lower Birkhead Formation reservoir net sandstone isopach.

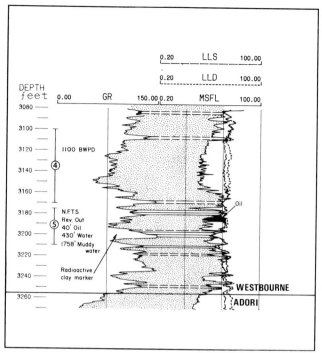

Figure 23. Tintaburra 2 Westbourne Formation log signature, lithology, and test results.

no acid-soluble carbonates are present. Formation damage is known to occur (Doolan et al., 1988) owing to kaolinite clay particle migration, and drilling/completion practices need to be optimized to avoid loss of well productivity.

Average depth to top of the reservoir is 3225 ft (984 m) and net oil pay varies from 5 ft (1.5 m) in flank wells to a maximum of 35 ft (11 m) on the crest of the structure. Average porosity in pay is 20%. Permeability in the oil zone ranges up to 3500 md and averages approximately 500 md. A porosity/permeability X-plot for Tintaburra 2, typical for the top Hutton, across the field is shown in Figure 25. Logs and capillary pressure data (Figure 26) indicate average connate water saturation in pay is 36%. The capillary pressure data indicate an oil-water transition zone of up to 20 ft (6 m); however, log and core data clearly indicate a transition zone of only 2–4 ft (0.6–1.2 m). The reason for the discrepancy is uncertain.

Gamma ray curve shapes used in conjunction with the dipmeter allow a reasonable analysis of depositional environment and imply a sediment transport direction to the southwest.

Fining-upward, medium- to coarse-grained, cross-bedded sandstone facies dominate the lower and middle Hutton sandstone, indicative of high-energy fluvial deposition. Coarsening-upward, coarse-grained, high-angle cross-bedded sandstone facies (Figure 24) predominate at the top of the formation, indicative of nearshore bar deposition.

High-energy sedimentation of both the above types is closely associated at the top of the Hutton, indicating a marginal marine/lacustrine environment rather than a fluvial one. The boundary between the top of the Hutton sandstone and Birkhead Formation is transitional in nature, reflecting gradually decreasing depositional energy associated with a transgression. Rapid facies variations occur at the top of the Hutton/base of the Birkhead, leading to dramatic variations in reservoir quality in the oil zone. These facies variations had a significant influence on the success of the appraisal drilling program.

A study of cores, logs, and dipmeter data was initiated after the first three wells, aimed at developing a predictive model for net reservoir distribution. The model consists of two stream mouth bar systems centered on wells 6 and 2 and prograding to the west-southwest. Figure 27 shows a classical coarsening-upward stream mouth bar facies sequence observed in all wells.

Coarse- to very coarse grained, high-angle cross-bedded sands form the main hydrocarbon reservoir and are best developed in wells 2 and 6. Wells 1, 3, and 4, lateral to the main depositional axis, have slightly poorer reservoir development. Variations of the top of the Hutton reservoir affect the net pay distribution (Figure 28).

The Birkhead transgression is marked by a cemented sandstone on top of the cross-bedded sandstone facies. Transgression from the west-

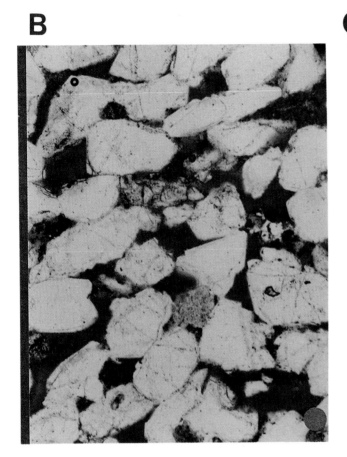
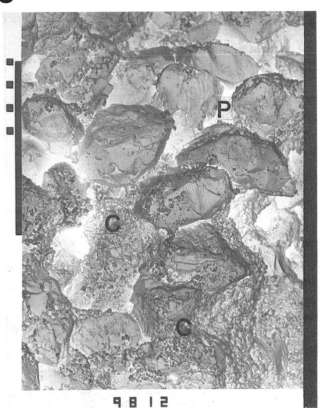

Figure 24. Top Hutton Sandstone oil reservoir. (A) Core slab showing medium- to coarse-grained, high-angle cross-bedded sandstone facies. (B) Photomicrograph of sample similar to that in (A) above showing upper medium-grained, well-sorted sandstone. Quartz (colorless) is the principal grain with minor feldspar (light brown). Red dot, 0.12 mm. (C) SEM micrograph of sample (B) above. Open pores (P) are about 100 microns in diameter. Clay (C) coats grains and may infill other pores. Bar length, 1 mm.

southwest marked a change in deposition to stacked upward-fining units, each representing a pulse of higher energy sedimentation into gradually increasing water depths on top of a subsiding bar.

Nearshore/interdistributary bay silt and shale nonreservoir sands were deposited laterally to the main depositional axis and transgressed over the fluvial system. A final high-energy pulse of sedimentation deposited net reservoir sands overlying either bay silts or bar top deposits and represents the top of the Hutton Formation. This model successfully predicted the development of these sands at Tintaburra 7.

The overlying lower Birkhead Formation shales represent offshore lacustrine facies distinct from sandy ripple laminated facies of the underlying bay environment. Notwithstanding facies variations at the top Hutton and possible near wellbore formation damage, there are no barriers to flow in the reservoir, which is not faulted within the pool.

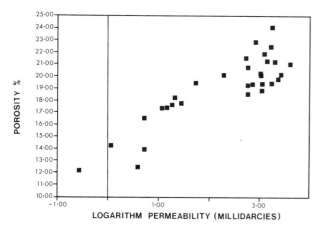

Figure 25. Representative top Hutton Sandstone porosity permeability X-plot.

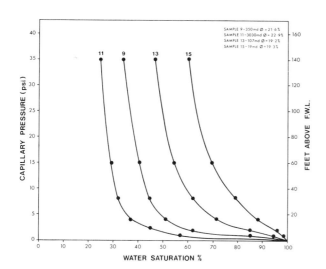

Figure 26. Tintaburra 7: capillary pressure vs. S_w plot.

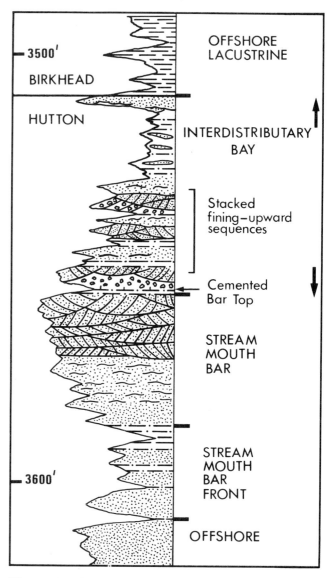

Figure 27. Characteristic lithological sequence at the top of the Hutton Formation.

Initial PIs (productivity indices) ranged from 1.65 to 5.0 BOPD/psi and averaged 3.0 BOPD/psi. PIs in a number of wells fell by up to 100% because of completion problems and/or near-wellbore formation damage owing to fines migration. Following installation of jet pumps on five of the six producing Tintaburra wells, field production increased from 300 BOPD to 700 BOPD from a total of six wells. Water cut increased rapidly during the early life of the field to stabilize at approximately 90%.

The oil column overlies approximately 300 ft (92 m) of Hutton Sandstone aquifer, and this provides an essentially infinite bottom water drive. The bottom water drive could be expected to result in recovery factors in excess of 30%. However, two factors negate this.

1. Reservoir permeability is very "streaky" and often greater near the OWC (fining-upward sequence), resulting in high initial water cuts.
2. Relative permeability data (Figure 29) clearly show K_{rw} (water relative permeability) end points of 0.4 to 0.5 which, combined with an adverse mobility ratio, means that advancing water breaks through the oil zone in fingers rather than pushing the oil zone ahead of it in a bank.

The economic constraints of producing the field at very high water cuts determine that ultimate primary recovery will be approximately 20%.

Oil recovery profiles per well dictate development well spacing, with 30 ac (12 ha) spacing currently proving to be economically the most viable.

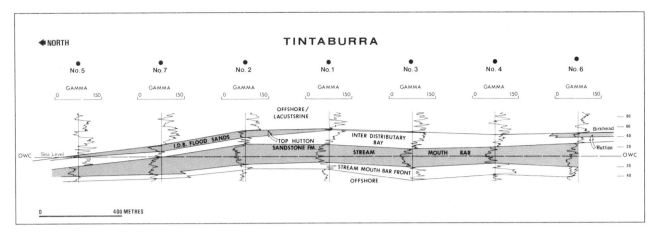

Figure 28. Structural cross section through Tintaburra field showing the top of Hutton/base of the Birkhead reservoir variations. For location of the wells see Figure 2.

Birkhead Formation

Reservoir quality sandstones within the Birkhead Formation only occur on the Toobunyah culmination. The reservoir comprises light brown, fine- to medium- and occasional coarse-grained, moderate- to well-sorted, lithic and arkosic sandstone. Kaolinite coats the grains and infills the pore spaces, severely reducing permeability. Lithics and feldspars form up to 35% of the framework grains. The reservoir is thinly interlaminated with siltstones. Primary intergranular porosity is dominant with minor secondary dissolution porosity after feldspars.

Gross pay ranges from zero feet on the Tintaburra culmination to a maximum of 50 ft (15 m) (22 ft or 6.7 m net pay). Average porosity in pay is 15% and permeability ranges up to 170 md, averaging approximately 5 md. Connate water saturation is uncertain because the reservoir is a low permeability, shaly sand. Connate water is fresh, and conventional log analysis is misleading.

Core facies analysis and log curve shape analysis suggest both the upper and lower Birkhead sandstones are progradational sand bodies of the distributary mouth bar type.

The lower Birkhead Formation sandstone, the main reservoir, is interpreted to have an easterly provenance (Figure 22).

Only Toobunyah 1 has been completed for production from the reservoir to date, and much is to be learned about the producing characteristics of the reservoir.

In excess of 2.5 MMSTB of oil is estimated in place; however, the recovery factor is uncertain because it depends on the extent (if any) of the pressure support from downdip water and the number and location of producing wells.

The very low GOR and absence of a significant water drive indicate primary recovery factor could be less than 5%.

Secondary recovery methods will be considered when more data are available.

Mooga Formation—Murta Member

The Murta Member comprises interbedded and interlaminated fine- to very fine grained sandstone, siltstone, and minor claystone units that can be readily correlated across the field. The Murta is absent in the central part of the Tintaburra/Toobunyah field owing to erosion by Neocomian submarine channeling (Figure 3). Depositional energy is low, and a lacustrine to brackish marine environment with a minor deltaic component is inferred. Quartz forms the dominant framework grain in the sandstones with rock fragments and feldspar each ranging up to 17%.

A significant amount of primary intergranular porosity is retained with minor secondary dissolution porosity. Ductile grain deformation reduces porosity. Authigenic kaolinite is the dominant cementing agent with minor silica and chlorite. Approximately 10 ft (3 m) of net oil pay is interpreted in the crestal well (Tintaburra 2) within a 40 ft (12 m) gross oil column. Field OWC is at 2068 ft (631 m) ss. Average porosity in pay is 18% and average permeability in the oil zone is less than 3 md. Because of the thin oil column and low productivity, economics have ensured the zone has never been put on production.

Cadna-owie Formation

The late Neocomian Cadna-owie Formation forms a major progradational sequence of interlaminated siltstones and claystones that coarsen upward into fine-grained, clay-choked sandstones.

The Wyandra sandstone member at the top of the sequence consists of interlaminated fine- to medium-grained sandstone and siltstone and represents the final stage of Neocomian deltaic progradation. Deposition energy, however, was low at Tintaburra/Toobunyah and reservoir quality is fair at best. Feldspar and rock fragments constitute up to 40% of the framework grains. Authigenic clay and calcite cement are the main diagenetic components.

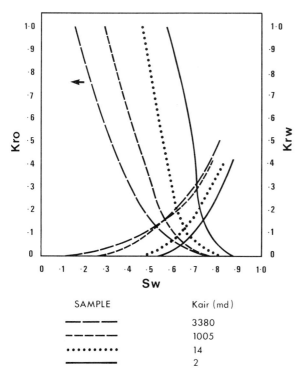

Figure 29. Representative oil-water relative permeability plots.

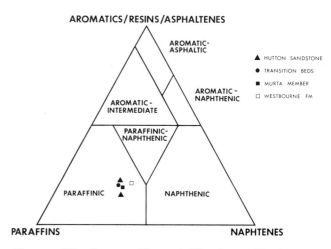

Figure 30. Composition of Tintaburra/Toobunyah crudes according to classification scheme of Tissot and Welte (1978).

The Cadna-owie Formation has been removed by Neocomian submarine channel erosion at Tintaburra 3, 4, 5, and 6.

A maximum 8 ft (2.4 m) of net oil pay is interpreted at Tintaburra 1 within a 40 ft (12.2 m) gross oil column (OWC 1883 ft or 574 m ss). Porosity in pay averages 17%, but permeability is very "streaky" and averages only 2 md.

Hydrocarbon Characteristics

Oils recovered from the Jurassic and Cretaceous sequence at Tintaburra all exhibit a similar C_{12+} bulk composition. All are low sulfur paraffinic crude oils (Figure 30).

Physical properties and hydrocarbon compositions indicate a consistent difference between Hutton/Birkhead crudes and those from stratigraphically higher reservoirs (Table 2). The low gasoline yield of the Hutton crude accounts for its low API gravity and high pour point.

Representative whole oil gas chromatograms of Murta, Birkhead, and Hutton crudes are shown in Figure 31.

All Tintaburra crudes have been affected by water washing, hence their low relative concentrations of low molecular weight aromatics. Notwithstanding small maturity and source differences, severe water washing of the Hutton crude is primarily responsible for its differing physical properties. Water washing preferentially removes the low molecular weight aromatics, then cyclic alkanes, followed by low molecular weight alkanes. Depletion of the low molecular weight alkanes results in the low gasoline yield for the Hutton crude.

Producing GOR for the Hutton is less than 20 SCF/bbl and at original reservoir pressure of 1730 psi (11,928 kPa), formation volume factor was 1.05 RB/STB.

Faults

The high-angle reverse fault on the eastern side of the Tintaburra anticline is clearly recognizable on seismic data (Figures 12 and 17), as is the smaller northwest–southeast-trending antithetic fault to the south of the Toobunyah culmination. Trend of the major faults is well-defined on the close seismic grid. The faults are not considered seals, and fault-independent closure is considered a prerequisite for success in the Eromanga basin sequence. The lack of fault seal is undoubtedly a function of the very sandy nature of the Jurassic sequence and the absence of thick laterally extensive sealing units.

Throw on the main north-south high-angle reverse fault is up to 300 ft (92 m) at the basement level with the throw dying out to zero in the more incompetent Wallumbilla Formation mudstones.

Seismic data clearly show postdepositional listric faulting within Wallumbilla sediments. The listric faults sole out within sometimes overpressured channel fill sediments. A listric fault intersects Tintaburra 1 and 3 with at least 50 ft (15 m) of the Wallumbilla Formation faulted out including the distinctive Correna gamma ray marker at Tintaburra 1 (Figure 3). Listric faulting also produces marked thickness variations within the Cadna-owie Formation and Murta Member on the Toobunyah culmination. Fault intersects are clearly identified on dipmeter data (Rostenburg, 1985).

Table 2. Physical properties and compositional analysis of Tintaburra/Toobunyah oils.

Crude	API Gravity (°)	Pour Point (°C)	Gasoline Yield (%)	Sulfur Yield (ppm)
Wyandra	50.l	0	46	75
Murta	48.5	6	49	210
Westbourne	47.0	5	44	210
Birkhead	45.0	15	—	—
Hutton	43–44	18	11–18	140–180

SOURCE

Despite extensive geochemical effort, the origin of oils recovered within the Eromanga basin sequence is enigmatic.

Early workers believed the underlying Permian-Triassic Cooper basin sequence was the sole source for Eromanga basin crudes. Geochemical data certainly support this in some cases, and the geographic distribution of oil fields along the Permian subcrop edge is strong supporting evidence. Numerous workers (Vincent et al., 1985; McKirdy, 1985; Newton, 1986; Alexander et al., 1988) now believe in an intra-Eromanga source for many of the oil fields.

The source for oil within the top Hutton sandstone at Tintaburra is thought to be organic-rich shales of the basal Jurassic and Birkhead formations.

TOC and pyrolysis data indicate the basal Jurassic and Birkhead formations have a moderate to high source richness, with TOC averaging 1.0% and ranging up to 4%. North of Tintaburra, in the Bodalla South-Kenmore area (Figure 1B), the basal Jurassic has a TOC content averaging in excess of 2.5% (Analabs, 1987). Pyrolysis yields an average of approximately 35 mg/g TOC.

The organic matter is entirely of land plant origin. Pyrolysis indicates (Figure 32) a mixed type II/III kerogen, and this is confirmed by actual examination of the dispersed organic matter (DOM) in which vitrinite and liptinite are observed in approximately equal quantities. Sporonite and liptodetrinite form the major liptinite macerals of the waxy, oil-prone kerogen. Algal organic matter forms a minor part of the dispersed organic matter in both the Birkhead and basal Jurassic.

Vitrinite reflectance data at Tintaburra indicate the basal Jurassic and Birkhead formations to be marginally mature (R_o 0.41–0.53) by conventional standards. The low productivity indices (SI/SI+S2) support this. These maturation levels are probably sufficient for oil generation from more labile liptinite macerals (Snowden and Powell, 1982).

The Murta Member of the Cretaceous Mooga Formation has a fair to moderate oil source potential across a large portion of the Eromanga basin (Cook, 1982; Kantsler et al., 1983). Average TOC content is 0.8% ranging up to 2.0%. Pyrolysis (Figure 32) and organic petrology indicate a mixed type II and III/IV kerogen. Liptinite (type II) makes up between 30% and 40% of the DOM with inertinite (type III/IV) making up the remainder. Sporonite is again the major liptinite maceral. Pyrolysis yields average 25 mg/g TOC. Approximately 200 ft (61 m) of potential source rocks are present within the Cadna-owie Formation and Murta Member, in which fair to moderate quality source rocks are interlaminated with poor source rock units. Present depth to the top of the sequence is 2900 ft (885 m).

Oils recovered in the Murta and Wyandra reservoirs at Tintaburra are interpreted to have been from sources in the surrounding Cadna-owie and Murta Member shales at maturation levels with a vitrinite reflectance equivalent of R_o 0.04–0.48.

Source quality of Jurassic and Cretaceous source rocks is thought to have been enhanced by extensive bacterial reworking of the primary land plant input (Thomas, 1982). McKirdy (1984) suggested bacterial activity during early diagenesis produced a lipid-rich bacterial biomass of increased oil source potential. The C_{12} and C_5-C_7 gasoline range hydrocarbon data from Tintaburra crudes indicate bacterial contribution is greatest in Cretaceous crudes (Figure 33).

Figure 34 indicates the Tintaburra crudes were derived primarily from terrestrial organic matter (C_{29}/C_{27} sterane >1.5) that accumulated below an oxic water column (pristane/phytane >2).

The presence of biomarkers in the Hutton crude specific to the Mesozoic implies an *intra-Eromanga source* for the Hutton crude. The distribution of oil on the northern Thargomindah shelf throughout the Jurassic and Cretaceous sequence, often in areally restricted sandstone bodies encased in potential source rocks, suggests an Eromanga origin for the crudes, and in this respect, biomarker geochemical data are entirely consistent.

Should it be necessary to consider oil migration from deeper, thermally more mature source rocks, Tintaburra is well situated on the northern end of the Thargomindah shelf to receive hydrocarbons migrating from the Nappamerri trough to the southwest (just south of Jackson field, Figure 1B) and Cretaceous depocenters to the north and northwest.

The present-day geothermal gradient on the Thargomindah shelf is greater than 6°C/100 m, almost the highest in the Eromanga basin and well above the world average of 3.0°C/100 m (Spicer, 1942).

The high geothermal gradient is thought to be a product of the Tertiary tectonism. This Tertiary

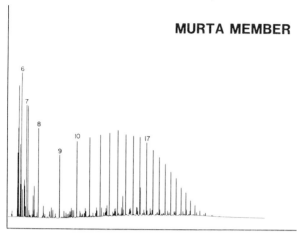

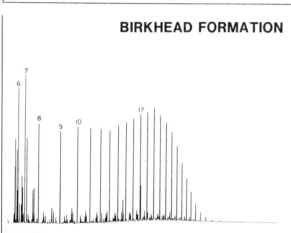

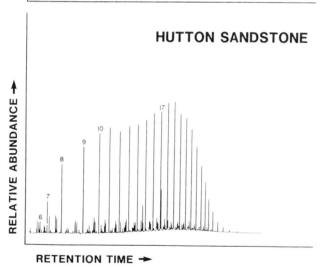

Figure 31. Representative whole oil gas chromatograms of Murta, Birkhead, and Hutton crudes.

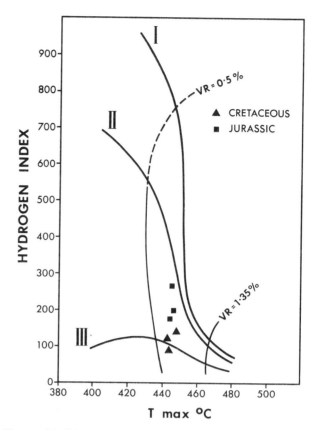

Figure 32. Kerogen type and maturity based on Rock-Eval, T_{max}, and hydrogen index data.

EXPLORATION AND DEVELOPMENT CONCEPTS

The exploration play successfully tested at Tintaburra/Toobunyah is not unique in the Eromanga basin; however, it has not been successful everywhere. The play—four-way dip closure at the top of porosity on Tertiary anticlines—has proven very successful on the Thargomindah shelf with ten fields producing from six different reservoirs.

The discoveries made to date are located along a number of prominent structural trends including the Harkaway, Tintaburra, and Tallyabra trends (Figure 20). South-bounding antithetic faults that set up closure in the critical southerly direction are common to a number of the fields.

The question of why the play works on the Thargomindah shelf and apparently does not in other areas of the basin cannot be answered with generalities. Each success and failure needs to be critically analyzed with respect to structure, seal, reservoir, and source/migration. Drilling results on the northern Thargomindah shelf have highlighted a number of important factors that have significant impact on the success of an exploration test.

1. The Jurassic and Cretaceous (Neocomian) sequence is essentially sandstone dominated, and

tectonic phase that set up the structural trap at Tintaburra also produced the heating pulse which matured the Mesozoic source rocks.

Both the timing of structuring and the heating event indicate hydrocarbon generation and migration to be post mid-Tertiary. Lopatian modeling incorporating a Tertiary heating phase confirms this.

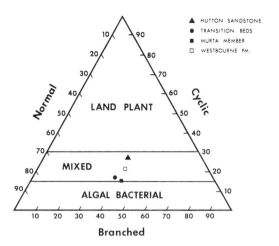

Figure 33. Source character based on C_5–C_7 alkanes.

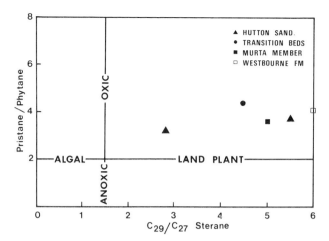

Figure 34. Source affinity of Tintaburra oils.

lateral continuity of seal over a structural target cannot be assumed. Fault-independent closure appears to be mandatory.
2. Rapid facies variations that occur at the tops of reservoir/base seal horizons are generally not resolvable with conventional seismic. Defining the time crest of a structural closure is not enough; every effort must be made to locate top of porosity beneath a laterally continuous seal.
3. The abnormally high heat flow on the Thargomindah shelf, a product of the Tertiary tectonic episode, has rapidly matured potential source rocks within the Eromanga basin sequence. These source rocks are only marginally mature by conventional standards, yet significant volumes of oil have obviously been generated.
4. The observation that every reservoir at Tintaburra/Toobunyah within fault-independent closure and overlain by laterally continuous seals contains oils suggests that sourcing is not a problem in the area.

Appraisal drilling at Tintaburra provided the first hard data on the nature of Necomian submarine channel fill and identified a number of channel-related plays. Associated plays include erosional topographic remnants and yet-to-be-defined basin floor mounded facies at channel mouths, possibly deepwater fan deposit.

At Tintaburra, the use of well data as seismic velocity restraints resulted in a useful measure of the uncertainty inherent in depth estimation by means of the amount of discrepancy between depth estimates and actual values; this is a departure from the usual practice of forcing depth maps to fit well intersections.

ACKNOWLEDGMENTS

Hartogen Energy Limited and the joint venture participants of ATP 299P(2) permitted publication of this paper. The author is grateful to the technical personnel of Hartogen Energy Limited whose work forms the basis of this paper. Special thanks to Doug Battersby, Dennis Morton, and Dr. Dave King whose advice and assistance in completing the study are appreciated.

Stephen Greaves has been responsible for seismic interpretation in the permit, and his work and suggestions have been invaluable in preparing this study. The author is indebted to Hartogen for providing the necessary drafting services and to the Drafting Department for their usual high standard. Last but certainly not least, thanks to Kerri Rich for typing and assistance.

Technical opinions expressed herein are those of the author and not necessarily those of the joint venturers in ATP 299P(2). Much of this paper is drawn from a previous publication by the same author (Newton, 1986).

REFERENCES CITED

Alexander, R., A. V. Larcher, R. I. Kagi, and P. L. Price, 1988, The use of plant derived biomarkers for correlation of oils with source rocks in the Cooper-Eromanga Basin system, Australia: APEA Journal, v. 28, n. 1, p. 310-324.

Analabs, 1987, Petroleum geochemistry, Eromanga-Cooper Basins, Queensland, Australia: Commercially available report.

Blackburn, G. J., 1985, Depth conversion—a comparison of method: Exploration Geophysics, v. 17, n. 2, p. 67-73.

Brown, L. F., 1985, Facies analysis and its role in seismic stratigraphic interpretation: PESA Distinguished Lecture Series.

Cook, A. C., 1982, Organic facies in the Eromanga Basin, in P. S. Moore and T. J. Mount, compilers, Eromanga Basin Symposium, Summary Papers: Geological Society of Australia and Petroleum Exploration Society of Australia, Adelaide.

Denmead, A. K., 1960, Occurrence of petroleum and natural gas in Queensland, in Geological Survey of Queensland, Publication 299, 78.

Doolan, P., P. D. Griffiths, and S. R. Welton, 1988, The successful recovery of well productivity in the Bodalla South oil field: APEA Journal, v. 28, n. 1, p. 7-18.

Kantsler, A. J., T. J. C. Prudence, A. C. Cook, and M. Zwigulis, 1983, Hydrocarbon habitat of the Cooper/Eromanga Basin: APEA Journal, v. 23, n. 1, p. 75-92.

McKirdy, D. J., 1984, Source rock evaluation, Tintaburra 1, Eromanga Basin, Queensland. Unpublished.

McKirdy, D. J., 1985, Biomarker geochemistry of three oils from Tintaburra 2, ATP 299P(2), Eromanga Basin. Unpublished.

Newton, C. B., 1986, The Tintaburra oil field: APEA Journal, v. 26, n. 1, p. 334-352.

Price, P. L., et al., 1985, Late Palaeozoic and Mesozoic palynostratigraphical units: CSR Oil and Gas Division, Palynology Facility, unpublished.

Rostenburg, J., 1985, Tintaburra oil field dipmeter report for Hartogen Energy Limited. Unpublished.

Senior, B. R., 1978, Landform developments, weathered profiles and Cainozoic tectonics in southwest Queensland: PhD Thesis, University of N.S.W. (unpublished).

Senior, B. R., and J. A. Mabbutt, 1979, A proposed method of defining deeply weathered rock units based on regional geological mapping in southwest Queensland: Geological Society of Australia Journal, v. 26, p. 237-254.

Snowdon, L. R., and T. G. Powell, 1982, Immature oil and condensate, modification of hydrocarbon generation model for terrestrial organic matter: American Association of Petroleum Geologists Bulletin, v. 66, p. 775-788.

Spicer, H. C., 1942, Observed temperatures in the earth's crust: GSA Special Paper 36, p. 281-292.

Tissot, B. P., and D. H. Welte, 1978, Petroleum formation and occurrence: Berlin-Heidelberg, Springer-Verlag, 538 p.

Thomas, B. M., 1982, Land plant source rocks for oil and their significance in Australian basins: APEA Journal, v. 22, n. 1, p. 164-178.

Vincent, P. W., I. R. Mortimore, and D. M. McKirdy, 1985, Hydrocarbon generation, migration and entrapment in the Jackson-Naccowlah area, ATP 259P, south western Queensland: APEA Journal, v. 25, n. 1, p. 62-84.

SUGGESTED READING

Biederstadt, K. C., 1988, Hydraulic jet pumping at Tintaburra oil field: APEA Journal, v. 28, n. 1, p. 19-27. Jet pumping is utilized to produce the Hutton sandstone oil reservoir at Tintaburra. This paper looks at some of the principles of jet pumping and explains why and how it is used at Tintaburra.

Habermehl, M. A., 1986, The Great Artesian Basin, Australia: BMR Journal of Geology and Geophysics, v. 5, p. 9-38.

Moore, P. S., and T. J. Mount, 1982, Eromanga Basin Symposium Summary Papers, Geological Society of Australia and Petroleum Exploration Society of Australia, Adelaide. Although a little dated, a useful compilation of papers on all aspects of the Eromanga basin.

Senior, B. R., N. F. Exon, and D. Burger, 1975, The Cadna-owie and Toolebuc formations in the Eromanga Basin, Queensland: Queensland Government Mining Journal, December 1975, p. 445-455.

Smythe, M., et al., 1984, Birkhead revisited—petrological and geochemical studies of the Birkhead Formation, Eromanga Basin: APEA Journal, v. 24, n. 1, p. 230-242.

Watts, K. J., 1987, The Hutton sandstone-Birkhead Formation transition, ATP 269P(1), Eromanga Basin: APEA Journal, v. 27, n. 1, p. 215-229. ATP 269P(1) covers the area north of ATP 299P(2) and includes the Bodalla South oil field. More information on the nature of the top Hutton/base Birkhead facies variations.

Appendix 1. Field Description

Field name .... Tintaburra field

Ultimate recoverable reserves .... 1.1 MMBO (Tintaburra culmination only)

Field location:
- Country .... Australia
- State .... Queensland
- Basin/Province .... Eromanga basin

Field discovery:
- Year first pay discovered .... Middle Jurassic Hutton Sandstone 1983
- Year second pay discovered .... Lower Cretaceous Wyandra sandstone Mbr of Cadna-owie Fm 1983
- Year third pay discovered .... Lower Cretaceous Murta Sandstone of Mooga Fm 1983
- Year fourth pay discovered .... Middle Jurassic Birkhead Fm sandstones 1985

Discovery well name and general location:
- First pay .... Tintaburra No. 1, Hutton Sandstone, Authority to Prospect 299P(2) area
- Second pay .... Tintaburra No. 1 Wyandra sandstone
- Third pay .... Tintaburra No. 1, Murta Member, Mooga Formation
- Fourth pay .... Toobunyah Member No. 1, Birkhead sandstone

Discovery well operator .... Hartogen Energy Limited

IP:
- First pay .... 300 BOPD from Tintaburra No. 1 (Hutton Sandstone)
- Second pay .... 30 BOPD from Tintaburra No. 1 (Wyandra sandstone)
- Third pay .... Murta sands never put on production
- Fourth pay .... 50 BOPD from Toobunyah No. 1 (Birkhead sandstone)

All other zones with shows of oil and gas in the field:

Age	Formation	Type of Show
Jurassic	Westbourne	Oil recovery on DST

Geologic concept leading to discovery and method or methods used to delineate prospect

Surface geology and Landsat data indicated a number of parallel north-northwest to south-southeast anticlinal trends on the northern Thargomindah shelf. Seismic data shot to investigate these trends confirmed a four-way dip closure at Tintaburra on the upthrown side of a high-angle reverse fault. Tintaburra No. 1 was designed to test the hydrocarbon potential of the Jurassic Hutton Sandstone sealed by the overlying Birkhead Formation shale and mudstones within fault-independent closure on the Tintaburra anticline. The structure was thought to be optimally located with respect to both potential Permian-Triassic and Jurassic-Cretaceous source rocks.

Structure:

Province/basin type .... Bally 121, Klemme IIIA

Tectonic history

Compressional deformation at the end of the Triassic followed by erosion of much of the Permian-Triassic sequence preceded deposition of the Eromanga basin sequence. The Jurassic-Cretaceous was a period of gentle cratonic downwarp and relative tectonic quiescence; however, oblique continent-island arc collision between Australia and Papua New Guinea in the late Tertiary resulted in compressional deformation on the Australian craton and is manifested in the Eromanga basin sequence by subtle basement fold and high-angle, reverse-faulted anticlines.

Regional structure

Basement involved compressional deformation forms the dominant structural style on the northern Thargomindah shelf and is manifested by a number of north-northwest to south-southeast-trending anticlines. The Tintaburra anticline parallels the prominent Harkaway anticline 4.5 mi (7 km) to the east.

Local structure

Tintaburra is a culmination on the western upthrown side of a Tertiary reverse-faulted anticline, the Tintaburra anticline, which plunges gently to the north-northwest.

Trap:

Trap type(s)

1. Anticline trap with multiple pays. The main pay, the top Hutton Sandstone, is dominantly structurally controlled, the accumulation being full to fault-independent closure.
2. Murta and Wyandra accumulation are partly stratigraphically controlled in that seals to these reservoirs are breached by Neocomian submarine channeling.
3. Birkhead reservoir pay on the southern (Toobunyah) culmination has a strong stratigraphic component. The reservoir shales out onto the southern end of the Tintaburra culmination and a greater oil column is present in the Birkhead reservoir sandstone than in the Hutton reservoir sandstone.

Basin stratigraphy (major stratigraphic intervals from surface to deepest penetration in field):

Chronostratigraphy	Formation	Depth to Top in ft (m)
Cretaceous Albian–Cenomanian	Winton	Surface
Cretaceous Albian	Mackunda/Allaru	300 (92)
Cretaceous Albian	Toolebuc	1400 (427)
Cretaceous Aptian–Albian	Wallumbilla	1440 (439)
Cretaceous Neocomian	Cadna-owie	2380 (726)
Cretaceous Neocomian	Mooga	2580 (787)
Jurassic Oxfordian–Tithonian	Westbourne	3000 (915)
Jurassic Oxfordian	Adori	3260 (994)
Jurassic Callovian	Birkhead	3350 (1022)
Jurassic Aalenian–Callovian	Hutton	3500 (1068)
Ordovician	Basement	3900 (1190)

Reservoir characteristics:

Number of reservoirs 4

Formations
(1) Cadna-owie Formation, Wyandra sandstone member; (2) Mooga Formation, Murta Member; (3) Birkhead (Toobunyah culmination only); (4) Hutton Sandstone

Ages .... Wyandra and Murta, Neocomian; Birkhead and Hutton, Callovian

Depths to tops of reservoirs .... Wyandra, 2374 ft (724 m); Murta, 2596 ft (792 m); Birkhead, 3473 ft (1059 m); Hutton, 3504 ft (1069 m)

Gross thickness (top to bottom of producing interval)
Gross pay thickness in the Wyandra, Murta, and Hutton at Tintaburra is 124 ft (38 m); an additional 71 ft (22 m) of gross oil column is present in the Birkhead Formation on the Toobunyah culmination

Net thickness—total thickness of producing zones
Average .... 40 ft (12 m)
Maximum .... 64 ft (20 m)

Lithology
Hutton Sandstone: well-sorted, medium-grained quartzose sandstone; monocrystalline quartz forms 60–80% of framework grains
Birkhead reservoir: fine- to medium-grained, lithic and arkosic sandstone, moderate to abundant authigenic kaolinite cement
Murta reservoir: fine- to very fine grained quartzose sandstone interlaminated with siltstone and claystone; authigenic kaolinite cement, rare calcite cement
Wyandra reservoir: fine- to very fine grained arkosic sandstone; authigenic kaolinite cement

Porosity type .... Hutton sandstone, primary intergranular porosity is dominant, interconnected with minor secondary dissolution porosity

Average porosity .... Hutton Sandstone, 20%; Wyandra sandstone, 17%; Murta Member, 18%; Birkhead (Toobunyah culmination), 15%

Average permeability .... Hutton Sandstone, 500 md; Wyandra sandstone, 2 md; Murta Member, 3 md; Birkhead (Toobunyah culmination), 5 md

Seals:

Hutton reservoir:
- **Upper**
  - **Formation, fault, or other feature** .... Birkhead
  - **Lithology** .... Shale/mudstone
- **Lateral**
  - **Formation, fault, or other feature** .... Downdip oil-water contact
  - **Lithology** .... Sandstone

Murta and Wyandra reservoirs:
- **Upper**
  - **Formation, fault, or other feature** .... Cadna-owie and Wallumbilla formations, respectively
  - **Lithology** .... Shale
- **Lateral**
  - **Formation, fault, or other feature**   OWC in part and Neocomian channeling with channel fill in part
  - **Lithology** .... Shale, sandstone

Birkhead reservoir:
- **Upper**
  - **Formation, fault, or other feature** .... Birkhead Formation shale
  - **Lithology** .... Shale
- **Lateral**
  - **Formation, fault, or other feature** .... Facies changes
  - **Lithology** .... Shales and siltstones

Source:

Hutton and Birkhead reservoirs:
- **Formation and age** ..... Basal Jurassic and Birkhead Formation shales of Lower to Middle Jurassic Age
- **Lithology** .... Shale/mudstone
- **Average total organic carbon (TOC)** 1.0%
- **Maximum TOC** 4.0%
- **Kerogen type (I, II, or III)** .... Mixed type II and III
- **Vitrinite reflectance (maturation)** .... $R_o = 0.53$
- **Time of hydrocarbon expulsion** .... Late Tertiary to present
- **Present depth to top of source** .... 3900 ft (1190 m)
- **Thickness** .... Average 20 ft (6 m)
- **Potential yield** .... Unknown

Murta and Wyandra reservoirs:
- **Formation and age** .... Cadna-owie Formation and Murta Member shales of Neocomian age
- **Lithology** .... Shale/mudstone
- **Average total organic carbon (TOC)** 0.8%
- **Maximum TOC** 2.0%
- **Kerogen type (I, II, or III)** .... Type III/IV, minor type II
- **Vitrinite reflectance (maturation)** .... $R_o = 0.45$
- **Time of hydrocarbon expulsion** .... Late Tertiary to present
- **Present depth to top of source** .... 2900 ft (884 m)
- **Thickness** .... Average 100 ft (30 m)
- **Potential yield** .... Unknown

Appendix 2. Production Data

Data for Hutton Sandstone as of 31 July 1987

Field name .... Tintaburra field

TINTABURRA

Field size:

Proved acres	500 ac (203 ha)
Number of wells all years	6
Current number of wells	5
Well spacing	30 ac (12 ha)
Ultimate recoverable	1.1 MMBO
Cumulative production	0.36 MMBO
Annual production	0.15 MMBO
Present decline rate	22% per year on jet pump
Initial decline rate	80% on natural flow
Overall decline rate	30%
Annual water production	2.0 MMBW
In place, total reserves	3.3 MMBO
In place, per acre foot	650 bbl/ac-ft
Primary recovery	1.1 MMBO
Secondary recovery	None
Enhanced recovery	None
Cumulative water production	2.8 MMBW

Drilling and casing practices:

Amount of surface casing set .... Surface casing set at 500 ft (151 m) in each well

Casing program
8⅝-in. surface casing set to approximately 500 ft (150 m) KB; 5½-in. production casing set to TD (approx. 3900 ft [1189 m]) and 2⅜-in. tubing run to 30 ft (9 m) above perforations

Drilling mud .... Surface casing shoe to TD; low solids KCl polymer

Bit program .... 12¼-in. 1-1-4 (IADC) to surface casing shoe and then 8½-in. 5-1-7 to TD

High pressure zones .... Artesian water sands below 2400 ft (732 m) require 10.4 ppg mud

Completion practices:

Interval(s) perforated .... Main pay (Hutton) perforated at 6 S.P.F. (TCP guns) (60# phasing) to 10 ft above OWC

Well treatment .... Perforate 1250 psi (8619 kPa) under-balance with oil in hole

Formation evaluation:

Logging suites
(1) DLL-MSFL-GR-CAL: TD to surface casing shoe; (2) SLS: TD to surface casing shoe; (3) LDL-CNL: TD to 200 ft (61 m) above Cadna-owie Formation; SHDT run on selected wells

Testing practices .... All zones with porosity and hydrocarbon indications tested on penetration or at TD after logging

Mud logging techniques
Mud logging service from surface casing shoe to TD; samples caught at 10 ft intervals, continuous monitoring of mud gas, drill rate, and all rig functions

Oil characteristics:

Type	Low sulfur paraffinic crude oil
API gravity	44°
Initial GOR	Less than 50 SCF/bbl
Sulfur, wt%	0.002
Viscosity, SUS	29
Pour point	18°C (64°F)
Gas-oil distillate	Gasoline yield 15%

Field characteristics:

Average elevation	540 ft (165 m)
Initial pressure	1730 psi (11,928 kPa)

Present pressure	1730 psi (11,928 kPa)
Pressure gradient	0.494 psi/ft (1.62 kPa/m)
Temperature	175°F (79°C)
Geothermal gradient	0.062°C/m (0.034°F/ft)
Drive	Strong bottom water drive
Oil column thickness	64 ft (20 m)
Oil-water contact	3010 ft (918 m) subsea
Connate water	36%
Water salinity, TDS	2400 mg/L (Hutton Sandstone)
Resistivity of water	2.4 ohm at 25°C (77°F)
Bulk volume water (%)	0.072%

Transportation method and market for oil and gas:
The oil is trucked 50 mi (80 km) to the head of the Jackson oil field to Brisbane pipeline. Pipeline distance to Brisbane is approximately 620 mi (1000 km). To December 1987, all domestically produced crude was taken by refiners under the Australian crude oil allocation system. From January 1988 a free market environment exists.

La Cira-Infantas Field—Colombia
Middle Magdalena Basin

PARKE A. DICKEY
Consultant
Owasso, Oklahoma

FIELD CLASSIFICATION

BASIN: Middle Magdalena
BASIN TYPE: Backarc
RESERVOIR ROCK TYPE: Sandstone
RESERVOIR ENVIRONMENT
  OF DEPOSITION: Fluvial Channels

RESERVOIR AGE: Oligocene
PETROLEUM TYPE: Oil
TRAP TYPE: Lenticular Sandstones over
  Domal Structure

TRAP DESCRIPTION: Downthrown anticline and dipping beds truncated against high-angle reverse fault

LOCATION

The La Cira-Infantas field is the first oil field to be discovered in Colombia, South America, and was the largest until the discovery of Caño Limon in the Eastern Plains in 1984. It is in the Magdalena River Valley, near the center of the country, at 7°04′N, 73°47′W, about 279 mi (450 km) from the Caribbean coast at Barranquilla and 155 mi (250 km) north of Bogotá, the capital of Colombia (Figure 1).

There are many smaller oil and gas fields in the vicinity of La Cira-Infantas, notably Casabe, 12 mi (20 km) west, and Provincia, 40 mi (65 km) north. These fields produce similar oil from the same geologic intervals but are not described in this paper.

At the latitude of La Cira-Infantas, the Andes Mountains consist of three separate ranges, called the Western Cordillera, the Central Cordillera, and the Eastern Cordillera. The Magdalena Valley lies between the Central and Eastern Cordilleras (Figure 1).

The cumulative production of La Cira-Infantas to December 1986 is 683 million barrels (MMBO).

La Cira and Infantas are two distinct fields with separate closures although both are on the same structure (Figure 2). The Infantas field was discovered first. It is a reverse-faulted anticline with oil on the downthrown (west) side of the fault. The La Cira field is a sort of half-dome north and west of Infantas, also on the west side of the same fault. They have always been considered separate fields although they are contiguous and produce from the same formations.

HISTORY

Pre-Discovery

In April 1536 the Spanish explorer Gonzalo Jimenez de Quesada left Santa Marta on the north coast of Colombia and started up the Magdalena River (Oviedo, 1959). In October they had progressed about 279 mi (450 km) and reached an Indian village called Latora. Some of the force had poled their way upstream in small boats (*bergantins*) and some, with horses, followed the banks. These were covered with thick vegetation, were sometimes swampy, and were often broken by tributary streams from the mountains.

They renamed the village Barrancas Bermejas ("vermilion banks") because of the red color of the banks of the river. From here they sent out scouting parties up the tributary rivers. They found an Indian with a load of salt in one of these tributaries, the Opon, and from that deduced that there must be civilization up in the mountains to the east.

They also heard of an oil seep up the Colorado River, near its confluence with the Oponcito. They later reported to Oviedo, the historian, that

> there is a spring of bitumen [*betun*] which is a pool that boils and runs out of the ground, and it is [located] entering into the forest at the foot of the hills, and there is a large quantity of thick liquid. And the Indians bring it to their houses and anoint themselves with this bitumen because they find it good to take away

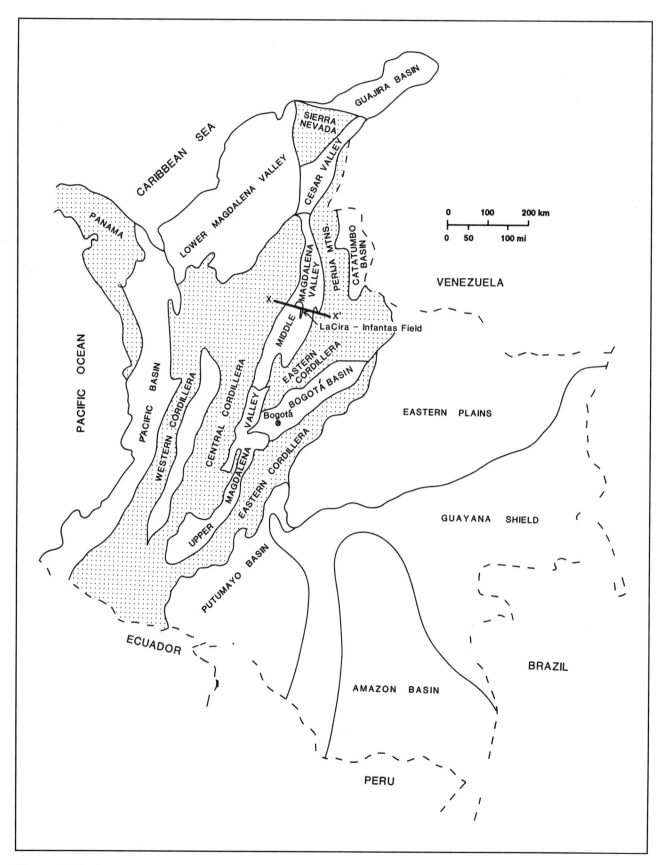

Figure 1. Sedimentary basins of Colombia showing location of La Cira-Infantas field.

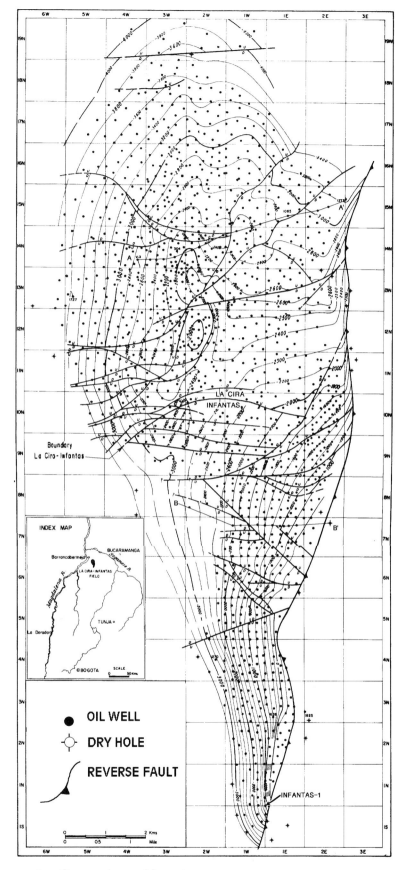

Figure 2. Structure map, top C zone lower Mugrosa Formation, La Cira-Infantas field. Contour interval, 100 and 200 ft (30 and 60 m). (After Taborda, 1965.)

tiredness and to strengthen the legs, and the Christians use this black fluid, with the smell of fish or worse, to tar (caulk) their boats.

The Spaniards named the place of the seep *Infantas*, that is, "princesses," after the two young daughters of the King of Spain. This reference to an oil seep in the New World is 200 years earlier than the first reference to an oil seep in New York State.

The De Mares Concession

In 1903 Jose Joaquin Bohorquez arrived in Barranca Bermeja, which by that time had become a river port (Santiago Reyes, 1986). Coffee was brought by mule train from the cities in the mountains on a trail through the jungle to the river, where it was loaded on steamboats. Bohorquez started a small transportation company. Besides handling shipments of coffee, he exploited the rubber and vegetable ivory that were abundant in the tropical rain forest covering the valley of the Magdalena. The climate of the valley is hot and humid, with 150 in. of rainfall yearly. Malaria was endemic, and until very recently few people wanted to live there. Most Colombians live in the cool and fertile valleys in the mountains.

Bohorquez took samples of the oil to Barranquilla and Cartagena but found little interest. Roberto de Mares saw the samples and came to Barranca to investigate further. In an informal partnership with Bohorquez, he proposed obtaining a concession to exploit the oil. He was a friend of the President of Colombia and was granted the concession in November 1905. Apparently, Bohorquez was cut out of the deal at some point, became impoverished, and was granted a small pension by the government in 1939.

The area of the concession is about 1976 mi^2 (5120 km^2) (Donoghue, 1951). The area was not known exactly until about 1928 when it was surveyed accurately by J. F. B. O'Sullivan of the Tropical Oil Company.

The contract required work to be started in 18 months, but this requirement was extended several times. In 1909 the Ministry of Public Works pronounced the concession canceled. It was restored in May 1915, with the requirement that work start within 12 months. De Mares finally interested the firm of Benedum and Trees of Pittsburgh, Pennsylvania, who had been very successful wildcatters in the United States and Mexico. According to Benedum, De Mares happened to meet a geologist named John Leonard who was taking a vacation to South America. Leonard told Benedum about De Mares's concession. He considered it an immense challenge, which it was. The area was almost unpopulated, was forested, and was plagued with malaria. Access was only by river boat. The Magdalena had regular steamboat traffic, but the Colorado was small and navigable only by canoes. To bring equipment from the United States would take months.

In October 1915, a geological party with John Leonard and a Colombian engineer resident in the United States named Luciano Restrepo arrived at Infantas and started a clearing. They also started to remove the logs and snags from the Rio Colorado. In February 1916, a group consisting of Benedum and Trees and others from Pittsburgh visited the area. At the end of February, De Mares asked approval of the government to transfer the concession to Benedum and Trees. In May 1916, they organized the Tropical Oil Company in Pittsburgh as a Delaware corporation with initial capital of $50,000,000.

The transfer of the concession was first denied but the company continued work. (The concession was not approved until 20 June 1919.) On 24 June 1916, ten days before the time to start work had expired, the Chief of Police of Barranca and representatives of De Mares, the government, and the Tropical Oil Company certified that work had begun. This ceremony was called the *Acta de San Vicente*.

Discovery

Although the government had still not approved the transfer of the concession to the Tropical, they proceeded with the development. Three cable-tool drilling rigs were found up the river, abandoned after an earlier attempt to exploit an oil seep. They were hauled to Barranca on the Magdalena and up the Colorado to Infantas by canoe. Infantas No. 1 was started 14 June 1916, 21 days before the expiration of the reinstated contract. It was not completed until 11 November 1918 with an initial production of 2000 bbl per day (BOPD) at 1943 to 2280 ft (553 to 695 m) (Donoghue, 1951). The discovery well is usually considered Infantas No. 2, which was spudded 12 December 1917, 1.2 mi (2 km) north of Infantas No. 1, and completed 29 April 1918, at 1531 to 1580 ft (467 to 481 m), with an initial production of 800 to 1000 BOPD. A third well about 984 ft (300 m) southwest of No. 2 was completed 19 December 1918 for 600 BOPD. Infantas No. 1 was located close to the fault and the oil seep, and less than 1 km from the southern limit of the field. Infantas No. 1 is the origin for the kilometer square grid of Figure 2.

The three successful wells aroused much interest in the oil possibilities of Colombia. In 1919, a new petroleum law was passed. Its requirements seemed onerous, but by the end of 1919 over 40 companies had taken out concessions.

In 1919, Benedum and Trees offered the De Mares concession to Standard Oil Co. (N.J.). A. V. Hoenig was sent to investigate (Gibb and Knowlton, 1956). He reported that the properties might be worth $5,000,000. At the same time, Standard sent James Flanagan on a confidential mission to review not only the properties but also the geographical and political problems. Flanagan had served the company in a political capacity in Peru. Many of the directors felt that Benedum would ask too much money, and the project was difficult and risky. The president, Walter Teagle, was an influential leader. He said, "There are very few sure things in the oil business, especially in the producing end, and the individual or corporation that does not take some chances never gets very far." The concession was finally purchased by International Petroleum Co. for 1,804,000 shares of International stock, worth $33,000,000. International Petroleum Co. had been organized a few years earlier to take over the La Brea and Parinas Estate properties in Peru from the London and Pacific Oil Co. International Petroleum was a subsidiary of Standard's Canadian affiliate, Imperial Oil Ltd. There was political hostility toward the United States in Colombia because of its support of the secession of Panama in 1903, so it was decided to make Tropical Oil Company an affiliate of the Canadian subsidiary. Many of the personnel sent to Colombia were Canadians.

Development of the Field

The years 1920 to 1924 were spent in preparation. A camp with bunk-houses, mess halls, and even a hospital were constructed at Infantas. A truck road was built from Barranca. Additional drilling rigs were brought in, and drilling started in 1921. By 1924, there were 17 producing wells. It was found that rotary tools were faster and cheaper than cable tools. By the end of 1926 when the pipeline was finished, there were 171 producing wells.

It was evident by this time that the field would extend northward from Infantas, so it was decided to transfer the camp and shops to El Centro, 5 mi (8 km) north of Infantas, and make it the center of operations. A meter-gauge railroad was built from Barranca Bermeja to El Centro and the truck road was abandoned. A small refinery was built at Barranca that went on stream in 1922.

Because of delays in getting equipment, partly caused by World War I, the government postponed the date at which the 30-year concession started to run. It was finally fixed as 25 August 1921.

In 1919, the Andean National Corporation Ltd. was formed as another affiliate of International Petroleum Co. in order to build a pipeline to the coast. James W. Flanagan was made manager, and he had difficulties coping with the political problems. Apparently he concealed the true ownership of the company, representing it as Canadian, which, of course, it partly was. The concession for the pipeline was granted in September 1922. Flanagan stated that it had been "the cleanest business of its kind that has ever been accomplished in this country." Actual construction of the line started in 1925 and the first oil reached the terminal at Mamonal, near Cartagena, on 10 June 1926.

The line was 10 in. (25.4 cm) in diameter and there were ten pump stations. The capacity at first was 30,000 BOPD, which was later increased by looping to 50,000 BOPD. The Andean line was replaced by a new line in 1985.

During the years 1922 to 1926, dwellings, shops, an electric generating plant, and other facilities were built at El Centro. By the end of 1927, a sum of $23,521,000 had been spent on the Colombian enterprise, but in that year the net income was $6,943,000.

Geological Mapping

During those years the whole concession was mapped by surface geology. It was entirely covered by tropical rain forest, inhabited by a few Indians. There was a mule trail from San Vicente de Chucuri, a town in the foothills of the Eastern Cordillera, east of the concession, through the forest to Barranca, and there were a few other small settlements on the west flank of the mountains. Surface geology was difficult because trails had to be cut. Outcrops were only found in the rivers and small streams. Although there were many dangerous animals, including poisonous snakes, the principal hardship was caused by swarms of insects, especially mosquitoes and sand flies called *jejenes* (pronounced "hay-hens"). Fresh food such as locally grown plantains, bananas, and cassava (yucca) were hard to transport by mules in the wet forest where it rained more than 150 in. per year. Canned food was also difficult to obtain and transport. The geologists and their Colombian helpers lived mostly on dried (jerked) beef, rice, coffee, and the hard cakes of brown sugar called *panela*.

The boundaries of the concession were surveyed by J. F. B. O'Sullivan, a Canadian born in Ireland. He measured a base line where the Sogamoso River comes out of the mountains at the northeast corner of the concession. He then proceeded to triangulate southward, occupying stations (called *banderas*) on the peaks of the foothills and in the adjacent forest until he arrived at a settlement called Marsella, near Landazuri at the foot of the mountains, where he measured another base line. So accurate was his work that the second base line agreed with the distance carried from the first by triangulation within 10 cm.

In spite of the difficulties and hardships, the entire surface of the concession had been mapped by 1930. Many American and Canadian geologists were involved. The most distinguished of these was O. C. Wheeler, who came to Colombia in 1921. He measured the section of Cretaceous in the gorge of the Rio Sogamoso and established the type section that is still in use. He became chief geologist in 1929 and in 1930 was transferred to Toronto where he was in charge of exploration in Colombia and Peru. He visited the area frequently. T. A. Link, later chief geologist of Imperial Oil Ltd. of Toronto, mapped the mountain front, which was the east boundary of the concession. It was marked by a large, high-angle, reverse fault, which brought up the Cretaceous along the east side of the concession. Other pioneer geologists were W. W. Waring and A. K. McGill.

These reconnaissance studies revealed a number of anticlinal structures. La Cira was discovered by Wheeler when he noted a fault in a railroad cut. South of the Colorado River, sharp anticlines were found and drilled. Mugrosa and Colorado produced light oil but were considered noncommercial. San Luis is another sharp anticlinal structure near the mountain front east of Infantas. The area between Infantas and the mountain front and north to the Sogamoso River was mapped in detail in 1935 by Parke Dickey, Oscar Haught, and Edward La Tour. Several gentle structures in this area have since proved productive.

The years 1922 to 1930 were also spent employing and training workmen. They had to be brought in from the mountains and especially from the coastal savannas. The coastal areas are cattle country, and the cowboys were used to physical labor and the hot climate. Very few could read or write, and they were totally ignorant of anything mechanical. The story is told of an American driller who got along well with only three words of Spanish: *No, otro lado!* ("No, the other side!").

The relations between the expatriates and the Colombians were generally excellent. With few exceptions, the Americans learned Spanish and devoted themselves to teaching the Colombians. Great mutual respect and affection developed. By 1932, all the drillers were Colombian, as well as the truck drivers, pumpers, and other roustabouts. When the concession reverted to the government oil company in 1951, it was the pumpers and roustabouts who kept the oil flowing while new Colombian superintendents and engineers learned their jobs.

The pay was about twice the going rate for unskilled labor in Colombia, so that jobs with the Tropical were eagerly sought. There were occasional labor troubles, including strikes, but these were the result of outside agitators and organizers. In government circles, it was always popular to denounce the American "imperialists." Once, on the floor of congress, a senator described conditions in the oil fields as miserable and said that the workmen were held in peonage. When the company representative in Bogotá invited the senator to visit the field he replied that he knew his accusations were untrue, but that he was not in politics for his health!

Medical Problems

The Standard Oil Co. (N.J.) organized a medical department in 1919 (Gibb and Knowlton, 1956). Dr. Alvin W. Schoenleber first went to Mexico and then Colombia. His reports showed appalling sanitary conditions, stating that all the people in Colombia in the vicinity of the concession were sick. Malaria, hookworm, and amoebic dysentery were endemic, and many other tropical diseases were prevalent. There were few doctors and no modern hospital facilities in the entire country. After his arrival, steps were taken to improve sanitation and to construct first a temporary hospital and finally a base hospital in El Centro. The wet tropical climate and the crowding resulting from the boom activity made sanitation measures difficult. In one month in 1921, the malaria admission rate was 1300 per 1000 employees, and the amoebic dysentery rate was 750 per 1000. Sanitary measures and medical assistance were sometimes impeded by the necessity to get on with the drilling. Finally, as many as 15 doctors were on the staff and a modern hospital was built in El Centro. The achievements of the Tropical became a model for Standard affiliates elsewhere in the world. The incidence of malaria and other diseases was drastically reduced, and health care became generally recognized as important to industry. Still, however, in 1937 when a road was built to the Lisama wildcat north of El Centro, the malaria rate was 150% of the number of workmen.

The Reversion of the De Mares Concession to the Government

The reversion of the De Mares concession to the Colombian government was a very noteworthy historical event (Mendoza, 1987). It was characterized by strict faithfulness to the terms of the contract on the part of the government, and a wholehearted willingness on the part of the company to turn over a functioning operation with all possible help to keep it functioning. It was marked also by evidence of mutual respect and affection between the Americans, Canadians, and Colombians.

The original contract between the government and Roberto De Mares specified that the term of the agreement was 30 years dating from the time exploitation work started. This date was postponed several times. In 1919, the International Petroleum Co. took over the operation and started "large scale exploitation." Accordingly, by an official resolution of 13 June 1921, the government specified that the start of operations was 25 August 1921.

However, the Chamber of Deputies in 1937 asserted that the start of operations was 14 June 1916, when the *Acta de San Vicente* was signed. The reversion to the government therefore should take place on 14 June 1946. The Supreme Court finally decided that the date of reversion was to be 25 August 1951.

Finally, at 12 P.M. midnight, 25 August 1951, the ceremony of signing the reversion took place in the big hall of the Club Internacional in El Centro. Present were the Minister of Development; the president of the Tropical Oil Co.; L. P. Maier, President of International Petroleum Co. of Toronto; and many other dignitaries of the local area, the government, and the company. In his speech, L. P. Maier praised the faithfulness with which both sides had lived up to their agreements and stated that while the equipment in the oil field was of great value to the country, the greatest gift was the staff of technical people and skilled workmen that the company had trained. The Minister in his speech pointed out that the country had obtained in a congenial and orderly manner the nationalization which in other countries had been premature and politically acrimonious. Both hoped that it would be an example to other countries. It is unfortunate that it has not been.

In a continuing effort to get the Colombian oil company off to a good start, the Tropical assigned American and Canadian advisers to some of the new department heads. This arrangement did not last long, for there was a tendency for the Colombians to give the responsibility to the foreign engineers.

It was, of course, assumed that the operation of the company by the government would enrich the treasury. Although no figures are available, this may be doubted. The labor laws and the unions maintained the labor force larger than really necessary, in spite of the field's declining production. At one point, men who had no work to do were trucked in the morning to a waiting zone where they spent the day, and then trucked back in the afternoon (Roberto Sarmiento, former government geologist, personal communication, 1982). In 1959, there were 30 chemical engineers in the refinery when six would have been enough (Frederick Wellington, refinery manager, personal communication, 1959). The real contribution to Colombia was, as Mr. Maier said, the training of engineers and workmen to the point that they could operate an oil field with little outside assistance.

Operations Since the Reversion

Following the reversion, there was a steady decline in production rate until 1960, when additional drilling for water injection maintained steady production of La Cira until 1975 (see decline curve, Figure 17). Because many of the wells had been drilled in the 1920s and 1930s before the introduction of electric logging, many oil sands had been shut off behind pipe. Some of these sands were located by gamma-ray logging and gun perforated. Others were opened by the infill drilling associated with water injection. By opening these untapped sands, Ecopetrol was able to maintain the production in spite of the fact that the field was fully drilled up.

Ecopetrol continued the development of the Galan field, which was an extension of Shell's Casabe field, discovered in 1941. Ecopetrol also resumed development of the Colorado field, discovered by Tropical in 1921, and Lisama, discovered by Tropical in 1937. Additional small fields have been discovered since, but while they are on the De Mares concession, they are on different structures from Infantas and La Cira and will not be considered in this paper.

DISCOVERY METHOD

The individuals and companies involved in the discovery have been mentioned above under *History*. They include Colombians Jose Joaquin Bohorquez and Roberto De Mares, American geologist John Leonard, and Pittsburgh wildcatter Mike Benedum.

John Leonard had been active in Mexico and had received a percentage for assembling a group of leases acquired by Benedum and Trees about 1911 (Owen, 1975). A search of the remaining files of Benedum and Trees was made by Roger Smith of Pittsburgh in March 1990, but no early reports were found.

There is no doubt that the primary discovery method was to drill on the huge oil seep at Infantas. Leonard had acquired the leases in Mexico on the basis of nearby oil seeps. How much he knew about the structure is not now known. The immediate terrain is broken and there are good outcrops, mostly in watercourses. The early geologists must have recognized the anticline, but whether they knew it was faulted and that the first wells were on the downthrown side of the fault is not known.

O. C. Wheeler is reported to have noticed a fault (and possibly an east dip) in a cut along the railroad from Barranca to El Centro. He realized that this might indicate another field west of Infantas, and La Cira was discovered. Later drilling showed that La Cira was a large dome on the west side of the Infantas structure. The upthrown east flank of the structure was drilled in several places but is nowhere productive. The C zone of the field outcrops and smells of oil. The Infantas structure plunges both south and north. There are several other anticlinal structures south of Infantas on the De Mares concession. They were drilled in the 1920s but oil shows were considered small and noncommercial.

STRUCTURE

Tectonic History

Pre-Cretaceous

During the Paleozoic and through the early Mesozoic, there must have been a depositional basin in northwestern South America. Outcrops of Devonian and other Paleozoic rocks are found in the centers of uplifts, sometimes metamorphosed and sometimes not metamorphosed at all. In the Jurassic and Triassic there is a series of red beds with extensive intrusions of diabase, resembling the Newark series of the eastern United States. Presumably this indicates a pulling apart of continental plates.

Cretaceous

During most of the Cretaceous, there was a miogeosyncline in eastern Colombia and a eugeosyncline in western Colombia, separated by what is now the Central Cordillera (Campbell, 1968) (Figure 3). About this time a large chunk of Pacific crust started to migrate northward and then eastward to become the Caribbean plate (Burke, 1988), sliding along the right lateral Romeral fault (Figure 3).

In eastern Colombia, sedimentation continued, derived mainly from the Precambrian Roraima quartzite and the metamorphics of the Guayana

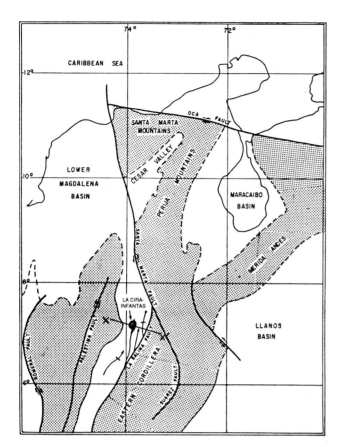

Figure 3. Tectonic map of northwestern South America showing great wedge of sediment driven northwestward along Santa Marta and Oca faults. Note offset of mountain ranges. (After Vasquez and Dickey, 1972.)

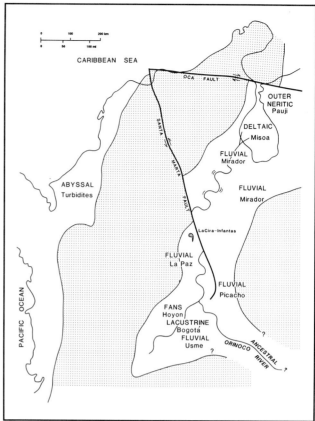

Figure 4. Facies map of middle Eocene showing approximate location of a wide river valley where the Eastern Cordillera is now located. (After Forero, 1974.)

shield to the east. In the south, much sand was deposited; in central Colombia, mostly shale; and north into the Maracaibo basin of Venezuela, mostly limestone. During much of this time, there were anoxic conditions with the deposition of huge thicknesses of black, organic shale, and in the Upper Cretaceous an especially organic shaly limestone with cherts, the La Luna.

This geosyncline contains sediments so similar to those in the Cretaceous geosyncline in Colorado, New Mexico, and Alberta that it is easy to conclude that it may have been continuous with them.

Tertiary

Sedimentation was continuous from the Cretaceous into the Paleocene, but in the Middle Magdalena Valley, continental sediments start to appear, characterized by coal beds. At the end of the Paleocene (Danian), there was uplift, and the Cretaceous was eroded over the Central Cordillera and locally along the Eastern Cordillera (Forero, 1974). During the Eocene, there was a broad valley where the Cordillera now lies (Figure 4). It must have been occupied by a large river, possibly the ancestral Orinoco. To the south, in the vicinity of Bogotá, the Eocene is largely lacustrine. In the Middle Magdalena Valley, it is entirely fluvial. River channels are embedded in oxidized continental mudstones. Farther north, this river deposited a thick succession of deltas in the Maracaibo basin. During the rest of the Tertiary, the Andean orogeny continued, reaching a climax in the Miocene. The Magdalena Valley was filled with as much as 30,000 ft (9000 m) of fluvial sediment. This valley filling is still continuing as the rainy season causes the muddy river to overflow its banks.

Regional Structure

During the Oligocene, the depression in the area of the Eastern Cordillera continued, but deposition was continental. Beginning in the Miocene and continuing into the Pliocene, there were tremendous movements in northern South America. There is still much controversy about the direction and timing of the motions (Burke, 1988). The Eastern Cordillera was uplifted with thrust faults along both east and west margins, but with a relatively horizontal and undisturbed zone in the center of the uplift. A triangular piece of crust (Figure 3) containing the Santa Marta massif and the Maracaibo basin was driven northwest something more than 60 mi (100

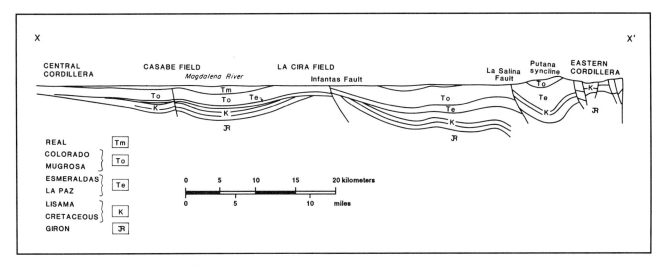

Figure 5. Cross section X-X' across the Middle Magdalena Valley. Location shown on Figures 1 and 3. (After Morales et al., 1958.)

km). The Caribbean plate moved several hundred kilometers eastward with respect to South America along the Oca fault and its projections. The Santa Marta fault sheared off the northern part of the Middle Magdalena basin to form the Cesar basin (Figure 3). What happened at the south end of the Santa Marta fault is not clear. Apparently, it became a series of northeast-trending faults, parallel to the strike of the Eastern Cordillera (Case and Holcombe, 1980). The La Salina fault is the westernmost of these offshoots. It is nearly vertical, with the east side upthrown several thousand meters. It may well have been formed by left lateral transcurrent movement, but this is hard to prove. A series of lesser faults forming north-plunging anticlines break away from the La Salina fault at a low angle. The Infantas anticline is one of these.

The Middle Magdalena Valley is a half-graben (Figure 5). The east side is formed by the La Salina and related faults which throw the Cretaceous against Miocene. The west side is an onlap of younger Miocene and Pliocene sediments on the metamorphic rocks of the Central Cordillera.

Local Structure

The principal structural feature of the La Cira-Infantas oil field is the Infantas anticline (Figure 6, B-B'). Along the crest is a reverse fault with the east side upthrown. The dip is about 60° to the east. The throw is from 800 to 2000 ft (250 to 600 m). To the south, the fault swings southwest and dies out. To the north, it continues along the east side of the domal La Cira structure and disappears when the associated anticline plunges northward.

The La Cira structure is a gentle dome whose crest is about 2.5 mi (4 km) west of the Infantas fault (Figure 6, A-A'). Both the La Cira and Infantas anticlines are cut by cross faults at angles of 60° to 90° from the axis. The throw of these faults is small, less than 200 ft (60 m); but the water-oil contact is usually displaced across the faults so they are apparently sealing.

There is an older structure under Infantas and La Cira (Figure 7). During the early Andean orogeny at the end of the Paleocene, the Cretaceous was uplifted into two rather symmetrical anticlines. It was eroded on the crest through the Cretaceous down to the Giron formation (Figure 8).

The Infantas fault was formed during the late Andean orogeny, probably during the Pliocene. All the Tertiary formations, including the Real series, are folded. The faulting occurred after the oil accumulated. The producing formations are exposed and eroded on the east flank of the anticline, but they still smell of oil.

STRATIGRAPHY

Pre-Cretaceous

The oldest formation penetrated in La Cira-Infantas is the Giron Formation (Figure 8). Its age is not determined exactly but is very likely Jurassic-Triassic. It is named for a town near Bucaramanga and is exposed in the gorge of the Lebrija River west of that city. It consists of gray, red, and green shales and sandstones. It contains frequent sills and flows of diabase or basalt. It is remarkably similar to the Newark formation of the eastern United States.

Cretaceous

The type locality for the Cretaceous is the gorge of the Sogamoso River at the northeast corner of the De Mares concession. The section was mapped by

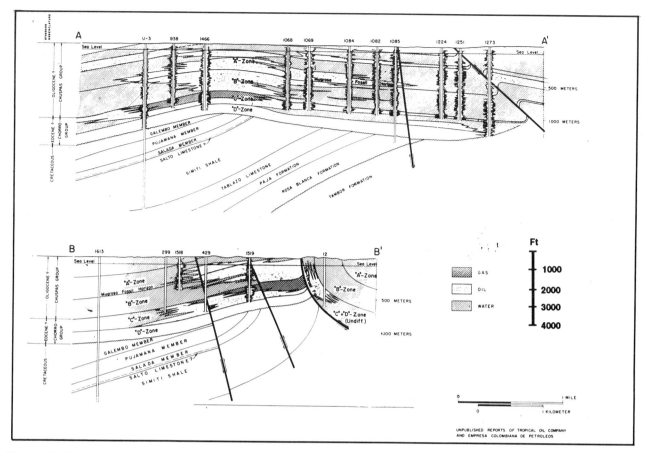

Figure 6. Cross sections A-A' and B-B', La Cira and Infantas fields. Locations shown on Figure 2. (After Morales et al., 1958.)

O. C. Wheeler in the 1920s, but the first published description was by Pilsbry and Olsson in 1935.

The lowermost formation believed to be Cretaceous is exposed in the Lebrija gorge above the Giron, where it was named the Tambor. It consists of red shales, sandstones, and conglomerates. It was penetrated in Infantas 1613.

The Rosa Blanca consists of massive dark gray bituminous limestones at the outcrop, but in the wells it also contains oolites. It is 900 ft (270 m) thick at Infantas. This richly fossiliferous formation is Hauterivian in age.

The Paja Formation consists of black organic shale, in places limy or silty. It is 2050 ft (320 m) thick at the outcrop but about 500 ft (150 m) thick at Infantas. Its age is lower Aptian to Barremian.

The Tablazo Limestone is a bluish-gray, massive limestone with marly beds at the outcrop where it is 500 ft (150 m) thick. At Infantas it contains shaly beds and is 800 ft (240 m) thick. Its age is upper Aptian to Albian.

The Simiti Shale was named for its outcrop at Cienaga Simiti on the west side of the Magdalena River, north of the De Mares concession. It consists of shale with occasional thin beds of limestone and sandstone. It is about 2000 ft (610 m) thick at Infantas. Its age is middle to lower Albian.

The Salto limestone is not recognized on the De Mares concession.

The La Luna formation was named for a sequence of argillaceous limestones near Maracaibo, Venezuela. In the De Mares concession, Wheeler named three formations: the Salada, which is limey shale; the Pujamana, which is a richly organic shale containing fish bones and scales; and the Galembo, which is black shale and limestone with fish scales and bones and several beds of chert. Oil shows are common. These formations are now considered members of the La Luna, which at Infantas is about 1700 ft (520 m) thick. Its age is Turonian to Santonian. It has been frequently analyzed for its organic content, which is often 10%. In the Maracaibo basin, it is considered to be the principal source rock, although there it is only 330 ft (100 m) thick. There is little doubt that the La Luna is one of the principal source rocks of the oil at La Cira-Infantas.

The uppermost Cretaceous unit is the Umir, which consists of gray shales with frequent coal beds. Its thickness is unknown but is about 6000 ft (1830 m) at the outcrop east of the De Mares concession. It

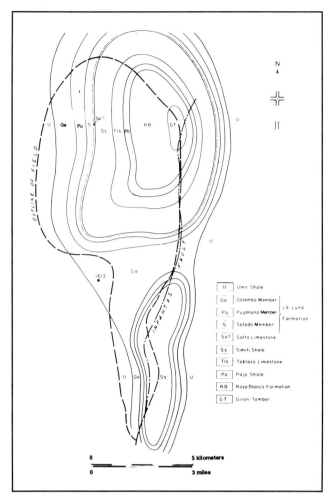

Figure 7. Pre-Tertiary geologic map of La Cira-Infantas fields. (After Morales et al., 1958.)

has been eroded away on the La Cira-Infantas structure. Fossil pollen from the Umir indicates that it is Maastrichtian in age, at least in part.

Tertiary

The Lisama Formation is conformable and gradational on the Umir. It consists of massive gray and reddish-brown shales with frequent beds of fine sand and occasional coals. Its age has been determined by pollen as Paleocene. Its maximum thickness is 3600 ft (1100 m) at the outcrop, but at La Cira-Infantas it has been removed by erosion at the mid-Eocene unconformity.

The Lisama is overlain unconformably by the La Paz formation. At its type locality of Cerro La Paz at the northeast corner of the De Mares concession, it consists of massive crossbedded arkosic sandstones and conglomerates, separated by beds of gray shale. It is here about 4000 ft (1220 m) thick. At La Cira, it consists of gray to brown and mottled shales with occasional fine sandstones about 600 ft (180 m) thick.

At the base of the La Paz at the Sogamoso gorge and also at Galan in the subsurface, there is a hard gray to pink shale, originally called the "Altered Shale" but now called the Toro member. It occurs mainly in the De Mares concession and reaches a maximum thickness of 200 ft (60 m). It has been considered as possibly a pyroclastic or a weathered zone at the unconformity.

The Esmeraldas Formation at La Cira consists of light gray to greenish shales with some sandstones. It contains the D zone, which is unproductive, probably because the sands do not make good reservoirs. It is between 550 and 750 ft (170 and 230 m) thick. Its top is marked by a zone of small gastropods called the Los Corros fossil horizon, which are brackish water fossils of upper Eocene age.

The Mugrosa Formation consists of gray to grayish-green sandstones, usually fine but sometimes pebbly, with interbedded gray and green shales in the lower part. This is the productive C zone. The middle part consists of blue and brown massive mottled shales with occasional thin interbeds of fine sandstone, which is the lower B zone. The upper third consists of gray fine- to coarse-grained sandstones, sometimes pebbly. This is the productive B zone. The thickness of the Mugrosa at La Cira is 1600 ft (488 m).

The top of the Mugrosa Formation is marked by the "Mugrosa fossil horizon," which consists of dark gray shales about 10 ft (3 m) thick containing gastropod shells and occasional fish bones and turtle plates. Their age is Oligocene. The Mugrosa fossil horizon is an extremely useful marker bed, coming as it does in the middle of a thick section of continental sediments, all very similar. It can be followed in the outcrop all over the De Mares concession because the snail shells are ferruginous and weather out to the surface. The shell molds show up in drill cuttings, and the horizon can be recognized on electric logs.

The lower 500 ft (150 m) of the Colorado Formation consists of fine-grained gray sandstones interbedded with mottled shales. Above this lie 700 ft (210 m) of gray, medium to coarse sandstones with yellowish-brown claystones mottled with violet. This is the productive A zone. Overlying this is 550 ft (168 m) of gray and red mottled claystones with thin-bedded, pebbly sandstones that are productive at Galan, west of La Cira. Overlying this are 600 ft (180 m) of brightly mottled shales and coarse sandstones. Overlying this is 250 ft (76 m) of dull gray to black, well-bedded carbonaceous shales that contain brackish water fossils, the La Cira fossil horizon. At La Cira-Infantas, the Mugrosa Formation is about 3000 ft (915 m) thick. Its age from the fossils is upper Oligocene to lower Miocene.

The Real Formation is exposed on the flanks of the Infantas-La Cira anticline where it is about 1600 ft (490 m) thick. It is much thicker and sandier to the southeast. It consists of sandstones with some interbedded shales. Its age is probably Miocene but may be partly Pliocene.

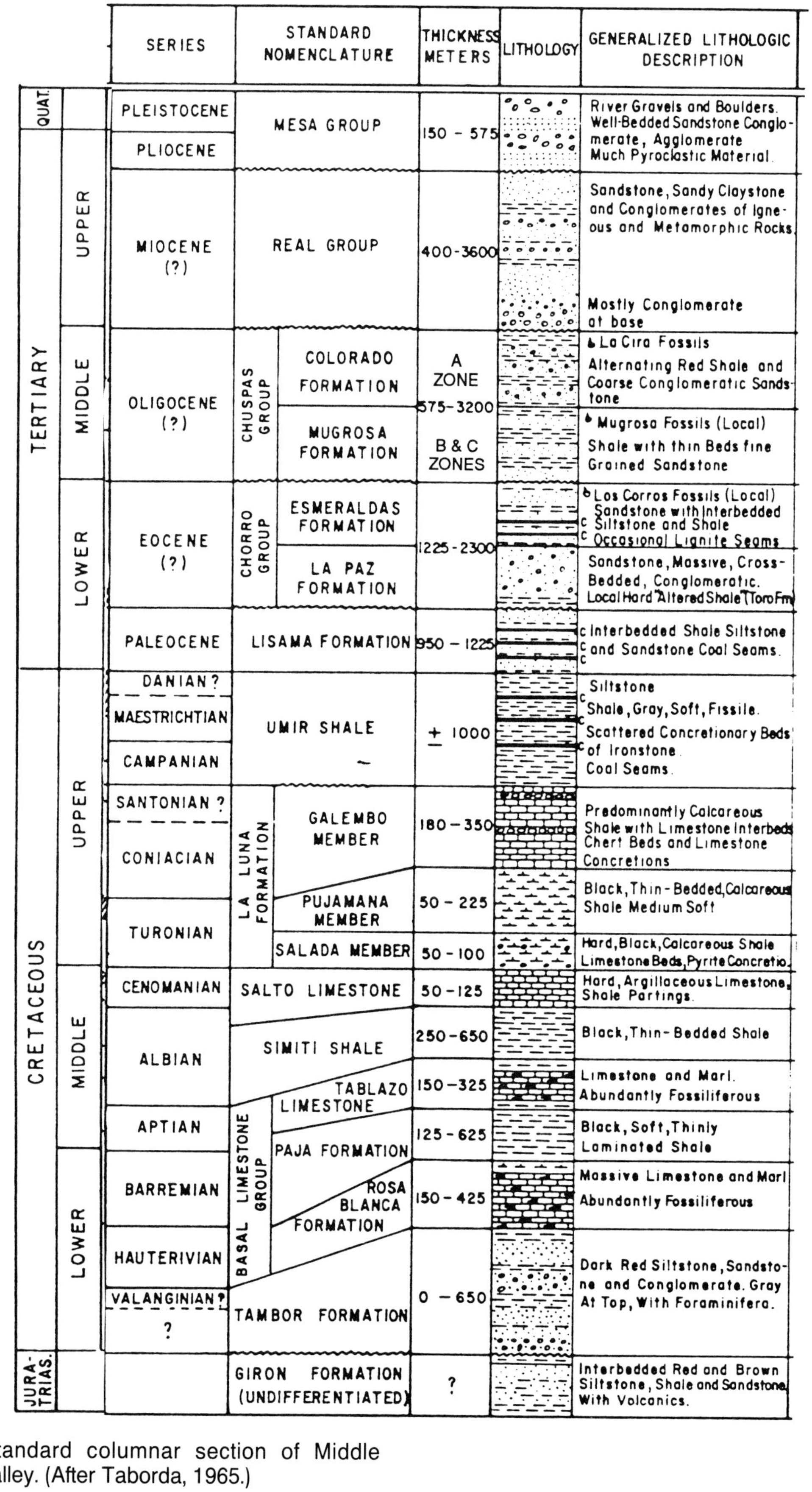

SERIES		STANDARD NOMENCLATURE		THICKNESS METERS	LITHOLOGY	GENERALIZED LITHOLOGIC DESCRIPTION
QUAT.	PLEISTOCENE	MESA GROUP		150 – 575		River Gravels and Boulders. Well-Bedded Sandstone Conglomerate, Agglomerate Much Pyroclastic Material
	PLIOCENE					
TERTIARY UPPER	MIOCENE (?)	REAL GROUP		400–3600		Sandstone, Sandy Claystone and Conglomerates of Igneous and Metamorphic Rocks. Mostly Conglomerate at base
TERTIARY MIDDLE	OLIGOCENE (?)	CHUSPAS GROUP	COLORADO FORMATION	A ZONE 575–3200		La Cira Fossils. Alternating Red Shale and Coarse Conglomeratic Sandstone
			MUGROSA FORMATION	B & C ZONES		Mugrosa Fossils (Local) Shale with thin Beds fine Grained Sandstone
TERTIARY LOWER	EOCENE (?)	CHORRO GROUP	ESMERALDAS FORMATION	1225–2300		Los Corros Fossils (Local) Sandstone with Interbedded Siltstone and Shale Occasional Lignite Seams
			LA PAZ FORMATION			Sandstone, Massive, Cross-Bedded, Conglomeratic. Local Hard Altered Shale (Toro Fm)
	PALEOCENE	LISAMA FORMATION		950–1225		Interbedded Shale Siltstone and Sandstone Coal Seams.
CRETACEOUS UPPER	DANIAN?	UMIR SHALE		±1000		Siltstone Shale, Gray, Soft, Fissile. Scattered Concretionary Beds of Ironstone Coal Seams.
	MAESTRICHTIAN					
	CAMPANIAN					
	SANTONIAN?	LA LUNA FORMATION	GALEMBO MEMBER	180–350		Predominantly Calcareous Shale with Limestone Interbeds Chert Beds and Limestone Concretions
	CONIACIAN					
	TURONIAN		PUJAMANA MEMBER	50–225		Black, Thin-Bedded, Calcareous Shale Medium Soft
			SALADA MEMBER	50–100		Hard, Black, Calcareous Shale Limestone Beds, Pyrite Concretio.
CRETACEOUS MIDDLE	CENOMANIAN	SALTO LIMESTONE		50–125		Hard, Argillaceous Limestone, Shale Partings.
	ALBIAN	SIMITI SHALE		250–650		Black, Thin-Bedded Shale
		BASAL LIMESTONE GROUP	TABLAZO LIMESTONE	150–325		Limestone and Marl. Abundantly Fossiliferous
	APTIAN		PAJA FORMATION	125–625		Black, Soft, Thinly Laminated Shale
CRETACEOUS LOWER	BARREMIAN		ROSA BLANCA FORMATION	150–425		Massive Limestone and Marl Abundantly Fossiliferous
	HAUTERIVIAN		TAMBOR FORMATION	0–650		Dark Red Siltstone, Sandstone and Conglomerate. Gray At Top, With Foraminifera.
	VALANGINIAN?					
JURA-TRIAS.		GIRON FORMATION (UNDIFFERENTIATED)		?		Interbedded Red and Brown Siltstone, Shale and Sandstone With Volcanics.

Figure 8. Standard columnar section of Middle Magdalena Valley. (After Taborda, 1965.)

The Mesa Formation consists of poorly consolidated sandstones and conglomerates with some igneous pebbles. The sands contain hornblende and magnetite and were apparently derived from the Central Cordillera. It is unconformable on the Colorado. Its age is probably Pliocene to Pleistocene. It is 800 ft (245 m) thick on the De Mares concession.

All of the Tertiary formations above the Lisama are fluvial and completely oxidized. It has always been hard for this author to imagine how thousands of meters of shales could be deposited under oxidizing conditions. Once while he was flying over the lower Magdalena in the rainy season, the explanation appeared. The whole area was under water from the flood of the Magdalena River, except for the natural levees. This area is in the tropical wet-dry climate where there is a pronounced rainy season followed by several months when it hardly rains at all. During the flood of the river, a few millimeters of mud was deposited over the entire valley. Then in the dry season the water table may have dropped a meter or more below the surface. The muds were oxidized. Plants grew and penetrated the mud with their roots. After burial, these roots provided a small zone of reducing material that changed the color of the iron oxides, causing the mottling. Huge thicknesses of mottled clays are found in Venezuela and Colombia, but very seldom in more temperate climates. It was pointed out many years ago by P. D. Krynine that mottled clays are typical of tropical climates.

The Tertiary sediments thin from east to west and become finer and less pebbly in that direction. The pebbles are mostly vein quartz and chert. This suggests that they were derived from the Eastern Cordillera, which consists mostly of Cretaceous sediments. It must have been rising all through the Tertiary, although the greatest orogenic paroxysms did not occur until the upper Miocene.

TRAP

La Cira and Infantas together constitute a faulted anticline, with oil production on the downthrown (west) side (Figure 2). The fault forms the lateral seal on the east side. There are asphalt seeps at several places along the fault, so it has been believed that the oil is leaking up along the fault plane, even though it is obviously the seal. The seeps might be explained by the exposure of the productive formations at the surface. Even at a considerable distance east of the fault, the B and A zone sands smell of oil.

The upper seals consist of the thick shales or rather mottled mudstones that separate the A, B, and C productive sands. The sand bodies themselves are stream channels, so there is considerable lateral sealing in the middle of the reservoirs. The western limits of production are downdip oil-water contacts (OWC) that step down northward. They are about 3000 ft (915 m) below sea level in the southern part of Infantas and 3800 ft (1160 m) in the northern part of La Cira. This indicates that there has been northward tilting since the oil accumulated. The productive A, B, and C zones are exposed at the surface on the east side of the fault, so there is no production on the east flank of the anticline.

The structure must have first formed during the post-Paleocene–pre-Eocene orogeny. During the time it was exposed, the Cretaceous was truncated (Figure 5). There is thus a great unconformity below the lowest Eocene productive zone. It seems likely that the oil migrated along this unconformity from source areas to the east. How it ascended into the producing zones that are separated by hundreds of meters of shale is not clear at all. The oil must have entered the structure before the faulting occurred, because the producing formations exposed on the east flank still smell of oil. It is possible that the structure was buried under thousands of meters of Miocene sediment and then uplifted and faulted in late Miocene or Pliocene. This would account for the peculiar character of the oil, described below.

Infantas

Infantas and La Cira are really two different fields although they are on the same anticline and are contiguous (Figure 2). Infantas was discovered first where the structure was cut by the Colorado River. The structure was plunging to the south, so that there is no production south of the river. The structure rises and widens northward so that at 5 mi (8 km) north of Infantas No. 1, it is 2 mi (3 km) wide. The dip ranges from 10° to 17° to the west, and there is about 1000 ft (305 m) of closure from the fault to the OWC.

La Cira

The La Cira field was discovered several kilometers farther northwest on what was believed to be a separate structure. It was only several years later that its development to the east showed that it was contiguous with Infantas. La Cira is a gentle dome on the west flank of the anticline. The crest of the structure is at 12°N 3°W, where the field is about 4 mi (7 km) wide. There is a shallow saddle east of the crest that separates it from Infantas. The field is about 6 mi (10 km) long north and south. There is about 1000 ft (305 m) of closure from the crest to the western oil-water contact. The OWC on the southwest is about -3400 ft (-1035 m) and on the north tip about -3800 ft (-1160 m). The east closure is the continuation of the Infantas fault to the north.

Reservoirs

In La Cira the A zone (Figure 8) of the Oligocene Colorado Formation consists of about 1200 ft (365 m) of interbedded sands and shales. The sands range from very thin to 100 ft (30 m) thick. They are clearly fluvial channels. The uppermost group, called the "58 Sands,"

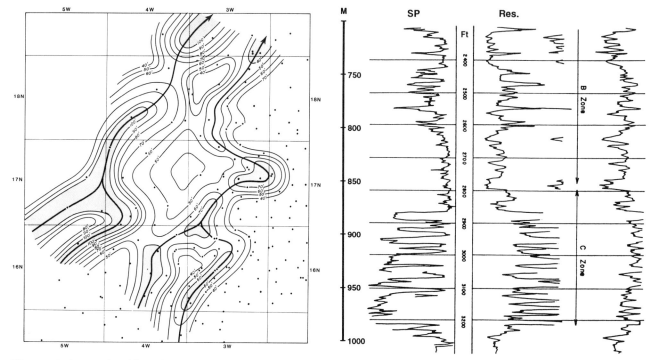

Figure 9. Isopach of "58 Sands" in northwest portion of La Cira field. Contour interval, 10 ft (3 m). Squares are 1 km (0.6 mi on a side). (After Valderrama, 1984.)

Figure 10. Electric log of well La Cira 1793 showing B and C zone sands. (Unquantified log to show relative positions of lithologic units and characteristic log signatures only.)

has been mapped by Valderrama (1984) (Figure 9). There seem to be parallel sets of channels with a northeast direction. The "58 Sands" are about 180 ft (55 m) thick.

Below them is the "116 Sands" group, which is about 400 ft (120 m) thick. There are more than 20 individual sand bodies, which are the best producers in the A zone. The lower sands are usually water-bearing.

The A zone is not developed at Infantas.

The B zone (Figure 8) is about 1800 ft (550 m) thick and is present in both Infantas and La Cira. The best sands are in the lower part and were quite prolific in north Infantas where they were named the "13 Sands" after Well Infantas 13. The Mugrosa fossil horizon is the top of the B zone and is about 500 ft (152 m) below the base of the water-bearing A zone sands.

The C zone (Figure 8) is the most prolific in both Infantas and La Cira. There is about 250 ft (76 m) of shale between the lower B zone sands and the top of the C zone (Figure 10), which consists of 400 to 500 ft (120 to 150 m) of section, about 50% sand. There are 20 or more individual sand bodies that come and go locally so they are difficult to correlate. The sands are soft and tend to flow into the wells. Sometimes this leaves a cavity, and the casing collapses.

It was recognized early that the sands must be river channels because of their restricted extent. Before electric logging it was impossible to tell where the individual pay sands were by examination of the cuttings. Consequently, many sands were shut off behind pipe.

Beginning in the late 1920s, efforts were made to correlate using heavy minerals. These efforts were not very satisfactory for detailed correlation, but several zones were distinguished. In 1933, Parke Dickey and Waldo Waring started electric logging using house-wiring wire and equipment borrowed from the electrical department. In 1935, a Blau-Gemmer logging truck was borrowed from the sister company, Humble Oil and Refining Co. In 1937, a gamma-ray logging device was acquired that identified the sands behind pipe. Later conventional logs were run by Schlumberger.

Porosity is variable but quite high, averaging 25.9% in the A zone, 27.2% in the B zone, and 27.9% in the C zone. Granulometric analyses have shown 8.52% coarse, 37.17% medium, and 54.31% fine sand and silt. Permeability varies up to 1500 md; it is difficult to determine because the sands are often unconsolidated.

Figure 11 shows two photomicrographs of a La Cira C zone sandstone. Note the abundance of altered feldspar, the poor sorting, and the extremely angular grains. The large pores in the lower figure may be secondary. Authigenic clays in the pores are abundant and have caused extensive formation damage.

Figure 11. Photomicrographs of a C zone sandstone, 2444 ft, La Cira 1880 well. Note poor sorting, extremely angular grains, abundant altered feldspar, and, possibly, secondary porosity. Scale bar, 1 mm. (Courtesy of Ecopetrol.)

Table 1. Oil gravities at La Cira-Infantas.

	Infantas	La Cira
A zone	25.9	21.4
B zone	27.2	24.1
C zone	27.9	23.8

Table 2. Bulk compositional data for oils (from Illich, 1983).

	Infantas	La Cira
Gravity, ° API	26.9	21.2
Depth, ft (m)	914–1067 (279–325)	1067–1219 (325–372)
Total sulfur	0.66%	0.90
Distillation data		
Gasoline (%)	18	6
Kero-lube (%)	58	63
Residuum (%)	24	31
Elution Chromatography		
Saturates (%)	56.4	58.9
Aromatics (%)	23.2	24.9
Others (%)	20.3	16.3
Arom/Sat	0.41	0.42

Table 3. Analyses of La Cira and Infantas oil by molecular type.

	Infantas (%)	La Cira (%)
C_{15} fraction	17.9	6.5
C_{15} plus fraction	82.1	93.5
Asphaltenes	3.7	2.0
Paraffins	56.4	58.9
Aromatics	23.2	24.9
Polars	8.3	8.5
Noneluted fraction	8.3	5.8

Character of the Oil and Gas

The API gravity of the oil ranges from 27.9° at Infantas to 21.4° at La Cira as shown in Table 1. The chemical composition of the oil is shown in Tables 2 and 3. Figures 12 and 13 are gas chromatograms of La Cira and Infantas oils. The Infantas oil is from Illich (1983). The La Cira oil was run by David Wavrek at the University of Tulsa. The proportions of individual compounds that the La Cira sample showed, normalized to methylcyclohexane, are shown in Table 4.

Source Rocks

The source of the La Cira-Infantas oil is clearly the underlying Cretaceous shales and limestones, as suspected by Hedberg in 1931. Geochemical analyses of their source potential have been made by Zumberge (1984).

The La Luna Formation, at its outcrop about 62 mi (100 km) north of Infantas-La Cira, is 490 to 1970 ft (150–600 m) thick. The Pujamana and Salada members are the richest. They consist of black calcareous shales with some interbedded limestone and chert. Foraminifera and other pelagic fossils suggest that they were deposited in moderately deep water with restricted circulation. Total organic carbon averages 3.51% in the Pujamana and 4.51% in the Salada. Cavities and fractures are often filled with oil or asphalt. The kerogen is mainly amorphous and the hydrogen-carbon and oxygen-carbon ratio indicate that it is type II. Vitrinite reflectance, spore coloration, and pyrolysis indicate that it is within the principal zone of oil generation. There are abundant C_{27} steranes relative to C_{29} steranes, and low quantities of C_{19} and C_{20} tricyclic diterpanes. Hopane triterpanes derived from bacteria are present. These indicate marine organic matter.

Migration and Degradation

The oils from La Cira and Infantas have a peculiar chemical composition. They appear to have been degraded and then reheated.

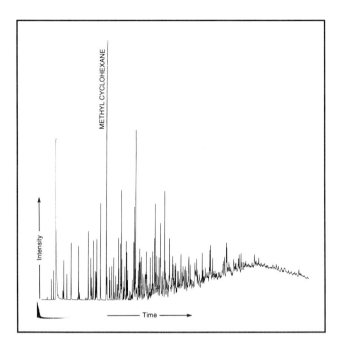

Figure 12. Gas chromatogram of La Cira oil. The proportion of individual compounds (normalized to methylcyclohexane) is shown in Table 4.

The oil appears to have been degraded because the higher normal alkanes (C_{10} plus) and the isoprenoids are almost missing. It has probably also been subjected to water washing because benzene and toluene, the most water-soluble of the hydrocarbons, are also almost missing. Water-washing always accompanies degradation.

On the other hand, the oils contain abundant light hydrocarbons, and these are mostly normal rather than branched. As a result, the API gravity is abnormally high for a severely degraded oil. It seems that after being degraded the oil was subjected to high temperatures and light hydrocarbons were formed by thermal cracking.

The sulfur content (0.9%) is low for a degraded oil. The introduction of light hydrocarbons would have caused precipitation of asphaltenes. Most of the sulfur is in the asphaltenes, so their precipitation would have reduced the sulfur content.

The geological history of the area explains how the degradation and reheating might have occurred. Figure 14 is a series of diagrammatic cross sections showing the successive periods of sedimentation broken by uplift and erosion.

Cretaceous (Figure 14-1). During the Cretaceous, there was a broad shelf on which shale and limestone were deposited. The west coast of South America is today, and may have been during the Cretaceous, a zone of upwelling and rich organic productivity. It sometimes became anoxic, so that large amounts of organic matter were preserved. West of the Middle Magdalena, there are turbidite deposits, while to the south and east, there are thick sandstones derived from the Guayana shield to the east. The total thickness of the Cretaceous is about 13,000 ft (4000 m). The highly organic La Luna formation in the upper part of the Cretaceous is about 2300 ft (700 m) thick.

Paleocene (Figure 14-2). During the Paleocene, the depositional environment became more continental and there are frequent coal beds. As much as 6600 ft (2000 m) may have been deposited. This may have been enough to mature the underlying Cretaceous source rocks and start migration.

Eocene uplift and erosion (Figure 14-3). At the end of the Paleocene, there was extensive uplift and erosion. In some localities, for example La Cira, the Paleocene and Cretaceous were completely eroded down to the basement Giron. The unconformity surface became a permeable zone through which oil could later migrate. Meteoric water may have penetrated deeply into the outcropping strata.

Eocene (Figure 14-4). During the Eocene, there was a great river valley located about where the Eastern Cordillera now is (Forero, 1974; Figure 4). Several thousand meters of fluvial sediments were deposited. Although the Eocene at La Cira is mostly shale, there are thick sandstones and conglomerates 30 mi (50 km) to the northeast at Cerro La Paz. During the Eocene, the oil may have been expelled from the Cretaceous and may have migrated along the unconformity surface, becoming degraded as it went.

Oligocene-Miocene (Figure 14-5). During the Oligocene and Miocene, the Magdalena Valley subsided and was filled with an enormous thickness of fluvial sediments derived from the slowly rising Eastern Cordillera. As much as 30,000 ft (9000 m) may have been deposited. These sediments lap westward onto the metamorphic rocks of the Central Cordillera. Probably additional oil was expelled from the source rocks. Any oil reservoired in the Eocene on the east side of the valley would have been reheated. On the west side of the valley, it would not have been reheated because the onlap of younger Tertiary sediments on the Central Cordillera resulted in a much thinner section.

Late Miocene-Pliocene (Figure 14-6). In the later Miocene and Pliocene, the great Andean orogeny took place. The Eastern Cordillera was uplifted, with complex thrust faults on both east and west margins. The Infantas and other anticlines in the center of the valley were uplifted and eroded down to the Eocene strata. Farther east, in the foothills, all the Oligocene and Miocene were removed. As much as 20,000 ft (6000 m) of fluvial sediments of the Oligocene and early Miocene were removed by erosion.

The burial history of the Middle Magdalena Valley is reconstructed in Lopatin diagrams (Waples, 1980). Figure 15 is based on the stratigraphic section at La Cira. Here the Upper Cretaceous and Paleocene formations were removed by erosion. Consequently, the La Luna source rock did not experience peak oil generation and expulsion until late Miocene. However, it is most unlikely that the oil moved straight up from the source rock to the reservoir.

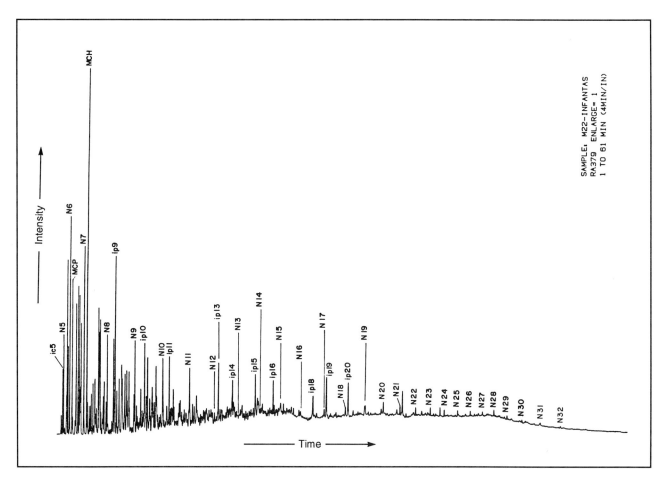

Figure 13. Gas chromatogram of Infantas oil. (From Illich, 1982.)

Table 4. Chromatogram of La Cira oil, compounds normalized to methylcyclohexane.

n-Butane	1.36
2 Methyl butane	9.52
n-Pentane	13.47
2 Methyl pentane	16.80
3 Methyl pentane	10.88
n-Hexane	23.13
Methylcyclopentane	20.07
Benzene	1.02
Cyclohexane	26.60
2 Methyl hexane	17.01
3 Methyl hexane	25.71
1-*cis*-3 Dimethyl cyclopentane	12.24
1-*trans*-3 Dimethyl cyclopentane	11.90
1-*trans*-2 Dimethyl cyclopentane	25.51
n-Heptane	45.60
Methylcyclohexane	100.00
Toluene	10.20

Farther east there was less erosion at the end of the Paleocene, and the Eocene is much thicker (Figure 16). The Umir and Lisama formations are preserved, and there is a great thickness of Eocene La Paz sandstone. Oil generation and expulsion from the La Luna may have started in the Eocene.

After migrating and becoming degraded, the oil was trapped in fluvial sands of the Eocene. But the Eocene reservoirs here reached the oil window in the late Miocene, suggesting that the reservoired oil may have been thermally cracked. The process of cracking degraded oil in the reservoir may have required less temperature and time than generation of oil in the source rock. It seems certain that the oil was generated from the source rock somewhere east of La Cira–Infantas and migrated westward toward the flank of the deeply subsided Magdalena Valley during the early Tertiary.

Water

The edge water on the west flank of the field showed very little encroachment; at most one or two locations. There was some fingering but no noticeable effect in enhancing production. The average water cut for the whole field increased to 8.61% in 1934 and 39.38% in 1940. Since 1942 the water cut has declined. This must indicate that the pressure in the aquifer has been taken off so that it will no longer advance.

A chemical laboratory was set up in 1927 to analyze and identify the source of water produced with the

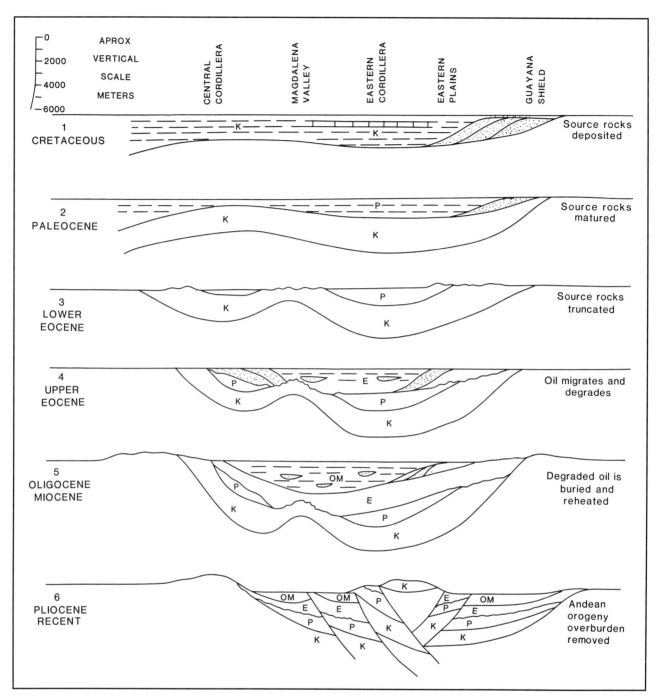

Figure 14. Series of diagrammatic cross sections of eastern Colombia showing successive vertical motions resulting in filling of basins with sediments. Vertical motions can exceed 6 mi (10 km). Horizontal motions are not shown but can exceed 60 mi (100 km).

oil. The differences in the chemical composition of the water in the different sands made it possible to ascertain from which sand the water was coming. Sometimes a C zone well would go to water, but an analysis showed it was coming from the A zone through a leaky casing.

The waters in the La Cira A zone are quite variable. They are low in total solids, low in calcium and magnesium, but high in bicarbonate. This indicates that they are partly meteoric in origin (Dickey, 1966). This evidence of meteoric origin of the water may explain why some of the La Cira oils are highly degraded. The B zone waters are more concentrated and contain more calcium and magnesium and much less bicarbonate. They are thus mostly connate in origin. The C zone waters are still more concentrated, ranging up to 68,000 mg/L total solids. Calcium is much higher than in the B zone, but magnesium is about the same. They are typical connate waters chemically. At Infantas the waters often have

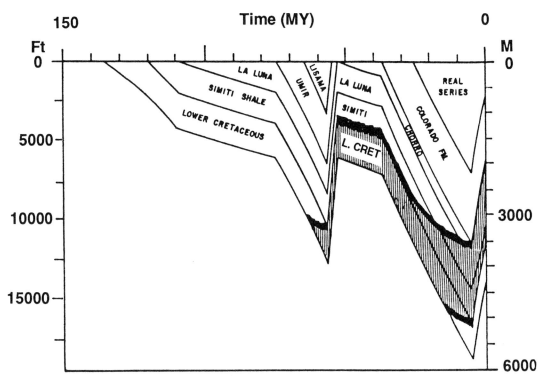

Figure 15. Lopatin diagram showing burial history at La Cira. The La Luna source rock did not experience peak oil generation until late Miocene. The oil must have been generated somewhere else, probably to the east.

different compositions from one fault block to another. Table 5 shows the composition of some typical waters from La Cira.

Conclusions Regarding Origin and Migration of Fluids

Beginning 53 Ma, at the end of the Paleocene, eastern Colombia ceased being a stable, passive shelf in whose waters marine life grew abundantly. A series of crustal movements began that involved as much as 6 mi (10 km) of vertical movement and more than 125 mi (200 km) of horizontal movement. Mountains were uplifted possibly as high as Mount Everest (29,000 ft; 8800 m) and then eroded, filling the ancestral Magdalena trough with 30,000 ft (9000 m) of sediment. The petroleum that formed from the organic matter deposited during the Cretaceous was squeezed out into reservoirs. With the tremendous crustal motions the reservoirs broke and spilled. Some of the oil trickled along the unconformity westward toward the Central Cordillera. Before reaching the outcrop of the unconformity the oil was covered with 20,000 ft (6000 m) of Oligocene and Miocene sediments derived from the great mountains to the east. This heated it to the point that it cracked and formed gasoline. The La Cira oil started as normal crude oil, but during its long and possibly tortuous journey it became degraded and then was later reheated.

Development and Production

Development

Infantas was developed on a well spacing of 10.9 ac (4.4 ha) per well. La Cira was first developed on a spacing of 8.56 ac (3.5 ha) per well in a triangular pattern. The spacing was later changed to 17.5 ac (7.1 ha) per well. Studies by Hughes in 1941 showed no increase in ultimate recovery on the closer spacing.

Wells were originally completed with 11-in. blank casing and 8½-in. slotted liner through the sand. Later blank casing was run through the sand and gun-perforated. Tubing was 2-in., 2½-in., and 3-in. By 1940 all wells were pumping. Most wells had electric motors, but a few were attached by rod lines to a central pumping power called by the workmen a *Catalina* after Saint Catherine's wheel.

Gas was reinjected starting in 1928. Condensate was extracted from the gas and added to the crude oil to reduce its viscosity. Channeling was severe, especially in the B zone. Injection wells were mostly halfway down the flanks of both fields. It was estimated by Hughes in 1941 that 10 million bbl of additional oil was recovered by the gas injection. Later estimates were that at La Cira 8 million bbl of additional oil could be ascribed to the gas injection and at Infantas 44 million. Secondary gas caps developed in both fields.

Original gas-oil ratio was 50 to 300 ft^3/bbl (8.8 to 52.8 m^{3}gas/m^{3}oil). The original viscosity at La Cira C zone was 15 cp, which has increased to 32, and

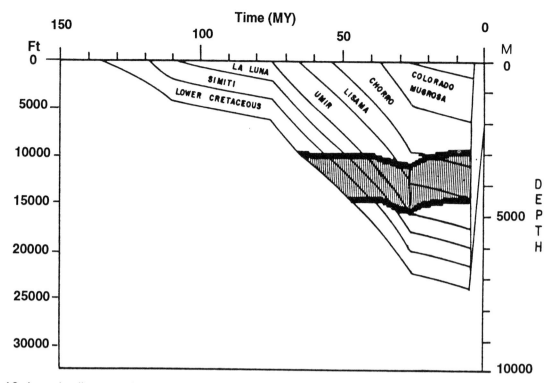

Figure 16. Lopatin diagram showing burial history at foothills of Eastern Cordillera, 30 mi (50 km) east of La Cira. Umir and Lisama formations are present and there is a great thickness of Eocene La Paz.

at Infantas 9 cp, which has increased to 15 cp. This increase is probably due to the stripping of light ends by the gas cycling. Sulfur ranges from 0.7 to 1.0%. The average formation temperature is 100°F (38°C). Original reservoir pressures were as shown in Table 6. These were estimated later because the bottom-hole pressure measuring device had not been invented when the field was discovered. Indeed, most of what we now know about reservoir behavior was still to be developed. When the Amerada bomb was introduced in the middle 1930s, the geologists had to draw the bottom-hole pressure maps because the petroleum engineers had never drawn contours!

Production Data

The production of oil from Infantas and La Cira for 60 years, from 1926 to 1986, has been 683 MMBO. Yearly production is shown in the decline curves, Figure 17. The summary of data is given in Table 7.

Infantas—The Infantas field was pretty well drilled up by 1934. It produced originally from C zone and later from B zone wells. Gas injection had started in 1928. At first, the decline rate at Infantas was very steep, about 10% per year. If this decline were projected, the field would be producing only 1 MMBO/yr in 1951, with a cumulative ultimate production of 147 million bbl. However, the curve flattened notably in 1942, probably as a result of a secondary gas cap forming at the crest of the structure, which gave a true gas-cap drive with gravity drainage. Predictions of future production in 1941 by Hughes and in 1951 by Bily (Lipstate and Bily, 1952) were much too low. Infantas had actually produced 223 million bbl to December 1986 and was still producing about 1 MMBO/yr.

La Cira—Development at La Cira continued until 1940. In 1942 and 1943, World War II interrupted tanker traffic, so export almost stopped and production was cut back from 18 to 7 MMBO/yr. Water injection into the aquifer west of the field was undertaken between 1946 and 1949. The increase in production between 1948 and 1951 may have been caused by drilling the additional wells even though the edge water injection had little effect. From 1951 to 1958, the production declined from 10 to 6 MMBO/yr, a rate of 6%/yr. Present production is about 3 MMBO/yr.

Waterflooding

The first attempt at waterflooding was made in 1946. Seven wells were drilled and perforated downdip from the La Cira field in the aquifer. Three wells were perforated inside the field. The aquifer wells had no apparent effect. From 1946 to 1949, the pool wells recovered 54,000 bbl of oil as a result of injecting 770,000 bbl of water.

In 1955 the Forest Oil Co. of Bradford, Pennsylvania, undertook waterflooding. They drilled and cored four wells in La Cira, which showed great lateral variation in thickness and quality of the sands. Swelling clays were also noted. The oil was 23° API with a viscosity of 30 cp. Residual oil after flooding the cores was 18%. They proposed a nine-year development plan with a five-spot pattern on 16 ac

Table 5. Water analyses from La Cira field (mg/l) (Dickey, 1933).

Zone	Well No.	Na	Ca	Mg	SO₄	Cl	HCO₃
A	213	7107	130	215	—	10,292	2454
A	138	2040	49	85			1974	2220
B	221	8208	977	429	—	15,512	431
C	588		6320	672	346	37,200	108

Table 6. Original reservoir pressures.

Field	Zone	Datum, feet (m) subsea	Pressure psi (bar)
Infantas	C	2000 (610 m)	875 (60.4)
La Cira	A	500 (152 m)	350 (24.2)
La Cira	B	1200 (366 m)	750 (51.8)
La Cira	C	3000 (915 m)	1450 (100.0)

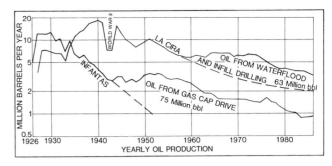

Figure 17. Production decline curves from 1926 to 1986, La Cira and Infantas fields.

spacing. In 1957, 280 ac were developed by 17 five-spots: 24 injectors and 18 producers. From 1957 to 1982, a total of 241 injectors were drilled, some on a five-spot and some on a seven-spot pattern. From 1966 to 1971, a polymer thickener was added to the water. It reduced the injection rate and gave rise to fracturing.

After 25 years of experience, it is obvious that the injection of water has not been very effective. The heterogeneities of the sand were such that only 60 to 80% of the sand was swept. The channels and faults left areas unaffected. Some wells watered out very early, supposedly through channel sands or fault planes. The viscosity ratio was adverse.

A summary made in March 1987 indicated 61,650,000 bbl of oil additional recovery from the waterflood. The estimation naturally depends on the projected natural decline without water injection. Some of the increased production is doubtless due to infill drilling that opened up zones previously untapped. A cumulative total of 801,000,000 bbl of water had been injected.

Certainly no additional attempts at improved recovery should be made without a detailed study of the shape and character of the individual sand channels. This would involve subdividing the producing zones into thin time-rock units and isopaching the individual sand channels within each unit. Pressure measurements should be made using the repeat formation tester in newly drilled wells and interference tests in old wells.

EXPLORATION CONCEPTS

In 1915 when De Mares and Leonard met Benedum, American interest in foreign oil had become widespread. Many companies were operating in Mexico (Owen, 1975). Shell discovered the Mene Grande field in Venezuela in 1914. The United States was not yet in the European War, although it was later said that "the Allies floated to victory on a sea of American oil." Leonard had led Benedum to substantial discoveries in Mexico in 1913. It was, therefore, fashionable to prospect for oil in Latin America.

The Mexican fields were mostly discovered by drilling on or near asphalt seeps, of which there were very many. The first discovery in Venezuela was made by drilling on the enormous Mene Grande seep in 1914. In 1915, therefore, oil was discovered in Latin America by drilling near seeps. Few geologists were employed, and then mostly on a consulting basis. Leonard is said to have received a 10% interest in the Mexican properties from Benedum and Trees. The value of anticlines was first obvious in Oklahoma and California about 1912. It may be that the fact that the Infantas seep was on an anticline visible at the surface made it doubly easy to sell.

It is hard for us to imagine now what was the role of geology in the oil business in 1918. There were only three published geological papers on Colombia (Karsten, 1858; Sievers, 1888; Hettner, 1892), all in German, and it would be surprising if any of the geologists were familiar with them. After the discovery in the 1920s, several papers were published in both German and English.

Apparently some wells had been drilled in the Magdalena Valley before Infantas No. 1. It was said in one account that Benedum and Trees brought a drilling rig from a previous wildcat up the river. A seep at Guataqui, near Girardot, may have been the locality.

There are dozens of oil seeps on both sides of the Magdalena Valley, besides a large field of very heavy oil (Cocorna). Probably this part of Colombia contained oil fields comparable to the giant fields of nearby Venezuela before the late Miocene orogeny breached and eroded most of them.

Between 1918 and 1980, 252 exploratory wells have been drilled in the Middle Magdalena basin, of which 40 discovered commercial fields and 11 discovered noncommercial fields. This is a success ratio of about 16%. Total recoverable reserves are estimated as 2 billion bbl of oil and 234 BCFG. (Ecopetrol unpublished report, 1986). Only five fields contain more than 100 million bbl. Two of these (Velasquez and

Table 7. Summary of production data of La Cira-Infantas fields.

Proved acres	19,000 (7695 ha)	
Number of wells, all years	1671	
Current number of wells	960	
	La Cira	Infantas
Well spacing, acres per well.	8.56 and 17.5 (3.5 and 7.1 ha)	10 (4.1 ha)
Ultimate recoverable oil (million bbl)	503	234
Cumulative production through 1986	459.9	223.2
Annual production (million bbl/yr)	3.1	1.0
Present decline rate(%/y)	4.5	3.5
Original oil in place (million bbl)	2659	1163
*Secondary recovery (million bbl)	69 (water)	8 (gas)
Recovery factor (%)	17.8	

*Above are the accepted values for secondary recovery. Review of the decline curves suggests that the high permeability combined with the steep dip makes it seem possible that the early gas injection at Infantas gave rise to a good gas-cap gravity drainage drive, resulting in as much as 75 million barrels of extra oil.

Provincia) are on central basin structures like La Cira-Infantas, and three (Cocorna, Cantagallo, and Casabe) are near the west margin.

ACKNOWLEDGMENT

The assistance of the geologists and engineers of Ecopetrol in the compilation of this report is acknowledged. I am especially grateful to Antonio Parada and Cesar Santiago.

REFERENCES CITED

Burke, K., 1988, Tectonic evolution of the Caribbean: Annual Review of Earth and Planetary Sciences, v. 16, p. 201-230.
Campbell, C. J., 1968, The Santa Marta wrench fault of Colombia and its regional setting: Transactions Fourth Caribbean Geological Congress, p. 247-261.
Case, J. E., and T. L. Holcombe, 1980, Geologic-tectonic map of the Caribbean: U. S. Geological Survey.
Dickey, P. A., 1933, Additional information on the chemistry of the oilfield waters from La Cira: Tropical Oil Company Report.
Dickey, P. A., 1966, Patterns of chemical composition in deep subsurface waters: AAPG Bulletin, v. 50, p. 2472-2478.
Donoghue, D., 1951, The De Mares Concession: Petroleum Engineer, March, p. A-39-A-40.
Ecopetrol, 1986, Distrito de Produccion, El Centro, 142 p.
Forero E. O., 1974, The Eocene of Northwestern South America: M. S. Thesis, University of Tulsa.
Gibb, G. S., and E. K. Knowlton, 1956, The resurgent years, 1911-1927: New York, Harper and Brothers, 754 p.
Hedberg, H. D., 1931, Cretaceous limestone as petroleum source rock in northwestern Venezuela: AAPG Bulletin, v. 15, p. 229-246.
Hettner, A., 1892, Die Kordillere von Bogota: Ergzh zu Permans Mittteilungen, Gotha.
Hughes, R. V., 1941, Pool studies: Tropical Oil Company Report.
Illich, H. A., 1983, Stratigraphic implications of geochemistry of oils from Middle Magdalena Valley, Colombia (abs.): AAPG Bulletin, v. 67, p. 487.
Lipstate, P. H., and C. Bily, 1952, Notes on the development and behavior of the La Cira and Infantas fields: Ecopetrol Company Report.
Karsten, H., 1858, Uber die geognostinschen Verhaltnisse des westlichen Columbien, der heitigen Republiken Neu-Granada und Ecuador: Versammlungen des Deutches Naturforschungen Gesellschaft in Wien, 1956, Vienna.
Mendoza, F., 1987, La reversion de la concesion De Mares: El Tiempo, Bogotá, August 15.
Morales, L. G., and the Colombian Petroleum Industry, 1958, General geology and oil occurrences of Middle Magdalena Valley, Colombia, in L. G. Weeks, ed., Habitat of oil: AAPG, p. 641-695.
Oviedo, G. F. de, 1959, Historia general y natural de las Indias: Biblioteca de Autores Espanoles, Madrid, v. III, p. 94.
Owen, E. W., 1975, The trek of the oil finders, history of the exploration for petroleum: AAPG Memoir 6, 1619 p.
Pilsbry, H. A., and A. A. Olsson, 1935, Tertiary freshwater molluscs of the Magdalena Embayment, Colombia, *with* Tertiary stratigraphy of the Middle Magdalena Valley, by O. C. Wheeler: Proceedings of the Academy of Natural Sciences of Philadelphia, v. 87, p. 7-39.
Santiago Reyes, M. A., 1986, Cronica de la Concesion De Mares: Ecopetrol, Bogotá, 127 p.
Sievers, W., 1888, Die Sierra Nevada de Santa Marta und die Sierra de Perija: Zeitschrift der Gesellschaft fur Erdkunde, Beil. Bd. 23, p. 1-159.
Taborda, B., 1965, The geology of the de Mares Concession, *in* Colombian Society of Geologists and Geophysicists, Field Trips Guide Book 1959-1978, Bogotá, p. 119-160.
Valderrama R. R., 1984, Atrapamiento estratigrafico de hidrocarburos en el Valle Medio del Magdalena: Ecopetrol Unpublished Report.
Vasquez, E. E., and P. A. Dickey, 1972, Major faulting in northwest Venezuela and its relation to global tectonics: Transactions Sixth Caribbean Geological Conference, p. 191-202.
Waples, D. W., 1980, Time and temperature in petroleum formation: application of Lopatin's method to petroleum exploration: AAPG Bulletin, v. 64, p. 916-926.

Appendix 1. Field Description

Field name .... *La Cira-Infantas*
Ultimate recoverable reserves .... *737,000,000 bbl*
Field location:
    Country .... *Colombia*
    State .... *Santander*
    Basin/Province .... *Middle Magdalena Valley*
Field discovery:
    Year first pay discovered .... *Oligocene Colorado and Mugrosa formations 1917*
Discovery well name and general location:
    First pay .... *Infantas No. 2 (in C zone) approx. location NW corner 2N-1E*
        (Infantas No. 1, completed later, located between 1N and 1S and 1W and 1E)
    Second pay .... *B zone (well name and location not available)*
    Third pay .... *A zone (well name and location not available)*
Discovery well operator .... *Benedum and Trees*
    Second pay .... *NA*
    Third pay .... *NA*
IP:
    First pay .... *800-1000 BOPD (Infantas No. 1 completed in 1918 had IP 2000 BOPD)*
    Second pay .... *NA*
    Third pay .... *NA*

Geologic concept leading to discovery and method or methods used to delineate prospect
Large asphalt seep.

Structure:
    Province/basin type .... *Rifted convergent margin, Bally 332, Klemme III Bc*
    Tectonic history
Marine geosyncline during Cretaceous. Uplift and erosion during Eocene. Uplift of Andes Mountains in Miocene.

    Regional structure
Field is on the central basin uplift in a half-graben with a large fault on east side of basin, onlap on west side.

    Local structure
Anticline broken by high-angle reverse fault. Oil trapped on downthrown side, oil-bearing strata exposed on upthrown side.

Trap:
    Trap type(s)
La Cira: lenticular sandstones over domal structure
Infantas: faulted anticline with lenticular sandstones

Basin stratigraphy (major stratigraphic intervals from surface to deepest penetration in field):

Chronostratigraphy	Formation	Depth to Top in ft (m)
Miocene	*Real*	*Exposed*
Oligocene	*Colorado*	*Exposed*
	Mugrosa	*1000 (305)*
Eocene	*Esmeraldas*	*3000 (915)*
Cretaceous	*Various*	*3000+ (915+)*

Reservoir characteristics:

- **Number of reservoirs** 3
- **Formations** .... Oligocene Colorado Formation A zone and Mugrosa Formation B and C zones
- **Ages** .... Oligocene
- **Depths to tops of reservoirs** .... A zone, 600 ft (185 m); B zone, 1100 ft (335 m); C zone, 2200 ft (670 m)
- **Gross thickness (top to bottom of producing interval)** .... B zone, 1050–1250 ft (320–380 m); C zone, 680 ft (205 m) average
- **Net thickness—total thickness of producing zones**
  - Average .... 250 ft (75 m)
  - Maximum .... 500 ft (150 m)
- **Lithology**
  Fine to medium sandstone, subarkosic, sometimes argillaceous, in north-south fluvial channels
- **Porosity type** .... Intergranular, some secondary
- **Average porosity** 20–30%
- **Average permeability** .... 2 to 1500 md, average 270 md

Seals:

- **Upper**
  - Formation, fault, or other feature .... Oligocene shales and mudstones
  - Lithology .... Shale and mudstones
- **Lateral**
  - Formation, fault, or other feature .... Fault in part, facies changes, OWC downdip
  - Lithology .... Various

Source:

- **Formation and age** .... La Luna and Simiti, Cretaceous
- **Lithology** .... Carbonate and shale
- **Average total organic carbon (TOC)** .... NA
- **Maximum TOC** .... >10%
- **Kerogen type (I, II, or III)** II
- **Vitrinite reflectance (maturation)** .... NA
- **Time of hydrocarbon expulsion** .... NA
- **Present depth to top of source** .... 1000 ft (300 m)
- **Thickness** .... Variable—up to 1000 ft (300 m)
- **Potential yield** .... NA

Appendix 2. Production Data

- **Field name** .... La Cira-Infantas
- **Field size:**
  - Proved acres 19,000
  - Number of wells all years 1671
  - Current number of wells 960
  - Well spacing .... Infantas, 10.9 ac (4.4 ha); La Cira, 8.56 (3.47 ha) and 17.5 (7.09 ha)
  - Ultimate recoverable .... Infantas, 234 million bbl; La Cira, 503 million bbl
  - Cumulative production .... Infantas, 223 million bbl; La Cira, 460 million bbl (to December 1986)
  - Annual production .... Infantas, 1.0 million bbl; La Cira, 3.1 million bbl
  - Present decline rate .... Infantas, 3.5%; La Cira, 4.5%

Primary recovery .... *Secondary operations initiated before primary complete*
Secondary recovery .... *69+ to date, La Cira (water injection); 8 to date, Infantas (gas injection)*

Enhanced recovery .... *NA*
Cumulative water production .... *NA*

Drilling and casing practices:

Amount of surface casing set .... *NA*

Casing program
Originally 11-in. casing with slotted liner. Later blank casing perforated. Severe sand flowage. Some collapsed casing.

Drilling mud and bit program
Early wells drilled with cable tools; later rotary wells with "natural mud"; more recent wells NA

High pressure zones .... *None in producing field*

Completion practices:

Interval(s) perforated .... *Sandy zones*
Well treatment .... *Aluminum hydroxychloride to stabilize clays*

Formation evaluation:

Logging suites .... *Gamma-ray, dual induction, double spaced neutron*

Oil characteristics:

Type .... *Naphthenic*
API gravity .... *Infantas, 25.9–27.9; La Cira, 21.4–24.1*
Initial GOR .... *50–300 ft³/bbl*
Sulfur, wt% *1.0*
Viscosity, SUS .... *Infantas, 32 cp (originally less viscous); La Cira, 32 cp*

Field characteristics:

Average elevation .... *Surface 300 ft (90 m) MSL*
Initial pressure .... *1450 psi at 3000 ft (100 bar at 915 m)*
Present pressure .... *Varies*
Pressure gradient .... *Normal*
Temperature .... *NA*
Geothermal gradient .... *0.84°F/100 ft (1.53°C/100 m)*
Drive .... *Dissolved gas; gravity drainage at Infantas*
Oil-water contact .... *La Cira, -3400 to -3700 ft (-1036 to -1128 m); Infantas, -2500 ft (-762 m) (tilted north)*

Connate water .... *35% est.*
Water salinity, TDS .... *10,000–40,000 mg/L*
Resistivity of water .... *NA*

Transportation method and market for oil and gas:
Pipeline to coast, local refinery; 1926–1950 mostly export; since, mostly domestic